Motor pool vehicle use analysis (3.1)
Number of calls to a fire station (4.2)
Maintenance of highway department tractors
(5.4)
Number of accidents reported to police (6.2, 6.4)

HOME ECONOMICS

Estimating consumer appeal of a new product
(1.1)
Analysis of monthly telephone bills (4.3)
Estimating breaking strength of threads (6.1)
Examining sugar content of potatoes (7.2)
Differences in market basket prices
(13.1, 13.2, 13.3)

MEDICINE

Displaying heights of high school seniors (2.1)
Survey of weights of people (5.4)
Estimating effectiveness of burn medication (6.3)
Examining drug effectiveness (7.1)
Medication to control high blood pressure (8.1)
Using the same individual pre and post treatment
(8.1)
Effect of smoking on lung cancer (10.2)
Weight loss attributable to experimental drug
(12.5)

PRODUCTION

Distribution of number of accidents (6.2)
Sick leave taken on various days of the week
(9.1, 9.2, 10.3)
Time between successive accidents (10.3)
Number of machine breakdowns (10.3)
Analysis of production rate of assembly line
workers (12.3)

PSYCHOLOGY

Sampling to determine employee job satisfaction
(1.1)
Determining psychological behavior patterns (1.1)
Estimating individual susceptibility to hypnosis
(6.1)
Peer ranking by employees (11.2)

QUALITY CONTROL

Estimating proportion of defective items (1.1, 7.3)
Number of computer malfunctions (5.3)
Time to failure of electronics components
(5.5, 6.4)
Analysis of the number of defective shirts
manufactured (10.1)

SOCIOLOGY

Examination of results of a political poll (1.1)
Results of analysis of a Customer Warranty Card
(1.4)
Collecting responses in a poll (3.2)
Distribution of charitable contributions (6.4)
Effect of an exercise program on TV viewing
habits (9.3)

MISCELLANEOUS

Biology—Growth rate of bacteria (3.2)
Genetics—Genotypes of plants (5.1)
Music—Numbers of flat and sharp band
instruments (5.1)
Athletics—Points scored in best quarter (5.1)
Flight Control—Time between airplane arrivals
(5.5)
Postal Service—Time distribution for mail
delivery (6.4)

A MODERN APPROACH TO STATISTICS

WILEY SERIES IN PROBABILITY
AND MATHEMATICAL STATISTICS

ESTABLISHED BY WALTER A. SHEWHART AND SAMUEL S. WILKS
Editors

Ralph A. Bradley, J. Stuart Hunter, David G. Kendall,
Rupert G. Miller, Jr., Geoffrey S. Watson

Probability and Mathematical Statistics

DUNN and CLARK • Applied Statistics: Analysis of Variance and Regression

ELANDT-JOHNSON • Probability Models and Statistical Methods in Genetics

ELANDT-JOHNSON and JOHNSON • Survival Models and Data Analysis

FLEISS • Statistical Methods for Rates and Proportions, *Second Edition*

FRANKEN, KÖNIG, ARNDT, and SCHMIDT • Queues and Point Processes

GALAMBOS • The Asymptotic Theory of Extreme Order Statistics

GIBBONS, OLKIN, and SOBEL • Selecting and Ordering Populations: A New Statistical Methodology

GNANADESIKAN • Methods for Statistical Data Analysis of Multivariate Observations

GOLDBERGER • Econometric Theory

GOLDSTEIN and DILLON • Discrete Discriminant Analysis

GROSS and CLARK • Survival Distributions: Reliability Applications in the Biomedical Sciences

GROSS and HARRIS • Fundamentals of Queueing Theory

GUPTA and PANCHAPAKESAN • Multiple Decision Procedures: Theory and Methodology of Selecting and Ranking Populations

GUTTMAN, WILKS, and HUNTER • Introductory Engineering Statistics, *Third Edition*

HAHN and SHAPIRO • Statistical Models in Engineering

HALD • Statistical Tables and Formulas

HALD • Statistical Theory with Engineering Applications

HAND • Discrimination and Classification

HARTIGAN • Clustering Algorithms

HILDEBRAND, LAING, and ROSENTHAL • Prediction Analysis of Cross Classifications

HOAGLIN, MOSTELLER, and TUKEY • Understanding Robust and Exploratory Data Analysis

HOEL • Elementary Statistics, *Fourth Edition*

HOEL and JESSEN • Basic Statistics for Business and Economics, *Third Edition*

HOLLANDER and WOLFE • Nonparametric Statistical Methods

IMAN and CONOVER • Modern Business Statistics

JAGERS • Branching Processes with Biological Applications

JESSEN • Statistical Survey Techniques

JOHNSON and KOTZ • Distributions in Statistics
          Discrete Distributions
          Continuous Univariate Distributions—1
          Continuous Univariate Distributions—2
          Continuous Multivariate Distributions

JOHNSON and KOTZ • Urn Models and Their Application: An Approach to Modern Discrete Probability Theory

JOHNSON and LEONE • Statistics and Experimental Design in Engineering and the Physical Sciences, Volumes I and II, *Second Edition*

JUDGE, HILL, GRIFFITHS, LÜTKEPOHL and LEE • Introduction to the Theory and Practice of Econometrics

JUDGE, GRIFFITHS, HILL and LEE • The Theory and Practice of Econometrics

KALBFLEISCH and PRENTICE • The Statistical Analysis of Failure Time Data

KEENEY and RAIFFA • Decisions with Multiple Objectives

LAWLESS • Statistical Models and Methods for Lifetime Data

LEAMER • Specification Searches: Ad Hoc Inference with Nonexperimental Data

McNEIL • Interactive Data Analysis

MANN, SCHAFER and SINGPURWALLA • Methods for Statistical Analysis of Reliability and Life Data

MARTZ and WALLER • Bayesian Reliability Analysis

A Modern Approach To Statistics

RONALD L. IMAN
Sandia National Laboratories

W. J. CONOVER
Texas Tech University

JOHN WILEY & SONS
New York · Chichester Brisbane
Toronto Singapore

Library of Congress Cataloging in Publication Data:

Iman, Ronald L.
    A modern approach to statistics.

    Includes index.
    I. Conover, W. J.   II. Title.
QA276.15.I43   1983   519.5   82-8344
ISBN 0-471-09667-9   AACR2

Printed in the United States of America

10 9 8 7 6 5 4 3 2 1

ACKNOWLEDGMENTS

We express our appreciation to the following individuals whose reviews of this book contributed greatly to smoothing and clarifying its presentation.

Eileen Boardman
Colorado State University
Fort Collins, Colorado

Francisco Samaniego
University of California, Davis
Davis, California

John Boyer
Kansas State University
Manhattan, Kansas

Chanchal Singh
St. Lawrence University
Canton, New York

Gerry Hobbs
West Virginia University
Morgantown, West Virginia

Vidya S. Taneja
Western Illinois University
Macomb, Illinois

Dallas E. Johnson
Kansas State University
Manhattan, Kansas

Thanks are also due to Mike Connors for drawing the cartoons. He demonstrated a great deal of innovation in bringing the cartoons to life after being presented with only a punch line and very rough sketches. Lastly, we are grateful to the professional dedication of the staff of John Wiley & Sons. In particular, the efforts of Robert Pirtle, editor for this book, Judith Watkins, and Alej Longarini are greatly appreciated. Their dedication to review and quality has led to the final form of this book.

R. L. I.
W. J. C.

ABOUT
THE AUTHORS

Ronald L. Iman is a statistical consultant at Sandia National Laboratories. He holds a Ph.D. in statistics from Kansas State University. He previously worked for The Upjohn Company and has several years of university teaching experience. His research papers are in applied nonparametric statistics and sensitivity analysis techniques for use with computer models. Dr. Iman is a Fellow of the American Statistical Association and has served on the ASA Council. He is chairman of the Steering Committee for the DOE Statistical Symposium, an associate editor of *The American Statistician*, and a member of the Editorial Board of *Selected Tables in Mathematical Statistics*.

W. J. Conover is Horn Professor of Statistics and Associate Dean of Research and Graduate Programs, College of Business Administration, Texas Tech University. He holds a Ph.D. in mathematical statistics from the Catholic University of America. Professor Conover is also a consultant at Sandia National Laboratories and a visiting staff member at Los Alamos National Laboratory. An earlier book written by him is *Practical Nonparametric Statistics*, Second Edition, Wiley, 1980. His research papers include both theoretical and applied nonparametric statistics. He is a Fellow of the American Statistical Association.

PREFACE

To the Student

As you begin this course you may well wonder what lies in store for you. At this time in your life with limited job experience it is difficult to be familiar with situations where the application of statistical methods would be appropriate. Thus, the authors have tried to provide a variety of problem settings, examples, and exercises that will allow you to realize some of the potential uses of statistics and how statistics can be important to you in your future or present employment. Additionally, even though you master the material and gain an appreciation for the potential use of statistics while taking this course, it may be difficult at some time in the future to recall all of the details of some particular statistical procedure. Thus, a great deal of effort has gone into making this text readable and organizing the various methods in a form that would permit this book to serve as a ready reference at some time in the future when the need arises.

RONALD L. IMAN   W. J. CONOVER

PREFACE
To the Instructor

In general the approach used in this book is conventional in many ways, but also unconventional in many other ways. Both authors have had extensive experience teaching and consulting, and this book reflects the authors' experiences, especially regarding the inadequacies and incompleteness of the presentation made in most basic statistics textbooks. In this vein, throughout the text the emphasis is on training the reader to go from a data-oriented situation to the proper statistical method. Most topics are introduced by means of a realistic problem setting, and then the method of solution unfolds in the subsequent discussion where the authors attempt to communicate how to analyze a real situation.

The problems, examples, and exercises further emphasize realistic settings in which these methods are useful. The problem settings and exercises are based on actual applications in the fields of biology, business, economics, psychology, sociology, education, home economics, agriculture, pharmacology, engineering, physical education, physiology, genetics, actuarial science, and others.

This book is intended as an introduction to modern statistical techniques in a one-semester course with the topics divided into lesson-size sections. Each section concludes with a set of exercises; review exercises occur at the end of each chapter. In all, the book contains approximately 540 exercises and numerous worked-out examples. Answers to selected exercises are given at the end of the text.

Each methodology in the book is presented in a self-contained format, complete with assumptions, explanation of notation, statements of hypotheses, and decision rules. This format is used for the convenience of the reader in referring back to the various statistical procedures. It is the authors' experience that most texts present methodology and assumptions in a helter-skelter fashion that makes it extremely difficult if not impossible for the reader to decide when to use each methodology. These same texts almost always fail to provide any guidance as to what to do if the assumptions are not satisfied. Thus, a question of ethics arises. Should the assumptions be considered? Should texts make some attempt to deal with these issues? The authors do not feel that the reader should be left on his or her own, assuming that all is well if the mechanics are known, but that these issues are quite important and should be addressed.

In order to avoid these shortcomings, the sections presenting testing procedures clearly state the assumptions required for validity of the testing procedure. Many of these testing procedures rely on the underlying assumption of normality. This assumption is easily checked using graphs specifically developed for this book for use with the Lilliefors test for normality. Thus, the cumbersome chi-square goodness-of-fit test for normality is avoided, enabling the assumption of normality to be checked easily thorughout the text. For situations where the normality assumption is not satisfied, guidance toward the appropriate nonparametric test is provided. This means that the parametric and nonparametric tests are presented side-by-side throughout the text. However, the transition to the nonparametric tests is made easy by presenting them as analogues to parametric tests. This is done by applying the parametric procedure to rank transformed data. The result is a test statistic that is functionally equivalent to the usual presentation of the nonparametric test statistic. This way of presenting nonparametric tests makes them easier to learn than when the standard method is used, and it eases the burden on instructors who may not have a strong nonparametric background. This approach was described and verified in recent research papers by the authors. A tutorial paper on the subject, with appropriate references, appeared in *The American Statistician* in August 1981.

The first chapter concentrates on sampling procedures and includes a section on sample surveys reprinted with the permission of the American Statistical Association. Due to the ease with which graphs are used, it is a temptation for many instructors to skim through techniques for displaying sample data such as occur in Chapter 2. The authors hope the instructor will not yield to this temptation since most students really do not know how to display sample data, and these techniques will be used throughout the text to aid in displaying and understanding the data as well as checking the reasonableness of many decisions. This is particularly true of scatterplots and the empirical distribution function. The graph of the empirical distribution function simplifies many of the ideas that usually give students difficulty, such as working with normal probabilities, grasping the concept of power, finding sample quantiles, and performing goodness-of-fit tests.

Years of study can be devoted to the fascinating study of probability. This book is intended to teach statistics rather than probability, so in Chapter 3 only the topics in probability that are necessary to teach statistics are included.

Descriptive sample statistics such as the sample mean, standard deviation, median, mode, and proportion and the associated population parameters are given in Chapter 4. Also included in this presentation are the empirical distribution function, quantiles, the sample correlation coefficient, and the rank correlation coefficient, since these statistics are integrated into the presentation throughout the text.

The commonly used binomial and normal distributions are introduced in

Chapter 5 along with instructions on how to use their corresponding Tables A1 and A2. These tables as well as others presented in this book are all based on cumulative distributions and therefore tie in closely with the authors' repeated emphasis on empirical and cumulative distribution functions. The table of the cumulative normal distribution has the advantage over the density function approach (i.e., area under a curve) emphasized in most statistics books because students grasp its use almost immediately—something that cannot be said of the tables based on the density function, as anyone who has spent countless hours explaining how to use it knows only too well. The Lilliefors test for normality is introduced in Chapter 5. Its use requires plotting an empirical distribution function of standardized sample values in the graphs developed for this text. The Lilliefors test provides an easy and accurate test for normality and is used throughout the text to challenge the normality assumption. Sampling distributions and the Central Limit Theorem are covered in Chapter 5 as are the closely related exponential and Poisson distributions. A companion Lilliefors test for exponential distributions is given.

Point and interval estimates are considered in Chapter 6 as well as properties of estimators. In addition to the usual confidence interval for the mean of a normal population, a procedure for finding a confidence interval for the median of a population is given. This procedure may be more appropriate in nonnormal populations. Confidence intervals are also provided for the population proportion. The framework for hypothesis testing is given in Chapter 7. Hypothesis testing is demonstrated in sections covering each of the population mean, proportion, and median. Power is explained in terms of the cumulative distribution function.

The matched pairs problem is presented in Chapter 8 from the standpoint of being a good experimental design to control unwanted variation. The paired t-test is presented as the parametric method of analysis. The normality assumption is checked with the Lilliefors graphs and the Wilcoxon signed ranks test is given as the alternative nonparametric procedure when the normality assumption is not satisfied. The Wilcoxon signed ranks test is presented as a rank transform test. That is, the parametric paired t-test is applied to rank transformed differences. The result is a test statistic that is functionally equivalent to the standard presentation of the Wilcoxon signed ranks test statistic; however, the transition is quite easy for the student as all that is needed is to apply the just learned parametric paired t-test to the ranks. The critical values are approximated by the Student's t-distribution, which is quite accurate even for small sample sizes. The basis for this approach is contained in a number of research papers by the authors.

The case of two independent samples is considered in Chapter 9. Large sample techniques are given first followed by the two-sample t-test. Assumptions of normality and equal variance are checked. Graphs of the empirical distribution functions are presented in the examples to aid the student in understanding the two-sample problem and resultant decisions. The Wilcoxon rank sum test is presented as a rank transform test obtained from the two-sample t-test.

Chi-square test procedures are considered in Chapter 10, including 2 × 2 and $r \times c$ contingency tables, and goodness-of-fit tests. Chapter 11 reviews correlation on raw and rank transformed data prior to the study of regression techniques in Chapter 12.

Methods for linear regression are given in Chapter 12, including the usual least-squares computation procedures as well as a study of the model for linear regression and hypothesis testing. A nonparametric test for slope based on rank correlation is also given. Methods of monotone regression are presented as linear regression computations applied to rank transformed data. These methods are appropriate for either linear regression or monotonic nonlinear regression.

The completely randomized design and associated one-way analysis of variance are presented in Chapter 13. These techniques represent a generalization of the two-sample procedure of Chapter 9. Fisher's least significant difference is given as a multiple comparisons procedure; research (including that by the authors) has shown this procedure to be robust and powerful compared to other multiple comparisons techniques. The Kruskal–Wallis test is given as the rank transform analogue to the one-way analysis of variance for the completely randomized design. Computer output is utilized for the analyses in this chapter.

Depending on the emphasis desired by the instructor, a number of sections would be omitted in a 3- or 4-semester-hour course without loss of continuity. In particular, any one of the following groups of sections could be eliminated totally independent of the elimination or inclusion of any of the other groups: (1.2, 1.3, and 1.4); (4.6, 11.2, 12.5, and 12.6); 5.5; (6.4 and 7.4); and (10.1, 10.2, and 10.3).

The authors hope you will find it a refreshing experience to teach from this text and would welcome any correspondence regarding strengths or weaknesses of the book that would affect any possible future editions.

RONALD L. IMAN   W. J. CONOVER

CONTENTS

xvii

APPENDIX A

Tables

APPENDIX B

Solutions to Selected Exercises

INDEX

A MODERN APPROACH TO STATISTICS

1

The Relationship Between Sampling and Statistics

It has been stated, "You don't have to eat the whole ox to know that the meat is tough." This statement is one way of paraphrasing what the science of statistics is all about. On the basis of a serving, an inference is made regarding the quality of the meat of the entire animal, or indeed even about the quality of oxen meat in general. However, in statistics the serving and the ox are replaced respectively with the terminology **sample** and **population**.

POPULATION *A **population** consists of a collection of individual units, which may be persons or experimental outcomes, whose characteristics are to be studied.*

SAMPLE *A **sample** is a portion of the population that is studied in order to learn about the population.*

For example, populations and samples are involved in each of the following situations.

1. **Polls.** A candidate for political office hires a polling firm to assess his chances in the upcoming election. The population consists of all voters in the candidate's district. The sample consists of all persons contacted by the polling firm.
2. **Quality Control.** The manufacturer of flashbulbs wants to be sure that the proportion of defective bulbs does not become too high. The population here is all flashbulbs made by the manufacturer. But since the test of the flashbulbs is by its very nature destructive, only a sample of the bulbs is examined.

1

3. **General Business.** Royalties for music played on the radio are determined by the frequency with which each piece of music is played. The population consists of all music played by all radio stations in the country. A sample of radio stations provides the information upon which the royalties are based.

WHAT WILL YOU LEARN FROM THIS BOOK

Clearly if valid conclusions are to be drawn with respect to the previous populations, the sampling cannot be done in a haphazard manner. A classic example of poor sampling occurred in the 1936 presidential election when the publishers of *Literary Digest* conducted a survey and concluded that Republican Alf Landon would defeat Democrat Franklin Roosevelt. The results of the 1936 election showed Landon carrying only two states! The reason for this fiasco could have been that the individuals polled came from lists of telephone owners, magazine subscribers, and car owners, most of whom were Republicans, or the reason could have been that the people who responded were not representative of the people polled.

In this book you will learn proper methods for the collection, display, and analysis of sample data. Specifically you will learn:

1. proper methodology for obtaining samples;
2. how to summarize data and display data graphically as an aid to the correct interpretation of the data;
3. how to analyze data and make estimates of unknown population characteristics;
4. how to choose the correct statistical procedure for testing conjectures (hypotheses) about the population.

The statistical methods presented have proven their usefulness in application after application. This book will serve as a handy reference because many of the statistical methods you are most likely to encounter in your work, in your own projects, or in reports written by colleagues in your field, are treated here. The authors hope you find the explanations easier to understand than those found elsewhere as every effort has been made to present this technical subject in a way that is understandable to individuals who may not have a technical background, and to do so without sacrificing the accuracy and correctness of the material.

1.1
Populations and Samples

Successful quantitative decisions are directly related to one's experience or knowledge of the situation at hand. For example, farmers who use a new insecticide on their fruit trees may find that undesirable side effects may cause extra cleaning expenses before their crops can be marketed. A television network that emphasizes situation comedies when the audience is interested in live sports coverage finds it difficult to obtain premium prices for its commercials. Many a political candidate has misread the pulse of the public and never been heard from again.

THE IMPORTANCE OF SAMPLING

When it is impossible to survey the entire population a properly obtained sample from the population can reveal a wealth of information about its makeup. The Gallup Poll, the Harris Poll, and the Nielson Ratings all rely on properly obtained samples to provide information about the wants, needs, and opinions of the population of people represented by the samples. The following example provides an illustration of how a sample could be obtained and how the information in it could be used.

SAMPLING IN A SMALL BUSINESS

Suppose you and a small group of college friends are planning to start a small business to pick up some extra money while going to school. You have decided to make fresh sandwiches, starting about 7 o'clock every evening, and circulate through the residence halls on campus between 10 and 11 that evening selling these delicious home-made sandwiches to hungry students. You have distributed a flyer advertising the types of sandwiches, their prices, and the approximate times they will be available, beginning on the following Monday.

As Monday draws nearer, you begin to wonder how many sandwiches you should make. If you make more than you can sell, you will lose money on every unsold sandwich. On the other hand, if you don't make enough sandwiches you will be faced with lost profits and dissatisfied potential customers who will learn to rely on other sources for their evening snacks. What kind of sandwiches should you make? If you make a large number of tuna salad sandwiches, but everyone wants chicken salad, you've made another costly error in judgment. Of course, after a week or two of operations you should know your market pretty well, but in the meantime poor planning could cost you a lot of money and upset many potential customers.

Having just read your assignment in statistics—the one about the value of sampling—you decide to conduct your own sample survey. One evening, between 10 and 11, you and your friends visit the residence hall floors where you will soon be selling sandwiches. You cannot very well stand in the hall yelling "sandwiches" yet, so you do the next best thing. You knock on three of the thirty doors on that floor, introduce yourself, and tell them that you are here to find out what kind of a product they will want next week when your service begins. If you were selling sandwiches tonight, what kind (if any) would they be likely to buy? You assure them that by answering your questions honestly and accurately they can help you supply the type and quantity of sandwiches they would like.

Back in your room, the survey results are tabulated. The number of potential orders is multiplied by ten to reflect the fact that only one-tenth of the rooms were used in the sample. This should provide you with some idea of how many sandwiches to make for the first night of sales. Of course, you may wish to increase these sales estimates by 10 percent, to make sure you have enough sandwiches for everyone, or you may want to make fewer sandwiches than the sales estimate, to lessen the chances of having sandwiches left over. You, as a manager and decision maker, need an accurate sales estimate for the first day of business, but you shouldn't let the sales estimate be your boss. Decisions should not be made by statistics. Rather, statistics provide information on which a better decision can be made.

DESIGNING A SAMPLE SURVEY

This example illustrates the four steps in designing (i.e., planning) and conducting a sample survey.

1. Planning. You decide which question to ask, how many and which rooms to include in your survey, and what time of day to conduct your survey. You need to plan for the unexpected also, such as what to do if no one is home, or if the room occupants refuse to cooperate with the survey. In general, planning also includes selection of the population and specification of the issues to be studied. In this case the population is the group of potential customers, and the issues to be studied are the types of sandwiches that will be in demand.

2. Data Collection. You and your friends actually went out, asked the questions, and got the answers, following the plans as closely as possible. Collecting the data can be a very informative activity. Since you were involved with collecting the data, you learned many things that will not show in the numbers. That is, several people may have suggested sandwich types that were not on your proposed menu. Or the manner in which some answers were supplied may cause you to doubt the accuracy of those particular answers. The data

"24 BANANA SANDWICHES? RIGHT!"

collection step may be the most important and most neglected step in a sampling study.

3. Data Analysis. The number of students who responded favorably is simply multiplied by ten to estimate sales. Other methods are available for stating how close this sales estimate can be expected to be to the actual sales figure. These other methods appear in subsequent chapters of this book.

4. Conclusions. The conclusions were stated as sales estimates. Other surveys may have more complicated objectives, requiring more elaborate conclusions. Often a survey will result in several recommendations for alternative courses of action, with an analysis of the likely consequences of each. The decision maker should consider these recommendations before making the final decision, but should not relegate the decision-making responsibility to the results of a survey.

STEPS IN THE DESIGN OF A SURVEY AND ANALYSIS OF SURVEY DATA

The four steps necessary for a sample survey are:
1. Prior planning and design of the survey.
2. Collection of the data in the sample.
3. Analysis of the data collected.
4. Conclusions based on the data analysis.

WHY USE A SAMPLE

There are several standard situations where a sampling study is preferable to a *census*, or complete study of the population.

1. A census may not be cost effective. In a marketing survey in which free samples are given out in order to estimate consumer response to a new product in a large city, little or no additional information is obtained by giving free samples to everyone in the city rather than to only a portion of the population.

2. There may not be enough time to obtain more than a sample. In political polls, information on voter preferences must be obtained on short notice in order to be useful in the campaign.

3. A carefully obtained sample may be more accurate than a census. In a large project such as a census, certain **biases** often appear systematically throughout the census, because there are not enough resources available to give each person as much attention as he or she deserves. (A bias is a tendency to make errors in one direction or the other.) For example, in a large inventory census or in a complete audit, errors due to fatigue or carelessness on the part of the census taker may introduce a serious bias in the results. This may result in less accuracy than if a sample had been carefully examined with an accurate report on the sample.

4. In destructive testing of products, a sample has to suffice. If the quality control test involves dismantling a car battery to see if it was made properly, exploding a firecracker to see if it works, or breaking a rope to check its strength, it is necessary to include only a few items in the sample. A census would under these situations result in no products available for marketing.

5. Sometimes a census is impossible. If a new medicine is being tested, the population of interest with regard to its efficacy is not only all people alive now, but all future generations as well. Clearly, a sample must suffice in this case. Another example is the enumeration of the wildlife population, where a census is not a feasible alternative.

CENSUS	A **census** is a complete enumeration of the entire population, as opposed to a sample, which consists of only a portion of the population.

BIAS	A statistical **bias** refers to the systematic tendency of a sample or method of analysis to give estimates of population characteristics that are either larger on the average (positive bias), or smaller on the average (negative bias), than the true quantity being estimated.

EXAMPLES OF SAMPLING

Here are some typical applications of sampling.

Industry / In order to find out how employees feel about certain management practices, such as Saturday overtime, or procedures for determining merit pay increases, a sample of employees may be selected and interviewed.

Politics / Elected representatives sample their constituents to get a reading of public opinion on pending legislation.

Quality Control / A sample of a manufactured product is tested to see if the manufacturing process is in control.

Agriculture / The national production of corn, soybeans, wheat, and many other crops is estimated prior to harvest time by examining many selected fields in a carefully designed survey. Foreign production is estimated from satellite photos.

Psychology / Behavior patterns for the entire population of people are determined from careful study of a sample of paid volunteers.

Education / The effectiveness of a new teaching method is estimated by trying it on several groups of students.

Home Economics / Product appeal is determined by letting a sample of consumers try the product, where that product may be a new fabric design, a different recipe, or perhaps a microwave oven.

Exercises*

1.1   Explain briefly the difference between a sample and a population.

* Solutions to Exercise numbers indicated in color will be found in Appendix B (pp. 451 to 490).

1.2 Give an example of a situation where a sample must be used due to the impossibility of examining the entire population.

1.3 Suppose you are placed in charge of designing a sample survey to determine the number of hours the average student spends viewing television in one of the dormitories on your campus. Explain what would be involved in each of the four steps of conducting a sample survey.

1.4 Why would you want to use a sample instead of a census in the following situations?

(a) Estimating the potential market for a new brand of soap by giving out free samples.

(b) Testing the effectiveness of a new flu vaccine.

(c) Estimating the nicotine content in a brand of cigarettes.

1.5 One of the popular shows on radio is the playing of the week's top 40 hits based on sales throughout the country. How would you guess that the determination of the top sellers is made?

1.2
Types of Sampling Techniques

PROBLEM SETTING 1.1

Several years ago the League of Women Voters in Manhattan, Kansas wanted to assist the city in applying for federal funds for urban renewal. As a part of the application the federal government required information concerning the location and extent of dilapidated housing in the city. Housing was considered dilapidated if and only if it met certain federal guidelines, which considered the appearance of the house, the number of bathrooms, the condition of the foundation, and other factors too numerous to mention. By carefully inspecting a residence and asking certain questions of its occupants, a League member could determine whether or not the house qualified for the term "dilapidated" according to federal guidelines.

Because of the time required to determine the condition of each residence, it is easy to see why the League of Women Voters, a volunteer organization, found it impossible to examine all 8000 residences in Manhattan, Kansas. They decided to take a sample.

TARGET POPULATIONS AND SAMPLED POPULATIONS

The **target population** of the League's survey was the collection of all residences within the city limits. In general the target population is the population about which it is desired to obtain information.

The **sampled population** was the list of all households given by the current city directory. The city directory, for those who are not familiar with this useful book, is available to businesses for a modest price, and usually may be found in the public library. It lists households, usually geographically street by street, along with information such as the names of the inhabitants and their occupations, if known. The city directory was a valuable source of information in this project.

Sometimes the sampled population is called the **sampling frame**. In this case the sampled population matches well with the target population, because the city directory is up to date and includes most, if not all, of the households in the city. Households that may be outside the city limits can be omitted from the study by checking addresses against a map containing the city limits.

TARGET POPULATION AND SAMPLED POPULATION	The **target population** is the population that is the ultimate object of study. The **sampled population** is the population from which the sample was obtained. The validity of the study depends on the target population and the sampled population being similar in the characteristics being studied.

One should be aware that the target population and the sampled population do not always agree. For example, university research projects often use students as subjects. Consider a research project that examines the effect of lack of sleep on the ability to concentrate. Subjects are obtained by selecting from students who have responded to an ad in the campus newspaper. The sampled population consists of young men and women who are above average in intelligence, who read the campus newspaper, and who happen to have more time than money. This may be quite unlike the target population that may be people of all ages, from all walks of life.

SIX TYPES OF SAMPLES

There are several different types of sampling procedures the League of Women Voters could have used, although they used only one. Please keep in mind that there is a precise mathematical definition for each of these sampling procedures, but the procedures are explained here only by example in the belief that this will lead to an easier understanding. Many books are devoted to sampling techniques and are available in any technical library.

1. Convenience Sample. By including each League member's household in the sample, and throwing in the households of their friends, they could easily obtain information on several hundred residences. Convenience samples are commonly used in "man on the street" broadcasts, and supermarket surveys. The results usually cannot be used to make valid conclusions about the target population, because the sampled population is likely to be quite different from the target population.

2. Representative Sample. A search is conducted for several "typical" middle income family dwellings, several "typical" low income apartment households, and so on, until a sample is constructed that contains representatives of each of the household types likely to be encountered in the city. Without some estimate of the total number of each of these household types, such information does not answer the questions in the application for urban renewal funds. Additionally, the method of determining which households to include or exclude from the sample invites personal bias to enter the selection process. The bias inherent in a sampling method such as this one is called a *sampling bias*. The following methods use **randomization** to remove the sampling bias.

RANDOMIZATION	**Randomization** *is the name given to any process which assures that each possible sample in a given class of samples is equally likely to be obtained.*

3. Random Sample. Sometimes called a "simple random sample" to distinguish it from the other variations given below, this is the first of the sample types presented that enables valid projections to the entire sampled population to be made with confidence. A sample with *n* observations in it is considered to be a simple random sample of size *n* if it is chosen in such a way that every possible collection of *n* units from the sampled population is equally likely to be selected. Notice that "random" refers to the method of selecting the sample. *It is not possible to tell whether a sample is random or not by looking at the sample itself.* One must examine the method by which the sample was obtained. The methods for obtaining a random sample are not easy. Two possible methods as they apply to the household survey by the League of Women Voters will be described.

Information on each of the 8000 or so households can be put on a card, one household per card, and thoroughly mixed. Then the desired number of cards (households) are drawn from the pile without first looking at the information on the card. If the mixing process is thorough, the sample will be random. A similar method was used by the Selective Service one year, when each of the 366 possible birthdates was put in a plastic capsule, the capsules were mixed in a drum, and the order in which the capsules were drawn determined the order in which young men with those birthdays were called into the armed forces. Unfortunately, the mixing of the capsules was not very thorough the first year this was tried, and the capsules with December dates tended to be the first ones drawn. Figure 1.1 shows a frequency tabulation of birthdates in the lottery. It is worth repeating: a truly random sample is difficult to obtain.

Another way of obtaining a random sample is with the aid of a table of

Months	Number of Selections in the First 183	Number of Selections in the Last 183
Jan.–Feb.	25	35
Mar.–Apr.	20	41
May–June	28	33
July–Aug.	32	30
Sept.–Oct.	32	29
Nov.–Dec.	46	15

FIGURE 1.1. Numbers of Birthdates Selected in the First and Last Half of the 1970 Draft Lottery, Grouped by Time of Year

random numbers. This is much easier and less expensive then the method just described. A table of random numbers (more precisely called random digits) is a table filled with digits 0 through 9, in which the digits follow no particular pattern whatsoever. These tables are usually obtained from well-tested computer programs designed to generate such numbers.

One way to use a random number table to obtain a random sample of households is to number all of the households consecutively from 0001 through 8000, or however far is necessary. Any order for numbering the households is satisfactory. Then a starting place in the table of random numbers, such as in Figure 1.2, is selected "at random," such as by closing one's eyes and pointing "blind" to some number in the table (using a 4-digit group as a number in this case) to be the starting number. Suppose the number 1484 from the fourth column and sixth row of Figure 1.2 is selected. Then the household with the same number is included in the sample. The number below 1484 in the random number table is then selected, say 2801, and household number 2801 is included in the sample. By reading vertically down the column, and continuing at the top of the next column, households are included until the sample is as large as desired. Of course, once the starting point is obtained, additional random numbers may also be found by reading horizontally across the row and continuing on to the next row. It should be noted that any number greater than 8000, or a number already used, may be discarded because the corresponding households are not available for inclusion in the sample.

Most of the statistical methods presented in this book are designed for use on a random sample, or several random samples. Probabilistic statements can be made about the population being sampled only when randomization procedures are used in obtaining a sample. In actual practice other types of sampling may be used, in which case some modification of the methods presented here must be made.

4. Systematic Sample. The League of Women Voters could have started at the beginning of the city directory, selected one household at random from the first 80 households listed, and then counted from that household, including every 80th household in their sample. In this way a sample of 100 households (8000/80 = 100) would be drawn. If only 50 households were needed for the sample, every 160th household, from a household randomly selected from the first 160, would be included. This method of sampling is easy to use when the sampled population is listed, such as accounts in a file drawer or names on a list. Quality control sampling from an assembly line often uses a systematic sample. It forces the sample to span the entire population in a systematic manner, but prevents two items that are close together in the list from both being included in the sample. This can be good in some situations or undesirable in others, depending on the situation. For example, a systematic housing survey may select only houses on the corner of each block, and corner houses may have different characteristics than the rest of the neighborhood. This would be undesirable. On the other hand, a systematic sample from an alphabetical list of college students to determine home states would ordinarily exclude brothers and sisters with the same surname, which might, but need not, be desirable.

81080	67493	23666	22251	17616	60716	77125	18653
83272	18379	46498	60045	80649	35179	03185	57068
82844	85553	16852	57931	84063	57516	46529	47030
33097	46244	16769	48531	56618	90035	88363	04097
48477	33067	76572	84835	96208	68558	23560	89245
61186	63971	20547	14846	77137	62636	88927	34322
92545	83866	06895	28019	08547	04275	79277	28833
05172	25637	13665	86725	45970	42670	35291	22685
73850	99275	97475	11064	93492	05362	57562	99582
77978	42899	65518	48688	96755	83554	76916	15224
16463	00350	44697	94868	22697	33740	60701	04034
56564	40277	66044	78417	52968	52982	82340	92970
26355	51841	01235	15986	65898	74181	51391	11313
87582	80276	88583	30633	50721	65017	48735	04476
15659	86285	09579	07969	17850	88197	14309	25013

FIGURE 1.2. A Brief Table of Random Numbers

5. Stratified Sample. This is the sampling method the League of Women Voters used in this problem. Boundary lines were drawn on a map of the city, outlining several relatively homogeneous sections of town that were known to be somewhat similar. One of the regions thus defined was the area around the business district, which was the area the League had in mind for urban renewal. Within each of these regions, called ''strata'' (singular = stratum), a random sample of households was selected using a random number table in the manner previously described. The sections believed to contain a higher proportion of dilapidated residences were sampled much more heavily than the newer sections of town. In this way proportionately more time was spent in the neighborhoods that contain most of the dilapidated housing and less time was spent in the more affluent neighborhoods. The numbers obtained were weighted appropriately to reflect these proportions, and accurate estimates were obtained for the entire city.

Stratified sampling improves accuracy when a heterogeneous population is divided into strata that are relatively homogeneous. The Gallup Poll, the Harris Poll, and most political polls attempt to use stratified sampling whenever possible, because of the greater accuracy from smaller sample sizes, as compared with simple random sampling.

6. Cluster Sample. Once the League of Women Voters decided which households to include in its sample, it could have tripled its sample size by telling each person to include the two households closest to the household selected as well. Thus the observations would be in clusters of three each.

An advantage of cluster sampling is ease in obtaining additional observations, especially when distances between observations are large and time required to take the observations is short. In the League's survey, the distance from one observation to the next was only a few blocks, and a minor inconvenience compared to the time and effort required to examine a household. Therefore cluster sampling was not used. In a survey of farm houses a different decision may have been reached.

A disadvantage of cluster sampling is that often units that are located close together are similar. Hence these additional units do not contribute much different information than the first unit examined.

SAMPLING ERROR AND NONSAMPLING ERROR

It is obvious that all of these sample types have one characteristic in common. None of the samples will look quite like the entire population would look in a census. Therefore a difference between the estimate furnished by the sample and the corresponding population value will most likely exist. If this difference can be attributed to the fact that there is a natural but inherent difference between a sample and the population, then that difference is said to be due to **sampling error**, or **the error due to sampling**. Statistical methods can supply an estimate of the amount of the sampling error. Sampling error refers to the natural variation from one sample to another. It does *not* imply a mistake on the part of anyone.

SAMPLING ERROR AND NONSAMPLING ERROR	**Sampling error** *is the name given to natural variability inherent among samples from a population. It is always present when samples are obtained.* **Nonsampling error** *is the name given to inaccuracies and errors that can and should be avoided by using sound experimental techniques.*

However, if the difference between the sample and the population is due to other causes, such as workers inventing data to fill in blank spots where real data are difficult to obtain, misleading answers supplied in a survey, or unintentional errors in transcribing or retranscribing the data, then the differences between the sample estimates and the true population values are said to be due to **nonsampling error**. Even a census may contain nonsampling errors. This source for error can be avoided through careful planning and sound sampling procedures.

An illustration of sampling error and nonsampling error will now be given. Suppose it is desired to estimate the proportion of students who smoke marijuana. Each member of your statistics class obtains a random sample of 10 students and reports the proportion who smoke marijuana. One student may report .3, another reports .2, and so on. The variability from one random sample to the next is sampling error. However, there may be a tendency for students in each sample to answer the question "Do you smoke marijuana?" with less than complete honesty. This introduces nonsampling error. Also some students in your class may have been too busy or too shy to complete the assignment, and may have fabricated some data, leading to another source of nonsampling error.

Exercises

1.6 A bank wishes to select a random sample of size 10 from 100 bank accounts for a sample audit. Assume these accounts can each be identified by 2 digits such as 00, 01, 02, 03, . . . , 99. Use a random starting point in the random number table of Figure 1.2 to select the accounts to be used in the audit.

1.7 How could a systematic sample of size 10 be obtained from the bank accounts for the sample audit in Exercise 1.6?

1.8 Why do you think that stratified sampling adds to the greater accuracy of political polls such as the Gallup Poll and the Harris Poll?

1.9 Consider the first 10 columns of Figure 1.2 and record the frequency with which each of the 10 digits 0, 1, 2, . . . , 9 occurs. Each column contains 15 digits so you will be summarizing 15 × 10 or 150 digits. It would be reasonable to expect each digit to appear the same number of times (i.e., 15 times). Your frequency tabulation represents the type of random variation you could expect to find in a sample due to sampling error.

1.10 A congressional committee wishes to examine the effect of proposed legislation on the nation's high schools. It randomly selects five high schools from the Washington D.C. area and conducts a study on those five schools.

(a) What is the target population?

(b) What is the sampled population?

1.11 A list of 100 drivers licenses has been made available for sampling. These licenses are numbered from 00 to 99, and a sample of size five is drawn. For each of the following samples, identify the sample as most likely to be a simple random sample, a convenience sample, a systematic sample, a cluster sample, or a stratified sample.

(a) License numbers 18, 38, 58, 78, and 98.

(b) License numbers 00, 01, 02, 03, and 04.

(c) License numbers 07, 16, 43, 58, and 81.

(d) License numbers 13, 36, 44, 77, and 90.

(e) License numbers 12, 13, 75, 76, and 77.

1.3

What is a Survey*

Characteristics of Surveys

The Need

Any observation or investigation of the facts about a situation may be called a survey. But today the word is most often used to describe a method of gathering in-

* This section consists of a report on survey sampling that was prepared for the American Statistical Association and is reprinted here with permission.

formation from a number of individuals, a "sample," in order to learn something about the larger population from which the sample has been drawn. Thus, a sample of voters is surveyed in advance of an election to determine how the public perceives the candidates and the issues. A manufacturer makes a survey of the potential market before introducing a new product. A government agency commissions a survey to gather the factual information it needs in order to evaluate existing legislation or draft new legislation. For example, what medical care do people receive, and how is it paid for? Who uses food stamps? How many people are unemployed?

It has been said that the United States is no longer an industrial society but an "information society." That is, our major problems and tasks no longer focus merely on the production of the goods and services necessary to our survival and comfort. Rather, our major problems and tasks today are those of organizing and managing the incredibly complex efforts required to meet the needs and wishes of nearly 220 million Americans. To do this requires a prompt and accurate flow of information on preferences, needs, and behavior. It is in response to this critical need for information on the part of the government, business, and social institutions that so much reliance is placed upon surveys.

Surveys come in many different forms and have a wide variety of purposes, but they do have certain characteristics in common. Unlike a census, they gather information from only a small sample of people (or farms, businesses or other units, depending on the purpose of the study). In a bonafide survey, the sample is not selected haphazardly or only from persons who volunteer to participate. It is scientifically chosen so that each individual in the population has a known chance of selection. In this way, the results can be reliably projected to the larger public.

Information is collected by means of standardized questions so that every individual surveyed responds to exactly the same question. The survey's intent is not to describe the particular individuals who by chance are part of the sample, but to obtain a statistical profile of the population. Individual respondents are never identified and the survey's results are presented in the form of summaries, such as statistical tables and charts.

The sample size required for a survey will depend on the reliability needed, which, in turn, depends on how the results will be used. Consequently, there is no simple rule for sample size that can be used for all surveys. However, analysts usually find that a moderate sample size is sufficient for most needs. For example, the well-known national polls generally use samples of about 1,500 persons to reflect national attitudes and opinions. A sample of this size produces accurate estimates even for a country as large as the United States with a population of over 200 million.

When it is realized that a properly selected sample of only 1,500 individuals can reflect various characteristics of the total population within a very small margin of error, it is easy to understand the value of surveys in a complex society such as ours. They provide a speedy and economical means of determining facts about our economy and people's knowledge, attitudes, beliefs, expectations, and behavior.

Who Does Surveys

We all know of the public opinion polls that are reported in the press and broadcast media. The Gallup Poll and the Harris Survey issue reports periodically, describing national public opinion on a wide range of current issues. State polls and metropolitan area polls, often supported by a local newspaper or TV station, are reported regularly in many localities. The major broadcasting networks and national news magazines also conduct polls and report their findings.

But the great majority of surveys are not exposed to public view. The reason is that, unlike the public opinion polls, most surveys are directed to a specific administrative or commercial purpose. The wide variety of issues with which surveys deal is illustrated by the following listing of actual uses:

1. The U.S. Department of Agriculture conducted a survey to find out how poor people use food stamps.

2. Major TV networks rely on surveys to tell them how many and what types of people are watching their programs.

3. Auto manufacturers use surveys to find out how satisfied people are with their cars.

4. The U.S. Bureau of the Census compiles a survey every month to obtain information on employment and unemployment in the nation.

5. The National Center for Health Statistics sponsors a survey every year to determine how much money people are spending for different types of medical care.

6. Local housing authorities make surveys to ascertain satisfaction of people in public housing with their living accommodations.

7. The Illinois Board of Higher Education surveys the interest of Illinois residents in adult education.

8. Local transportation authorities conduct surveys to acquire information on people's commuting and travel habits.

9. Magazine and trade journals utilize surveys to find out what their subscribers are reading.

10. Surveys are used to ascertain what sort of people use our national parks and other recreation facilities.

Surveys of human populations also provide an important source of basic social science knowledge. Economists, psychologists, political scientists, and sociologists obtain foundation or government grants to study such matters as income and expenditure patterns among households, the roots of ethnic or racial prejudice, comparative voting behavior, or the effects of employment of women on family life. (Surveys are also made of nonhuman populations, such as of animals, soils, and housing; they are not discussed here, although many of the principles are the same.)

Moreover, once collected, survey data can be analyzed and reanalyzed in many different ways. Data tapes with identification of individuals removed can be made available for analysis by community groups, scientific researchers, and others.

Types of Surveys

Surveys can be classified in a number of ways. One dimension is by size and type of sample. Many surveys study the total adult population, but others might focus on special population groups: physicians, community leaders, the unemployed, or users of a particular product or service. Surveys may be conducted on a national, state, or local basis, and may seek to obtain data from a few hundred or many thousand people.

Surveys can also be classified by their method of data collection. Thus, there are mail surveys, telephone surveys, and personal interview surveys. There are also newer methods of data collection by which information is recorded directly into computers. This includes measurement of TV audiences carried out by devices attached to a sample of TV sets that automatically record in a computer the channels being watched. Mail surveys are seldom used to collect information from the general public because names and addresses are not often available and the response rate tends to be low, but the method may be highly effective with members of particular groups—for example, subscribers to a specialized magazine or members of a professional association. Telephone interviewing is an efficient method of collecting some types of data and is being increasingly used. A personal interview in a respondent's home or office is much more expensive than a telephone survey but is necessary when complex information is to be collected.

Some surveys combine various methods. Survey workers may use the telephone to "screen" for eligible respondents (say, women of a particular age group) and then make appointments for a personal interview. Some information, such as the characteristics of the respondent's home, may be obtained by observation rather than questioning. Survey data are also sometimes obtained by self-administered questionnaires filled out by respondents in groups—for example, a class of school children or a group of shoppers in a central location.

One can further classify surveys by their content. Some surveys focus on opinions and attitudes (such as a pre-election survey of voters), while others are concerned with factual characteristics or behavior (such as a survey of people's health, housing or transportation habits). Many surveys combine questions of both types. Thus, a respondent will be asked if s(he) has heard or read about an issue, what s(he) knows about it, his (her) opinion, how strongly s(he) feels and why, interest in the issue, past experience with it, and also certain factual information that will help the survey analyst classify the responses (such as age, sex, marital status, occupation, and place of residence).

The questions may be open-ended ("Why do you feel that way?") or closed ("Do you approve or disapprove?"); they may ask the respondent to rate a political candidate or a product on some kind of scale; they may ask for a ranking of various alternatives. The questionnaire may be very brief—a few questions taking five minutes or less—or it can be quite long, requiring an hour or more of the respondent's time. Since it is inefficient to identify and approach a large national sample for only a few items of information, there are "omnibus" surveys that combine the interests of several clients in a single interview. In such surveys, the respondent will be asked a dozen questions on one subject, half a dozen more on another subject, and so on.

Because changes in attitude or behavior cannot be reliably ascertained from a single interview, some surveys employ a "panel design," in which the same respondents are interviewed two or more times. Such surveys are often used during election campaigns, or to chart a family's health or purchasing pattern over a period of time. They are also used to trace changes in behavior over time, as with the social experiments that study changes by low income families in work behavior in response to an income maintenance plan.

What Sort of People Work on Surveys

The survey worker best known to the public is the interviewer who calls on the phone, appears at the door, or stops people at a shopping center. Though survey interviewing may occasionally require long days in the field, it is normally part-time occasional work and is thus well suited for individuals who do not seek full-time employment or who wish to supplement their regular income. Previous experience is not usually required for an interviewing job. Most research companies will provide their own basic training for the task. The main requirements are an ability to approach strangers, to persuade them to participate in the survey, and to conduct the interview in exact accordance with instructions.

Behind the interviewers are the in-house research staff who design the survey, determine the sample design, develop the questionnaire, supervise the data collection, carry out the clerical and computer operations necessary to process the completed interviews, analyze the data, and write the reports. In most survey research agencies, the senior people will have taken courses in survey methods at the graduate level and will hold advanced degrees in sociology, statistics, marketing, or psychology, or they will have the equivalent in business experience. Middle-level supervisors and research associates frequently have similar academic backgrounds, or they have advanced out of the ranks of clerks, interviewers, or coders on the basis of their competence and experience.

Are Responses Confidential

The privacy of the information supplied by survey respondents is of prime concern to all reputable survey organizations. At the U.S. Bureau of the Census, for example, the confidentiality of the data collected is protected by law (Title 13 of the U.S. Code). In Canada, the Statistics Act guarantees the confidentiality of data collected by Statistics Canada, and other countries have similar safeguards. Also, a number of professional organizations that rely on survey methods have codes of ethics that prescribe rules for keeping survey responses confidential. The recommended policy for survey organizations to safeguard such confidentiality includes:

1. Using only code numbers for the identity of a respondent on a questionnaire, and keeping the code separate from that of the questionnaires.

2. Refusing to give names and addresses of survey respondents to anybody outside of the survey organization, including clients.

3. Destroying questionnaires and identifying information about respondents after the responses have been put onto computer tape.

4. Omitting the names and addresses of survey respondents from computer tapes used for analysis.

5. Presenting statistical tabulations by broad enough categories that individual respondents cannot be singled out.

How a Survey is Carried Out

As noted earlier, a survey usually has its beginnings when an individual or institution is confronted with an information need and there are no existing data that suffice. A politician may wish to tap prevailing voter opinions in his district about a proposal to build a superhighway through the county. A government agency may wish to assess the impact on the primary recipients and their families of one of its social welfare programs. A university researcher may wish to examine the relationship between actual voting behavior and expressed opinion on some political issue or social concern.

Designing a Survey

Once the information need has been identified and a determination made that existing data are inadequate, the first step in planning a survey is to lay out the objectives of the investigation. This is generally the function of the sponsor of the inquiry. The objectives should be as specific, clear-cut, and unambiguous as possible. The required accuracy level of the data has a direct bearing on the overall survey design. For example, in a sample survey whose main purpose is to estimate the unemployment rate for a city, the approximate number of persons to be sampled can be estimated mathematically when one knows the amount of sampling error that can be tolerated in the survey results.

Given the objectives, the methodology for carrying out the survey is developed. A number of interrelated activities are involved. Rules must be formulated for defining and locating eligible respondents, the method of collecting the data must be decided upon, a questionnaire must be designed and pretested, procedures must be developed for minimizing or controlling response errors, appropriate samples must be designed and selected, interviewers must be hired and trained (except for surveys involving self-administered questionnaires), plans must be made for handling nonresponse cases, and tabulation and analysis must be performed.

Designing the questionnaire represents one of the most critical stages in the survey development process, and social scientists have given a great deal of thought to issues involved in questionnaire design. The questionnaire links the information need to the realized measurement.

Unless the concepts are clearly defined and the questions unambiguously phrased, the resulting data are apt to contain serious biases. In a survey to estimate the incidence of robbery victimization, for example, one might want to ask "Were you robbed during the last six months?" Though apparently straightforward and clear-cut, the question does present an ambiguous stimulus. Many respondents are unaware of the legal distinction between robbery (involving personal confrontation of the victim by the offender) and burglary (involving breaking and entering but no confrontation), and confuse the two in a survey. In the National Crime Survey, conducted by the Bureau of the Census, the questions on robbery victimization do not mention "robbery." Instead, several questions are used that, together, seek to capture the desired reponses by using more universally understood phrases that are consistent with the operational definition of robbery.

Designing a suitable questionnaire entails more than well-defined concepts and distinct phraseology. Attention must also be given to its length, for unduly long questionnaires are burdensome to the respondent, are apt to induce respondent fatigue and hence response errors, refusals, and incomplete questionnaires, and may contribute to higher nonresponse rates in subsequent surveys involving the same respondents. Several other factors must be taken into account when designing a questionnaire to minimize or prevent biasing the results and to facilitate its use both in the field and in the processing center. They include such diverse considerations as the sequencing of sections or individual questions in the document, the inclusion of check boxes or precoded answer categories versus open-ended questions, the questionnaire's physical size and format, and instructions to the respondent or to the interviewer on whether certain questions are to be skipped depending on response patterns to prior questions.

Selecting the proper respondent in a sample unit is a key element in survey planning. For surveys where the inquiry is basically factual in nature, any knowledgeable person associated with the sample unit may be asked to supply the needed information. This procedure is used in the Current Population Survey, where the sample units are households and any responsible adult in a household is expected to be able to provide accurate answers on the employment-unemployment status of the eligible household members.

In other surveys, a so-called "household" respondent will produce erroneous and/or invalid information. For example, in attitude surveys it is generally accepted that a randomly chosen respondent from among the eligible household members produces a more valid cross section of opinion than does the nonrandomly selected house-hold respondent. This is because a nonrandomly selected individual acting as household respondent is more likely to be someone who is at home during the day, and the working public and their attitudes would be underrepresented.

Another important feature of the survey planning process is devising ways to keep response errors and biases to a minimum. These considerations depend heavily on the subject matter of the survey. For example, memory plays an important role in surveys dealing with past events that the respondent is expected to report accurately, such as in a consumer expenditure survey. In such retrospective surveys, therefore, an appropriate choice of reference period must be made so that the respondent is not forced to report events that may have happened too long ago to remember accurately. In general, attention must be given to whether the questions are too sensitive, whether they may prejudice the respondent, whether they unduly invade the respondent's privacy, and whether the information sought is too difficult even for a willing respondent to provide. Each of these concerns has an important bearing on the overall validity of the survey results.

Sampling Aspects

Virtually all surveys that are taken seriously by social scientists and policy makers use some form of scientific sampling. Even the decennial Censuses of Population and Housing are sampling techniques for gathering the bulk of the data items, although 100 percent enumeration is used for the basic population counts. Methods of sampling are well grounded in statistical theory and in the theory of probability. Hence, reliable and efficient estimates of a needed statistic can be made by surveying a carefully constructed sample of a population, as opposed to the entire population, provided of course that a large proportion of the sample members gives the requested information.

The particular type of sample used depends on the objectives and scope of the survey, including the overall survey budget, the method of data collection, the subject matter, and the kind of respondent needed. A first step, however, in deciding on an appropriate sampling method is to define the relevant population. This target population can be all the people in the entire nation or all the people in a certain city, or it can be a subset such as all teenagers in a given location. The population of interest need not be people; it may be wholesale businesses or institutions for the handicapped or government agencies, and so on.

The types of samples range from simple random selection of the population units to highly complex samples involving multiple stages or levels of selection with stratification and/or clustering of the units into various groupings. Whether simple or complex, the distinguishing characteristics of a properly designed sample are that all the units in the target population have a known, non-zero chance of being included in the sample, and the sample design is described in sufficient detail to permit reasonably accurate calculation of sampling errors. It is these features that make it scientifically valid to draw inferences from the sample results about the entire population, which the sample represents.

Ideally, the sample size chosen for a survey should be based on how reliable the final estimates must be. In practice, usually a trade-off is made between the ideal sample size and the expected cost of the survey. The complexity of a sample plan often depends on the availability of auxiliary information that can be used to introduce efficiencies into the overall design. For example, in a recent Federal Government survey on characteristics of health-care institutions, existing information about the type of care provided and the number of beds in each institution was useful in sorting the institutions into "strata," or groups by type and size, in advance of selecting the sample. The procedure permitted more reliable survey estimates than would have been possible if a simple random selection of institutions had been made without regard to size or type.

A critical element in sample design and selection is defining the source of materials from which a sample can be chosen. This source, termed the sampling frame, generally is a list of some kind, such as a list of housing units in a city, a list of retail establishments in a county, or a list of students in a university. The sampling frame can also consist of geographic areas with well-defined natural or artificial boundaries, when no suitable list of the target population exists. In the latter instance, a sample of geographic areas (referred to as segments) is selected and an interviewer canvasses the sample "area segments" and lists the appropriate units—households, retail stores or whatever—so that some or all of them can be designated for inclusion in the final sample.

The sampling frame can also consist of less concrete things, such as all possible permutations of integers that make up banks of telephone numbers, in the case of telephone surveys that seek to include unlisted numbers. The quality of the sampling frame—whether it is up-to-date and how complete—is probably the dominant feature for ensuring adequate coverage of the desired population.

Conducting a Survey

Though a survey design may be well conceived, the preparatory work would be futile if the survey were executed improperly. For personal or telephone interview surveys, interviewers must be carefully trained in the survey's concepts, definitions, and procedures. This may take the form of classroom training, self-study, or both. The training stresses good interviewer techniques on such points as how to make initial contacts, how to conduct interviews in a professional manner, and how to avoid influencing or biasing responses. The training generally involves practice interviews to familiarize the interviewers with the variety of situations they are likely to encounter. Survey materials must be prepared and issued to each interviewer, including ample copies of the questionnaire, a reference manual, information about the identification and location of the sample units, and any cards or pictures to be shown to the respondent.

Before conducting the interview, survey organizations frequently send an advance letter to the sample member explaining the survey's purpose and the fact that an interviewer will be calling soon. In many surveys, especially those sponsored by the Federal Government, information must be given to the respondent regarding the voluntary or mandatory nature of the survey, and how the answers are to be used.

Visits to sample units are scheduled with attention to such considerations as the best time of day to call or visit and the number of allowable callbacks for no-one-at-home situations. Controlling the quality of the field work is an essential aspect of good survey practice. This is done in a number of ways, most often through observation or rechecking of a subsample of interviews by supervisory or senior personnel, and through office editing procedures to check for omissions or obvious mistakes in the data.

When the interviews have been completed and the questionnaires filled out, they must be processed in a form so that aggregated totals, averages or other statistics can be computed. This will involve clerical coding of questionnaire items that are not already precoded. Occupation and industry categorizations are typical examples of fairly complex questionnaire coding that is usually done clerically. Also procedures must be developed for coding open-ended questions and for handling items that must be transcribed from one part of the questionnaire to another.

Coded questionnaires are keypunched, entered directly onto tape so that a computer file can be created, or entered directly into the computer. Decisions may then be needed on how to treat missing data and "not answered" items.

Coding, keypunching, and transcription operations are subject to human error and must be rigorously controlled through verification processes, either on a sample basis or 100 percent basis. Once a computer file has been generated, additional computer editing, as distinct from clerical editing of the data, can be accomplished to alter inconsistent or impossible entries, for example, a 6-year-old grandfather.

When a "clean" file has been produced, the survey data are in a form where analysts can specify to a computer programmer the frequency counts, cross-tabulations, or more sophisticated methods of data presentation or computation that are needed to help answer the concerns outlined when the survey was initially conceived.

The results of the survey are usually communicated in publications and in verbal presentations at staff briefings or more formal meetings. Secondary analysis is also often possible to those other than the survey staff by making available computer data files at nominal cost.

Shortcuts to Avoid

As we have seen, conducting a creditable survey entails scores of activities, each of which must be carefully planned and controlled. Taking shortcuts can invalidate the results and badly mislead the user. Four types of shortcuts that crop up often are failure to use a proper sampling procedure, no pretest of the field procedures, failure to follow up nonrespondents, and inadequate quality control.

One way to ruin an otherwise well-conceived survey is to use a convenience sample rather than one based on a probability design. It may be simple and cheap, for example, to select a sample of names from a telephone directory to find out which candidate people intend to vote for. However, this sampling procedure could give incorrect results since persons without telephones or with unlisted numbers would have no chance to be reflected in the sample, and their voting preferences could be quite different from persons who have listed telephones. This is what happened with the *Literary Digest* presidential poll of 1936 when use of lists of telephone owners, magazine subscribers, and car owners led to a prediction that President Roosevelt would lose the election.

A pretest of the questionnaire and field procedures is the only way of finding out if everything "works," especially if a survey employs a new procedure or a new set of questions. Since it is rarely possible to foresee all the possible misunderstandings or biasing effects of different questions and procedures, it is vital for a well-designed survey plan to include provision for a pretest. This is usually a small-scale pilot study to test the feasibility of the intended techniques or to perfect the questionnaire concepts and wording.

Failure to follow up nonrespondents can ruin an otherwise well-designed survey, for it is not uncommon for the initial response rate to most surveys to be under 50 percent. Plans must include returning to sample households where no one was home, attempting to persuade persons who are inclined to refuse, and, in the case of mail surveys, contacting all or a subsample of the nonrespondents by telephone or personal visit to obtain a completed questionnaire. A low response rate does more damage in rendering a survey's results questionable than a small sample, since there is no valid way of scientifically inferring the characteristics of the population represented by the nonrespondents.

Quality control, in the sense of checking the different facets of a survey, enters in at all stages—checking sample selection, verifying interviews, and checking the editing and coding of the responses, among other things. In particular, sloppy execution of the survey in the field can seriously damage the results. Without proper quality control, errors can occur with disastrous results, such as selecting or visiting the wrong household, failing to ask questions properly, or recording the incorrect answer. Insisting on proper standards in recruitment and training of interviewers helps a great deal, but equally important is proper review, verification, and other quality control measures to ensure that the execution of a survey corresponds to its design.

Using the Results of a Survey

How Good is the Survey

The statistics derived from a survey will rarely correspond exactly with the unknown truth. (Whether "true" values always exist is not important in the present context. For fairly simple measurements—the average age of the population, the amount of livestock on farms, and so on—the concept of a true value is fairly straightforward. Whether true values exist for measurements of such items as attitudes toward political candidates, I.Q.'s, and so forth, is a more complex matter.)

Fortunately, the value of a statistic does not depend on its being exactly true. To be useful, a statistic need not be exact, but it does need to be sufficiently reliable to serve the particular needs. No overall criterion of reliability applies to all surveys since the margin of error that can be tolerated in a study depends on the actions or recommendations that will be influenced by the data. For example, economists examining unemployment rates consider a change of 0.2 percent as having an important bearing on the United States economy. Consequently, in the official United States surveys used to estimate unemployment, an attempt is made to keep the margin of error below 0.2 percent. Conversely, there are occasions when a high error rate is acceptable. Sometimes a city will conduct a survey to measure housing vacancies to determine if there is a tight housing supply. If the true vacancy rate is very low, say 1 percent, survey results that show double this percentage will not do any harm; any results in the range of 0 to 2 or 3 percent will lead to the same conclusion—a tight housing market.

In many situations the tolerable error will depend on the kind of result expected. For example, during presidential elections the major television networks obtain data on election night from a sample of election precincts, in order to predict the election results early in the evening. In a state in which a large difference is expected (pre-election polls may indicate that one candidate leads by a substantial majority and is likely to receive 60 percent of the vote), even with an error of 5 or 6 percent it would still be possible to predict the winner with a high probability of being correct. A relatively small sample size may be adequate in such a state. However, much more precise estimates are required in states where the two candidates are fairly evenly matched and where, say, a 52–48 percent vote is expected.

Thus, no general rule can be laid down to determine the reliability that would apply to all surveys. It is necessary to consider the purpose of the particular study, how the data will be used, and the effect of errors of various sizes on the action taken based on the survey results. These factors will affect the sample size, the design of the questionnaire, the effort put into training and supervising the interview staff, and so on. Estimates of error also need to be considered in analyzing and interpreting the results of the survey.

Sources of Errors

In evaluating the accuracy of a survey, it is convenient to distinguish two sources of errors: (1) sampling errors, and (2) nonsampling errors, including the effect of refusals and not-at-homes, respondents providing incorrect information, coding or other processing errors, and clerical errors in sampling.

Sampling errors   Good survey practice includes calculation of sampling errors, which is possible if probability methods are used in selecting the sample. Furthermore, information on sampling errors should be made readily available to all users of the statistics. If the survey results are published, data on sampling errors should be included in the publication. If information is disseminated in other ways, other means of informing the public are necessary. Thus, it is not uncommon to hear television newscasters report on the size of sampling errors as part of the results of some polling activity.

There are a number of ways of describing and presenting data on sampling errors so that users can take them into account. For example, in a survey designed to produce only a few statistics (such as the votes that the candidates for a particular office are expected to get), the results could be stated that Candidate A's votes are estimated at 57 percent with the error unlikely to be more than 3 percent, so that this candidate's votes are expected to fall in the range of 54–60 percent. Other examples can be found in most publications of the principal statistical agencies of the United States Government, such as the Bureau of the Census.

Nonsampling errors   Unfortunately, unlike sampling errors, there is no simple and direct method of estimating the size of nonsampling errors. In most surveys, it is not practical to measure the possible effect on the statistics of the various potential sources of error. However, in the past 30 or 40 years, there has been a considerable amount of research on the kinds of errors that are likely to arise in different kinds of surveys. By examining the procedures and operations of a specific survey, experienced survey statisticians will frequently be able to assess its quality. Rarely will this produce actual error ranges, as for sampling errors. In most cases, the analyst can only state that, for example, the errors are probably relatively small and will not affect most conclusions drawn from the survey, or that the errors may be fairly large and inferences are to be made with caution.

Nonsampling errors can be classified into two groups: random types or errors whose effects approximately cancel out if fairly large samples are used, and biases, which tend to create errors in the same direction and thus cumulate over the entire sample. With large samples, the possible biases are the principal causes for concern about the quality of a survey.

Biases can arise from any aspect of the survey operation. Some of the main contributing causes of bias are:

1. Sampling Operations. There may be errors in sample selection, or part of the population may be omitted from the sampling frame, or weights to compensate for disproportionate sampling rates may be omitted.

2. Noninterviews. Information is generally obtained for only part of the sample. Frequently there are differences between the noninterview population and those interviewed.

3. Adequacy of Respondent. Sometimes respondents cannot be interviewed and information is obtained about them from others, but the "proxy" respondent is not always as knowledgeable about the facts.

4. Understanding the Concepts. Some respondents may not understand what is wanted.

5. Lack of Knowledge. Respondents in some cases do not know the information requested, or do not try to obtain the correct information.

6. Concealment of the Truth. Out of fear or suspicion of the survey, respondents may conceal the truth. In some instances, this concealment may reflect a respondent's desire to answer in a way that is socially acceptable, such as indicating that s(he) is carrying out an energy conservation program when this is not actually so.

7. Loaded Questions. The question may be worded to influence the respondents to answer in a specific (not necessarily correct) way.

8. Processing Errors. These can include coding errors, data keying, computer programming errors, and so on.

9. Conceptual Problems. There may be differences between what is desired and what the survey actually covers. For example, the population or the time period may not be the one for which information is needed, but had to be used to meet a deadline.

10. Interviewer Errors. Interviewers may misread the question or twist the answers in their own words and thereby introduce bias.

Obviously, each survey is not necessarily subject to all these sources of error. However, a good survey statistician will explore all of these possibilities. It is considered good practice to report on the percent of the sample that could not be interviewed, and as many of the other factors listed as practicable.

Budgeting a Survey

We have seen from the preceding sections that many different stages are involved in a survey. These include tasks such as planning, sample design, sample selection, questionnaire preparation, pretesting, interviewer hiring

and training, data collection, data reduction, data processing, and report preparation. From a time point of view, these different stages are not necessarily additive since many of them overlap. This is illustrated in the attached diagram, which portrays the sequence of steps involved in a typical personal interview survey. Some steps, such as sample design and listing housing units in the areas to be covered in the survey, can be carried out at the same time a questionnaire is being revised and put into final form. Although they are not additive, all of these steps are time consuming, and one of the most common errors is to underestimate the time needed by making a global estimate without considering these individual stages.

How much time is needed for a survey? This varies with the type of survey and the particular situation. Sometimes a survey can be done in two or three weeks, if it involves a brief questionnaire, and if the data are to be collected by telephone from a list already available. More usually, however, a survey of several hundred or a few thousand individuals will take anywhere from a few months to more than a year, from initial planning to having results ready for analysis.

A flow diagram for a particular survey is very useful in estimating the cost of such a survey. Such a diagram ensures that allowance is made for the expense involved in the different tasks, as well as for quality checks at all stages of the work. Thus, among the factors that enter into an expense budget are the following:

1. Staff time for planning the study and steering it through the various stages.
2. Labor and material costs for pretesting the questionnaire and field procedures.
3. Supervisory costs for interviewer hiring, training, and supervision.
4. Interviewer labor costs and travel expense (and meals and lodging, if out-of-town).
5. Labor and expense costs of checking a certain percentage of the interviews (by reinterviews).
6. Cost of preparing codes for transferring information from the questionnaire.
7. Labor and material costs for editing, coding, and keypunching the information from the questionnaire onto computer tape.
8. Cost of spot-checking to assure the quality of the editing, coding, and keypunching.
9. Cost of "cleaning" the final data tapes—that is, checking the tapes for inconsistent or impossible answers.

STAGES OF A SURVEY

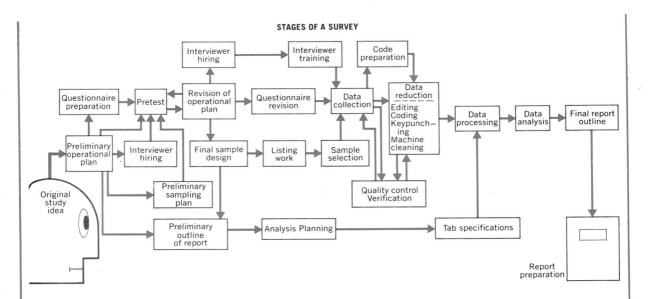

10. Programming costs for preparing tabulations and special analyses of the data.

11. Computer time for the various tabulations and analyses.

12. Labor time and material costs for analysis of the data and report preparation.

13. Telephone charges, postage, reproduction, and printing costs.

An integral part of a well-designed survey, both in terms of time and of costs, is allowance for quality checks all along the way. For example, checks have to be made that the sample was selected according to specifications, that the interviewers did their work properly, that the information from the questionnaires was coded accurately, that the keypunching was done correctly, and that the computer programs used for data analysis work properly. For these reasons, a good survey does not come cheap, although some are more economical than others. As a rule, surveys made by personal interview are more expensive than by mail or by telephone; and costs will increase with the complexity of the questionnaire and the amount of analysis to be carried out. Also, surveys that involve more interviews tend to be cheaper on a per interview basis than surveys with fewer interviews. This is particularly so where the sample size is less than about a thousand because "tooling up" is involved for just about any survey, except one that is to be repeated on the same group.

Where to Get More Information

Several professional organizations have memberships heavily involved in survey research. They also frequently have workshops or sessions on surveys as parts of their regional and annual meetings. The principal organizations are the following:

1. The *American Statistical Association* is concerned with survey techniques and with general application of survey data. It has a separate Section on Survey Research Methods, which sponsors sessions on surveys at the annual meetings of the association. The many chapters of the association in the various parts of the country also periodically have meetings and workshops on survey methods, and its publications, the *Journal of the American Statistical Association* and the *American Statistician*, carry numerous articles about surveys.

2. The *American Marketing Association* is concerned, among other things, with the application of survey methods to marketing problems. Like the American Statistical Association, it sponsors sessions on survey methods at its annual meetings, and still other sessions are sponsored by its local chapters. Its publications, the *Journal of Marketing* and the *Journal of Marketing Research*, frequently contain articles on surveys.

3. The *American Association for Public Opinion Research* focuses on survey methods as applied to social and media problems. Its journal, the *Public Opinion Quarterly*, regularly carries articles on survey techniques and on the application of survey methods to political and social problems.

A number of other professional associations in North America place emphasis periodically on survey methods—for example, the Statistical Society of Canada, the American Sociological Association, the American Political Science Association, the Association for Consumer Research, the American Public Health Association, the American Psychological Association, and the Canadian Psychological Association. There are also various business oriented associations such as the Advertising Research Foundation and the American Association of Advertising Agencies, that give attention to survey methods as applied to business. These and other associ-

ations publish a number of journals that carry a great deal of material on survey methods.

There are many good books on survey methods written for nontechnical readers. A few of these are:

1. Tanur, Judith, et al., *Statistics: A Guide to the Unknown*. San Francisco: Holden-Day Pub. Co., 1972.
2. Hauser, Philip, *Social Statistics in Use*. New York: Russell Sage Foundation, 1975.
3. Williams, William H., *A Sampler on Sampling*. New York: John Wiley & Sons, 1978.

For further information contact . . .
Executive Director
American Statistical Association
806 15th Street, N.W.
Washington, D.C. 20005

Exercises

1.12 In a household survey is it better to (a) ask questions of the person answering the door or (b) obtain a list of household residents from the person answering the door and select one at random to interview?

1.13 In a household survey if no one is home at one house is it better to (a) return to that home a second time, or third time, if necessary, until someone is home or (b) go to the next door, or next house, if necessary, until finding someone at home?

1.14 In a household survey if the respondent does not understand the question, is it better to (a) repeat the question, using essentially the same wording that is in the questionnaire or (b) use your own words in repeating the question, using a different vocabulary than is in the questionnaire?

1.15 When recording a response to a household survey is it better for the interviewer to (a) rephrase the response in the interviewer's words to correct bad grammer and maintain respondent anonymity or (b) record the response in the respondent's exact words, as nearly as possible?

1.16 Which of the following situations represent sampling error and which represent nonsampling errors?

(a) A pollster interviews 100 voters and finds 57 percent support for a political candidate. The next day the pollster interviews 100 more voters and finds 46 percent support for the candidate.

(b) In a household survey an interviewer fills in some of the answers left blank by the respondent.

(c) In a household survey the interviewer rewords the questions.

1.17 Which of the 10 types of nonsampling errors are represented by the following?

(a) An error is made in compiling the results of a survey.

(b) An interviewer finds no one home, so the interviewer goes next door to ask the neighbor to respond to questions for the person not at home.

(c) The respondent is afraid of revealing information and does not tell the interviewer the truth.

1.4
Meaning and Role of Statistics

There is more than one meaning for the word "statistics," and more than one meaning for the word "statistician." When you listen to a broadcast of a sports event, the announcer frequently reviews the game's "statistics," mentioning such numbers as batting averages, numbers of free throws attempted, total yardage by the star fullback, and so on. This is one use of the word statistics that refers to its plural form. In its second meaning, the word statistics means the science of data summary and analysis. A third use of the word statistics occurs in Chapter 4.

TWO MEANINGS OF THE WORD STATISTICS

Statistics *(plural) are numbers, usually used as summaries of larger sets of data.*
Statistics *(singular) is the science of data analysis; that is, Statistics is concerned with scientific methods for collecting, organizing, summarizing, presenting, and analyzing sample data as well as drawing valid conclusions about population characteristics and making reasonable decisions on the basis of such analysis.*

QUANTITATIVE AND QUALITATIVE DATA

Statistics as numbers will be considered first. To understand some of the many forms statistics can take, consider a registration card that a customer might be asked to fill out following the purchase of an appliance (see Figure 1.3). This card has several types of data. First there are **qualitative**-type data, such as the name and address. Some of the data, such as the price of the item, or the age of the main user, indicate "how much" or "how many" and are called **quantitative**-type data. Other data are not so easy to classify, such as the serial number or the product model number. If the serial number indicates *how many* items were manufactured up to that point, it could be quantitative. But if the serial number serves merely as an identification number for that particular appliance, it is a qualitative item of information.

TWO TYPES OF DATA

Quantitative *data represent information that answers the questions "how much" or "how many" and is measured on a numerical scale.*
Qualitative *data represent information that identifies or names some quality, category, or characteristic, but is not quantitative.*

```
┌─────────────────────────────────────────────────────────────────────────┐
│ CUSTOMER REGISTRATION                                                     │
│ Mr.                                                                       │
│ Mrs. _____ Date Purchased _____         │
│ Ms.                                                                       │
│ Address _____ Apt. # _____               │
│ City _____ State _____ Zip Code _____       │
│ Product                                                                   │
│ Model No. _____ Serial No. _____ Price _____      │
│ Dealer's Name/City _____           │
│ Selected Product                      Type of Store                       │
│ as Result of:                         Where Purchased                     │
│                                                                           │
│ 1. _____ Previous Ownership        1. _____ Discount Store          │
│ 2. _____ Store Display             2. _____ Department Store        │
│ 3. _____ Catalog                   3. _____ Hardware Store          │
│ 4. _____ Magazine                  4. _____ Appliance Store         │
│ 5. _____ Newspaper                 5. _____ Other _____        │
│ 6. _____ Other _____ The age of the Main User is _____     │
│ TO HELP US SERVE YOU BETTER, PLEASE MAIL THIS CARD IMMEDIATELY            │
└─────────────────────────────────────────────────────────────────────────┘
```

FIGURE 1.3 A Typical Customer Registration Card

On the customer registration card the only quantitative data are the price of the appliance and the age of the main user. The other information is qualitative. Note that some of the qualitative data are numerical, such as Zip Code and Serial Number, while others are nonnumerical, such as Name. Some qualitative data may be either numerical or nonnumerical, such as "Selected Appliance as Result Of" or "Type of Store Where Purchased," depending on whether the verbal description is used or the numerical designation for each category is used. The numbers 1 through 6 (in the case of "Selected Product As Result Of") merely serve as surrogate names for the six different categories.

Quantitative data, on the other hand, must be numerical, because they answer either "how much" or "how many." Most of this book is devoted to handling quantitative data, because statistics is generally a quantitative science. But many important and useful statistical methods use qualitative data, as you will find throughout this book.

PREPARING A QUESTIONNAIRE

Consider once again the Customer Registration card. The idea of the card is for you, the customer, to fill it out and send it in. Now the actual card this example was copied from had requested much more information. Most of the questions have been left out for the sake of simplicity. The company wants to know all about you, the paying customer. What inspired you to buy their product? How can they get you to buy more of their products? How can they get others to buy their products? Which of their marketing techniques are the most effective?

If the company requests too much information from you, you might simply throw the card (which might now resemble a five-page questionnaire) into the nearest wastebasket. So the company keeps the questionnaire brief and easy to answer. (Notice the multiple choice questions.) And by including a blank space for "Serial No." many people will assume that if they don't fill out the card in its entirety and send it in their warranty will not be valid. Thus the company may get as many as 80 or 90 percent of its customers to fill out cards and send them in. This can be compared to responses to questionnaires that often are as low as 10 or 20 percent.

Sampling questionnaires, such as this one, require careful planning in order to be successful. A successful questionnaire is usually the result of a team of people, with skills in fields such as marketing, psychology, and (you guessed it) statistics. Many of the types of questions a company hopes to get answered by such a questionnaire can only be answered using statistical methods.

The use of statistical methods requires the right kind of data. The questions need to be written with the statistical method in mind. Many times a survey turns out to be a wasted effort when the person in charge of the survey brings the data, already collected, to a statistician and says "Analyze these data," but the data are not capable of answering the questions the survey was intended to answer.

SUMMARIZING DATA FROM A QUESTIONNAIRE

The Customer Registration Card in Figure 1.3 appears to be the result of good planning on the part of the company that issued it. Now suppose that you and many others like you have sent in their Customer Registration Cards (attn: Warranty Registration Dept.) and the company now has stacks of cards available for study. One of the first things the company probably does is bring in their "information systems" specialist to put all these data into their computer in such a way that all the interesting facts and figures can be extracted quickly and accurately.

After the cards are collected and compiled, many more quantitative measurements are available. Consider the following questions, whose answers involve quantitative measurements.

1. How many people sent in cards for Model No. TM-151?
2. Of the TM-151 cards, what percentage were purchased in discount stores?
3. What is the average price of the TM-151's that were sold in appliance stores?

Clearly there are many other measurements of a quantitative nature that are now available from this collection of information.

It is always necessary to summarize the results of surveys such as this one, because there are simply too much data to make any sense out of in its present form. Suppose the company wants to know how many appliances (of all types) are sold in the six major geographic regions of the United States. The states are assigned to various geographic regions and Figure 1.4 is constructed.

In Figure 1.4 the region represents a qualitative variable and the number of appliances sold is a quantitative variable. Often a pictorial representation of this information is easier for the reader to understand, and is especially helpful to readers who don't have much time to spend studying the numbers. Several graphical methods of displaying such information are given in the next chapter.

Region	No. of Appliances Sold
Northeast	1436
Southeast	1724
Central	1125
Midwest	838
Northwest	175
Southwest	1840

FIGURE 1.4 The Number of Appliances Sold by Geographic Region, as Reported by the Customer Registration Cards Received During 1980

Exercises

1.18 What is the difference between quantitative data and qualitative data?

1.19 Identify each item of numerical information asked for on an employee questionnaire as either quantitative or qualitative.

(a) Social Security Number

(b) Date of Birth

(c) Age

(d) Weight

(e) Height

(f) Employee Number

(g) Phone Number

(h) Street Address

(i) Number of Dependents

(j) Years of Professional Experience

1.20 Given below is a reproduction of a product information card that accompanied an appliance. Which of the pieces of information requested represent quantitative data and which represent qualitative data?

Model Number ———————————————— Date Purchased————

Name ———————————————— Address ————————————————

City ———————————————— State ———————————— ZIP ————

Name of Store ————————————————

1. IS THIS APPLIANCE A REPLACEMENT?
   ☐ YES   ☐ NO

2. WHO WAS THE APPLIANCE PURCHASED BY?
   ☐ MAN   ☐ WOMAN   ☐ BOTH

3. IF YOU PURCHASED THIS APPLIANCE DID YOU SEE IT DEMONSTRATED?
   ☐ YES   ☐ NO

4. AGE OF HEAD OF HOUSEHOLD
   ☐ 25-34   ☐ 45-54   ☐ OVER 54
   ☐ 35-44   ☐ UNDER 25

5. IF GIFT, WHAT WAS OCCASION?
   ☐ BIRTHDAY   ☐ OTHER
   ☐ CHRISTMAS   ☐ GRADUATION
   ☐ MOTHER'S DAY

6. WHICH FACTORS MOST INFLUENCED YOUR PURCHASE?
   ☐ MAGAZINE   ☐ NEWS-AD
   ☐ PRICE   ☐ FEATURES
   ☐ RECOM-MENDED   ☐ WARRANTY
   ☐ PREV. OWNER   ☐ RADIO/TV
   ☐ BRAND NAME
   ☐ SALES PERSON

1.21   The appliance company had a specific purpose in mind for each of the questions on the product information card of Exercise 1.20 when it designed the card. A summary or tabulation will be made of the responses on this card as they are received from the customer. Suppose you are responsible for making recommendations aimed at improving future sales based on such a tabulation. How do you see each of the questions 1 to 6 as providing specific help to you in making such recommendations?

1.5
Review Exercises

1.22   The following is a list of deceased U.S. presidents, the age at which they were inaugurated for the first time, and the age at which they died.

Name	Inaugurated	Died	Name	Inaugurated	Died
1. Washington	57	67	19. Hayes	54	70
2. J. Adams	61	90	20. Garfield	49	49
3. Jefferson	57	83	21. Arthur	50	56
4. Madison	57	85	22. Cleveland	47	71
5. Monroe	58	73	23. Harrison	55	67
6. J. Q. Adams	57	80	24. McKinley	54	58
7. Jackson	61	78	25. T. Roosevelt	42	60
8. Van Buren	54	79	26. Taft	51	72
9. Harrison	68	68	27. Wilson	56	67
10. Tyler	51	71	28. Harding	55	57
11. Polk	49	53	29. Coolidge	51	60
12. Taylor	64	65	30. Hoover	54	90
13. Fillmore	50	74	31. F. Roosevelt	51	63
14. Pierce	48	64	32. Truman	60	88
15. Buchanan	65	77	33. Eisenhower	62	78
16. Lincoln	52	56	34. Kennedy	43	46
17. A. Johnson	56	66	35. L. Johnson	55	64
18. Grant	46	63			

Use a table of random numbers to obtain a random sample of size 5 of the ages in which the U.S. presidents were first inaugurated, from this list.

1.23   Use a table of random numbers to obtain a systematic sample of size 7 from the names of the presidents given in Exercise 1.22.

1.24   Use a table of random numbers to obtain a stratified sample of seven ages at which U.S. presidents died, from the list in Exercise 1.22. Divide the list of presidents into seven strata of equal size according to the order in which they appear.

1.25   Describe how you might set up an approximate random-sampling method for drawing a sample of:

(a) Ten employees out of 147 employees in a company.

(b) Forty parts, out of a shipment of metal parts, where the shipment consists of 83 boxes, and each box has 144 parts in it.

(c) Twenty students currently taking a statistics course, where the course has 15 sections of students, and the number of students per section ranges from 23 to 48.

(d) Twenty-five businesses in a city that has about 500 businesses.

(e) Thirty accounts in a department store that has over 1500 accounts.

1.26 Restaurants often have comment cards available for their customers to use. Comment on the advantages and disadvantages of this method of obtaining a sample.

1.27 A newsletter by a conservative women's organization requested that the readers send their opinions regarding a proposed bill to their local Congressman. Comment on the advantages and disadvantages of this method of obtaining a sample.

1.28 Describe a method that the U.S. Department of Agriculture could use in July to estimate the total amount of the U.S. fall wheat harvest.

1.29 A government safety inspector is in charge of seeing that all offshore oil rigs adhere to certain regulations. She does not have the staff or the resources to inspect every rig every month, so she sets up a stratified sampling scheme, where the strata are geographical areas, selected at random, and the oil rigs within each stratum are selected at random. All site visits are unannounced. Discuss the advantages and disadvantages of this method over a systematic census where each oil rig is visited on a regular basis according to a published schedule.

Bibliography

Additional material on the topics presented in this chapter can be found in the following publication.

American Statistical Association (1980). *What Is A Survey?*, Publication of the subcommittee of the section on survey research methods.

2

Displaying Sample Data

PRELIMINARY REMARKS

One of the major concerns of individuals who are confronted with **quantitative** data is how to **summarize** and **display** the data in a manner that will aid in a correct interpretation for both themselves and others who may see only the summary or display. For example, meaningful graphical displays are needed to communicate the information contained in the quantitative data for the following situations:

1. corn yields on 200 Nebraska farms;
2. the frequency distribution of genotypes resulting from the dihybrid cross of fruit flies (Drosophila melanogaster);
3. typing speeds of students in a typing class;
4. classifications of company employee hirings for the past 5 years to demonstrate compliance with equal opportunity hiring policies;
5. the number of stocks that have increased in price on the daily stock market for each of the last 30 days;
6. energy consumption and gasoline prices since 1972;
7. percentage improvement in strength during one semester for members of a weightlifting class;
8. census counts of persons by age and sex;
9. vital capacity of lungs by age of smoker.

These situations, as well as others, provide many opportunities for innovative ways of displaying quantitative data. In this chapter some basic and easily used **graphical** techniques are presented. Many of these graphical procedures will be used throughout this text to display sample data prior to an analysis of the data. Such persistent usage of these methods should aid the reader in attaining some perspective regarding the appropriate application of these various types of graphs. A rough rule for application of these graphs is that if the general form of the group performance of the data is of interest, the methods of the first two sections of this chapter are more appropriate. If the primary interest concerns the relative location or size of each particular observation as compared with the group, then the methods of Section 2.3 are more appropriate. Showing the relationship between two variables calls for the use of the plots of Section 2.4.

2.1
Frequency Distributions

PROBLEM SETTING 2.1

A typing instructor has tried a new typing instruction technique with her students. As part of the evaluation process the students are given typing tests. For each of these tests the instructor records each student's scores indicating speed (i.e., net words per minute) and accuracy (i.e., total errors). Twenty students made the following scores for typing speeds.

68, 72, 91, 47, 52, 75, 63, 55, 65, 35
84, 45, 58, 61, 69, 22, 46, 55, 66, 71

THE STEM AND LEAF PLOT

A casual examination of these data by the instructor shows the highest score to be 91 and the lowest to be 22; however, it is difficult for even a trained eye to determine much else from these data. Therefore, at this point a technique for displaying quantitative data—called a **stem and leaf plot**—will be presented. This procedure, in addition to being an easy and quick way of displaying the data, is also extremely useful for arranging the observations from smallest to largest; there will be many opportunities to use it throughout this book. The **stem and leaf plot** is formed by starting with the **stem** and then putting on the **leaves**. In this case, since the numbers have only 2 digits, the first digit of each number, the tens digit, is the **stem**, and the second digit, the ones digit, is the **leaf**. A vertical line is drawn with the **stem** on the left of the line and the **leaves** on the right of the line. Figure 2.1a shows the placing of the stems for the previous typing scores, while Figure 2.1b shows how the first score (68) is placed. Each new typing score added to the stem and leaf plot results in one

(a) Stems Only	(b) Stems With First Observation (68)	(c) Complete Stem and Leaf Plot
2	2	2 | 2
3	3	3 | 5
4	4	4 | 7 5 6
5	5	5 | 2 5 8 5
6	6 | 8	6 | 8 3 5 1 9 6
7	7	7 | 2 5 1
8	8	8 | 4
9	9	9 | 1

FIGURE 2.1 Construction of the Stem and Leaf Plot for Typing Scores

"THIS ISN'T EXACTLY WHAT I HAD IN MIND WHEN I ASKED FOR A 'STEM AND LEAF' PLOT!"

new leaf. The completed stem and leaf plot for the typing scores is given by Figure 2.1c.

To read the typing scores from Figure 2.1, start at the first row and read the score 22. The second row contains 35 while the third row contains three scores: 47, 45, and 46, and so on for the other rows. Note that the number of leaves must be equal to the number of observations. From Figure 2.1, the largest (91) and smallest (22) scores can readily be seen. In addition, an entire picture of how the scores are distributed (or scattered) emerges. For example, it is readily apparent that there are more scores in the sixties than in any other group; only five scores are less than 50, and only two scores are above 80. Additionally, some of the numbers on the stem may have no corresponding leaves. That is, in Figure 2.1 the stem position "3" would have no corresponding leaf if the observation "35" were removed from the data set.

CONSTRUCTING A HISTOGRAM FROM A STEM AND LEAF PLOT

An interesting result is obtained by rotating the stem and leaf plot 90 degrees counterclockwise so that it appears as in Figure 2.2. If each of the leaves is thought of as a building block, the stem and leaf plot resembles a graph known as a **histogram**. In general, a histogram is a graphical method for presenting data, where the observations are located on the horizontal axis (usually grouped into intervals) and the frequency of those observations is depicted in

FIGURE 2.2 Stem and Leaf Plot Rotated to Resemble a Histogram

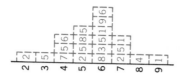

FIGURE 2.3 A Histrogram Constructed from a Stem and Leaf Plot

some way along the vertical axis. In fact, a stem and leaf plot is about the easiest way to generate a histogram. The stem and leaf "building block" histogram would appear as in Figure 2.3.

CONSTRUCTING A HISTOGRAM WITH EQUAL CLASS WIDTHS

The stem and leaf plot used to construct the "building block" histogram of Figure 2.3 is quite easy to use but there are settings where a satisfactory histogram cannot be derived immediately from a stem and leaf plot. For example, if the leading decimal place assumes only two or three different values, a stem and leaf plot would not be very helpful. There are also situations where only a summary of the data is available and a stem and leaf plot couldn't be used at all. The construction of a more general type of histogram is now considered. First, however, several terms associated with histograms need to be defined.

TERMS ASSOCIATED WITH HISTOGRAMS

*A **class** is an interval containing sample observations. Each sample observation is classified into one and only one class.*
*The **class boundaries** are the endpoints or limits for each class.*
*The **class width** is the distance between the class boundaries of a class.*
*The **class mark** is the midpoint of a class and is found midway between the class boundaries.*
*The **frequency** of a class is the number of sample observations associated with a class.*
*The **relative frequency** of a class is its frequency divided by the total number of observations in the sample.*

To illustrate each of these terms see Figure 2.4, which appears in the example that follows. There are six classes in the histogram in that example. The first class has class boundaries 147.5 and 152.5, the second class has boundaries

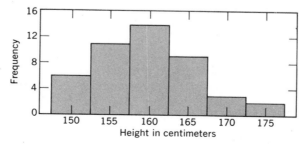

FIGURE 2.4 A Frequency Histogram for the Heights of 45 Female High School Seniors

152.5 and 157.5, and so on. The class widths are each 5.0. The class marks are noted in the figure as 150, 155, and so on, and lie exactly midway between the class boundaries. The frequencies of the six classes are 6, 11, 14, 9, 3, and 2, respectively, while the relative frequencies are obtained by dividing these numbers by the total number of observations, 45, to get .13, .24, .31, .20, .07, and .04. The relative frequencies always sum to 1.00, except for roundoff error.

| HISTOGRAM WITH EQUAL CLASS WIDTHS | A **histogram** for classes with **equal class widths** is a graphical presentation of the sample data using classes on the horizontal axis and either frequency or relative frequency as the vertical axis. |

Formal rules for the construction of a histogram are stated for the case where all class widths are the same. In the next section, the case of unequal class widths is considered. Many computer programs are available for constructing histograms and stem and leaf plots.

RULES FOR GRAPHICAL DISPLAY OF QUANTITATIVE DATA IN A HISTOGRAM

1. If the data set is of a reasonable size (say 200 observations or less), a stem and leaf plot can be quickly constructed by hand and may provide a good starting point for constructing a histogram. For larger data sets, a computer could be used to make a stem and leaf plot.
2. Decide upon the number of classes to be used in grouping the data, usually from 5 to 20, depending upon the number of observations. (See the suggested rule in this section.)
3. Decide upon class boundaries such that each class has the same width and so that every observation can be classified uniquely into exactly one class.
4. In a frequency histogram, the height of each vertical bar in the histogram represents the number of observations in each class. In a relative frequency histogram, the height of each vertical bar in the histogram represents the proportion of observations in each class. For example, a class with 10 observations should be twice as high as a class with 5 observations, but only a tenth as high as a class with 100 observations.

NUMBER OF CLASSES

The rules for displaying sample data in a histogram are not very specific with respect to the number of classes to use in the construction of a histogram. The reason for this vagueness is that a certain amount of artistry is involved and there are no hard and fast rules that are universally accepted by statisticians. Rather, statisticians are usually aiming at an accurate representation of the data. That is, a histogram should not misrepresent the data. Using too few classes gives an inaccurate picture by smoothing out too many details. (An example of this situation is given in Figure 2.5.) Too many classes present too many details and obscure the overall view. (An example of this situation is given in Figure

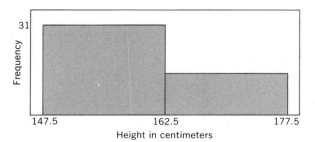

FIGURE 2.5 A Frequency Histogram Using Only Two Classes

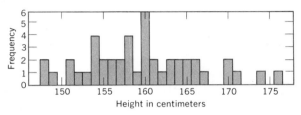

FIGURE 2.6 A Frequency Histogram with Too many Classes

2.6.) With this thought in mind, a rule is offered, which although somewhat arbitrary in nature does allow some uniformity to be applied in the construction of histograms. For some types of data, such as integer-valued data, the integers themselves may be logical classes to use, and rules for the number of classes to use may be inappropriate.

RULE FOR FINDING THE NUMBER OF CLASSES TO USE IN THE CONSTRUCTION OF A HISTOGRAM

The number of classes, k, to be used in the construction of a histogram for sample data is the smallest integer value of k such that $2^k \geq n$, where n is the number of observations in the sample. Examples follow.

Number of Observations n	Number of Classes k
8 or less	3
9 to 16	4
17 to 32	5
33 to 64	6
65 to 128	7
129 to 256	8
257 to 512	9
513 to 1024	10
more than 1024	The smallest value of k such that $2^k \geq n$

EXAMPLE

Display the heights of 45 female high school seniors in a histogram where the heights are recorded in centimeters.

170	151	154	160	158
154	171	156	160	157
160	157	148	165	158
159	155	151	152	161
156	164	156	163	174
153	170	149	166	154
166	160	160	161	154
163	164	160	148	162
167	165	158	158	176

From the previous rule 6 classes are used with 45 observations. The smallest observation is 148 and the largest is 176 with a difference of $176 - 148 = 28$. The class width is found as $28/6 = 4.67$. This number is nearly equal to 5, which is more convenient to use. In order to make each class mark a convenient multiple of 5, the lower bound if defined as 147.5 for the first class and 177.5 is the upper bound for the last class. A frequency summary of the data is helpful in constructing the histogram.

Class	Tally	Frequency
At least 147.5 but less than 152.5	ЖІ I	6
At least 152.5 but less than 157.5	ЖІ ЖІ I	11
At least 157.5 but less than 162.5	ЖІ ЖІ IIII	14
At least 162.5 but less than 167.5	ЖІ IIII	9
At least 167.5 but less than 172.5	III	3
At least 172.5 but less than 177.5	II	2
		45

The frequency histogram is easily constructed from this frequency summary and is given in Figure 2.4. See also Figures 2.5 and 2.6.

Exercises

2.1 The amount of food consumed by laboratory rats has been recorded in grams for each of the last 40 days. Make a stem and leaf plot for these data. (*Hint:* The stem should consist of the first 2 digits of these numbers and the leaf will be the last digit.)

186	121	143	159	180
125	178	215	166	158
187	148	151	162	153
128	133	188	170	168
153	123	134	184	208
178	186	162	202	160
201	200	175	126	150
174	218	165	166	185

2.2 Summarize the data in Exercise 2.1 in a histogram.

2.3 Pregnant laboratory rats were injected with an experimental drug prior to the birth of their litters. The number of live births was recorded for each of the 35 rats.

Number of Survivors	Number of Litters
0	3
1	7
2	14
3	8
4	2
7	1

That is, one litter had 7 survivors, two litters had 4 survivors, and so on. Represent these data by a histogram.

2.4 In one university, seniors are required to pass a grammar test prior to receiving approval for graduation. The frequencies of errors made by 230 seniors are summarized in the accompanying histogram. Answer the following questions by referring to the histogram.

(a) How many classes were used in the summary?

(b) What are the class boundaries for each class?

(c) What is the class width?

(d) What is the class mark for each class?

(e) What is the frequency of each class?

(f) What is the relative frequency of each class?

(g) What percent of the seniors made 10 or fewer errors?

(h) What percent of the seniors made more than 20 errors?

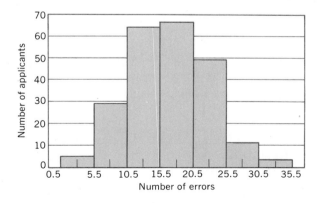

2.5 A discount store wishes to put in a fast checkout lane for shoppers who purchase only a small number of items. The store manager observes 102 customers selected at random and counts the number of items that each has. The sample provides the following frequency summary.

Number of Items	Frequency	Number of Items	Frequency
1	4	14	7
2	1	15	4
3	9	16	2
4	7	17	1
5	6	18	2
6	10	21	1
7	6	22	1
8	5	23	1
9	12	25	1
10	7	26	1
11	3	29	1
12	4	36	1
13	4	42	1

Construct a histogram for these data.

2.6   The earnings per share for 25 stocks randomly selected from a recent issue of *Standard and Poor's Stock Guide* are given below. Summarize these data in a stem and leaf "building block" histogram. (*Hint:* Use the first digit 0, 1, 2, 3, 4, 5, 6, 7 as the stem and the last two digits as the leaves.)

$5.44	$1.25	$0.68	$4.70	$2.30
0.95	6.85	0.41	1.34	0.45
3.01	1.60	2.11	6.79	2.57
3.55	0.20	1.80	7.30	2.07
1.60	1.43	2.34	4.76	1.31

2.7   A traffic safety officer is trying to make the city residents more aware of the danger of accidents. A survey of the last 50 weeks shows the number of serious traffic accidents per week to be as follows.

0	5	9	8	2	10	4	12	14	8
3	8	4	4	0	9	3	10	10	9
1	1	2	14	4	9	1	8	2	7
2	6	2	9	8	5	0	18	8	11
4	6	12	1	12	8	6	6	6	11

(a) Display these data in a relative frequency histogram with five classes of equal width.

(b) What are the class boundaries for the middle class in your histogram?

(c) What is the class mark for the middle class in your histogram?

2.2
More on Frequency Distributions

HISTOGRAMS WITH UNEQUAL CLASS WIDTHS

In the previous section it is assumed that all class widths are to be equal. Frequently this is not convenient since there are settings where equal class widths are not appropriate as well as settings where only a summary of the sample data is available. Such is the case with the projected population of the United States for the year 2000, which appears in Figure 2.7. In Figure 2.7

the projections are broken down by age groups where the groupings follow natural areas of interest such as pre-school, grade school, high school, college, adults in four age groups covering 10 years each, and retirement age. Note that the last interval has no upper limit indicated; that is, it is open ended. In order to plot these data as a histogram it will be necessary to modify the definition of a histogram.

FIGURE 2.7 Population Projection for the United States in the Year 2000. (Source: Department of Commerce, Bureau of the Census.)

Age Group	Projected Number (millions)
Under 5 years	17.9
5 to 13 years	35.1
14 to 17 years	16.0
18 to 24 years	24.7
25 to 34 years	34.4
35 to 44 years	41.3
45 to 54 years	35.9
55 to 64 years	23.3
65 and over	31.8
	260.4

HISTOGRAM WITH UNEQUAL CLASS WIDTHS

A **histogram** for classes with **unequal class widths** is a graphical presentation of the sample data that consists of rectangular boxes plotted on each class interval where the **areas** of the rectangles are proportional to the number of observations in each class. The height of a rectangle for a particular class is found by dividing that class frequency by its width.

EXAMPLE

Make a histogram for the population projections given in Figure 2.7. First the summary of Figure 2.7 is reorganized to reflect the class widths. This means that an arbitrary upper bound must be assumed for the last class, such as 100.

Class	Frequency (millions)	Class Width	Height = Frequency/Class Width
Less than 5	17.9	5	3.58
5 but less than 14	35.1	9	3.90
14 but less than 18	16.0	4	4.00
18 but less than 25	24.7	7	3.53
25 but less than 35	34.4	10	3.44
35 but less than 45	41.3	10	4.13
45 but less than 55	35.9	10	3.59
55 but less than 65	23.3	10	2.33
65 to 100	31.8	35	.91
	260.4		

The histogram (Figure 2.8) is now constructed from this summary. Note that the vertical axis used in this histogram has no direct interpretation, and that the heights of the rectangular boxes have been constructed to accurately reflect the proportion of individuals in each class. This feature makes it easy for the eye to make accurate comparisons. For example, Figure 2.7 shows that there are about twice as many individuals in the 5 to 14 age group as there are in the groups on either side, but the class width is also about twice the size of the classes on either side. The histogram in Figure 2.8 is adjusted for this difference and shows that the population density in the 5–14 interval is about the same as in the adjacent classes. The scale used to construct the height can be inches, centimeters, or whatever is desired as long as it is consistent throughout. The histogram will always look the same.

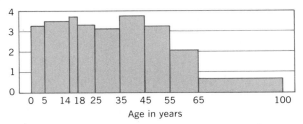

FIGURE 2.8 Histogram of Projected Population Distribution

A Warning. Histograms with unequal class interval widths are often misconstructed and misinterpreted. Sometimes the unequal interval widths are ignored so that higher bars are used to represent larger frequencies, giving the false impression of a greater density of observation in those intervals. Also the open ends usually have arbitrary endpoints selected, and the selection of endpoints influences the appearance of the graph.

FREQUENCY POLYGON

A useful variation of the histogram is known as a frequency polygon. If a histogram has been constructed it is easily converted to a frequency polygon by connecting the midpoints (class marks) of the tops of each rectangle by straight line segments. The frequency polygon is completed by adding two line segments—one at each extreme class in the histogram—which connect the class marks of these classes with the value of zero in the classes that would be adjacent to but outside of these extreme classes. This is done in order to reflect the fact that the frequency returns to zero outside of the intervals covered by the histogram. The selection of these endpoints is arbitrary in the case of histograms with unequal class interval widths, so caution must be observed when interpreting frequency polygons in those situations. The frequency polygon is demonstrated with the 20 typing scores given earlier in this chapter. These scores can be summarized in five classes as follows:

Typing Score	Frequency
22 but less than 36	2
36 but less than 50	3
50 but less than 64	6
64 but less than 78	7
78 but less than 92	2

The histogram with two extra classes added to accommodate the frequency polygon is given in Figure 2.9. Of course the histogram does not need to be part of the graph. The frequency polygon appears by itself in Figure 2.10.

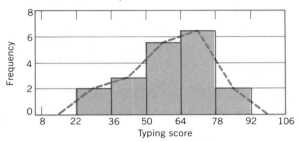

FIGURE 2.9 Histogram of Typing Scores with Frequency Polyon

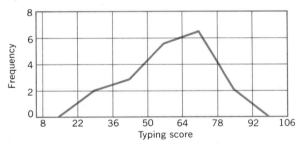

FIGURE 2.10 Frequency Polygon of Typing Scores

GRAPHING TIME SERIES

Many data sets in business and economics represent observations made at regular time intervals, such as yearly population figures or monthly sales data. These numbers are called **time series** data. Figure 2.11 represents a time series, the population of the United States as measured by the U.S. Census Bureau from 1790–1970. This graph is called a **vertical bar graph** because the rectangles do not touch each other. It is easy to see not only what the population size has been in the past years, but also how the growth rate of the population has changed. For example, the growth rate in the 1800s appears to be almost exponential, because the tops of the bars exhibit a sharp upward curvature. But the growth rate in the 1900s appears to be much more linear in appearance.

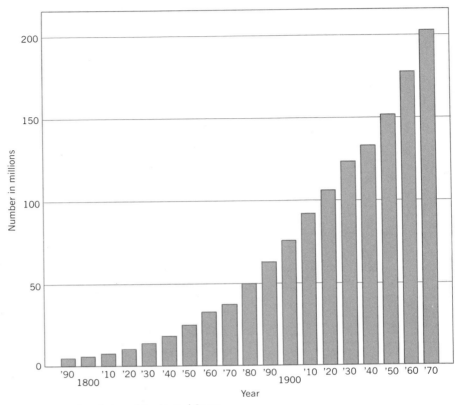

FIGURE 2.11 Population of the United States

INDIVIDUALIZED GRAPHICAL METHODS

Graphs may be used in innovative ways to depict several types of information simultaneously. Figure 2.12 is such a graph. It shows the age distribution of the 1970 United States population, for both Males and Females. This figure presents information on one qualitative variable (sex) and two quantitative variables (age, and number of persons) all at the same time.

Side by side graphs in Figure 2.13 present several types of information also. One graph has the 1981 sales figures of several leading manufacturers of an electronic product, while the other represents a time series of growth rates for 11 years.

An example of artistic innovation appears in Figure 2.14. The population growth rates of Southwestern states are depicted in two different ways, one for a quick visual impression and the other for a more accurate numerical reading on each state. These figures represent only a few of the many attractive and innovative ways of presenting data in a clear and understandable manner.

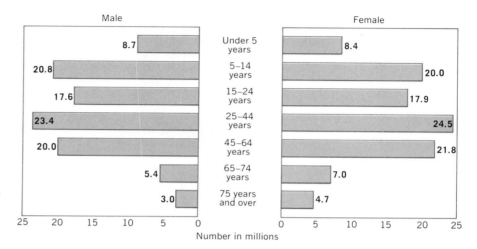

FIGURE 2.12 The Number of Persons by Age and Sex in the United States in 1970

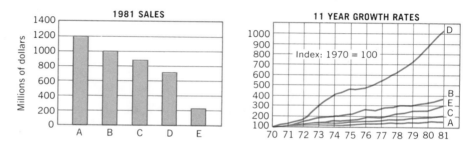

FIGURE 2.13 A Comparison of Absolute Sales, and Growth Rate of Several Leading Manufacturers of an Electronic Product, Showing Company D is Fourth in Sales in 1981, but First in 10 Year Growth Rates

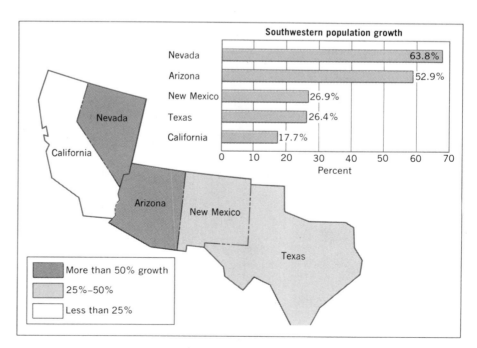

FIGURE 2.14 Population Growth Rates of the Five Southwestern States, 1970–1980

Exercises

2.8 Magazine advertisements frequently use histogram-type representations to compare a client's sales with the competition. However, some of these graphs can be misleading in the way they are constructed. What is misleading about the following graph?

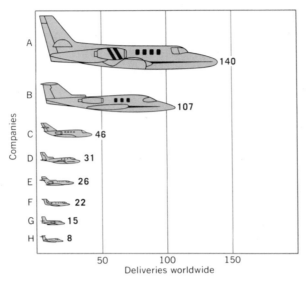

2.9 The histograms given below compare salaries for executives with four different functions. The rectangles have been constructed so that their areas are proportional to the percentages they represent. However, there is one basic fault involved in the construction of the histogram. What is it?

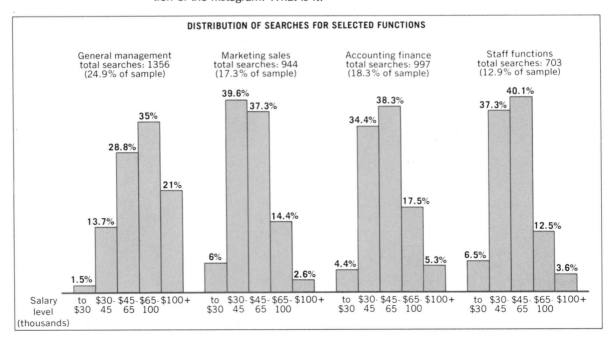

2.10 A frequency distribution summary of adjusted gross income compiled from 1976 income tax returns is given below. Summarize these data in a histogram.

Adjusted Gross Income (thousands of dollars)	Number of Taxpayers (in thousands)
0 to 3	15,015
3 to 5	8,837
5 to 10	19,891
10 to 15	14,182
15 to 20	11,182
20 to 25	6,662
25 to 30	3,611
30 to 50	3,632
50 to 100	945
100 to 500	221
500 to 1000	4
1000 and over	1

2.11 Construct a frequency polygon for the data in Exercise 2.10.

2.12 Given below are the population projections for the year 1985. Summarize these data in a histogram similar to the one in Figure 2.8 and comment about any possible differences in the relative frequency distributions for the years 1985 and 2000.

Age Group	Projected Number (millions)
Under 5 years	18.8
5 to 13 years	29.1
14 to 17 years	14.4
18 to 24 years	27.9
25 to 34 years	39.9
35 to 44 years	31.4
45 to 54 years	22.8
55 to 64 years	21.7
65 and over	27.3
	232.9

2.13 Construct a frequency polygon for the data in Exercise 2.12.

2.14 Prior to putting a new tire on the market a tire company conducts tread life tests on a random sample of 100 of these new tires. The test results are summarized below. Construct a frequency polygon for these data.

Number of Miles	Frequency
10,000 but less than 15,000	6
15,000 but less than 20,000	21
20,000 but less than 25,000	38
25,000 but less than 30,000	19
30,000 but less than 35,000	16

2.15 Refer to Figure 2.11 and determine the population of the United States in 1880 and 1940. In what year did the population first exceed 100,000,000?

2.16 How many of the seven age groups in Figure 2.12 show more females than males? How many years of age are covered in each of these seven groups?

2.17   What is the basic idea being communicated by Company D in Figure 2.13?

2.18   Which of the Southwestern states depicted in Figure 2.14 shows the largest growth? Which one shows the largest growth rate?

2.19   Convert the following "pie graph" to a "vertical bar" graph.

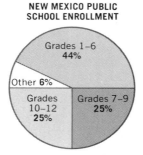

2.20   Represent the following breakdown of a company's sales in a "pie graph" like the one shown in Exercise 2.19.

Packaging	36.0%
Retail	27.7%
Lumber	10.5%
Pulp and Paper	14.7%
Printing	5.1%
Other	6.0%

2.3
Cumulative Frequency Distributions

CONSTRUCTING AN OGIVE

The graphs of the previous section were based on either the premise of frequency or relative frequency indicating either the frequency of observations or the proportion of observations in each class. In this section methods are presented for displaying sample data in a cumulative frequency graph.

> **CUMULATIVE (RELATIVE) FREQUENCY**   The **cumulative frequency** *is the total number of observations less than or equal to a given number or class.* The **cumulative relative frequency** *is the cumulative frequency divided by the total number of observations.*

Cumulative frequency and cumulative relative frequency are illustrated by returning to the histogram displayed in Figure 2.9. In that figure the frequencies of observations in each class from left to right are 2, 3, 6, 7, and 2 for a total of 20 observations. The relative frequencies are found by dividing each of these frequencies by the sample size 20. For a plot based on cumulative frequency

Typing Speed	Frequency	Relative Frequency
22 but less than 36	2	.10
36 but less than 50	3	.15
50 but less than 64	6	.30
64 but less than 78	7	.35
78 but less than 92	2	.10
Typing Speed	Cumulative Frequency	Cumulative Relative Frequency
21 or less	0	.00
35 or less	2	.10
49 or less	5	.25
63 or less	11	.55
77 or less	18	.90
91 or less	20	1.00

FIGURE 2.15 Typing Scores Organized for a Cumulative Frequency Graph

rather than relative frequency the typing scores need to be summarized as in Figure 2.15.

The first method for displaying a cumulative relative frequency is called an **ogive**. The ogive is useful for estimating the proportion of observations in the sample that are less than or equal to any value. It may also be used to visualize where each observation lies relative to the rest of the sample.

> OGIVE
> An **ogive** *is a plot of cumulative relative frequency versus the upper class boundary for each class. The successive points in this plot are then connected with straight lines.*

A plot of the cumulative relative frequencies from Figure 2.15 is given in the ogive in Figure 2.16.

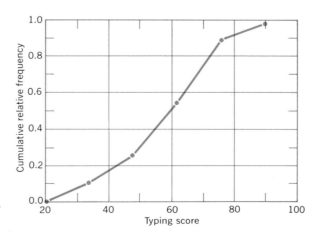

FIGURE 2.16 An Ogive for Typing Scores

REAL NUMBER LINE PLOT

The typing scores of Problem Setting 2.1 are used to illustrate another simple but informative way to display the data graphically. For this approach a simple concept is borrowed from mathematics and the data are plotted on a real number line as illustrated in Figure 2.17. This plot is easily made and makes clear how the spacing differs between pairs of scores. Clearly the scores of 22 and 35 are inferior to the others since they fall well below the otherwise tightly clustered group in the middle. At the other extreme the scores of 84 and 91 are well above their nearest competitors. The addition of a second dimension to a real number line plot to make information more readily discernible will now be illustrated.

FIGURE 2.17 Real Number Line Plot of Typing Scores

THE EMPIRICAL DISTRIBUTION FUNCTION

Notice that the ogive depends on the class boundaries selected. Another type of graphing technique somewhat akin to the cumulative frequency plot but that does not depend on class boundaries is based on the **empirical distribution function** (e.d.f.). The word empirical means the same as the word sample; hence an e.d.f. is a plot made from sample data. The e.d.f. is graphed by adding a second dimension to a real number line plot. The second dimension is formed by building a step function or "staircase" from left to right on the real number line, adding another step to the "staircase" every time a sample data point is encountered. The height of the step at each of the data points is the reciprocal of the sample size. For a sample of size 20 each step is of height $1/20 = .05$. In the event there are two identical values in the sample data, the

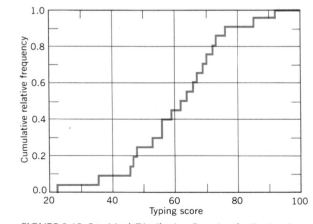

FIGURE 2.18 Empirical Distribution Function for Typing Scores

step height at that value is twice as high as an ordinary step. Likewise, steps with heights three times the ordinary step height would indicate three identical sample data values and so on. This means that the vertical axis of an e.d.f. plot will always be labeled from 0 to 1. The "staircase" will now be added to the real number line plot of Figure 2.17 to generate an e.d.f. This e.d.f. appears in Figure 2.18.

Although the vertical axis for the e.d.f. has the same label as the ogive, the e.d.f. differs from the ogive not only because it is a step function, but also because each of the data points is represented in the e.d.f. rather than being grouped together in classes as in the ogive. Extensive use of the e.d.f. will be made throughout this book, and it will be treated more formally in the next chapter. However, some of its usefulness may readily be seen. For example, the top (or bottom) 5 percent, 10 percent, and so on scores are easily obtained from reading the e.d.f. plot, as are other percentages of interest. Formal rules for plotting an e.d.f. are now given.

EMPIRICAL DISTRIBUTION FUNCTION	An **empirical distribution function** *is a function of* x, *which equals the proportion of the sample less than or equal to* x, *for all* x *from* $-\infty$ *to* $+\infty$. *In graphical form it is a graph of the cumulative relative frequency of a sample.*

RULES FOR GRAPHICAL DISPLAY OF DATA IN AN EMPIRICAL DISTRIBUTION FUNCTION

1. The sample data points must first be ordered from smallest to largest. A stem and leaf plot can be quite helpful in ordering the sample data. A real number line plot can also be used to order the data.
2. The empirical distribution function looks like a stairstep function as it proceeds from left to right, having a value of 0 to the left of the smallest data point and a value of 1 to the right of the largest data point. In between, the cumulative step heights will start at $1/n$ at the smallest data point and increase to $2/n$, $3/n$, and so on to $n/n = 1$ as additional sample data points are encountered from left to right along the horizontal axis. In the event of identical sample values, the step heights should be increased to indicate exactly the multiplicity of the data point.

Exercises

2.21   Construct an ogive for the tread life data given in Exercise 2.14.

2.22   Construct an ogive for the population prediction data for the year 1985, which was presented in Exercise 2.12.

2.23   Construct an ogive for the litter survival data given in Exercise 2.3.

2.24   Summarize the data given in Exercise 2.1 in an e.d.f. plot. Use the e.d.f. to determine the percent of the time that the food consumed exceeded 200 grams.

2.25   Refer to the data given in Exercise 2.5 and summarize these data in an e.d.f. If the discount store desires to have 10 percent or less of its customers eligible for the fast checkout lane, what is the maximum number of items a person can have and still use the fast lane?

2.26   Use an e.d.f. to summarize the data given in Exercise 2.6. What percent of the companies would you estimate to have earnings of less than $1.00 per share?

2.27   Construct an ogive for the accident survey data given in Exercise 2.7.

2.4

Scatterplots

UNIVARIATE AND BIVARIATE DATA

While the type of data considered so far is called **univariate** because it involves only one variable, much quantitative data are bivariate in nature; that is, a pair of measurements is recorded for each sample unit.

UNIVARIATE DATA AND BIVARIATE DATA	*Univariate sample data consist of values of a single variable measured on each unit in the sample.* *Bivariate sample data consist of values of a pair of variables measured on each unit in the sample.*

The following examples illustrate bivariate quantitative data.

1. Bank assets and deposits for last year in a city with 12 banks.
2. Spring and fall enrollments at 50 major universities.
3. Expenditures in each of two fiscal years by 22 federal agencies in a particular region of the country.
4. The number of tourists visiting each of several National Park Service areas for two consecutive years.
5. Height and weight measurements for a group of individuals.
6. Vital capacity of the lungs and age for a group of smokers.
7. Age at inauguration and age at death for U. S. Presidents (see Exercise 1.22).

In Problem Setting 2.1 the statement was made that the typing instructor had available, in addition to the typing scores, the number of errors that each student made. Hence, she has bivariate observations. The bivariate typing data that the instructor has available are given in Figure 2.19.

Student Number	Typing Scores	Number of Errors
1	68	8
2	72	2
3	35	9
4	91	14
5	47	9
6	52	13
7	75	12
8	63	3
9	55	0
10	65	14
11	84	0
12	45	14
13	58	14
14	61	12
15	69	2
16	22	2
17	46	5
18	55	5
19	66	13
20	71	2

FIGURE 2.19. Typing Scores and Number of Errors for Each of 20 Students

SCATTERPLOT

The methods of the previous two sections could also be used to display the errors, and this would probably be a worthwhile exercise. However, consider a joint display of these bivariate observations in a **scatterplot**. This means that a plot needs to be constructed where the horizontal axis represents the typing speeds and the vertical axis represents the number of errors. Each student is represented as a single point in the two-dimensional display given in Figure 2.20. The x-coordinate is the typing score and the y-coordinate is the number of errors.

The usefulness of the scatterplot is apparent when comparing the various students. Previously (see Figures 2.1, 2.2, and 2.3) there was no way to choose between the two students with typing speeds of 55. Now it is clear that one student had 5 errors while the other had none. Likewise, the student with a typing speed of 91 previously appeared to be far superior to all others. However, this student had 14 errors while the student with a speed of 84 had no errors. In this particular application auxiliary lines could be added to the scatterplot to make identification of the best typists easy. For example, if it is desired to identify only students with 2 or fewer errors and a speed of 65 or higher, a line drawn parallel to the horizontal axis and intersecting the vertical

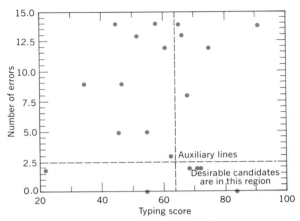

FIGURE 2.20 Scatterplot of Typing Score and the Number of Errors Made by the 20 Students

axis between 2 and 3 shows six students with 2 or fewer errors. Likewise, a vertical line at 65 shows nine students with speeds of at least 65. However, in the lower right-hand corner of the plot there are only four students satisfying both conditions simultaneously.

Clearly Figure 2.20 exhibits the relationship that exists between the two variables, and is more informative than the real number line plot of Figure 2.17. Moreover, the real number line plot for the typing scores, given in Figure 2.17, may be obtained from Figure 2.20 by moving each point vertically to the x-axis.

THE LOG TRANSFORMATION AS AN AID TO GRAPHING DATA

One major area of concern when making a graphical display of quantitative data is that all points be identifiable in the graph. For example, draw a real number line and try to plot the following numbers on it: 1, 10, 100, 1000, 10,000, and 100,000. After 100,000 is set as the upper limit on the scale on the graph, you will find that all of the numbers less than 1000 are so close together that they cannot be distinguished. A solution that is frequently offered in this situation is to plot the **logarithms** of the numbers, which are in this case 0, 1, 2, 3, 4, and 5. Recall that a logarithm, $\log_{10}X$, is the exponent to which 10 must be raised to equal X. If the logarithms of these numbers are plotted, each of these points is easily identifiable. This is known as **transforming** the data.

> **THE LOG TRANSFORMATION**
>
> *The **log transformation** of the sample data $X_1, \ldots, X_n$ replaces the sample data with their corresponding base 10 logarithms; that is, $\log_{10}X_1, \ldots, \log_{10}X_n$. The log transform can only be used when all sample values are greater than zero.*

As an example of the usefulness of logs in graphical display, consider data from a study of 12 Albuquerque, New Mexico banks given in Figure 2.21.

Information is available on the year end deposits (in thousands of dollars) of the banks as well as the percentage growth or decline of deposits over the previous year.

	1979 Deposits (in thousands)	Growth of Deposits 1978–1979 (percent)
Albuquerque National Bank	$675,709	5.1
First National Bank	457,085	12.0
Bank of New Mexico	242,682	7.1
American Bank of Commerce	99,615	−2.9
Rio Grande Valley Bank	78,947	22.3
Citizens Bank	39,603	−7.5
Fidelity National Bank	38,213	6.9
Southwest National Bank	42,813	13.5
Republic Bank	33,445	−2.3
Western Bank	30,271	5.1
Plaza del Sol National Bank	10,370	13.9
El Valle State Bank	5,635	−6.0

FIGURE 2.21 Deposit and Deposit Growth Data for 12 Albuquerque, New Mexico Banks

Plotting the deposit data presents some problems due to the large differences in the amount of deposits held by the different banks. To illustrate this point consider a real number line plot of deposits given in Figure 2.22

FIGURE 2.22 Real Number Line Plot of Bank Deposits

It is clear from observing Figure 2.22 that some points are so close together as to be indistinguishable. This is caused by the few extremely large values that create the scaling problem in the first place. Such values that stand out from the others in the sample are commonly referred to as **outliers**. One way to moderate the influence of outliers in the real number line plot is to replace the data with their logs. That is, use a log transform on the data. A graph of a real number line plot based on log transformed bank deposits is given in Figure 2.23. A comparison of Figures 2.22 and 2.23 shows clearly that the data points become more nearly distinguishable by using logs without completely losing a sense of distance or relative ordering among the points.

FIGURE 2.23 Real Number Line Plot of Log Transformed Bank Deposits

> **OUTLIERS**   *Observations that are much larger or much smaller than the rest of the observations in the sample are called **outliers**.*

LOG SCATTERPLOT

A bivariate scatterplot can be used to see if there might be an association between the amount of bank deposits and their growth rate. That is, do larger banks tend to grow faster than smaller banks or slower than smaller banks, or is there no discernible relationship? The log transform can be used with the deposit data in a scatterplot, but the percentage data contain negative numbers; hence the log transform cannot be used. The resulting scatterplot is given in Figure 2.24. There does not appear to be any tendency for the points to exhibit any relationship or pattern such as a straight line. An alternative method for graphing such data involves the use of special graph paper known as semi-log graph paper. Log-log graph paper is also available for cases when both variables can be log transformed.

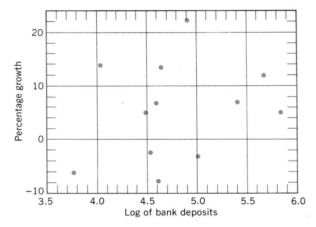

FIGURE 2.24 Scatterplot of Log Transformed Bank Deposits and Percentage Growth

THE RANK TRANSFORMATION

The log transformation cannot be used on the percentages associated with the bank deposit data since some of the numbers are negative. A transformation that works equally well on all types of numbers is the **rank transformation**. The rank transformation will be used extensively in this text and has many uses beyond simple plotting. The rank transformation merely replaces each of the data points by its respective **rank**. The rank of an observation is its relative position when the observations are arranged in order from smallest to largest. Consider the following data set and the corresponding set of ranks.

Data:	5.6,	7.2,	3.4,	11.8,	6.9,	8.1
Ranks:	2,	4,	1,	6,	3,	5

The rank transformation is easily performed and can be used on positive or negative numbers. There is only one additional item to remember and that is how to handle **ties** in the sampling data; ties are sample observations that are equal to one another. Ranks within each group of tied observations are determined by adding together the ranks that would have been assigned within that group had there been no ties, and then dividing this total by the number of tied observations and assigning the result to each tied observation in that group. This is called **the method of average ranks**.

THE RANK TRANSFORMATION	The **rank transformation** *for a sample of n observations replaces the smallest observation by the integer 1 (called the rank), the next smallest by* **rank** *2, and so on until the largest observation is replaced by* **rank** *n.*

Consider the following revision of the previous data set.

Data:	6.9,	7.2,	3.4,	11.8,	6.9,	8.1
Ranks:	2.5,	4,	1,	6,	2.5,	5

Here the previous data set has been changed to include two values of 6.9. Had these values differed only slightly they would have been assigned ranks of 2 and 3, so each is assigned a rank of $(2 + 3)/2 = 2.5$ to settle the tie. If the original data set is further modified so that the 7.2 is also a 6.9, then the tie is settled by assigning each tied sample value the rank of $(2 + 3 + 4)/3 = 3$.

USE OF THE STEM AND LEAF PLOT TO ASSIGN RANKS

A stem and leaf plot may be used to order the sample data from smallest to largest. This ordering makes the assignment of ranks easy. For example, refer back to the original stem and leaf plot of typing speeds and see how quickly the ranks can be assigned from this plot.

RANK TRANSFORM SCATTERPLOT

As an example of the usefulness of ranks in a graphical display of data, the banking data given in Figure 2.21 are used again. First the original data are replaced with their corresponding ranks.

Deposits	Percentage Growth	Ranks of Deposits	Ranks of Percentages
675,709	5.1	12	5.5
457,085	12.0	11	9
242,682	7.1	10	8
99,615	−2.9	9	3
78,947	22.3	8	12
39,603	−7.5	6	1
38,213	6.9	5	7
42,813	13.5	7	10
33,445	−2.3	4	4
30,271	5.1	3	5.5
10,370	13.9	2	11
5,635	−6.0	1	2

Note that the deposits are almost in descending order and that the percentages have two values of 5.1. These values are tied for positions 5 and 6 so each is assigned the rank of $(5 + 6)/2 = 5.5$. The bivariate scatterplot appears in Figure 2.25.

Figure 2.25 presents an easy to read graph with the points not clustered and there is no longer a scaling problem. Figure 2.25 makes it apparent that there is no discernible pattern present; that is, the points appear to be scattered at random in the plot. This random scattering is in contrast to other possible plots that might show some relationship such as one connecting the pairs (1, 1), (2, 2), (3, 3), . . . , (12, 12) and forming a strong straight line relationship from the lower left-hand corner of the plot to the upper right-hand corner; see Figure 2.26. The study of such relationships will take place in a later chapter.

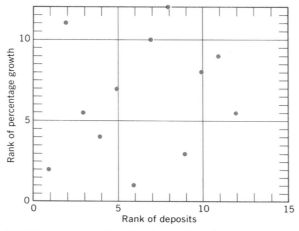

FIGURE 2.25 Scatterplot of Rank Transformed Bank Deposits and Rank Transformed Percentage Growth

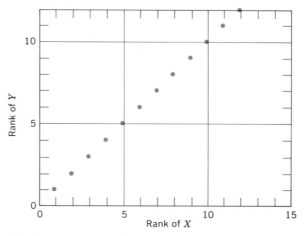

FIGURE 2.26 Rank Transform Scatterplot Showing a Straight Line Relationship

One reason for using scatterplots is to display visually points that simultaneously represent two variables. Examination of a scatterplot makes relationships between the two variables easy to see. The pattern in the scatterplot often aids in interpreting the results of a statistical analysis of such data and will be used throughout this book under the topics of paired data, correlation, and regression.

Warning. Changing the scale on graphs can distort the data. While log transforms and rank transforms can be useful, one must be careful to interpret them in terms of logs or ranks.

Exercises

2.28   The Bureau of Labor Statistics reported expenditures on clothing and personal care for low, intermediate, and high budget four-person families in the Northcentral portion of the United States for the autumn of 1979, as shown in the table.

| | Type of Budget | | | | | |
| | Low | | Intermediate | | High | |
	Cloth.	Per. Care	Cloth.	Per. Care	Cloth.	Per. Care
Chicago	797	326	1153	426	1684	586
Cincinnati	994	305	1430	401	2083	550
Cleveland	920	409	1323	544	1930	756
Detroit	830	339	1195	452	1750	619
Kansas City	927	391	1329	516	1940	724
Milwaukee	973	357	1390	465	2034	645
Minneapolis	872	357	1251	470	1820	651
St. Louis	820	334	1184	426	1747	573

Plot these data in a scatterplot taking care to use a different symbol ($*$, $\circ$, $\bullet$) with each budget so that the points are easily identified within the plot. What facts does the scatterplot make readily apparent? Do the values for Cincinnati and Milwaukee appear to fit the pattern within each budget type?

2.29   A summary of financial statistics of institutions of higher education compiled by the National Center for Education Statistics for the fiscal year 1977 gives a breakdown of revenues. Given below is a summary by different regions of the country of the percentage of revenue obtained from tuition and fees and government appropriations (mostly state government). Plot these percentages in a scatterplot. Does your plot indicate any relationship between tuition and fees and the size of government appropriations?

| | Percentage of Total Revenue | |
| | From Tuition and Fees | From Government Appropriations |
Region		
New England	33.0	17.1
Mideast	29.1	27.2
Great Lakes	22.8	33.3
Plains	19.7	35.4
Southeast	18.0	37.9
Southwest	13.6	45.8
Rocky Mountains	17.0	36.3
Far West	11.9	41.9

2.30 The National Center for Education Statistics has compiled data on the type of degree granted to males and females for 11 consecutive academic years. Data regarding the awarding of first-professional degrees are given below. Use these data to make a scatterplot of the number of first-professional degrees granted to males versus the number of first-professional degrees granted to females. What fact does the graph make apparent during the last few years of the study regarding the awarding of first-professional degrees?

| | Number of First-Professional Degrees | |
Academic Year	To Men	To Women
1966–67	31,064	1,429
1967–68	33,083	1,645
1968–69	34,069	1,612
1969–70	33,344	1,908
1970–71	35,797	2,479
1971–72	41,021	2,753
1972–73	46,827	3,608
1973–74	48,904	5,374
1974–75	49,230	7,029
1975–76	53,210	9,851
1976–77	52,668	12,112

2.31 Market share trends of the liquor industry are given below for the years 1959 and 1974. During this time demand shifted away from domestic bourbons and blends. In this setting it would be desirable to make a graph to show how this shift has affected preference for various liquors. That is, how does the ranking of liquors in 1959 compare with the ranking of liquors in 1974 in a scatterplot? Make this rank transform scatterplot and determine which liquors have made the greatest change in position in the market.

	Percent of Market	
	1959	1974
Whiskey types:		
Straights	25.9	16.3
Spirit blends	32.1	13.8
Scotch	7.7	13.7
Canadian	5.0	11.4
Bonds	4.6	1.0
Other	0.3	0.3
Vodka	7.3	16.9
Gin	9.2	10.0
Cordials	3.5	6.2
Brandy	2.4	4.0
Rum	1.5	3.7
Other	0.5	2.7

2.32 One way to evaluate a stock is to see how many shares are held by financial institutions—that is, banks, investment and insurance companies. Given below is such a listing for 10 discount variety chain stores along with their recent earnings per share as obtained from a recent issue of *Standard and Poor's Stock Guide*. A plot is needed that shows if the amount of institutional holdings is associated with the size of the stock's earnings, but the representation of the data listed below in a scatterplot is difficult due to the large variation in the size of the numbers of shares of stock held. However, the desired information can be obtained from a rank transform scatterplot. Make such a plot for these data.

Company	Number of Institutional Shares Held (thousands)	Earnings per Share
Danners Inc.	197	$1.28
Duckwall-Alco	24	2.20
Fed-Mart	99	4.80
Grand Central	2	1.22
Hartfield-Zodys	377	1.97
K-Mart	53,170	2.96
Thrifty Corp.	544	1.46
Wall-Mart	3,452	2.34
Woolworth	6,587	4.84
Zayre	632	2.64

2.33 Use a log transformation of the number of institutional shares held data in Exercise 2.32 and then make a scatterplot of these logs versus earnings per share. Compare your results with Exercise 2.32.

2.34 The monetarist school of economic thought states that all monetary expansion in excess of real economic growth is eventually transmitted into inflation. In other words, money supply (cash and bank deposits) should only be permitted to grow as fast as the output of goods and services grows. Any more rapid growth will merely create excess dollars to chase after the same supply of goods and services—and send prices soaring. Given below are 20-year summary data for 14 countries. Plot the anticipated inflation versus actual inflation in a scatterplot to see if the monetarist hypothesis is supported by these data.

| Country | Annual Rates of Change 1960–1979 | | | |
	(1) Money Supply Growth (percent)	(2) Real Economic Growth (percent)	(1) − (2) Anticipated Inflation (percent)	Actual Inflation (percent)
Brazil	44.7	9.1	35.6	32.1
Japan	16.8	8.4	8.4	6.1
Italy	16.5	4.3	12.2	9.1
Spain	16.2	6.3	9.9	10.5
Mexico	15.8	6.1	9.7	8.4
France	10.4	4.1	6.3	6.4
Netherlands	9.5	4.2	5.3	6.4
Sweden	9.1	3.1	6.0	6.6
Germany	8.7	3.8	4.9	4.4
U.K.	8.1	2.5	5.6	8.3
Canada	8.1	4.8	3.3	5.3
Switzerland	7.8	2.8	5.0	5.1
Belgium	7.1	4.2	2.9	5.2
U.S.A.	5.3	3.6	1.7	4.7

2.35   Convert the scatterplot of Exercise 2.34 to a rank transform scatterplot.

2.5

Review Exercises

2.36   Given below is a frequency summary of the weights of 114 students. Summarize these data in a histogram. Why is a stem and leaf plot not appropriate for these data?

Weight (lbs)	Frequency
104.5 to 114.5	1
114.5 to 124.5	5
124.5 to 134.5	16
134.5 to 144.5	29
144.5 to 154.5	24
154.5 to 164.5	19
164.5 to 174.5	12
174.5 to 184.5	4
184.5 to 194.5	3
194.5 to 204.5	1

2.37   Use a stem and leaf plot to represent the number of hours worked in one week in a manufacturing company by 50 employees.

47	50	52	55	46
50	53	49	43	48
51	54	52	47	49
45	51	53	49	47
52	61	50	51	53
49	50	45	55	51
52	46	47	50	48
57	50	48	55	49
53	49	54	44	47
51	45	51	50	52

2.38   Plot the data in Exercise 2.37 as an empirical distribution function.

2.39   Plot the data given below in a scatterplot for heights (X) and weights (Y) of 10 college students.

Height (X):	64	73	71	69	66	69	75	71	63	72
Weight (Y):	121	181	156	162	142	157	208	169	127	165

2.40   Make a rank transform scatterplot of the data given in Exercise 2.39.

2.41   Construct an ogive for the sample data given in Exercise 2.36.

2.42   Given below are the weight losses (in grams) of laboratory rats during the 24 hours after they were injected with an experimental drug. Summarize these data with a stem and leaf plot.

4.2	4.4	4.8	4.9	4.4
3.9	4.3	4.5	4.8	3.9
3.6	4.1	4.3	3.9	4.2
4.1	4.0	4.0	3.8	4.6
3.8	4.7	3.9	4.0	4.2
4.4	4.6	4.4	4.9	4.4
4.1	4.3	4.2	4.5	4.4
4.2	4.1	4.0	4.7	4.1
4.7	4.2	4.1	4.4	4.8
4.3	4.6	4.5	4.6	4.0

2.43   Summarize the sample data given in Exercise 2.42 in a histogram.

2.44   Construct an empirical distribution function for the sample data given in Exercise 2.42.

Bibliography

Additional material on the topics presented in this chapter can be found in the following publication.

Tukey, John W. (1977). *Exploratory Data Analysis*, Addison-Wesley, Reading, Mass.

3

Probability

The weather bureau has made probability a household word by including probabilities in the weather forecasts. Yet probability is often misunderstood, as evidenced by people who maintain that the weather bureau was wrong because it rained when the probability of rain was only 20 percent. Probability is only a measure of the uncertainty associated with events. It can indicate which events are more likely to occur than other events, but it does not enable the future to be predicted with certainty.

There is a solid and extensive mathematical development of the theory of probability. The discussion of probability often involves dealing cards, tossing coins, rolling dice, and other games of chance, since the theory of probability has its origins in the need to determine whether certain gambling games were favorable or unfavorable to the player. Not all probability lends itself to a rigorous mathematical development, however. In the business world probability often takes the form of an expert opinion, as when the financial officer

"WHY SO DISTRAUGHT CLAYBORNE? THE WEATHER FORECAST CALLED FOR A 50% CHANCE FOR RAIN."

estimates the probability of a bond issue selling out within the first week, or the marketing specialist assesses probabilities of a new product selling more than a given number of units.

The ability to interpret correctly results of statistical analyses requires some knowledge of probability, to which this chapter provides a brief introduction.

3.1
Sample Spaces, Events, and Probabilities

SAMPLE VARIABILITY

There is a lot of chance variability associated with collecting data. A woman taking a survey in a supermarket may find three out of the ten people questioned had never heard of Planter's Peanuts, for example. It is possible that those are the only three people out of the hundreds that could have been selected for interview, that had never heard of Planter's Peanuts, and they just happened to be selected for the interview. Such a chance occurrence may be considered "unusual." The very fact that something can be called "unusual" or "unlikely" indicates that there is a "more usual" pattern to be expected, even in selecting people at random for a survey. There is some order, or some pattern, that can be used as a mental image, against which phenomena are compared and determined to be "unusual" or "not unusual." That is, in any process that involves data collection the individual observations will be mostly unpredictable, but many observations collectively should tend to reveal some kind of a tendency or pattern. This leads into the concept of **probability**, which will be introduced formally later in this section. But first some useful terms will be presented.

When some people hear the word "experiment" they see a mental picture of a scientific laboratory with several technicians in white coats doing mysterious things with test tubes and chemicals. The word experiment is given a much broader interpretation in the sciences, however. In particular, making a marketing survey is an experiment, trying a new method for treatment of a disease and recording the results is also an experiment, and noting the length of time it takes for a child to recognize an error is an experiment too.

EXPERIMENT | An **experiment** is any planned process by which observations are made and/or data are collected.

SAMPLE SPACE

The result of an experiment is called the **outcome** of the experiment. An experiment results in one and only one outcome at a time. It is often useful to have some concept of the possible outcomes of an experiment. The collection of all possible outcomes of an experiment is called the **sample space** of that experiment. The list need not be a written list, but often is only a mental list of the possible outcomes. Several outcomes considered together are referred to as an **event**. The usual notation for events is A, B, C, and so on, while S is usually reserved for the sample space, which is the set of all outcomes and hence is the largest possible event.

OUTCOME	An **outcome** of an experiment is any possible result of the experiment.

SAMPLE SPACE	The **sample space** of an experiment is the collection (or set) of all possible outcomes of the experiment.

EVENT	An **event** is one or more outcomes considered as a group.

EXAMPLE

The transportation department for a university is responsible for supervising the use of university-owned cars by the various colleges on campus. For every request for a vehicle the dispatcher records the college making the request. The experiment consists of someone requesting a vehicle. The possible outcomes of the experiment are:

AS / Arts and Sciences
AG / Agriculture
BA / Business Administration
ED / Education
EN / Engineering
HE / Home Economics

The above list of all six possible outcomes is the sample space. Some events include

$A = \{AS, AG, HE\},$

which means "the request is from Arts and Science, or Agriculture, or Home Economics," or

$B = \{AG\}$

or

$C = \{HE, BA, ED, EN\}.$

The sample space may be written as

$S = \{AS, AG, BA, ED, EN, HE\}.$

PROBABILITY

If an experiment could be repeated many times, then the number of times each outcome occurs could be recorded. The **relative frequency** associated with each outcome will tend toward some number between 0 and 1 as the number of repetitions gets larger and larger. This long-term relative frequency would approach the **probability of the outcome** if the nature of the experiment remains unchanged. The probabilities associated with the outcomes in a sample space must satisfy two requirements:

1. The probability of each outcome must be a number between 0 and 1, inclusive.
2. The sum of the probabilities of all of the outcomes in the sample space must equal 1.0.

EXAMPLE

In the previous example the sample space for the vehicle requests consisted of the six colleges originating the requests:

$$S = \{AS, AG, BA, ED, EN, HE\}.$$

Suppose that based on previous experience the following probabilities have been determined for the outcomes:

$$P(AS) = .47, \qquad P(ED) = .04,$$
$$P(AG) = .13, \qquad P(EN) = .11,$$
$$P(BA) = .22, \qquad P(HE) = .03.$$

Note that the sum of the probabilities is 1.00, and that all probabilities are between 0 and 1.

The probability of an event equals the sum of the probabilities associated with the outcomes that constitute the event. The probability of an event A is denoted by $P(A)$. It follows that the probability of either of two events A or B occurring, including the case where both events occur, can be found by summing the probabilities attached to the outcomes comprising the two events. If two events share no outcomes, they are said to be **mutually exclusive**, and the probability of either one or the other occurring may be found by adding the probabilities of the two events.

PROBABILITY OF AN OUTCOME	The **probability of an outcome** is a number between 0 and 1 that represents the long-term relative frequency with which that outcome could be expected to occur, if the experiment could be repeated many times.
PROBABILITY OF AN EVENT	The **probability of an event** is the sum of the probabilities of the outcomes that comprise the event.
ADDITION RULE	Let A and B be the two events that are **mutually exclusive**; that is, they share no outcomes in common. Then the **probability of either A or B** occuring is the sum of the probabilities of the two events: $P(A \text{ or } B) = P(A) + P(B)$ (for mutually exclusive events).

EXAMPLE

Probabilities for the various events defined in the first example in this section are as follows:

$P(A) = P(\{AS, AG, HE\}) = .47 + .13 + .03 = .63,$
$P(B) = P(\{AG\}) = .13,$
$P(C) = P(\{HE, BA, ED, EN\}) = .03 + .22 + .04 + .11 = .40,$

and of course

$P(S) = 1.00.$

Since the events B and C are mutually exclusive, the probability of either B or C occurring is found using the addition rule

$P(B \text{ or } C) = P(B) + P(C) = .13 + .40 = .53.$

For events that are not mutually exclusive, such as

$A = \{AS, AG, HE\}$

and

$D = \{HE, BA\},$

the probability of A or D is found by considering the set of outcomes that comprise the two events:

$$P(A \text{ or } D) = P(\{AS, AG, HE, BA\})$$
$$= .47 + .13 + .03 + .22$$
$$= .85.$$

The interpretation of this probability is that the next vehicle request is much more likely to come from either Arts and Sciences, Agriculture, Home Economics, or Business Administration than from the remaining two colleges, with an 85 percent chance for the former group versus a 15 percent chance for the remaining colleges.

OTHER DEFINITIONS OF PROBABILITY

The definition of probability given in this section is called the **relative frequency** definition of probability. The most rigorous and logically consistent definition involves using a set of axioms to define probability. This **axiomatic** definition of probability is found in most mathematical texts on statistics, and is used in some applied texts also. Although it is more rigorous, it is less intuitive and more difficult to understand than the relative frequency definition used here.

If probabilities are obtained from the opinions of one or more people, then the probabilities are called **subjective** probabilities. Oddsmakers in Las Vegas attach subjective probabilities to the various outcomes of future sports events, and these are used to form betting odds customers may use for making wagers. Participants in corporate planning sessions are often asked to estimate the chances of particular events occurring in the future, such as, "What are the chances Jones will quit if we hire Smith?"

There is a danger in attaching too much confidence in the ability of past events to provide probabilities of future events occurring. For example, if taxes were raised in 5 out of the last 6 years, it would be dangerous to say that the probability of taxes being raised next year is 5/6. Conditions change over time, and other factors may need to be considered in estimating probabilities. The relative frequency approach to probability should be used with common sense when assigning probabilities to events. True probabilities are never known; they are either assumed or estimated.

PROBABILITY MODELS

Probabilities may be obtained on the basis of subjective estimates, or estimated from relative frequencies. They may even result from a set of assumptions. In any event, the result is called a probability **model**, which is used to provide numbers to use as probabilities. Later in this book procedures will be provided for checking the set of assumptions on which the probability model is based.

PROBABILITY MODEL	A **probability model** is a set of assumptions that leads to the assignment of precise probabilities to the outcomes in a sample space.

JOINT PROBABILITIES

Some types of data are characterized by more than one type of information. Job applicants may list age, social security number, years of experience, and so on. The customer registration card discussed in Chapter 1 provides a good example of such data. When probabilities are provided for sample spaces containing such data, the probabilities are called **joint probabilities**. The following example illustrates what can be done with **bivariate** outcomes—that is, with outcomes that contain two types of information, called **bivariate** data, as defined in Chapter 2.

JOINT PROBABILITY	The probability associated with a bivariate outcome of an experiment is called a **joint probability**. The joint probability of two events A and B is denoted by P(A and B).

EXAMPLE

Employees are classified according to the type of job they hold and their sex. The total number of employees, and the numbers that fall into each category, are given in the frequency table in Figure 3.1. Division of each number in Figure 3.1 by the total number of employees 706 gives the relative frequency table of Figure 3.2 If an employee is chosen at random, the numbers in Figure 3.2 represent the joint probabilities of obtaining each type of employee. For this reason the table in Figure 3.2 is often called a **joint probability table**.

	Male	Female	Total
Clerical	27	51	78
Administrative	24	12	36
Production	314	261	575
Other	14	3	17
Total	379	327	706

FIGURE 3.1 Frequency Table of Employees

	Male	Female	Total
Clerical	.038	.072	.110
Administrative	.034	.017	.051
Production	.445	.370	.815
Other	.020	.004	.024
Total	.537	.463	1.000

FIGURE 3.2 Joint Probability Table of Employees

MARGINAL PROBABILITIES

The probabilities associated with one outcome of one variable by itself irrespective of the values of the other variable(s) may be obtained by summing all of the probabilities associated with the value of that variable, where the sum is taken over all outcomes of the other variable(s) not being considered. This is easy to see in Figure 3.2. There the probability of selecting a clerical employee is .110, which is obtained by summing the two joint probabilities for male clerical employees and female clerical employees. Because these sums of probabilities typically appear in the margins of joint probability tables, as in Figure 3.2, they are called **marginal probabilities**. Another marginal probability from Figure 3.2 is the probability that a randomly selected employee will be male, .537. This marginal probability is found by summing the joint probabilities for male clerical employees, .038, male administrative employees, .034, male production employees, .445, and other male employees, .020. The collection of male employees in these four categories constitutes the entire group of male employees, so the sum of the probabilities from those mutually exclusive categories results in the probability of a randomly selected employee being male.

MARGINAL PROBABILITY	*The sum of several joint probabilities over all values of one or more variables is called a* **marginal probability**.

CONDITIONAL PROBABILITY

If some information about the sampling frame is given so that the sampled population is different from the original population, then this new **condition** changes the probabilities. For instance, suppose that an employee is selected at random, and the information is provided that the employee is a clerical employee. Now what is the probability that the employee is male?

It can be seen from Figure 3.1 that of all employees, there is a slight majority of males, so the overall probability of selecting a male employee is $379/706 = .537$, as stated in Figure 3.2. However, the condition is given that the employee is clerical. There is a total of 78 clerical employees and only 27 of them are male, or $27/78 = 34.6$ percent. Therefore, the probability is .346 of a randomly selected employee being male, given that the employee is clerical. This is quite a change from the marginal probability .537 given earlier as the probability of an employee being male.

CONDITIONAL PROBABILITY

The **conditional probability** of one event, given that another event or condition is true, is given by the joint probability of two events divided by the marginal probability of the given event. Notation for this equation is

$$P(A \mid B) = \frac{P(A \text{ and } B)}{P(B)},$$

where $P(A \mid B)$ is read "the probability of A given B."

INDEPENDENT EVENTS

Two events A and B are **independent** if the conditional probability of A given B is equal to the unconditional probability of A. Notation for this is

$$P(A \mid B) = P(A). \qquad (3.2)$$

Equivalently, two events are independent if their joint probability equals the product of their individual probabilities. Notation for this is

$$P(A \text{ and } B) = P(A) \cdot P(B). \qquad (3.3)$$

INDEPENDENCE

When the conditional probability is different from the unconditional probability, the two events are said to be **dependent**. Conversely, two events are **independent** if the conditional probability of one event given the other is the same as the unconditional probability of the one. The events "selecting a male employee" and "selecting a clerical employee" are dependent events because the conditional probability of the first event, selecting a male employee, given the second, selecting a clerical employee, is .346, while the unconditional probability of the first event is .537.

To see how Equation (3.3) works, consider the joint probability of "selecting a male employee" and "selecting a clerical employee," which is given in Figure 3.2 as .038:

P(male and clerical) = .038.

The product of the two individual probabilities

$$P(male) = .537 \quad and \quad P(clerical) = .110$$

is .537 (.110) = .059, which does not equal the joint probability of the same two events. Therefore the events are dependent.

The same conclusion is reached using Equation (3.2). The probability of a randomly selected employee being male (A) given that the employee is clerical (B) was found previously to be .346:

$$P(A \mid B) = .346.$$

This is not equal to the probability of an employee being male, which is .537:

$$P(A \mid B) = P(A) = .537.$$

Therefore the events A and B are not independent.

Exercises

3.1 A student takes a multiple choice exam consisting of 20 questions. The exam score (outcome) consists of the total number of points, where 5 points are awarded for every correct answer, and no points for incorrect answers. Describe the following: (a) the experiment, (b) two possible outcomes, (c) the sample space, and (d) the outcomes in the event "the grade is above 70."

3.2 Suppose the probability model assumed for the test scores in Exercise 3.1 is as follows:

$$P(100) = .01 \quad P(95) = .07 \quad P(90) = .15$$
$$P(85) \;= .22 \quad P(80) = .22 \quad P(75) = .16$$
$$P(70) \;= .10 \quad P(65) = .04$$

and probability .03 of getting less than 65.

(a) Find the sum of the probabilities assigned by the probability model for the above outcomes.

(b) Find the $P(75$ or more).

(c) Find the $P(A)$, where A is 95 or 100.

(d) Find the $P(70$ or 80 or 90 or 100).

3.3 A survey of all employees regarding the choice of three medical plans produces the following frequencies.

	Marital Status		
	Married	Single	Total
Plan 1	20	50	70
Plan 2	140	10	150
Plan 3	45	15	60
Total	205	75	280

Assume that one employee is selected at random from the 280 employees.

(a) Convert the frequency table to a joint probability table.

(b) Find the probability of Plan 1 being preferred by the employee selected.

(c) Find the probability that the selected employee is married.

(d) Is the preference for Plan 1 independent of being married?

(e) Find the conditional probability of Plan 1 being preferred given that the employee is married.

(f) Find the conditional probability of Plan 1 being preferred given the employee is single.

(g) Find the conditional probability of Plan 2 being preferred given the employee is married.

(h) Find the conditional probability of Plan 2 being preferred given the employee is single.

(i) Based on your answers in (e)–(h), do you think plan preference is independent of marital status?

3.4 Volunteers in a physical fitness study are classified as to how much change in general level of stamina they have experienced in the last 12 months, and which type of physical activity they have been engaged in. The results are given in the following table of frequencies.

Change in Stamina (percent)	Exercise			Total
	Swimming	Running	Other	
> 20	1	2		6
10 to 20	13	12		45
0 to 10	20	20		60
−10 to 0	2	2		20
<−10	—	—	—	12
Total	48	36	59	143

Assume that one person is selected at random from the 143 volunteers.

(a) Fill in the blanks in the frequency table and convert it to a joint probability table.

(b) Find the probability that the change in stamina has a gain of over 20 percent.

(c) Find the conditional probability that the exercise was "Other," given that stamina increased over 20 percent.

(d) Find the probability that stamina decreased.

(e) Find the conditional probability that swimming was the exercise, given that stamina decreased by more than 10 percent.

3.5 A men's clothing store has a custom of extending credit to its customers. A study of all of its accounts revealed that 12 percent of the accounts were delinquent, and the other 88 percent were not. Also revealed in this study was the fact that 66 percent of the accounts belonged to married men. Only 4 percent of the accounts were both delinquent and belonging to married men. There were 500 accounts in all.

(a) Construct a joint probability table for a randomly selected account.

(b) Find the joint probability of an account being non-delinquent and belonging to a non-married man.

(c) Find the conditional probability of an account being delinquent, given that the account belongs to a non-married man.

(d) Is account status independent of marital status?

3.6 Students withdrawing from school have been classified according to sex and reason for withdrawing. The reason for withdrawing can be lack of money (MON), low grades (GPA), needed at home (HOM), personal problems (PP), or a dislike for college life (DCL). The frequencies are as follows:

| | Type of Termination | | | | | |
	MON	GPA	HOM	PP	DCL	Total
Male	82	78	18	10	6	194
Female	48	17	7	0	15	87
Total	130	95	25	10	21	281

Assume that one person is selected at random from the 281 students.

(a) Convert the frequency table to a joint probability table.

(b) The student is known to have withdrawn for a DCL. Find the conditional probability that this student is a female.

(c) What is the probability of withdrawal by a reason other than MON or DCL?

3.7 A magazine vendor classifies 100 customers by type of magazine purchased and age. The following table of frequencies is a summary of these classifications:

| | Age Group | | | | |
	<30	30–40	41–50	>50	Total
News	12	10	11	14	47
Sport	10	7	8	6	31
Hobby	1	3	5	13	22
Total	23	20	24	33	100

Assume that one of the 100 customers is selected at random.

(a) Convert the frequency table to a joint probability table.

(b) Find the probability that the customer is over 40.

(c) Find the conditional probability that the customer is over 50 if a hobby magazine is sold.

(d) Find the conditional probability that the customer is over 40 if a hobby magazine is sold.

(e) Find the probability that a customer is under 41 and purchases a sport magazine.

3.8 If the joint probability of a resident taking both the morning paper and the evening paper is .2 and the marginal probability of taking the evening paper is .6, find the conditional probability of a resident taking the morning paper given that the resident takes the evening paper.

3.2
Random Variables and Their Distributions

RANDOM VARIABLES

The outcomes of an experiment are much easier to work with when they can be associated with numbers. However, outcomes by themselves are not necessarily numerical. In a poll the response of an individual being polled may be entirely verbal, consisting of an opinion regarding the quality of the public transportation system, for example. Usually an attempt is made to associate the response with a number, such as

1 = completely satisfied
2 = partially satisfied
3 = no opinion
4 = mildly dissatisfied
5 = strongly dissatisfied.

The numbers are easier to record, to tally, and to summarize in a report. Any rule for assigning numbers to outcomes, such as the rule just described, is called a **random variable**. Random variables are usually denoted by capital letters such as X, Y, or Z, with or without subscripts.

RANDOM VARIABLE	A **random variable** is a function that assigns real numbers to the outcomes of an experiment.

EXAMPLE

Several test tubes are used to examine the growth rate of a particular strain of bacteria. Let X be the growth rate, measured in percent increase per day. Then each test tube results in one observation (a number) for X, which is the random variable in this experiment.

DISCRETE RANDOM VARIABLES

The random variable in the opening paragraph of this section assigned only a limited number of values to the possible outcomes of the experiment—that is, 1, 2, 3, 4, and 5. There are other experimental situations where a random variable assigns only a limited number of values to the experimental outcomes. Examples include:

1. the number of customers per hour (0, 1, 2, 3, etc.);
2. the number of days sick leave taken by an employee (0, 1, 2, 3, etc.);
3. the proportion of voters, out of 43 surveyed, who prefer Candidate Jones ($0/43$, $1/43$, $2/43$, . . . , $43/43$);
4. the percent effectiveness (percent kill) on an insecticide, when applied to a finite group of insects;
5. the number of defective items in a lot.

Notice that in each of these examples the random variable assumes values that are all separate and distinct from one another. For example, in illustration (1) above the number of customers could be 1 or 2 but not any number between 1 and 2. In statistical terminology random variables that take on only separate and distinct values are called **discrete** random variables.

DISCRETE RANDOM VARIABLE	*A random variable is called* **discrete** *if all of its possible values are separate and distinct.*

PROBABILITY FUNCTION OF A DISCRETE RANDOM VARIABLE

A random variable associates numbers with the outcomes of an experiment. Sometimes those numbers are equally likely, but sometimes some of the possible values of the random variable are more likely to occur than others. The function that gives the probability associated with each possible value of the random variable is called a **probability function**. A graph of the probability function is often a useful device for displaying the relative likelihood of obtaining the various possible values of the random variable.

PROBABILITY FUNCTION	*The* **probability function** *of a discrete random variable is the function that associates probabilities with the various numerical values the random variable can assume.*

EXAMPLE

Company records indicate the number of college degrees earned by each of its employees. Out of 150 employees, the number of degrees earned by each is given as follows:

Number of Degrees	Number of Employees	Relative Frequency
0	75	.50
1	36	.24
2	18	.12
3	15	.10
4	6	.04
	150	1.00

Let X equal the number of college degrees earned by an employee who is randomly selected from the 150 employees. Then X is a discrete random variable that can take on the values 0, 1, 2, 3, or 4. The probabilities associated with each value are given by the relative frequencies listed above. These relative frequencies are used to form the probability function for X:

$P(X = 0) = .50$
$P(X = 1) = .24$
$P(X = 2) = .12$
$P(X = 3) = .10$
$P(X = 4) = .04$

A graph of the probability function for X is given in Figure 3.3.

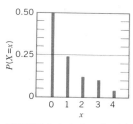

FIGURE 3.3 A Graph of the Probability Function of X.

THE CUMULATIVE DISTRIBUTION FUNCTION

Another way of expressing the probabilities associated with a random variable is by means of a **cumulative distribution function** (c.d.f.), which merely accumulates all of the probabilities of values less than or equal to the value of X of interest. The notation for a c.d.f. is $P(X \leq x)$, or sometimes simply $F(x)$. Since every c.d.f. is a cumulative probability function, it increases from zero for very small x to one for very large x.

The c.d.f. of a discrete random variable has a graph that identifies it immediately as being associated with a discrete random variable; the graph appears to be a series of stair steps, where the heights of the stairs are not necessarily equal. (See Figure 3.4.) The vertical lines in the graph do not actually belong in the graph of the c.d.f., but are usually included to make the graph easier to read. Incidentally the height of each step in the c.d.f. equals the probability associated with the value x at the step. Since the sum of all of the probabilities in a probability function equals 1.0, the c.d.f. also goes to 1.0 as x gets large, because the c.d.f. is the cumulative probability.

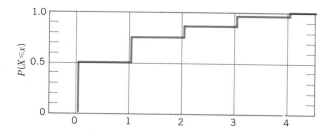

FIGURE 3.4 A Graph of the Cumulative Distribution Function of X

CUMULATIVE DISTRIBUTION FUNCTION	The **cumulative distribution function** of a random variable X is the function that gives the probability that X is less than or equal to x, written as P(X ≤ x).

EXAMPLE

The cumulative distribution function for the probabilities given in Figure 3.3 is given by

$$P(X \leq x) = 0 \text{ for } x < 0$$
$$= .50 \text{ for } 0 \leq x < 1$$
$$= .74 \text{ for } 1 \leq x < 2$$
$$= .86 \text{ for } 2 \leq x < 3$$
$$= .96 \text{ for } 3 \leq x < 4$$
$$= 1.00 \text{ for } x \geq 4$$

because, when x is greater than 4, the probabilities for $X \leq x$ include all of the individual probabilities associated with X. A graph of the cumulative distribution function of X is given in Figure 3.4.

CONTINUOUS RANDOM VARIABLES

Recall that the values possible for a discrete random variable are all separate and distinct, such as 1 or 2, but no numbers between 1 and 2. Not all measurements are of this type, however. For instance, measurements of time, such as the time required to serve a customer, may take the value 1 minute or 2 minutes, or any number between 1 and 2 minutes. There are no separate and distinct values in this case.

When actually recording the amount of time required to serve a customer, the measurement may be recorded as 1 minute or 2 minutes if the amount of time is rounded off to the nearest integer. Or the time may be recorded as 1.3 minutes if the measurement is rounded to the nearest tenth. Recorded measurements may look like measurements on a discrete random variable because the rounding off represents the interval within which that observation lies. That is, the recorded measurement 1.3 minutes represents all of the actual times between 1.25 and 1.35 minutes. The recorded measurement may appear to be discrete, but the actual time is not a discrete random variable.

Random variables that have no separate and distinct values are called **continuous** random variables; their possible values form one or more continuous intervals of numbers. Where the typical discrete random variable is counting something, the typical continuous random variable is measuring something, such as time, weight, height, volume, and so forth. Examples of continuous random variables include:

1. the time it takes to serve a customer;
2. the time an employee arrives at work;
3. the weight of a six-month-old baby;
4. the actual weight of the contents in a box of cereal;

5. the length of a steel reinforcement rod;

6. the actual volume of a 3 c.c. bottle of medicine.

The c.d.f. of a continuous random variable is distinctive because its graph has *no* steps in it, in contrast with the c.d.f. of a discrete random variable that graphs as a series of stair steps. That is, the graph of the c.d.f. of a continuous random variable appears as a smooth continuous function, such as the graphs in Figures 3.5 and 3.6.

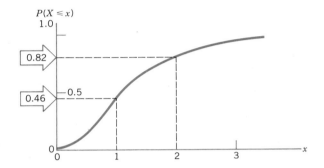

FIGURE 3.5 A Typical Cumulative Distribution Function for a Continuous Random Variable

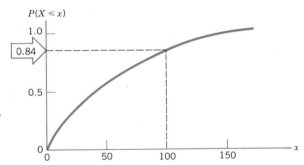

FIGURE 3.6 The Cumulative Distribution Function for Daily Percent Increase in Bacteria

---

CONTINUOUS RANDOM VARIABLE *A random variable is **continuous** if none of its possible values are separate and distinct.*

In Figure 3.5 probabilities may be read directly from the graph. The probability of X being less than or equal to 2 is obtained by reading the height of the c.d.f. directly above the point x = 2, which is shown to be .82 by the dotted lines in Figure 3.5. This is written as

$$P(X \le 2) = .82.$$

Similarly the probability that X is less than or equal to 1.0 is read from Figure 3.5 as .46,

$$P(X \le 1.0) = .46.$$

Probabilities over intervals of the x-axis, such as the probability that X is between 1 and 2, are found by measuring the corresponding interval on the vertical axis. The interval from $x = 1$ to $x = 2$ on the x-axis corresponds to the interval from .46 to .82 on the vertical (probability) axis, so the probability of X being between 1 and 2 is the distance between .46 and .82, or .36. Mathematically this is written as

$$P(1 < X < 2) = P(X < 2) - P(X < 1)$$
$$= .82 - .46$$
$$= .36.$$

Notice that no distinction is made between $X < 2$ and $X \leq 2$ with continuous random variables because the event $X = 2$ has zero probability attached to it. Notice also that the probability of exceeding a certain number, called an **exceedance probability**, is easily found using the method just described. For example, the probability of X exceeding 2 is given by the portion of the vertical axis that corresponds to $x > 2$, which is

$$P(X > 2) = 1 - P(X \leq 2)$$
$$= 1 - .82$$
$$= .18.$$

All c.d.f.'s are read in the way just described for continuous random variables, except that if the random variable is not continuous, distinction needs to be made between the probabilities $P(X \leq x)$ and $P(X < x)$ in those cases where $P(X = x)$ is not equal to zero. In general the latter two probabilities sum to the first,

$$P(X < x) + P(X = x) = P(X \leq x),$$

and this should be considered when finding probabilities from noncontinuous c.d.f.'s.

EXAMPLE

In the first example of this section the random variable X represented the daily percent increase in bacteria. Suppose that the c.d.f. of X is given by Figure 3.6. From the c.d.f. the probability of the daily increase in bacteria being less than 100 percent for any particular day is seen to be .84. The probability of the daily increase exceeding 100 percent is $1 - .84 = .16$. This means that in the long run about 16 percent of the days will involve percent increases greater than 100 percent.

EMPIRICAL DISTRIBUTION FUNCTIONS VERSUS CUMULATIVE DISTRIBUTION FUNCTIONS

In Chapter 2 the empirical distribution function was introduced for sample data. It is a function that increases in height from 0 to 1.0 as the value of x increases. In that way it resembles the c.d.f.'s introduced in this section. The difference between the two is like the difference between population and sample. The c.d.f. is a population curve that represents true probabilities of getting observations in any given range, while the e.d.f. is a sample curve that represents actual relative frequencies of sample data in any given range. Since relative frequencies tend to get closer to probabilities as the sample sizes get larger, the empirical distribution functions also tend to get closer to their respective cumulative distribution functions as the sample sizes get larger. This fact is the basis for using e.d.f.'s from sample data to estimate unknown c.d.f.'s in later chapters of this book.

EXAMPLE

A random sample of 20 days' records is obtained and the percentage increases are recorded. The empirical distribution function of these 20 sample data points is plotted in Figure 3.7, along with the c.d.f. Notice that the e.d.f. shows some sampling error; that is, it does not follow the c.d.f. as closely as is possible for a sample of size 20. Yet there is a tendency for the e.d.f. to follow along the general direction of the c.d.f. The e.d.f. of a larger sample will tend to follow the c.d.f. closer than that of a smaller sample. A sample of size 2000 for example will almost certainly produce an e.d.f. that will be virtually indistinguishable from the c.d.f.

Whenever the cumulative distribution function is associated with a population from which a sample is obtained for examination, that c.d.f. is also called the **population c.d.f.**, or sometimes the **population distribution function**. An example of a discrete population c.d.f. will now be given.

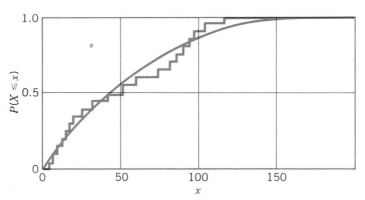

FIGURE 3.7 The Empirical Distribution Function From a Sample Size 20, and the Continuous Cumulative Distribution Function of the Population From Which the Sample was Obtained

POPULATION DISTRIBUTION FUNCTION	*The* **population distribution function**, *or* **population c.d.f.**, *is the cumulative distribution function of a population.*

EXAMPLE

Consider again the example in which the random variable is the number of degrees earned by each of 150 employees of a company. The random variable *X* equals 0, 1, 2, 3, or 4. A random sample of 20 employees is observed with the following results.

Number of Degrees	Number of Employees in Sample	Sample Relative Frequency	e.d.f.	Population Relative Frequency	c.d.f.
0	8	.40	.40	.50	.50
1	5	.25	.65	.24	.74
2	2	.10	.75	.12	.86
3	4	.20	.95	.10	.96
4	1	.05	1.00	.04	1.00
	20	1.00		1.00	

The empirical distribution function for these 20 sample data points appears in Figure 3.8, along with the population c.d.f. from Figure 3.4. Note that the e.d.f., with its multiplicity of points at $x = 0, 1, 2, 3$, and 4, has five steps, just like the population c.d.f. Since there are only five values possible in the population, the sample e.d.f. is forced to bear a strong resemblance to the population c.d.f. because the locations of the steps will coincide. As the sample size increases, the heights of the steps in the e.d.f. will also match up well with the heights of the steps in the c.d.f., so the two curves will tend to be virtually indistinguishable with large sample sizes.

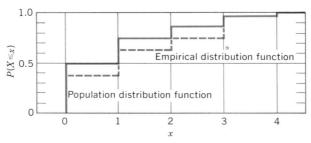

FIGURE 3.8 The Empirical Distribution Function from a Sample of Size 20, and the Discrete Cumulative Distribution Function of the Population from which the Sample was Obtained.

HISTOGRAMS VERSUS DENSITY FUNCTIONS

In Chapter 2 histograms were recommended as a way of displaying sample data graphically. Histograms are more appropriate for observations on continuous random variables than they are for observations on discrete random variables, although they can be used on either. (Sometimes vertical bar graphs are natural for discrete random variables.) Suppose a histogram is drawn for a large number of sample data points from a continuous population. Just as the empirical distribution function tends to approach the population distribution function as the sample gets large, the histogram tends to resemble more closely a population function called a **density function**. The derivation of the density function from the cumulative distribution function requires calculus. For the purposes of this book, the density function will be provided when it is needed, so no knowledge of calculus will be necessary.

DENSITY FUNCTION

The **density function** *of a continuous random variable is the function that a histogram of a random sample will tend to resemble as the sample size gets large and the class widths get small.*

EXAMPLE

A histogram is drawn for the 20 sample data points illustrated in the e.d.f. of Figure 3.7. Recall from Chapter 2 that a histogram of 20 observations suggests five classes, so the classes 0 to 25, 25 to 50, and so on are selected for this illustration. The histogram appears in Figure 3.9 along with a graph of the population density function that corresponds to the c.d.f. in Figure 3.7. Note that the same sampling variability is exhibited here for the histogram. Also note that no scale is given on the vertical axis, since the vertical scales tend to be different for histograms and density functions.

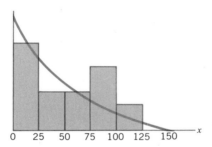

FIGURE 3.9 A Histogram of a Sample of Size 20, along with the Population Density Function.

As the sample size gets large, the number of classes in a histogram gets large also, and the histogram tends to become smoother and more similar in appearance to a density function. However, even for a sample of size 2000 the recommended number of classes is still only 11, and a graph consisting of 11 rectangles may be quite unlike a smooth density function. In general histograms tend to show more variability than empirical distribution functions do. Also they do not tend to converge as fast to the density function as the empirical distribution function tends to converge to the population distribution function.

INDEPENDENT RANDOM VARIABLES

A very useful concept is that of **independent random variables**, which follows very closely the definition of independent events in the previous section. Independence of random variables is a common assumption for the statistical procedures presented throughout this book. Examples of independent and dependent random variables occur in several places in subsequent chapters.

> Two random variables X and Y are **independent** if the values they assume satisfy the definition of independent events; that is, if
>
> $$P(X \le x, Y \le y) = P(X \le x)\, P(Y \le y)$$
>
> for all possible values of x and y.

INDEPENDENT RANDOM VARIABLES

Exercises

3.9 An employment agency gets calls for unskilled laborers in groups of 1, 2, 3, 4, 5, and more than 5. The probability model used for the number of laborers requested is .40, .23, .11, .09, .08, and .09 for each respective group size. Let the random variable $X =$ the number of laborers requested in a group and find the following probabilities for this discrete random variable.

(a) $P(X = 1)$ (d) $P(1 \le X \le 3)$

(b) $P(X \le 3)$ (e) $P(X > 3)$

(c) $P(1 < X < 3)$ (f) $P(X = 0)$

3.10 Use a vertical bar graph to show the probability function of the random variable X in Exercise 3.9.

3.11 Graph the cumulative distribution function for the random variable X in Exercise 3.9.

3.12 A sample of 10 calls received by the employment agency in Exercise 3.9 showed the following requested group sizes:

1, 3, 1, 1, 2, 5, 10, 1, 4, 1,

Plot the empirical distribution function for these data on the same graph with the c.d.f. shown in Exercise 3.11. How do the two graphs compare?

3.13 The length of time between telephone calls ($X =$ time in minutes) in an office has the following distribution function:

$$
\begin{aligned}
P(X \le x) &= 0 && \text{for} && x < 0 \\
&= \sqrt{x/10} && \text{for} && 0 < x \le 10 \\
&= 1 && \text{for} && x > 10.
\end{aligned}
$$

Use the distribution function to find the following probabilities for this continuous random variable.

(a) $P(X \le 10)$ (d) $P(X > 5)$

(b) $P(X \le 5)$ (e) $P(1 < X < 5)$

(c) $P(X < 5)$ (f) $P(X = 4)$

3.14 The density function for the random variable X in Exercise 3.13 is given as

$$
\begin{aligned}
f(x) &= .158/\sqrt{x} \text{ for } 0 < x \le 10 \\
&= 0 \qquad \text{for } x \le 0 \text{ or } x > 10.
\end{aligned}
$$

Construct a graph of this density function.

3.15 Graph the distribution function given in Exercise 3.13.

3.16 A sample of 10 calls arriving in the office mentioned in Exercise 3.13 showed the following lengths of time between calls in minutes:

6.3, 4, 0.2, 1.1, 7.1, 1.5, 9, 0.8, 0.6, 2.8

Plot the empirical distribution function for these data on the same graph with the c.d.f. shown in Exercise 3.15. How do the two graphs compare?

3.3
Review Exercises

3.17 A university has conducted a survey of 250 employees to obtain their opinion on collective bargaining by a teacher's union. The frequency summary is given in the following table.

Employee Type	Opposed	Undecided	In Favor	
		Opinion		
Staff	18	19		75
Faculty	63	12		125
Administrator				50
Total	100	35	115	250

Assume that one of the 250 employees is selected at random.

(a) Fill in the empty cells in the frequency table and then convert it to a joint probability table.
(b) What is the marginal probability of a favorable opinion?
(c) What is the probability of an opinion being favorable and from a faculty member?
(d) Find the conditional probability of an opinion being favorable given that the opinion was expressed by a faculty member.

3.18 If the joint probability of a motorist being both uninsured and having an accident is .3 and the marginal probability of being uninsured is .5, find the conditional probability of a motorist having an accident given that the motorist is uninsured.

3.19 The number of magazines sold each hour at a street corner newsstand is described by the probability model given below. Let X = the number of magazines sold each hour and find the probabilities below.

Number of Magazines Sold:	0	1	2	3	4	5	6
Probability:	.14	.27	.27	.18	.09	.04	.01

(a) $P(X \leq 1)$ (d) $P(X = 3)$
(b) $P(2 \leq X \leq 4)$ (e) $P(X > 4)$
(c) $P(2 < X < 4)$ (f) $P(X \geq 4)$

3.20 Use a vertical bar graph to construct a graph of the probability function of the random variable X in Exercise 3.19.

3.21 Graph the cumulative distribution function for the random variable in Exercise 3.19.

3.22   From past experience an insurance company knows that 1 check in 500 received for payment of premiums will be drawn on insufficient funds and will be postdated. Additionally, 1 check in 50 will be postdated. A postdated check is received by the company. What is the probability that the check will be drawn on insufficient funds?

3.23   A joint probability table is given below for colorblindness by sex. Use this table to find the conditional probability that an individual will be male given that the individual is colorblind. Also find the conditional probability that an individual will be female given that the individual is colorblind. Is colorblindness independent of sex?

	Normal Vision	Colorblind
Male	.475	.025
Female	.495	.005

3.24   A machine produces four parts during each hour of operation. Each manufactured part is then graded as either acceptable or not acceptable. Describe the sample space for one hour's production.

3.25   Two factories manufacture the same machine part. Each part is classified as having either 0, 1, 2, or 3 manufacturing defects. The joint probability distribution for this setting is as follows:

	Number of Defects			
	0	1	2	3
Manufacturer A	.1250	.0625	.1875	.1250
Manufacturer B	.0625	.0625	.1250	.2500

(a) A part is observed to have no defects. What is the conditional probability that it was produced by Manufacturer A?

(b) A part is known to have been produced by Manufacturer A. What is the conditional probability that the part has no defects?

(c) A part is known to have 2 or more defects. What is the conditional probability that it was manufactured by A?

Bibliography

Additional material on the topics presented in this chapter can be found in the following publication.

Mosteller, F., Rourke, R., and Thomas, G. (1970). *Probability with Statistical Applications,* 2nd ed., Addison-Wesley, Reading, Mass.

4

Descriptive Statistics and Population Characteristics

PRELIMINARY REMARKS

There is a popular riddle that goes something like this. One person is talking about a photograph, and says,

> "Brothers and sisters,
> I have none;
> But that man's father
> Is my father's son."

What is the relationship between the speaker and the person in the photograph?

Confusing, isn't it? Well, a report with too many numbers in it may be equally confusing. Just as the speaker could have made things clearer by stating that the picture was of his son, a report can be made clearer by summarizing the data. Chapter 2 described methods of summarizing the data with graphs. This chapter shows how to summarize the data with statistics.

The word **statistic** originally referred to numbers published by the state, where the numbers were the result of a summarization of data collected by the government. Thus some people think of a statistic as a number that is based on several numbers in a sample, the proportion of a population that is in a particular category, and so on. In this sense a statistic is just a number. However, if one stops to consider that the numbers being averaged may vary from one sample to the next or that the population may change from one year to the next, one can justify extending the idea of a statistic from being only a number to being a rule for finding the number. Then "the average of the numbers in the sample" is the statistic, and the actual average obtained in one sample is a value of the statistic. As a rule for obtaining a number, a statistic meets the requirements of being a random variable, a function that assigns numbers to the points in the sample space (for an appropriately defined sample space). A statistic also conveys the idea of a summarization of data, so usually a statistic is considered to be a random variable that is a function of several other random variables. Then a value assumed by the statistic is implicitly assumed to be the result of some arithmetic operations performed on other numbers (the data) that, in turn, are the values assumed by several random variables. Note that this is the third use of the word statistics, as alluded to in Section 1.4.

> **STATISTIC** A **statistic** *is a random variable that is a function of the observations in a sample.*

The statistics presented in this chapter are all standard, well known, and widely used methods of summarizing data. They include the mean, standard deviation, mode, percentiles, proportion, correlation coefficient, and other statistics closely related to these six main statistics. In addition to being used to summarize data, these statistics are also used to make inferences about the unknown characteristics of the populations from which the samples were obtained. Therefore, population characteristics are also introduced and discussed along with the descriptive statistics usually associated with them. Specific instructions on how to make these inferences are delayed until later chapters, however.

Anyone who has had a course in statistics is expected to know how to compute the mean, the standard deviation, the mode or modes, some percentiles, a proportion, and a correlation coefficient. These statistics will be presented in this chapter.

4.1
The Sample Mean

PROBLEM SETTING 4.1

A statistics instructor recently finished grading the first hour exam of the semester. Forty students took the exam, and these are the results:

77	68	86	84	95	98	87	71
84	92	96	83	62	83	81	85
91	74	61	52	83	73	85	78
50	81	37	60	85	100	79	81
75	92	80	75	78	71	64	65

Take a minute or two to study these numbers until they mean something to you. Every number is somebody's exam grade, which may represent several hours (or several minutes) of concentrated study, in addition to the one hour of intense activity during the exam. Were there many perfect papers (100 points)? Were there many very poor papers? Did the class as a whole do well on the exam? How can these data be presented so they can be more easily understood?

THE STEM AND LEAF PLOT

As you learned in Chapter 2, most data can be digested more easily after a stem and leaf plot has been made. These data are no exception; see Figure 4.1.

```
 3 | 7
 4 |
 5 | 0 2
 6 | 8 1 0 2 4 5
 7 | 7 5 4 5 8 3 1 9 1 8
 8 | 4 1 6 0 4 3 3 5 3 7 1 5 5 1
 9 | 1 9 2 6 5 8
10 | 0
```

FIGURE 4.1 A Stem and Leaf Plot of the 40 Exam Scores

It is clear from Figure 4.1 that the numbers range from 37, isolated at the low end, to 100, alone at the top of the class. More scores fall in the 80's than in any other class, with the 70's a clear second. The stem and leaf plot presents the entire data picture. No information is lost because all of the test scores are still present. To the contrary, some information is made clearer by the orderly arrangement of the numbers.

THE EMPIRICAL DISTRIBUTION FUNCTION

The empirical distribution function also presents a clear picture of the data without losing any information. The empirical distribution function for these data is given in Figure 4.2. Although it is not as easy to retrieve the individual test scores from the empirical distribution function as it is from a stem and leaf plot, the e.d.f. has some advantages, as will be seen in the next section.

Both the empirical distribution function and the stem and leaf plot are informative, but they are also inconvenient at times. Imagine asking someone who took that exam "How did the rest of the class do?" You would hardly expect to be treated to an immediate stem and leaf plot, or e.d.f. On the other hand, it seems like there should be something more informative than "Okay, I guess"

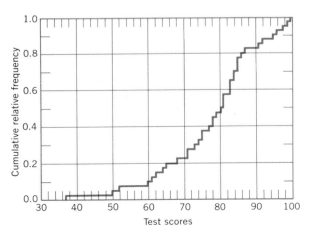

FIGURE 4.2 The Empirical Distribution Function for the Test Scores

for a response. There is clearly a need for one or two statistics that can be used to describe the data clearly and succinctly.

THE SAMPLE MEAN

Even a person accustomed to handling numbers will find 40 exam scores difficult to interpret all at once. Statistics to summarize these data would be helpful. A helpful statistic in this case is some measure of the "average" value. One technical term for "average value" is the "mean," or the "sample mean" in this case, since these 40 students may be thought of as a sample from the population of all students taking this course. The sample mean is obtained by adding all of the observations together and dividing by the sample size. The usual notation for the sample mean is $\overline{X}$ (pronounced "ex bar"):

$$\overline{X} = (77 + 68 + 86 + \cdots + 65)/40$$
$$= 3102/40$$
$$= 77.55.$$

Note how this sample mean 77.55 fits right into the middle of the stem and leaf plot in Figure 4.1, and the e.d.f. in Figure 4.2. This illustrates the concept that the sample mean is often a good statistic for locating the middle of a sample.

The equation just presented is awkward because there are so many numbers. The notation ". . ." is used to represent the numbers that are omitted for the sake of brevity. A much more convenient notation involves the summation notation (Σ) you have learned in a previous math course.

Let the individual test scores be denoted by X_1, X_2, X_3 and so on up to X_{40}. The test scores are represented by X_i, $i = 1, 2, \ldots, 40$. Then the formula for the sample mean is conveniently expressed as

$$\overline{X} = \frac{1}{n}(X_1 + X_2 + \cdots + X_n) = \frac{1}{n}\sum_{i=1}^{n} X_i$$

where n is the sample size. In this case $n = 40$. If the sample had been denoted by $Y_1, Y_2, \ldots, Y_{40}$, then $\overline{Y}$ would be used for the sample mean.

The **sample mean** $\overline{X}$ *is a measure of the center of a sample, and is given by the equation*

SAMPLE MEAN $$\overline{X} = \frac{1}{n}\sum_{i=1}^{n} X_i$$ (4.1)

where n is the sample size and $X_1, X_2, \ldots, X_n$ are the sample observations.

CALCULATING THE MEAN FROM GROUPED DATA

Often the data are grouped when first seen, because the complete data set may have been too lengthy to report and somebody has summarized the data for easier understanding by the reader. The grouped data might have the form shown in Figure 4.3 for the 40 test scores.

Grade	Interval	Number of Students
A	91–100	7
B	81–90	13
C	71–80	11
D	61–70	5
F	≤ 60	4

FIGURE 4.3 The Test Scores Presented as Grouped Data

How can the sample mean be found when only the grouped results in Figure 4.3 are known? The answer is that the exact value for the sample mean cannot be found but an approximate value can be found. The approximate value is obtained by assuming that all of the observations in an interval happened to fall right at the midpoint of that interval, called the class mark. The midpoint is selected because it is somewhat representative of all possible values in the interval. For instance, the seven A's could all be taken at the midpoint of that interval, 95.5, which is found by averaging the largest and the smallest possible numbers in the interval. Thus the A's convert to seven 95.5's, the B's become thirteen scores of 85.5 each, the C's are eleven values of 75.5, and the D's convert to five 65.5's. What should be done about the F's? If the standard procedure is used, the largest score, 60, and the smallest score, presumably zero, would be averaged to get a **class mark** (midpoint) of 30. However, intuition suggests that the three F's are more likely to be closer to 60 than to zero, and in fact it might be more accurate to treat that class as from 51 to 60, with a midpoint at 55.5. Remember, it is assumed that the only information available is the grouped data as they appear in Figure 4.3, and the actual test scores are not available for examination.

The end classes in grouped data often pose difficult problems for estimating the sample mean. There is no substitute for good judgment, and the judgment of the person doing the analysis is needed in situations like this. The only guideline offered here is "Always tell the reader the basis for your calculations, including your reasoning, if possible."

FINDING THE MEAN FROM GROUPED DATA

The equation for finding the mean from grouped data is as follows:

$$\bar{X} = \frac{1}{n} \sum_{i=1}^{k} f_i m_i \qquad (4.2)$$

where

$$n = \sum_{i=1}^{k} f_i$$

and

k = the number of intervals,
f_i = the frequency in the ith interval,
m_i = the midpoint of the ith interval.

The midpoint in each class interval will be selected as a basis for the calculations, except for the F's, which will be taken to equal 55.5, the midpoint of the interval from 51 to 60. Note that the use of Equation (4.1) for $\bar{X}$ will involve adding the same numbers several times, since each interval midpoint is assumed to occur several times. This procedure is simplified in Equation (4.2) by multiplying each interval midpoint by the number of times it occurs, called its frequency. Figure 4.4 illustrates the shortened calculations.

(1) Grade	(2) Interval	(3) Frequency (f_i)	(4) Midpoint (m_i)	(5) $(f_i)\,(m_i)$
A	91–100	7	95.5	668.5
B	81–90	13	85.5	1111.5
C	71–80	11	75.5	830.5
D	61–70	5	65.5	327.5
F	≤ 60	4	55.5	222.0
			Total	3160.0

$$\sum_{i=1}^{40} X_i \cong \sum_{i=1}^{5} f_i m_i \;\; = 3160$$

$$\bar{X} \cong \frac{1}{40} \,(3160) = 79.00$$

FIGURE 4.4 The Worksheet for Approximating $\bar{X}$ from Grouped Data

The term ΣX_i is approximated by $\Sigma f_i m_i$ at the bottom of column (5). The resulting sample mean is 79.00, slightly larger than the more accurate value 77.55, which was obtained from the original observations. This method of approximating the sample mean is not completely satisfactory because of the use of interval midpoints or other numbers instead of the original observations. But when the original observations are not available, this may be the best a person can do.

The previous calculations reveal that the sample mean from grouped data may be close to, but perhaps not exactly equal to, the true sample mean calculated from the original observations. There is one situation however where the equation for grouped data actually gives the **exact** sample mean. That situation is when the observations are often repeated in the data set. An illustration of that situation is given in the following example.

EXAMPLE

A survey of 25 faculty members in a College of Engineering was taken to study their vocational mobility. They were asked "In addition to your present position, at how many additional educational institutions have you served on the faculty?" Their responses were as follows:

1 2 1 0 0 1 0 2 1 0 0 1 2 1 3 0 1 1 1 0 2 0 1 1 2

These numbers may be summarized in a frequency distribution. Notice that the x_i are used to represent the observations rather than m_i since it is the exact observations x_i that are used in this example, instead of the class marks m_i of the grouped data.

x_i	f_i
0	8
1	11
2	5
3	1
	$n = 25$

The sample mean is easily computed using Equation (4.2):

$$\overline{X} = \frac{1}{n} \sum_{i=1}^{k} f_i x_i$$

$$= \frac{1}{25}[(8)(0) + (11)(1) + (5)(2) + (1)(3)]$$

$$= \frac{24}{25} = 0.96.$$

This is the exact sample mean, since the exact data values were used in the equation. Thus the equation for grouped data may also be used to compute the mean of a sample that is presented in the form of a frequency distribution.

Exercises

4.1   Pharmaceutical companies use large numbers of mice in experiments designed to screen new chemical compounds for effectiveness in treating some types of disease in hopes of developing a new drug to market. These experiments are carefully run and monitored and the mice are routinely weighed as part of the quality control process. Find the sample mean of the following data, which represent the weight in grams of three-week-old laboratory mice:

15.7   14.8   13.7   16.1   15.2

4.2   Production figures for a company for 12 consecutive quarters are given below. Find the average production for these 12 quarters.

Year	Quarter	Total Production (number of units)
1978	1	1048
	2	964
	3	833
	4	1265
1979	1	1117
	2	848
	3	769
	4	1306
1980	1	1082
	2	968
	3	812
	4	1240

4.3 A random sample of homes in a neighborhood revealed the length of time that each resident had lived in that particular neighborhood. These data were: 4 years, 8 months, 2 years, 27 years, and 14 months. Find the sample mean for the number of years lived in the neighborhood.

4.4 A sample of six graduate students selected at random reported the following monthly rents: $245, $195, $250, $280, $215, $345. Find the sample mean for these data.

4.5 A class was given a test with 10 questions on it. Two students got all 10 correct, 8 got 9 correct, and so on as in the following table.

Number Correct	Number of Students
10	2
9	8
8	16
7	12
6	5
5 or less	0

Find the mean number correct first using the formula for ungrouped data and then with the formula for grouped data. Compare your answers.

4.6 Find the average mileage based on tread life tests of a random sample of 100 new tires.

Number of Miles	Frequency
at least 10,000 but less than 15,000	6
at least 15,000 but less than 20,000	21
at least 20,000 but less than 25,000	38
at least 25,000 but less than 30,000	19
at least 30,000 but less than 35,000	16

4.7 Refer to the sample data presented in Exercise 2.5 and find the mean number of items purchased by shoppers.

4.8 The sales slips for the morning business in a neighborhood hardware store are as follows: $1.36, $18.40, $183.79, $2.65, $1.95, and $7.16. Find the sample mean for these sales data. Do you think the sample mean gives a meaningful measure of the center of this sample?

4.9 A frequency table of salary levels for all employed alumni attending a 10-year high school class reunion furnished the following grouped data.

Salary Group	Number of Alumni
$ 0– 5,000	0
5,000–10,000	27
10,000–15,000	73
15,000–20,000	48
20,000–25,000	13
25,000–30,000	8
30,000–35,000	1

Find the sample mean salary for these data.

4.2
The Population Mean

All of the sample statistics to be presented in this chapter can be computed on any set of numbers, whether or not that set of numbers represents some sort of a sample. In the latter case the sample statistics are useful as summary descriptions of the data. However, in the former case, if the observations represent a sample from some population, then those statistics can be used for more than just a summary description of the data. They may be used to gain some useful and valuable insights into the population from which the sample was obtained. While the previous section considered the sample mean and empirical distribution function, in this section their counterparts in the population will be discussed. Specific statistical methods for converting observable sample information into information about the unobservable population counterparts is the subject of subsequent chapters.

AN EXAMPLE

Consider the population of all freshmen at a university that requires an ACT score for admittance. At the end of their first semester both their GPA and ACT score could be obtained to see if a relationship exists between the two measurements. If there are 4800 such freshmen, obtaining this pair of measurements for each freshman could prove to be quite a chore. A random sample of size 10 is selected from the 4800 freshmen, and the ACT score and GPA are recorded in Figure 4.5, along with some sample statistics.

Student	ACT	GPA
1	32	4.0
2	30	3.1
3	29	2.6
4	27	2.3
5	26	2.4
6	26	2.1
7	25	2.4
8	24	1.9
9	24	1.8
10	23	1.0
Mean	26.6	2.36

FIGURE 4.5 Some Statistics on a Sample of University Freshmen

The **population mean** μ is computed from the formula

POPULATION MEAN FOR FINITE POPULATIONS

$$\mu = \frac{1}{N} \sum_{i=1}^{N} X_i \qquad (4.3)$$

for finite populations with N items in the population. Sometimes the notation E(X) is used instead of μ, where E(X) is read "the expected value of X."

*A **parameter** is a fixed numerical quantity that describes a population characteristic.*

THE POPULATION MEAN

If the mean ACT score had been computed for the entire population of 4800 freshmen, Equation (4.3) would be used. This is the same formula as used for the sample mean except for a different notation, using μ ("mu") instead of $\bar{X}$. Also it is customary to let N denote the number of items in the population if the population is finite, reserving the lower case n for the sample size. Greek letters are used to convey the idea of a fixed population characteristic, called a **parameter**. Unlike a statistic that is a random variable and changes from one sample to another, a parameter is a constant and does not change although its value may be unknown. In this case μ represents the **population mean**, a fixed population characteristic that represents the center (in some sense) of the population of values. You can see that $\bar{X}$ provides a useful estimate of the population mean in situations such as this where the population mean is very difficult or impossible to find. Just how useful the estimate $\bar{X}$ is depends largely upon how much $\bar{X}$ varies from sample to sample. The study of this variability plays an important role in this course and in all of statistics.

The population of ACT scores are all reported as integers. Thus, the population of values of this variable consists of a finite set of values, most of which occur several times. That is, if X denotes the ACT score, then $X = 25$ for many different freshmen, $X = 26$ for many others, and so on. Thus it is more convenient to use the formula

$$\mu = \frac{1}{N} \sum_i x_i f_i \tag{4.4}$$

to find μ, where $x_1, x_2, x_3, \ldots$ represent all possible values of X in the population, and f_i represents the frequency with which the event $X = x_i$ occurs in the total population. Note that the only difference between Equation (4.4) and Equation (4.2) is the use of N to represent all elements in the population and that the x_i are used to represent the observations rather than m_i.

Equation (4.4) can be rearranged slightly to look like

$$\mu = \sum_i x_i \left(\frac{f_i}{N} \right). \tag{4.5}$$

Now μ looks like a **weighted mean**; that is, the various values of $X = x_i$ in the population are each multiplied (weighted) by their relative frequencies (f_i/N) in the population.

*A **weighted mean** of several numbers x_i is the sum $\sum w_i x_i$, where weights w_i are positive numbers that sum to 1.0.*

| RANDOM SELECTION | An item is **randomly selected** from a population containing N items if the selection process is such that each item in the population has an equal probability 1/N of being selected. |

Random selection of an item from a population of size N means that each item in the population has an equal probability $1/N$ of being selected. Since exactly f_i of the items result in $X = x_i$, the probability that X equals x_i is just f_i times the probability $1/N$ of each item. Therefore the relative frequencies f_i/N in Equation (4.5) represent the probabilities that a **randomly selected** value of the population will equal the respective values x_i. So Equation (4.5) could be written as

$$\mu = \sum_i x_i p_i \tag{4.6}$$

where $p_i = f_i/N$ is the probability that a randomly selected member of the population will equal x_i.

| POPULATION MEAN | In populations where X equals the values x_1, x_2, x_3, . . . , with respective probabilities p_1, p_2, p_3, . . . , the **population mean** may be computed from the formula

$$E(X) = \mu = \sum_i x_i p_i \tag{4.7}$$ |

EXAMPLE

Consider the numbers of fires per day requiring assistance from a local fire substation. From their past records one day per month is randomly selected for the past 12 months to obtain 12 observations, as given in Figure 4.6, to estimate the population mean number of fires per day.

	No. of Fires		No. of Fires
Jan.	3	July	3
Feb.	1	Aug.	4
Mar.	4	Sep.	0
Apr.	2	Oct.	1
May	2	Nov.	4
Jun.	6	Dec.	3

FIGURE 4.6 Number of Fire Calls in One Day, for Each of 12 Months

Notice that the population is not clearly defined, since the sample could have covered the last 24 months just as easily as the last 12. The population really consists of all days, past, present, *and future*, for which assistance in fires may be required.

THE RELATIONSHIP BETWEEN THE SAMPLE MEAN AND THE POPULATION MEAN

The sample mean could be calculated without grouping the data, especially because the sample size is small, but it is necessary to group the observations in order to show how the relative frequencies are obtained. The grouped data appear in Figure 4.7. The formula for the sample mean for grouped data from Equation (4.2) is

$$\bar{X} = \frac{1}{n} \sum x_i f_i$$

which can be written as

$$\bar{X} = \sum x_i \left(\frac{f_i}{n}\right). \tag{4.8}$$

No. of Fires (x_i)	Frequency (f_i)	f_i/n
0	1	.083
1	2	.167
2	2	.167
3	3	.250
4	3	.250
5	0	.000
6	1	.083
	12	1.000

FIGURE 4.7 The Number of Fire Calls in One Day, in Grouped Form

This latter form is similar to Equation (4.5) and shows that each number of fires x_i is multiplied by its relative frequency of occurrence (f_i/n) in the sample. Either way

$$\bar{X} = 2.75 \text{ fires/day.}$$

When some sort of a probability model is assumed for the population, the probabilities have the same role as the relative frequencies have in the sample. For example, if a probability of .1 of having zero fire calls is assumed, then in the long run about .1 of the days will have no fire calls, or any particular day has about 1 chance in 10 of having no fire calls. The value .1 can be used in place of (f_i/n) in Equation (4.8) for the sample mean and it becomes the first term in Equation (4.7) for the population mean.

Suppose that on the basis of an extended study of past records and expert advice from the firemen, the probability model in Figure 4.8 is determined. Figure 4.8 differs from Figure 4.7 in that the population probabilities appear in Figure 4.8 while the relative frequencies obtained from a sample of observations is given in Figure 4.7. As the sample size gets larger, those relative frequencies should tend toward the population probabilities.

No. of Fires (x_i)	Probability (p_i)
0	.10
1	.15
2	.15
3	.20
4	.15
5	.15
6	.10
Total	1.00

FIGURE 4.8 A Probability Model for the Number of Fire Calls in One Day

The population mean computed from Equation (4.7) is

$$\mu = 0(.10) + 1(.15) + 2(.15) + 3(.20) + 4(.15) + 5(.15) + 6(.10)$$
$$= 3.0 \text{ fires/day.}$$

In a large sample the sample mean $\overline{X}$ will tend to be fairly close to the population mean, since in a large sample (f_1/n) will tend to be close to .10, (f_2/n) will tend to be close to .15, and so on.

THE POPULATION DISTRIBUTION FUNCTION

An empirical distribution function of the GPA data from Figure 4.5 is given in Figure 4.9, along with the population distribution function. The 4800 steps in the population distribution function appear as a smooth curve. It is evident that the empirical distribution function provides an estimate of the population distribution function, useful when the population distribution function is unknown.

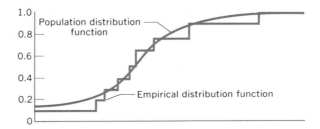

FIGURE 4.9 Graphs of the Empirical Distribution Function and the Corresponding Population Distribution Function

As an example of a discrete type of population distribution function, consider the example involving the number of fire calls per day. The random variable X (the number of fires) assumes only the values 0, 1, 2, 3, and so on, with respective probabilities .10, .15, .15, .20, and so on, as given in Figure 4.8. The population distribution function is a **step function**, with the heights of the steps equal to the respective probabilities, as illustrated in Figure 4.10. The sample observations are provided in Figure 4.6. Here again the empirical distribution function provides a close approximation to the population distribution function. The larger the sample, the better the estimate will tend to be.

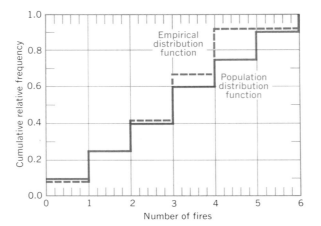

FIGURE 4.10 Graphs of the Empirical Distribution Function and the Corresponding Population Distribution Function Based on Figures 4.6 and 4.8, Respectively.

Exercises

4.10   *Consumer Reports* conducted a survey of insurance claims for 31 insurance companies based on 31,640 claims reported between January 1976 and June 1979. Even with a large sample such as this, the entire population is not accounted for; however, the estimates from this sample may be quite close to the true population values. As part of their summary, *Consumer Reports* gave the table below, which summarizes the dollar amount of claim settlements. If the relative frequencies (percents) reported in this table are treated as true population probabilities, find the population mean dollar amount for these claims. (*Hint:*   Use the class mark to represent each class. The arbitrary choice of 75,000 was made for the class mark in the last class.)

Dollar Amount of Claim Settlement	Class Mark (dollars)	Percent of Claims[a]
1–99	50	13.8
100–199	150	15.8
200–299	250	10.9
300–399	350	7.9
400–499	450	5.0
500–999	750	16.8
1000–1499	1250	6.9
1500–1999	1750	4.0
2000–2499	2250	3.0
2500–4999	3750	5.9
5000–9999	7500	4.0
10,000–49,999	30000	5.0
50,000 or more	75000	1.0

[a] These percentages were normalized to total 100 percent due to the rounding off in the *Consumer Reports* article

4.11   Planning for television broadcasts of events such as the World Series presents a number of scheduling and advertising problems for television networks due to the fact that the series winner is determined by the first team to win 4 games played between the two participating teams. This could take as few as 4 games or as many as 7 games. A study of the frequency distribution of the number of games required to determine a winner would provide some useful information to the networks, but only sample data are available since the population consists of all World Series played past, present, and future. Summary data are given below for all World Series from 1923 to 1980. Since there are 56 years of data, treat the relative frequencies as probabilities and find the mean for this population.

World Series 1923–1980	
Number of Games Played	Frequency
4	11
5	10
6	12
7	23
Total	56

4.12   Graph the population distribution function for the frequency distribution given in Exercise 4.11. Your graph should appear as a step function such as was given in Figure 4.10.

4.13   A restaurant has established the following probability model for groups of 8 or less requiring seating.

Group Size:	1	2	3	4	5	6	7	8
Probability:	.10	.30	.10	.20	.08	.11	.03	.08

Find the population mean for the group size.

4.14   The number of customers waiting for service in a barber shop is described by the probability model given below. Find the population mean for the number of customers waiting to be served.

Number Waiting:	0	1	2	3	4	5	6	7
Probability:	.10	.20	.25	.15	.10	.08	.06	.06

4.15   Graph the population distribution function for the probability model given in Exercise 4.13.

4.16   Graph the population distribution function for the probability model given in Exercise 4.14.

4.3
The Standard Deviation for the Sample and for the Population

WHAT IS THE SAMPLE STANDARD DEVIATION

While the first statistic usually computed on a set of data is the sample mean, the second statistic is usually the sample standard deviation. It is a measure of how much "spread" or variability is present in the sample. If all of the numbers in the sample are very nearly equal to each other, the standard deviation is close to zero. If the numbers are well dispersed, the standard deviation will tend to be large.

CALCULATION OF THE SAMPLE STANDARD DEVIATION

The computation of the standard deviation will be given first, followed by its interpretation. Since the standard deviation is a measure of the variability in a sample, it seems reasonable to start by noting how far each observation lies from the mean of the sample. The expression $(X_i - \bar{X})$ represents how far "out" each observation lies. These quantities are squared to make all of them positive, and the squared values are averaged to find the **sample variance**. It is customary to use $n - 1$ as the divisor in finding the sample variance instead of n, because n tends to underestimate the population variance (i.e., has a negative bias), and dividing by $n - 1$ eliminates this bias. More detail is given in Chapter 6. The square root of the sample variance is the **sample standard deviation**, and is denoted by s. Use of the square root allows the standard deviation to be expressed in the same units as the original observation. The sample standard deviation is a measure of distance; in particular, it is a measure of the distance from the observations in a sample to the middle of that sample. Consider the sample values 3, 4, 5, 6, and 7, which have $\bar{X} = 5$ and $s = 1.58$, while a more spread out group of numbers, 1, 2, 5, 8, 9, has the same sample mean $\bar{X} = 5$ but a larger sample standard deviation $s = 3.54$.

One formula for the sample standard deviation is

$$s = \sqrt{\frac{1}{n-1} \sum_{i=1}^{n} (X_i - \bar{X})^2} \ . \tag{4.9}$$

If there are several standard deviations involved in a discussion, one may be denoted by s_x, another by s_y, and a third by s_z, or perhaps s_1, s_2, and s_3 may be used as notation. Another formula for the sample standard deviation is usually used because it is easier to compute. It is exactly equivalent to the one just given.

$$s = \sqrt{\frac{1}{n-1} \left[\sum_{i=1}^{n} X_i^2 - \frac{(\Sigma X_i)^2}{n} \right]} \tag{4.10}$$

To use Equation (4.10) each observation X_i is squared, then summed to get ΣX_i^2. A common error occurs if the X's are first summed and then squared, the reverse of the correct order. Equation (4.10), though usually easier to use, is more susceptible to roundoff error than Equation (4.9), so the user of Equation (4.10) is cautioned to carry many significant digits in the calculations, especially if $\bar{X}$ is very large compared to the estimated value of s.

SAMPLE STANDARD DEVIATION   *The **sample standard deviation** s is a measure of the amount of variability within a sample. It may be computed using either Equation (4.9) or Equation (4.10).*

SAMPLE VARIANCE	The **sample variance** s^2 is the square of the sample standard deviation s.

EXAMPLE

Suppose your telephone bill for the last five months is given below in column (2).

(1) Month (i)	(2) Bill (X_i)	(3) X_i^2	(4) $X_i - \bar{X}$	(5) $(X_i - \bar{X})^2$
(1) Jan.	$12.37	153.02	−0.85	.7225
(2) Feb.	10.43	108.78	−2.79	7.7841
(3) Mar.	13.66	186.60	0.44	.1936
(4) Apr.	18.14	329.06	4.92	24.2064
(5) May	11.51	132.48	−1.71	2.9241
Totals	66.11	909.94	0.01 (roundoff error)	35.8307

$$\bar{X} = \frac{1}{5}(66.11) = \$13.22$$

$$s = \sqrt{\frac{1}{4}[909.94 - (66.11)^2/5]} = \sqrt{8.9584} = \$2.99$$

$$\text{or } s = \sqrt{\frac{1}{4}(35.8307)} = \sqrt{8.9577} = \$2.99$$

The sample mean is $13.22. Actually the sample mean is $13.222, but it has been rounded off to $13.22, because the sample mean is usually not given to more significant digits than the observations themselves.

The total for column (3) is used in Equation (4.10), along with the total for column (2), to get a standard deviation of $2.99, which is also rounded off to the nearest cent. This standard deviation represents, in a sense, the average amount of variability of these five observations.

Columns (4) and (5) are not needed, except to illustrate how Equation (4.9) is used. The total for column (4) will always be zero, except for roundoff error caused by using $13.22 for $\bar{X}$ instead of $13.222. The total for column (5) is used in Equation (4.9) to obtain a standard deviation of $2.99.

BACK TO THE TEST SCORES

The computations for the standard deviation of the 40 test scores given in Problem Setting 4.1 yield the following intermediate calculations:

$$\sum_{i=1}^{40} X_i^2 = [(77)^2 + (68)^2 + (86)^2 + \cdots + (65)^2]$$

$$= 247,714.$$

Equation (4.10) for the standard deviation gives

$$s = \sqrt{\frac{1}{39}\,[247{,}714 - (3102)^2/40]}$$
$$= \sqrt{183.4333}$$
$$= 13.54.$$

A USEFUL RULE OF THUMB

The sample standard deviation is a measure of how widely scattered the observations are from the sample mean. As a rule of thumb, often about two-thirds of the sample observations are within one standard deviation of the mean. That is, for these 40 test scores about 26.67 observations ($^2/_3$ of 40) can be anticipated as being between 64.01 and 91.09. Those values are obtained from the calculations

$$\overline{X} - s = 77.55 - 13.54 = 64.01$$

and

$$\overline{X} + s = 77.55 + 13.54 = 91.09.$$

Every test score that is within one standard deviation of the mean has been circled in the following diagram.

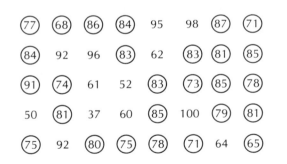

There are 27 observations between 64.01 and 91.09, in close agreement with the rule of thumb prediction.

Another rule of thumb says about 95 percent of the observations are within two standard deviations of the mean. Since

$$\overline{X} - 2s = 50.47$$

and

$$\overline{X} + 2s = 104.63,$$

it can be seen that all but two observations are between 50.47 and 104.63, or within two standard deviations of $\bar{X}$, namely, the two test scores 50 and 37. The 38 out of 40 test scores that are between 50.47 and 104.63 happen to be precisely 95 percent of the total of 40 observations.

Although the results agreed extremely well with the two-thirds rule and the 95 percent rule, these rules should be considered merely rules of thumb. Another rule says that about 99.7 percent of the observations tend to fall within three standard deviations of the mean. This third rule is not as useful as the other two, however, because the sample size needs to be very large before it becomes useful to discuss the concept of about .003 of the observations lying more than three standard deviations from the mean. For a sample with 40 observations, .003 converts to one-eighth of an observation. In this example no test scores are observed to lie outside the interval from $\bar{X} - 3s = 36.93$ to $\bar{X} + 3s = 118.17$.

These rules apply better to some types of data sets than to others. If the histogram has a bell-shaped appearance, these rules are quite accurate, but they are still surprisingly close for many other types of data sets.

A QUICK CHECK ON THE COMPUTATIONS

Since it is easy to make an error somewhere in the process, the computations should always be checked against the sample to see if $\bar{X}$ is near the middle of the observations and to see if about two-thirds of the observations are between $(\bar{X} - s)$ and $(\bar{X} + s)$. This simple test should reveal whether $\bar{X}$ or s contains a gross error in calculation.

THE COEFFICIENT OF VARIATION

The **coefficient of variation** is a single statistic that combines both $\bar{X}$ and s. It provides a comparison of the standard deviation in units of the sample mean. Thus it has the advantage of being free of dimensions (such as "dollars" or "pounds"). Also larger numbers tend to have more variability than smaller numbers. The weights of a sample of men, for example, tend to have a larger standard deviation than the weights of a sample of women. Division by $\bar{X}$ tends to equalize the measures in the two samples so that more meaningful comparisons can be made. This descriptive statistic is used only when all of the observations are positive (so $\bar{X}$ cannot be zero) as a single statistic that combines both s and $\bar{X}$. The coefficient of variation for the test score data is CV = 13.54/77.55 = .175.

THE COEFFICIENT OF VARIATION	*The ratio of the standard deviation to the mean is called the* **coefficient of variation** *and is often denoted by* CV.
	$CV = s/\bar{X}$ (4.11)

STANDARDIZED OBSERVATIONS, OR z-SCORES

An observation by itself does not indicate as much information as when the mean and the standard deviation of the entire sample are also given. In this way the observation can be placed relative to the other observations in the sample. There is a simple method for conveying all of this information in a single number. It involves **standardizing** the observations. Standardizing involves first subtracting the mean from each observation, and then dividing by the standard deviation. The resulting numbers are called z-scores.

$$Z = \frac{X - \bar{X}}{s} \qquad (4.12)$$

If X is greater than $\bar{X}$, then Z is positive. If X is less than $\bar{X}$, then Z is negative. The size of the z-score indicates how many standard deviations separate X and $\bar{X}$. If the z-score is greater than 2.0 for instance, this indicates that the X value is more than two standard deviations above the mean. The rule of thumb would suggest that the X value is in the outer 5 percent of the sample, or in the upper 2½ percent of the sample.

AN APPLICATION OF z-SCORES

Suppose one of your instructors said that every person in the class could drop their lowest hour exam score from consideration from the course grade. Also, one of the exams was unusually difficult and the class mean on that exam was low, only 50, with a standard deviation of 10. However, you had studied extra for that exam and got a score of 80, the highest grade in the class. Even though that 80 may be your lowest exam grade, it may represent your best effort. An examination of z-scores will show this:

$$Z = \frac{80 - 50}{10} = 3.0$$

which is your highest z-score. You may want to try to convince your instructor that it would be more fair to figure grades by averaging z-scores, and to drop the lowest z-score when computing averages. In this way, the grade being dropped would depend more on the individual's performance relative to the rest of the class, and less on the level of difficulty of the exam. This represents the philosophy of thought behind using z-scores instead of "raw scores," and should help to explain the popularity of z-scores in many applied sciences.

MORE ABOUT z-SCORES

Some interesting properties of z-scores are listed below.

1. The sample mean of the z-scores is always zero. This is because the sample mean $\bar{X}$ was first subtracted from the raw scores X_i.
2. The sample standard deviation of the z-scores is always one. This is because the raw scores are divided by the sample standard deviation s.

3. The empirical distribution function of the z-scores looks exactly like the empirical distribution function of the original sample, except for new "standardized" numbers along the horizontal axis.

CALCULATING THE STANDARD DEVIATION FROM GROUPED DATA

If the sample data are already grouped into intervals, and the individual observations are not available, the exact sample standard deviation cannot be obtained. However, an approximate value can be obtained by treating all of the observations in an interval as if they were equal to the midpoint of that interval, as was done for obtaining an approximate value of the mean in the previous section. This method will now be illustrated using the same example used at the end of Section 4.1 in Figure 4.3 with the aid of Equation (4.13) and Equation (4.14).

The grouped data are given again in Figure 4.11, with the calculations needed to find $\bar{X}$ as were presented in Figure 4.4, plus an additional column that is needed to compute the sample standard deviation from grouped data. The term ΣX_i^2 needed to find s is approximated by $\Sigma f_i(m_i)^2$, the total for the sixth column. This approximation results in a **standard deviation for grouped data** equal to 12.10, which is slightly smaller than the standard deviation for the individual test scores, 13.54, computed earlier in this section. In general the standard deviation of the grouped data may be larger or smaller than the standard deviation of the original observations. As was stated earlier for $\bar{X}$ computed from the grouped data, if only the grouped data are available, this may be the best method available for estimating the sample standard deviation.

The **sample standard deviation for grouped data** may be found from either

$$s = \sqrt{\frac{1}{n-1}\sum_{i=1}^{k} f_i(m_i - \bar{X})^2} \qquad (4.13)$$

or

SAMPLE STANDARD DEVIATION FROM GROUPED DATA

$$s = \sqrt{\frac{1}{n-1}\left[\sum_{i=1}^{k} f_i(m_i)^2 - (\Sigma f_i m_i)^2/n\right]}, \qquad (4.14)$$

where

$$n = \sum_{i=1}^{k} f_i = \text{the total number of observations,}$$

k = the number of intervals,

f_i = the frequency in the ith interval,

m_i = the midpoint of the ith interval.

Grade	Interval	Frequency (f_i)	Midpoint (m_i)	$f_i m_i$	$f_i(m_i)^2$
A	91–100	7	95.5	668.5	63,841.75
B	81–90	13	85.5	1111.5	95,033.25
C	71–80	11	75.5	830.5	62,702.75
D	61–70	5	65.5	327.5	21,451.25
F	≤60	4	55.5	222.0	12,321.00
	Totals	40		3160.0	255,350.00

$$\sum X_i \cong \sum f_i m_i = 3160$$

$$\bar{X} \cong \frac{1}{40}(3160) = 79.00$$

$$\sum X_i^2 \cong \sum f_i(m_i)^2 = 225,350$$

$$s = \sqrt{\frac{1}{39}[255,350 - (3160)^2/40]}$$

$$= 12.10$$

FIGURE 4.11 The Worksheet for Approximating $\bar{X}$ and s from Grouped Data

If the data appear to be grouped, but actually represent repeated observations of exact discrete values as in the final example of Section 4.1, then the formula for grouped data may be used. The result will be the exact value of s, because exact observations are used in the calculations.

THE POPULATION STANDARD DEVIATION

The **population standard deviation** is denoted by σ ("sigma" or "lower case sigma") and is given by Equation (4.15). The population standard deviation is a measure of the variability within the population. The sample standard deviation s is the statistic customarily used for estimating σ when σ is not known. The **population variance** σ^2 is simply the square of the population standard deviation σ.

POPULATION STANDARD DEVIATION FOR FINITE POPULATIONS

The **population standard deviation** σ for finite populations with N items is given by

$$\sigma = \sqrt{\frac{1}{N}\sum_{i=1}^{N} x_i^2 - \mu^2}, \tag{4.15}$$

which is slightly different than the formula for s. Besides using μ instead of $\bar{X}$, the divisor is N instead of (n − 1) as used when computing s.

When the same value appears repeatedly in a population, the formula for the standard deviation may be simplified just as the formula for the mean was simplified. Let x_1, x_2, x_3, ... represent the various different values of X in the population. If f_i represents the population frequency of the values x_i for X, then the population standard deviation becomes

$$\sigma = \sqrt{\sum_i x_i^2 \left(\frac{f_i}{N}\right) - \mu^2} \,. \tag{4.16}$$

The probability p_i of a randomly selected value X equaling x_i may be used in place of the relative frequency (f_i/N) in (4.16) to get

$$\sigma = \sqrt{\sum_i x_i^2 \, p_i - \mu^2} \,.$$

In populations where X assumes the values x_1, x_2, x_3, ... , *with respective probabilities* p_1, p_2, p_3, ... , *the* **population standard deviation** *may be computed using either of the equations*

POPULATION
STANDARD
DEVIATION

$$\sigma = \sqrt{\sum_i (x_i - \mu)^2 \, p_i} \tag{4.17}$$

or

$$\sigma = \sqrt{\sum_i x_i^2 \, p_i - \mu^2} \,. \tag{4.18}$$

EXAMPLE

A probability model for the number of fires X in one day was introduced in the previous section as follows.

Values of X:	0	1	2	3	4	5	6
Probability p_i:	.10	.15	.15	.20	.15	.15	.10

The population mean was found to be $\mu = 3$. The population standard deviation may be found using Equation (4.17) as

$$\sigma = \sqrt{\sum_i (x_i - \mu)^2 \, p_i}$$

$$= [(0 - 3)^2(.10) + (1 - 3)^2 \,(.15) + \cdots + (5 - 3)^2(.15) + (6 - 3)^2(.10)]^{1/2}$$

$$= \sqrt{3.30}$$

$$= 1.817.$$

Equation (4.18) could be used instead to get

$$\sigma = \sqrt{\sum_i x_i^2 p_i - \mu^2}$$
$$= [(0)^2(.15) + (1)^2(.15) + \cdots + (5)^2(.15) + (6)^2(.10) - (3)^2]^{1/2}$$
$$= \sqrt{12.30 - 9}$$
$$= \sqrt{3.30}$$
$$= 1.817,$$

which is identical with the previous result, as expected.

Exercises

4.17 Compute the standard deviation for the data in Exercise 4.1 using both formulas for ungrouped data (Equations (4.9) and (4.10)). Compare your answers for the two formulas.

4.18 Calculate the coefficient of variation for the sample data given in Exercise 4.1.

4.19 Refer to Exercise 2.1 and calculate the mean and standard deviation. How many observations do you find within one standard deviation of the mean? How many observations do you find within two standard deviations of the mean? How do these results compare with the rules of thumb given in this section?

4.20 Convert the sample data given in Exercise 2.1 to z-scores. Compute the mean and standard deviation for the z-scores. Plot an e.d.f. of the z-scores as well as an e.d.f. of the original data values. How do these e.d.f's compare to one another?

4.21 Calculate the z-scores for the data of Exercise 4.2. What fact do the z-scores make clear about the production in the second and third quarters of each year?

4.22 Use the formula for grouped data to calculate the sample standard deviation for the sample data in Exercise 2.14.

4.23 Use the formula for grouped data to calculate the sample standard deviation for the sample data in Exercise 2.5.

4.24 Refer to the data given in Exercise 4.5. Compute the standard deviation for these data first using the formula for ungrouped data and then using the formula for grouped data. Compare your answers.

4.25 Find the standard deviation for the data given in Exercise 4.9.

4.26 Find the population standard deviation for the frequency distribution reported in Exercise 4.10. Use 75,000 as the class mark for the last class.

4.27 Find the population standard deviation for the frequency distribution given in Exercise 4.11.

4.28 Find the population standard deviation for the probability model given in Exercise 4.13.

4.29 Find the population standard deviation for the probability model given in Exercise 4.14.

4.4
Modes, Quantiles, and Proportions

PROBLEM SETTING 4.2

The owner-manager of a shoe store is considering ordering a new style of ladies fashion boot, but she is concerned about overstocking since she is unsure about how popular that style of boot will be. To get some idea of what sizes to order, she examines sales records that indicate foot sizes for the 20 most recent sales of women's shoes. Those foot sizes are given in Figure 4.12. As she sorts through the records containing these foot sizes she automatically keeps a tally by first writing down the foot sizes she knows to be reasonable sizes, and then placing a check mark beside each size as it occurs. When she has gone through the records her final tally looks like Figure 4.13.

THE SAMPLE MODE

The store owner has obtained the *frequency distribution* of the sample of size 20. The most frequently occurring size, called the **sample mode**, is size 6½. If she were to order only one pair of this new style boot, she would probably want to order the *modal* size, which is 6½ in this sample, because past records indicate that it is the size most likely to fit a customer.

SAMPLE MODE	The **sample mode** is the observation that occurs the most frequently. The **modal class** is the class with the greatest frequency.

9½	7	5½	7½	9
5½	5	7	5½	8
7	6½	6½	6	6½
6½	9	7½	6½	6

FIGURE 4.12 Twenty Hypothetical Women's Shoe Sizes

Size	Tally	Frequency
5	√	1
5½	√ √ √	3
6	√ √	2
6½	√ √ √ √ √	5
7	√ √ √	3
7½	√ √	2
8	√	1
8½		0
9	√ √	2
9½	√	1
10		0

FIGURE 4.13 A Frequency Distribution of Shoe Sizes

A sample mode is usually considered to be the most frequently occurring value. However, sometimes a value that occurs more frequently than the values on either side of it is also called a mode. Thus size 5½, which occurs more frequently than the sizes 5 or 6 on either side of it in the sample, could be considered to be a mode. It would be called a **secondary mode** because its frequency is less than the frequency at size 6½, which is the **primary mode**. Do you see another secondary mode in the sample?

The term **mode** is also used with histograms, and graphs of density functions and probability functions. In these cases the mode of the graph is the highest point in the graph. Graphs can also have secondary modes. A graph with one mode is called **unimodal** while a graph with two modes is called **bimodal**.

THE SAMPLE MEDIAN

The store manager is not interested in ordering just one pair, however. She is more interested in ordering one dozen pairs of various sizes. She will have a better selection for her regular customers, she can always sell the leftover sizes in a clearance sale, and besides that she gets a substantial price break when she orders in lots of a dozen.

Which sizes should she order? She feels safest in ordering boots that are somewhere near the middle of the distribution of sizes. The exact middle value, when the observations are arranged from smallest to largest, is called the **sample median**. In this case there is no exact middle value since there is an even number of observations in the sample. When the sample size is an even number, the two middle observations are averaged to get the sample median. This is simple to do in this case since both middle observations are 6½. Thus the median size is $x_{.5} = 6½$.

SAMPLE MEDIAN	The **sample median** $x_{.5}$ is the middle observation when the sample is ordered from smallest to largest. If the number of observations is even, the median is the average of the two middle observations.

The sample median is a useful statistic for locating the middle of a sample. It will always be one of the observations in the sample, or halfway between two observations, unlike the sample mean, which will rarely equal one of the observations.

For example, suppose you have five uncles, and the number of children each has is as follows: 1, 0, 0, 4, 3. To find the sample median these numbers are arranged from smallest to largest: 0, 0, 1, 3, 4. The median is 1 child. The sample mean is 1.6 children, larger than the median. If your father is counted among the five uncles, and your father has six children, the sample median becomes $(1 + 3)/2 = 2$ children. Half of the families have more than 2 children and half have less. The sample mean is 2.3, which coincidentally has the same property.

INSENSITIVITY OF THE SAMPLE MEDIAN TO OUTLIERS

The sample median is not as sensitive as the sample mean to observations that are very large or very small compared with the rest of the sample. That is, one

CONNORS 1-81

"THE MEDIAN WEIGHT IS 168, BUT THE MEAN WEIGHT IS 260?!"

very large observation, when averaged in with the other observations, may increase the sample mean until it no longer is a reasonable measure of the middle of the sample, but that one large observation will not change the sample median as much. If your family, instead of having 6 children, as in the previous example, had 16 children, the sample median would remain unchanged. Half of the families would still have more than 2 children and half would have fewer than 2 children. But the mean is now increased to 4.0. Only one family is now larger than the mean, and that family is your family, which produced the "outlying" observation 16.

In some cases, however, the sample mean is preferred because it *does* reflect the total size of all of the observations. The choice of whether to use the mean or the median, or perhaps both, depends largely on personal preference and what happens to be customary in the particular field of application.

EXAMPLE

Several years ago the U.S. government started a policy of reporting median incomes rather than mean incomes. A median income of $20,000 implies that half of the incomes are less than $20,000 and half are greater than $20,000. If your income is $20,000, you know exactly how you rank with the rest of the population.

A mean income of $20,000 is more difficult to interpret, however, since a few extremely large salaries may be responsible for most of the $20,000 average, while the great majority of salaries may be considerably less than $20,000.

SKEWNESS

The bar graph for the shoe sizes is given in Figure 4.14. It is said to be **skewed to the right** because the bar graph extends farther to the right of the mode than

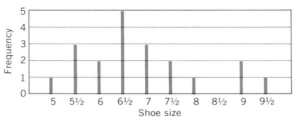

FIGURE 4.14 The Bar Graph for the Shoe Size Data, Illustrating Skewness to the Right

it does to the left. The modal size is 6½, and there are six sizes to the right of the modal size, but only three to the left. Examples of histograms skewed to the right are given in Figures 2.4 and 3.9.

SKEWNESS

A histogram, bar graph, or the graph of a probability function or density function, is said to be **skewed to the right** *if it extends farther to the right of the mode than it does to the left. It is said to be* **skewed to the left** *if the reverse is true.*

When graphs are skewed to the right, the sample mean is typically larger than the sample median, and both are typically larger than the mode. This rule holds in most cases but not always, as is illustrated in this data set where the mode and the median are both equal to 6½. The sample mean, incidentally, is 6.88, which does follow this rule, since it is larger than the median. For graphs that are skewed to the left, the opposite relationship usually holds.

SAMPLE PERCENTILES

Can you remember taking a standard achievement exam such as the ACT or SAT exams? In addition to your exam score, you were probably given a **percentile**. If you scored in the 87th percentile, your counselor may have explained that it meant 87 percent of the people taking the exam got scores lower than or equal to yours. **Percentiles** are also statistics that locate different portions of the sample. Percentiles generally range from the first percentile to the 99th percentile, although fractional percentiles such as the 87.3 percentile are sometimes used. The 50th percentile is the median. If the shoe store manager wanted to order boots from the middle part of the distribution of sizes, she might have found the 10th percentile and the 90th percentile from her sample and only ordered boot sizes between these two values. In this way she wouldn't be ordering any very small or very large boots that may be hard to sell.

SAMPLE PERCENTILE

If r is any whole number from 1 to 99, then the rth percentile $x_{r/100}$ for a sample is any value that has at most r percent of the observations less than that value, and at most $(100 - r)$ percent of the observations greater than that value.

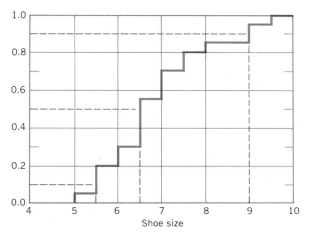

FIGURE 4.15 The Empirical Distribution Function for Shoe Sizes

The easiest way to find sample percentiles is from the empirical distribution function. The e.d.f. for the 20 sample observations is given in Figure 4.15. Any sample percentile may be found easily by locating the desired percentage on the vertical axis, moving horizontally to the e.d.f., and reading the corresponding value from the horizontal axis. The dotted lines in Figure 4.15 show the 10th percentile to be size 5½, the median to be 6½, and the 90th percentile to be size 9. Any number between 5½ and 6 may be used to represent the 20th percentile. For continuous data that are grouped, the ogive should be used instead of the e.d.f. to obtain approximate sample percentiles.

OTHER SAMPLE QUANTILES

Some percentiles are used more than others and are given their own names. The 25th percentile is called the **lower quartile** and the 75th is called the **upper quartile**. The 10th, 20th, 30th, and so on percentiles are called **deciles**. These percentiles, deciles, and quartiles are all special cases of **quantiles**, the more general term. Quantiles are associated with a number between 0 and 1. For example, the .50 quantile is the 50th percentile, the fifth decile, or the median. The ⅓ quantile is between the 33rd and 34th percentiles.

SAMPLE QUARTILES	*The 25th percentile $X_{.25}$ is called the lower quartile or* **first quartile**, *and the 75th percentile $X_{.75}$ is called the upper quartile or* **third quartile**.

SAMPLE QUANTILE	*A* **qth sample quantile** X_q, *where q is a quantity between 0 and 1, is a value that has at most a proportion $(1 - q)$ of the sample observations greater than X_q and at most a proportion q of the sample observations less than X_q.*

THE POPULATION MEDIAN

The population counterpart to the sample median is called, naturally, the **population median**. Just as the sample median is the middle observation in a sample, the median in a finite population is the middle observation in the population in the sense that half of the population is smaller and half of the population is larger than the median. Other quantiles for finite populations are defined in a manner analogous to the way sample quantiles are defined, since finite populations are similar in many respects to large samples.

The median for random variables in general is defined as the middle value of the random variable, in a probability sense. That is, if $\lambda_{.5}$ (where λ is the Greek letter "lambda") is the median of a random variable X, then the probability of $X < \lambda_{.5}$ is no larger than .5, and the probability of $X > \lambda_{.5}$ is also no larger than .5. In practice the population median is easily obtained from the population distribution function in the same way that the sample median is easily obtained from the empirical distribution function.

POPULATION MEDIAN

> The **median of a population** is that number $\lambda_{.5}$ (read "lambda sub .5") which has the property that a randomly obtained observation will be less than $\lambda_{.5}$ with probability no greater than .5, and will exceed $\lambda_{.5}$ with probability no greater than .5.

EXAMPLE

A shoe salesman is assisting the owner-manager of a shoe store in placing her order. He shows her a chart that contains the national distribution of women's shoe sizes. This chart is in the form of a cumulative distribution function and is given in Figure 4.16. The median is obtained by locating .5 on the vertical axis, and reading horizontally to see which shoe size corresponds to that cumulative

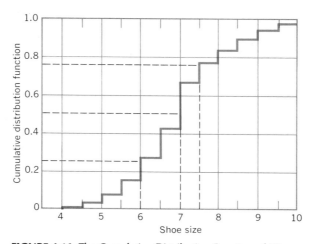

FIGURE 4.16 The Cumulative Distribution Function of Women's Shoe Sizes

probability. The dotted lines at the height .5, leading to the shoe size 7, illustrate this procedure and show that the median population shoe size is $\lambda_{.5} = 7$.

POPULATION QUANTILES

The qth **population quantile**, $0 < q < 1$, called λ_q, is most easily found by finding q on the vertical scale of the c.d.f. and reading horizontally to find the x-coordinate with that ordinate. The probability associated with values in the population less than λ_q is q or less, and the probability in the other direction is $(1 - q)$ or less. The terminology for population quantiles parallels that of sample quantiles, with $\lambda_{.25}$ and $\lambda_{.75}$ being the lower and upper population quartiles, respectively, $\lambda_{.43}$ being the 43rd population percentile, and so on. Figure 4.16 illustrates the method for finding the population quartiles, which are $\lambda_{.25} = 6$ and $\lambda_{.75} = 7\frac{1}{2}$ in the previous example.

POPULATION QUANTILE	The **qth population quantile** λ_q is that number that has probability q or less associated with population values less than λ_q, and probability $(1 - q)$ or less associated with population values greater than λ_q.

THE INTERQUARTILE RANGE

Just as the median may be used as a measure of location instead of the mean, quantiles may be used to measure the spread of the observations instead of the standard deviation. One way is to use the interquartile range, which is the distance from the lower quartile to the upper quartile. From the e.d.f. in Figure 4.15 the upper sample quartile is easily found to be $X_{.75} = 7\frac{1}{2}$ and the lower sample quartile is $X_{.25} = 6$, so the sample interquartile range is $1\frac{1}{2}$ sizes. By coincidence the population interquartile range, obtained from Figure 4.16, has the same value. The interquartile range will usually tend to be a little larger than the standard deviation. However, the sample interquartile range will always be less than or equal to the **sample range**, the largest observation in the sample minus the smallest observation, which is $9\frac{1}{2} - 5 = 4\frac{1}{2}$, in the previous examples.

INTERQUARTILE RANGE	The **interquartile range** is the distance from the lower quartile to the upper quartile, $\lambda_{.75} - \lambda_{.25}$ for populations and $X_{.75} - X_{.25}$ for samples.

EXAMPLE

The sample median and the sample interquartile range can be compared with the sample mean and the sample standard deviation for the shoe size data. Preliminary calculations give

$$n = 20, \quad \sum_{i=1}^{20} X_i = 137.5, \quad \text{and} \quad \sum_{i=1}^{20} X_i^2 = 974.75.$$

The sample mean is

$$\bar{X} = \frac{1}{n}\sum_{i=1}^{20} X_i = \frac{1}{20}(137.5) = 6.875,$$

which is fairly close to the sample median

$$X_{med} = 6.5$$

as would be expected because there are no **outliers** present in the sample. Outliers affect the sample mean and sample standard deviation much more than they affect the sample median and the sample interquartile range. The sample standard deviation is

$$s = \sqrt{\frac{1}{n-1}\left[\sum X_i^2 - (\sum X_i)^2/n\right]}$$

$$= \sqrt{\frac{1}{19}[974.75 - (137.5)^2/20]}$$

$$= \sqrt{1.5493}$$

$$= 1.2447,$$

which as expected is slightly smaller than the sample interquartile range 1.5. Again, the presence of outliers can greatly affect the sample standard deviation, making it much larger than the sample interquartile range.

THE SAMPLE PROPORTION

A statistic that is closely related to sample quantiles is the **sample proportion**. The sample proportion is the fraction of the sample that meets some stated criterion. If the store manager orders only boot sizes between the 10th percentile, size 5½, and the 90th percentile, size 9, but not including those sizes, she would be ordering sizes belonging to only 65 percent of the sample values, not 80 percent as one might have thought. This is because only 13 of the 20 sample values were greater than 5½ and less than 9, and $13/20 = 65$ percent. The sample proportion between sizes 5½ and 9 is 65 percent. The notation for sample proportion is $\hat{p}$. If the sizes 5½ and 9 are included in the calculations, the sample proportion becomes $\hat{p} = 18/20 = 90$ percent, which is larger than the 80 percent one would expect between the 10th and 90th percentiles.

SAMPLE PROPORTION	The **sample proportion** $\hat{p}$ is the fraction of the sample that meets some stated criterion.

Sample proportions often may be read from the e.d.f. The procedure is just the opposite of that for finding quantiles. To find the sample proportion less than or equal to any number, first find that number on the horizontal scale, then read up to the e.d.f., to the top of the step if a step is involved. The ordinate of that point, read from the vertical scale, is the desired proportion.

EXAMPLE

The proportion of sample shoe sizes less than or equal to size 5½ is the height at the top of the step above 5½ in Figure 4.15, which is $\hat{p} = .2$. The proportion less than *but not including* 5½ is read from the bottom of the step above 5½, which is $\hat{p} = .05$. The height of the step itself, .15, is the proportion of the sample that equals size 5½. Proportions of the sample *greater than* given values may be found by subtracting, from one, the sample proportion less than or equal to that number. The sample proportion greater than size 5½ is one minus .2, the sample proportion less than or equal to 5½, which comes out to be 80 percent.

POPULATION PROPORTION

Population proportions are defined and found in a way similar to sample proportions, only using the population distribution function instead of the e.d.f. For example, the population proportion of women having shoe size equal to 7 is found from Figure 4.16 as $.67 - .43 = .24$.

POPULATION PROPORTION	The **population proportion p** *is the fraction of the population that meets some stated criterion.*

Exercises

4.30   Find the mode for the test scores given in Problem Setting 4.1. Is the mode unique for these data or does more than one number satisfy the definition of the mode?

4.31   The data of Problem Setting 4.1 are summarized in a frequency distribution in Figure 4.3. What is the modal class for this summary?

4.32   The data of Problem Setting 4.1 are summarized in an e.d.f. in Figure 4.2. Refer to Figure 4.2 to find the following sample statistics: (a) the median, (b) the lower quartile, (c) the upper quartile, (d) the interquartile range, (e) the ninth decile, and (f) the 80th percentile.

4.33   Find the median for each of the following data sets.

Data set #1:   1   2   3   4   3   2   1
Data set #2:   1   2   3   3   2   1

4.34   Find the mean and median for the following data sets. Why does the median change very little from set to set while the mean changes a great deal?

Data set #1:   7.2   5.1   8.4   6.3   9.1
Data set #2:   7.2   5.1   8.4   6.3   9.1   70.8

4.35   The data of Exercise 2.32 for the number of institutional shares held (in thousands) for 10 discount variety stores are repeated below. For these data find the median and the mean as well as the interquartile range and the standard deviation. Why do each of these respective measures differ so greatly from one another?

Company	Number of Institutional Share Held (thousands)
Danners Inc.	197
Duckwall-Alco	24
Fed-Mart	99
Grand Central	2
Hartfield-Zodys	377
K-Mart	53,170
Thrifty Corp.	544
Wal-Mart	3,452
Woolworth	6,587
Zayre	632

4.36 Find the modal class for the data given in Exercise 4.9.

4.37 Construct an ogive for the frequency distribution reported in Exercise 4.10 and use it to find the population median. Use 75,000 as the class mark for the last class.

4.38 Find the population median for the frequency distribution reported in Exercise 4.11.

4.39 Determine the population median from the graph of the population distribution function made in Exercise 4.15.

4.40 Find the population interquartile range from the graph of the population distribution function made in Exercise 4.15. How does this value compare with the population standard deviation found in Exercise 4.28?

4.41 Find the population interquartile range for the graph of the population distribution function made in Exercise 4.16. How does this value compare with the population standard deviation found in Exercise 4.29?

4.5
The Correlation Coefficient

PROBLEM SETTING 4.3

A professor is interested in knowing if there is a relationship between the order in which an exam is handed in and the score made on the exam. When a student hands in the exam early—long before most students have finished their exams—the professor might think that the student must be very smart, finishing so quickly. On the other hand, the professor may be disappointed because the student made too many careless errors and did not attempt to answer some of the more challenging questions. This raises the question "Do the students who hand their exams in early tend to do better (or worse) than the rest of the group?"

In order to address this question, the order in which the exams were handed in and the score on each exam was noted for one group of students with 34 people in it. The maximum possible score was 200 points. See Figure 4.17.

Order	Score	Order	Score	Order	Score	Order	Score
1	182	10	174	19	115	27	77
2	99	11	140	20	132	28	95
3	193	12	91	21	114	29	188
4	183	13	108	22	83	30	141
5	92	14	164	23	160	31	143
6	125	15	119	24	140	32	134
7	179	16	165	25	133	33	157
8	59	17	135	26	117	34	155
9	100	18	174				

FIGURE 4.17 Test Scores from 34 Students and the Order in which They were Handed in

A SCATTERPLOT

While the previous sections of this chapter dealt with univariate data, this problem setting introduces bivariate data. It is difficult to see any pattern in the bivariate data in Figure 4.17 as they are listed. A scatterplot of these numbers might reveal some tendencies that are not yet apparent. The scatterplot is given in Figure 4.18.

The scatterplot in Figure 4.18 is very interesting to examine. The horizontal axis represents the order in which the exams were handed in, so the points on the left represent the earlier exams and the points on the right represent the exams handed in later. Now the picture becomes clearer. There still does not appear to be a tendency for the earlier exams to be better or worse as a group than the other exams. It looks like the earlier exams include some very good scores, some of the poorer scores and not many of the more average scores. However, the tendency in this direction is only slight and may be a chance happening in this case. At this point it can be said with some confidence that there does not seem to be much of a tendency for the exams handed in earlier to be any better (or worse) than the exams handed in later. Therefore, if you are one of the students, you can ignore the people who hand their exams in early,

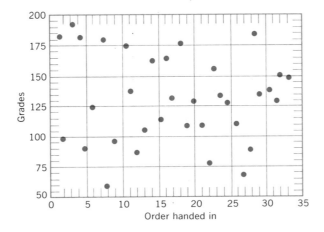

FIGURE 4.18 Test Scores from 34 Students and the Order in which They were Handed in

knowing that your exam still has the same chance of being better or worse than their exams. This illustration points out once more the usefulness of graphical methods for a quick visual inspection of a set of data.

THE SAMPLE CORRELATION COEFFICIENT

There is a convenient statistic that can be used to measure the strength of a linearly increasing or decreasing relationship—that is, a straight line relationship of one variable with another. This statistic is called the **sample correlation coefficient**, and is denoted by r. It is a statistic that tends to be close to zero in cases like the one just examined, where there is no tendency for one variable (exam scores in this case) to increase or decrease as the other variable (the order in which the exams were handed in) increases. The correlation coefficient is always between -1 and $+1$. It equals -1 only if all the points in the scatterplot are on a straight line with a downward slope (downward as viewed from left to right on the graph). It equals $+1$ only if all the points are on a line with an upward slope. The more the points scatter away from a linear relationship, the closer r comes to equaling zero. This does not imply that no relationship exists, it merely implies that the relationship is not linear.

POSSIBLE SCATTERPLOTS FOR THE SAMPLE CORRELATION COEFFICIENT

Although a single statistic like the sample correlation coefficient goes a long way in describing the relationship between two variables, it cannot convey the entire picture. Many very different looking scatterplots can yield the same value of r as the Figures 4.19, 4.20, and 4.21 indicate. Figure 4.19 shows four different scatterplots that share one property in common, there is no tendency

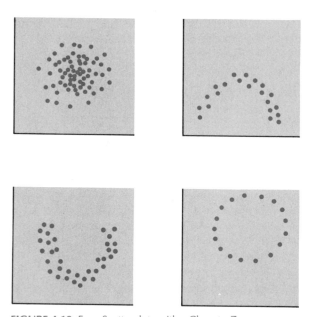

FIGURE 4.19 Four Scatterplots with r Close to Zero

 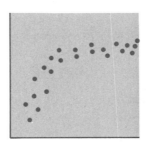 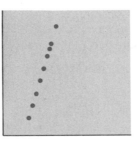 

FIGURE 4.20 Two Scatterplots with r Close to .5   **FIGURE 4.21** Two Scatterplots with $r = 1.0$

toward a **linear** formation and hence r equals zero. If there is a tendency toward following a straight line that is exactly horizontal, then r equals zero also. In Figure 4.20 one scatterplot has a tendency to follow a straight line, but the pattern is not well defined, while in the other the pattern is much clearer, but it is a nonlinear pattern. Finally, Figure 4.21 shows that the only way to get $r = 1.0$ is for all of the points to line up exactly on a straight line sloping upward. A straight line sloping downward would give $r = -1.0$.

HOW TO CALCULATE r

The formula for calculating r from sample data is given in Equation (4.19).

The **sample correlation coefficient** r *measures the strength of the* **linear** *relationship between two variables. It is computed from*

SAMPLE
CORRELATION
COEFFICIENT

$$r = \frac{\sum_{i=1}^{n} X_i Y_i - n \bar{X} \, \bar{Y}}{\sqrt{\left(\sum_{i=1}^{n} X_i^2 - n\bar{X}^2\right)\left(\sum_{i=1}^{n} Y_i^2 - n\bar{Y}^2\right)}}.$$

(4.19)

Many inexpensive calculators are programmed to compute r with a minimum of effort, and anyone who expects to do much computing of correlation coefficients should consider investing in one of these. The steps for computing r manually are given in the following example.

EXAMPLE

Let X represent the order in which the exams of Problem Setting 4.3 were handed in, from $X = 1$ for the first exam to $X = 34$ for the last exam. Let Y denote the scores on the exams. Intermediate calculations yield $\sum X_i$ and $\sum X_i^2$ as in computing the mean $\bar{X}$ and the standard deviation s_x of X, and $\sum Y_i$ and $\sum Y_i^2$ used in finding the mean $\bar{Y}$ and standard deviation s_y of Y. Also the term $\sum XY$ needs to be computed by taking each X value times the corresponding Y and summing. The worksheet appears in Figure 4.22. The correlation coefficient r is close to zero, with a slightly negative value. The nearly zero value of r provides objective testimony that there does not seem to be a tendency for a linear association between the scores of the exams and the order turned in. Most important, however, is that there is now an objective measure of that tendency, supplied by the value of r.

	X	Y	X²	Y²	XY
	1	182	1	33,124	182
	2	99	4	9,801	198
	3	193	9	37,249	579
	4	183	16	33,489	732
	. . .	. . .	. . .	. . .	. . .
	33	157	1,089	24,649	5,181
	34	155	1,156	24,025	5,270
Total	595	4,566	13,685	654,248	79,495

$$\bar{X} = 17.5, \quad \bar{Y} = 134.29, \quad s_x = 9.958, \quad s_y = 35.27$$

FIGURE 4.22 Worksheet for Calculating r for the Test Scores

$$r = \frac{79,495 - 34(17.5)(134.29)}{\sqrt{13,685 - 34(17.5)^2} \; \sqrt{654,248 - 34(134.29)^2}}$$

$$= -.035$$

CHANGING THE EXAMPLE

Suppose that the earlier exams handed in were better than the later exams. How would this affect r? It is instructive to pretend this had happened, to see how different values of r result from different pairings of X with Y. The most extreme case possible is that the exams were submitted in perfect order from the best exam (193) handed in first to the worst exam (59) handed in last. In order to arrange the exams from best to worst a stem and leaf plot is first made. See Figure 4.23. Now the test scores are easily arranged from largest to smallest. If the first exam handed in was the 193, the second exam handed in was the

```
19 | 3
18 | 2 3 8
17 | 9 4 4
16 | 4 5 0
15 | 7 5
14 | 0 0 1 3
13 | 5 2 3 4
12 | 5
11 | 9 5 4 7
10 | 0 8
 9 | 9 2 1 5
 8 | 3
 7 | 7
 6 |
 5 | 9
```

FIGURE 4.23 A Stem and Leaf Plot of the Exam Scores

188, and so on, then the scatterplot would look like Figure 4.24. The calculations worksheet is given in Figure 4.25.

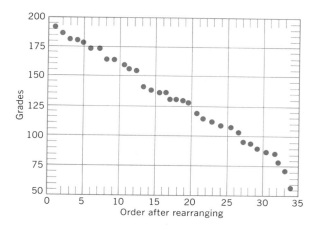

FIGURE 4.24 A Scatterplot of Exam Scores in the Revised Example

	X	Y	X²	Y²	XY
	1	193	1	37,249	193
	2	188	4	35,344	376
	3	183	9	33,489	549
	. . .	. . .	. . .	. . .	. . .
	33	77	1089	5,929	2541
	34	59	1156	3,481	2006
Total	595	4566	13,685	654,248	68,396

$$\bar{X} = 17.5, \quad \bar{Y} = 134.29, \quad s_x = 9.958, \quad s_y = 35.27$$

FIGURE 4.25 The Calculations Worksheet for r in the Revised Example

Notice that the sum of the first four columns is unchanged from before. Changing the order of the observations has no effect on their mean and standard deviation. The only term changed is the sum, ΣXY, which becomes 68,396 instead of its former value 79,495. Decreasing the term ΣXY makes the numerator in r smaller (more negative) and results in r getting closer to its minimum value, -1. This particular pairing of X with Y results in the smallest possible value for ΣXY because the largest values of X are multiplied by the smallest values of Y, and vice versa. The result is the smallest value of r possible with these numbers. From Equation (4.19) r is computed as follows:

$$r = \frac{68,396 - 34(17.5)(134.29)}{\sqrt{13,685 - 34(17.5)^2} \ \sqrt{654,248 - 34(134.29)^2}}$$

$$= -.993.$$

Of course the minimum possible value is $r = -1.0$, but this is possible only when all of the points lie in a straight line. The scatterplot shows that the points are not exactly lined up, even though they are arranged as "perfectly" as possible.

The opposite pairing of X's with Y's—where the first test handed in has the worst score, the second test has the second worst score, and so on until the last test has the best score—results in the maximum value for ΣXY since the largest values of X are multiplied by the largest values of Y. It can be shown that $\Sigma XY = 91,414$ for this arrangement, and $r = .993$, the same as before but with a positive sign instead of a negative sign. Reversing the order for one of the variables will always simply result in a change of sign for r. This value of r, almost equal to its maximum possible value 1.0, conveys the fact that the observations lie almost exactly along a straight line with an upward slope.

SAMPLE COVARIANCE

Another way of measuring the tendency of two variables to increase or decrease together is called the **sample covariance**. It is denoted by s_{xy} and given in Equation (4.20). While the sample covariance increases or decreases as r increases or decreases, it has a definite disadvantage over r in that the sample covariance is not standardized, so the relative size of s_{xy} is difficult to interpret.

SAMPLE COVARIANCE

The **sample covariance** between X and Y in a sample of paired observations (X_1, X_1), (X_2, Y_2), $\cdots$, (X_n, Y_n) measures the tendency for X and Y to increase or decrease together in the sample and is computed as

$$s_{xy} = \frac{1}{n-1}\left[\sum_{i=1}^{n} X_i Y_i - n\overline{X}\,\overline{Y}\right]. \tag{4.20}$$

It is possible to standardize s_{xy} by dividing by the two standard deviations s_x and s_y. In fact, it is easy to show with a little algebra that this furnishes an alternative formula for computing r,

$$r = \frac{s_{xy}}{s_x s_y}, \tag{4.21}$$

which is exactly equivalent to Equation (4.19).

EXAMPLE

For the test scores in Figure 4.22 the sample covariance is

$$s_{xy} = \frac{1}{33}[79,495 - 34(17.5)(134.29)]$$

$$= -12.350$$

from Equation (4.20). This is substituted into Equation (4.21), along with s_x and s_y from Figure 4.22 to get

$$r = \frac{-12.350}{(9.958)(35.27)} = -.035,$$

which is exactly the same value as before.

THE POPULATION COVARIANCE

Other sample statistics presented in this chapter have population counterparts. Likewise, the sample covariance has a counterpart in the population given by Equation (4.22).

POPULATION COVARIANCE

The **population covariance** *measures the tendency for X and Y to increase or decrease together in the population. For finite populations with N items it is given by*

$$\sigma_{xy} = \frac{1}{N} \sum_{i=1}^{N} x_i y_i - \mu_x \mu_y, \tag{4.22}$$

where μ_x and μ_y are the population means.

If the bivariate random variable assumes the values (x_1, y_1), (x_2, y_2), (x_3, y_3), $\cdots$ with probabilities p_1, p_2, p_3 $\cdots$, then the population covariance may be computed using

$$\sigma_{xy} = \sum_i x_i y_i p_i - \mu_x \mu_y. \tag{4.23}$$

THE POPULATION CORRELATION COEFFICIENT

The sample correlation coefficient has a population counterpart given by Equation (4.24).

POPULATION CORRELATION COEFFICIENT

The **population correlation coefficient** *between X and Y is*

$$\rho = \frac{\sigma_{xy}}{\sigma_x \, \sigma_y}, \tag{4.24}$$

where σ_{xy}, σ_x, and σ_y are the population covariance and standard deviations for X and Y, respectively.

The population correlation coefficient ρ (called ''rho'') is a measure of the strength of the linear relationship between two variables.

Exercises

4.42 In Exercise 2.29 a scatterplot was to be made to determine how tuition and fees are affected by the size of government appropriations. Calculate the sample correlation coefficient for these same data. What is the interpretation of the value you have calculated?

4.43 Calculate the sample correlation coefficient for the data presented in Exercise 2.31.

4.44 Find the sample correlation coefficient for the banking data presented in Figure 2.21.

4.45 Refer to the typing scores and error rates given for the 20 students in Figure 2.19 and calculate the sample correlation coefficient for these data. What interpretation do you associate with your answer?

4.46 Calculate the sample correlation coefficient for anticipated inflation and actual inflation using the data given in Exercise 2.34.

4.47 Calculate the sample correlation coefficient for the data given in Exercise 2.32.

4.48 Refer to the Bureau of Labor Statistics data given in Exercise 2.28. Calculate the sample correlation coefficients for each of the three budget types. Also, calculate the sample correlation coefficient for all 24 pairs of observations. How would you explain the difference between the correlation coefficients for each budget type and the one obtained for all data pooled together?

4.6
Rank Correlation

SPEARMAN'S RHO

The sample correlation coefficient r measures the amount of **linear** dependence between two variables. It is simply a measure of how closely the two variables follow a straight line relationship. However, outliers present in the data may have a strong, unwanted effect on the numerical value of r. Therefore, another measure of correlation, known as **Spearman's rho**, r_s, is also a popular statistic for describing a relationship between two variables. Spearman's rho measures the strength of an increasing (or decreasing) relationship between two variables. Also it treats all observations equally, so outliers have no more effect on its value than any other observations have. It may be possible that this relationship is quite strong (r_s is close to 1 or -1), but is not linear so that r is not close to 1 or -1.

SPEARMAN'S RHO

Spearman's rho r_s *is the rank correlation coefficient, obtained by computing the sample correlation coefficient on the ranks. It is computed from the equation*

$$r_s = \frac{\Sigma R_{xi} R_{yi} - C}{\{\Sigma R_{xi}^2 - C\}^{1/2} \{\Sigma R_{yi}^2 - C\}^{1/2}} \tag{4.25}$$

where $C = n(n + 1)^2/4$. *If there are no ties in the data, either of the following equations is simpler to use:*

$$r_s = \frac{\Sigma R_{xi} R_{yi} - C}{n(n^2 - 1)/12} \tag{4.26}$$

or

$$r_s = 1 - \frac{6 \sum_{i=1}^{n} [R_{xi} - R_{yi}]^2}{n(n^2 - 1)}, \tag{4.27}$$

where R_{xi} *is the rank of* X_i *and* R_{yi} *is the rank of* Y_i, *and n is the sample size. Equations (4.26) and (4.27) should never be used if there are ties.*

AN EXAMPLE WITH OUTLIERS

As an example of data with an outlier or two (there is seldom agreement among statisticians on whether or not a given observation should be called an outlier), recall the banking data used earlier in Chapter 2 and repeated in Figure 4.26. A scatterplot of data is difficult to present because of the relatively large deposits of the first two or three banks. However, the sample correlation coefficient may be computed as before. It gives $r = .1446$, indicating a slight, very slight, tendency for the percent of deposit growth to increase linearly with the total deposit size.

	1979 Deposits (thousands of dollars)	Growth of Deposits 1978–1979 (percent)
Albuquerque National Bank	675,709	5.1
First National Bank	457,085	12.0
Bank of New Mexico	242,682	7.1
American Bank of Commerce	99,615	−2.9
Rio Grande Valley Bank	78,947	22.3
Citizens Bank	39,603	−7.5
Fidelity National Bank	38,213	6.9
Southwest National Bank	42,813	13.5
Republic Bank	33,445	−2.3
Western Bank	30,271	5.1
Plaza del Sol National Bank	10,370	13.9
El Valle State Bank	5,635	−6.0

FIGURE 4.26 Deposit Data on Albuquerque Banks

To illustrate how one large observation can dominate the entire variable relationship, as measured by r, two numbers in Figure 4.26 will be interchanged to see how r changes considerably. Let the largest bank, Albuquerque National, be paired with the highest deposit growth, 22.3 percent, and move the 5.1 percent figure down to the Rio Grande Valley Bank. This raises the correlation coefficient from $r = .1446$ to a new value of $r = .6338$. That one change in the growth rate has changed a low value of r to a high value, changing a weak indication of correlation to a strong one, mostly because the size of the deposits of Albuquerque National Bank is many times larger than most of the other banks involved.

Spearman's rho is affected equally by all observations, so it will not change as much. It takes on a value $r_s = .1997$ for the original data (see Figure 4.27), close to the value for r. After the two growth rates are exchanged, r_s becomes only .3818, less than half of the change experienced by r. Spearman's rho is a popular measure of the increasing (or decreasing) relationship between two variables, and it is preferred when a few outliers tend to dominate the scene.

HOW TO CALCULATE SPEARMAN'S RHO

The calculation of Spearman's rho is very similar to the calculations for r, with a few minor simplifications. Basically Spearman's rho is r calculated on ranks instead of the original data. The values of X are replaced by the ranks, 1 for the smallest X, 2 for the second smallest X, and so on up to 12 for the deposits of the Albuquerque National Bank. The Y values are also replaced by their ranks from 1 for the smallest, -7.5, to 12 for the largest, 22.3. Average ranks are used in case of ties—that is, in case several observations are exactly equal to each other. The original observations, their ranks, and the calculations are given in Figure 4.27. A scatterplot of the ranks is given in Figure 2.25.

The formula for r_s as given by Equation (4.25) is the same as the formula for r, given by Equation (4.19), with two exceptions:

1. The ranks R_{x_i} and R_{y_i} of X_i and Y_i, respectively, are used for the calculations instead of the X's and Y's themselves.
2. The mean ranks are always equal to $(n + 1)/2$, so this is used to replace $\overline{X}$ and $\overline{Y}$ in the formula.

Calculation of Spearman's rho from Equation (4.25) for the bank data uses $C = 12(13)^2/4 = 507$ and gives

$$r_s = \frac{535.5 - 507}{(650 - 507)^{1/2}(649.5 - 507)^{1/2}}$$

$$= \frac{28.5}{[(143.0)(142.5)]^{1/2}}$$

$$= .1997.$$

Interchanging the growth rates of Albuquerque National Bank and Rio Grande Valley Bank affects only the sum of $R_x R_y$, increasing that sum to 561.5,

Bank (abbr.)	Deposits (X)	Growth (Y)	Rank of X (R_x)	Rank of Y (R_y)	R_x^2	R_y^2	$R_x R_y$
ANB	675,709	5.1	12	5.5	144	30.25	66
FNB	457,085	12.0	11	9	121	81	99
BNM	242,682	7.1	10	8	100	64	80
ABC	99,615	−2.9	9	3	81	9	27
RGVB	78,947	22.3	8	12	64	144	96
CB	39,603	−7.5	6	1	36	1	6
FNB	38,213	6.9	5	7	25	49	35
SNB	42,813	13.5	7	10	49	100	70
RB	33,445	−2.3	4	4	16	16	16
WB	30,271	5.1	3	5.5	9	30.25	16.5
PSNB	10,370	13.9	2	11	4	121	22
EVSB	5,635	−6.0	1	2	1	4	2
Totals			78	78	650	649.5	535.5

FIGURE 4.27 Worksheet for Calculating Spearman's Rho

and thereby increasing the covariance of the ranks to 4.9545, and increasing r_s to .3818.

AN INTERPRETATION OF SPEARMAN'S RHO

Spearman's rho describes the ranks in exactly the same way that r describes the data. Since r measures the strength of the linear (straight line) relationship between X and Y, r_s measures the strength of the linear relationship between the rank of X and the rank of Y. In cases such as examined earlier, where all of the exams were handed in in perfect order from the best to the worst, a scatterplot of the ranks would show that the ranks lie on a straight line exactly, except for ties, and Spearman's rho will equal -1.0 exactly if there are no ties. Ties will have a slight effect on r_s because the ranks will not lie exactly on a straight line. This is why Spearman's rho is considered to be a measure of the increasing (or decreasing) relationship between two variables, while the correlation coefficient r measures the strength of the linear relationship between two variables.

THE POPULATION COUNTERPART TO SPEARMAN'S RHO

For a finite population with N bivariate data points, each variate can be replaced by its ranks from 1 to N and r_s can be computed on those N data points just as r_s is computed for a sample. In this sense a population counterpart ρ_s to r_s exists for finite populations. It indicates the degree of linear relationship that exists between the ranks of X and Y in the population. However, if the population is defined in terms of possible bivariate observations (x_1, y_1), (x_2, y_2), (x_3, y_3), and so on, where these values have probabilities p_1, p_2, p_3, and so on, respectively, then a population counterpart to r_s is not so easily defined, so it is usually not considered in the study of statistics.

Exercises

4.49   Calculate Spearman's rho on the data of Exercise 2.29. Compare your answer with the sample correlation coefficient computed in Exercise 4.42.

4.50   Calculate Spearman's rho on the data of Exercise 2.31. Compare your answer with the sample correlation coefficient computed in Exercise 4.43.

4.51   Use both Equations (4.25) and (4.27) to calculate Spearman's rho for the job applicant data given in Figure 2.19. What explanation do you have for the difference in the answers using the two equations?

4.52   Calculate Spearman's rho for the anticipated inflation data and actual inflation data given in Exercise 2.34.

4.53   Compute Spearman's rho for the data given in Exercise 2.32. Compare your answer with the simple correlation coefficient computed for these data in Exercise 4.47. How do you explain the large difference in the two answers?

4.54   Rework Exercise 4.48 using Spearman's rho and compare your answers with those obtained when the simple correlation coefficient was computed.

4.7

Review Exercises

4.55 Calculate the sample correlation coefficient for the data given in Exercise 2.39.

4.56 Calculate the rank correlation coefficient for the data given in Exercise 2.39.

4.57 Calculate the sample mean and standard deviation for the data given in Exercise 2.37.

4.58 Convert the sample data given in Exercise 2.37 to z-scores. (See Exercise 4.57.)

4.59 Find the sample median and interquartile range for the data given in Exercise 2.37. (See Exercise 2.38.)

4.60 Find the sample mode for the data given in Exercise 2.37.

4.61 Calculate the sample mean and standard deviation for the data given in Exercise 2.42.

4.62 Convert the sample data given in Exercise 2.42 to z-scores. (See Exercise 4.61.)

4.63 Find the sample median and interquartile range for the data given in Exercise 2.42. (See Exercise 2.44.)

4.64 Find the sample mode for the data given in Exercise 2.42.

4.65 Find the sample mean and standard deviation for the data summarized in Exercise 2.36.

4.66 A popular sports magazine allocates its advertising space such that each page either contains no ads or has $\frac{1}{3}$, $\frac{2}{3}$, or the entire page devoted to advertising. The probability model below describes the allocation used by the magazine on their ads. Find the mean proportion of each page devoted to advertising.

Fraction of the Page Devoted to Ads:	0	$\frac{1}{3}$	$\frac{2}{3}$	1
Probability:	.408	.017	.025	.550

4.67 Find the population standard deviation for the probability model given in Exercise 4.66.

4.68 Graph the probability distribution function for the probability model given in Exercise 4.66 and use it to find the population median.

5

Some Useful Distributions

Thus far, much effort has been devoted to the discussion of *sample* data, display of *sample* data, and summarization of *sample* data. In addition, references have been made to *populations* with *population* means and *population* variances defined. One of the primary reasons for studying *sample statistics* is to obtain some information about the corresponding *population values*.

For example, the empirical distribution function for a random sample is useful because it approximates the distribution function of the population. That is, for reasonably large sample sizes the empirical distribution function resembles the population distribution function, and as the sample size gets larger the approximation tends to get better. Since the population distribution function (or mean, or variance) is the item of interest, and it is almost always unknown, the empirical distribution function (or sample mean, or sample variance) can be used to make inferences about the unknown population.

In this chapter some distributions are introduced that are used in many situations, as population distribution functions, as distribution functions of sample statistics, or as models of common random occurrences. There are countless numbers of different distribution functions. These are just a few of the more useful ones.

Much of this material relates to Chapter 3. Random variables are discussed freely throughout this chapter. Recall that a random variable is a rule for assigning numbers to the possible outcomes of an experiment. The terms "the mean of X" or "the standard deviation of X" are used interchangeably with the mean and standard deviation of the population distribution function associated with the random variable X.

EXAMPLE

Consider the number of babies born in one day in Cleveland, Ohio. Let X be a random variable denoting this number. Then the expression $X = 0$ means the number 0 is assigned to the outcome for a day on which **no** babies were born.

Likewise, $X = 9$ is used for a day on which exactly 9 babies were born. It may be helpful to think of the following expressions as saying the same thing.

"The number of babies born is 6"
      "$X = 6$"

Thus, the use of the letter "X" is a shortcut notation (for convenience in formulas) for the expression "the number of babies born." This random variable is a **discrete random variable**. Two discrete distributions that are useful as probability models for this type of random variable are the **binomial** and the **Poisson** distributions. Both are introduced in this chapter.

EXAMPLE

Let X be a random variable representing the amount of time (in hours) elapsed from the time a bulb is first turned on until it fails in a continuous test; that is, X is the failure time. Thus, $X = 11.89$ means a bulb burned for 11.89 hours before it failed. This is an example of a **continuous random variable**. Two popular continuous distributions introduced in this chapter are the **normal** and the **exponential** distributions.

5.1
The Binomial Distribution

PROBLEM SETTING 5.1

A lung association group ran an extensive ad campaign on local TV to get smokers to participate in a nationwide effort to give up smoking for one day. After the nonsmoking day was over, the lung association contacted smokers to see if they had given up smoking for one day. The smokers responded with either a simple yes or no. The tally of the responses enables the lung association to know how many smokers were influenced by the ads and they can then make a determination as to the effectiveness of the ads.

A BINOMIAL RANDOM VARIABLE

For this study the experiment is the process of asking all of the smokers if they quit smoking because of the TV ads. The random variable X is defined as the number of smokers who indicated they were influenced by the TV ads. Clearly the random variable X can assume only the values 0, 1, 2, 3, . . . , n, where n is the total number of smokers contacted, and therefore X is a discrete random variable.

Consider the characteristics that define this experiment. First, the experiment consists of n independent trials. Trials are repetitions under identical conditions. Each trial is one smoker's response, and the responses are assumed to be independent; that is, each smoker's response is not influenced by the responses of other smokers. Second, each smoker gives a dichotomous response—that is, either yes or no. Third, it can be assumed that the probability of a yes answer is the same from smoker to smoker. This is an example of an experiment in which the **binomial distribution** can be used as the probability distribution of X. A random variable whose distribution is the binomial distribution is called a **binomial random variable**.

A BINOMIAL RANDOM VARIABLE

*A random variable has the **binomial distribution** if the following conditions exist.*

1. *There are one or more trials. (The number of trials is denoted by n, and is a known number.)*
2. *Each trial results in one of two outcomes. (The two outcomes are usually called "success" and "failure" for convenience.)*
3. *The outcomes from trial to trial are independent. That is, the probability of an outcome for any particular trial is not influenced by the outcomes of the other trials.*
4. *The probability of success, denoted by p, is the same from trial to trial. (So is the probability of failure q, where q = 1 − p.)*
5. *The random variable equals the number of successes in the n trials. (Thus the random variable may equal any integer value from 0 to n.)*

EXAMPLE

If there is a total of $n = 4$ smokers in the sample in Problem Setting 5.1, and if X is the total number of affirmative responses, then the possible values of X are 0, 1, 2, 3, and 4. Further, if the TV ad influenced 25 percent of all smokers into quitting for one day, the probability of each person responding "yes" is $p = .25$. Note that p is usually unknown. The probability that the first smoker says "no" is

$$P(no) = q = .75.$$

The joint probability of the first smoker saying "no" and the second smoker saying "yes" is

$$P(no, yes) = P(no) \cdot P(yes) = q \cdot p = (.75)(.25)$$

because of the assumed independence between responses. The probability of a "no," then "yes," and then "no" from the third smoker is

$$P(no, yes, no) = q \cdot p \cdot q = (.75)(.25)(.75).$$

And finally the probability of only the second smoker saying "yes," with the first, third, and fourth smokers responding "no," is

$$P(no, yes, no, no) = q \cdot p \cdot q \cdot q = p \cdot q^3 = (.25)(.75)^3.$$

How many other ways are there to get $X = 1$, or one "yes" and three "no" responses? The first smoker could be the one to respond "yes," with probability

$$P(\text{yes, no, no, no}) = p \cdot q \cdot q \cdot q = p \cdot q^3 = (.25)(.75)^3.$$

The "yes" could come from the third smoker,

$$P(\text{no, no, yes, no}) = q \cdot q \cdot p \cdot q = p \cdot q^3 = (.25)(.75)^3.$$

Finally the "yes" could be from the fourth smoker,

$$P(\text{no, no, no, yes}) = q \cdot q \cdot q \cdot p = p \cdot q^3 = (.25)(.75)^3.$$

Thus there are four different ways in which there could be exactly one "yes" response, so the probability of $X=1$ is the sum of the four probabilities,

$$P(X = 1) = 4pq^3 = 4(.25)(.75)^3 = .422.$$

Probabilities for the other possible values of X are given in Figure 5.1. Later in this section it will be illustrated how to get these probabilities from Table A1, so these calculations will not be necessary in most cases.

$$
\begin{aligned}
f(0) &= \binom{4}{0}(.25)^0(.75)^4 = .316 = P(X=0) \\
f(1) &= \binom{4}{1}(.25)^1(.75)^3 = .422 = P(X=1) \\
f(2) &= \binom{4}{2}(.25)^2(.75)^2 = .211 = P(X=2) \\
f(3) &= \binom{4}{3}(.25)^3(.75)^1 = .047 = P(X=3) \\
f(4) &= \binom{4}{4}(.25)^4(.75)^0 = \underline{.004} = P(X=4) \\
& \qquad\qquad\qquad\qquad\text{Total} = 1.000
\end{aligned}
$$

FIGURE 5.1 Binomial Probabilities for $n = 4$, $p = .25$

THE BINOMIAL PROBABILITY FUNCTION

When a random variable satisfies the requirements to be a binomial random variable, it takes one of the possible values 0, 1, 2, up to n, the number of trials. The probability associated with each possible value x is denoted by $f(x)$, and is given by

$$f(x) = \binom{n}{x} p^x q^{n-x} \qquad \text{for } x = 0, 1, 2, \ldots, n. \tag{5.1}$$

The term $\binom{n}{x}$, called the **binomial coefficient**, is computed using the formula,

$$\binom{n}{x} = \frac{n!}{x!(n-x)!} \tag{5.2}$$

derived in the appendix to this chapter, where $n! = n(n-1)(n-2) \cdots (2)(1)$ for $n \geq 1$ and $n! = 1$ for $n = 0$. The term p^x in Equation (5.1) represents the probability of x successes, in x trials, the term q^{n-x} represents the probability of $(n-x)$ failures in $n-x$ trials, and they are multiplied together because the trials are independent. The binomial coefficient represents the number of different orders in which n trials can result in x successes and $(n-x)$ failures. The function $f(x)$ is called the **probability function**. This formula for $f(x)$ is just another way of assigning probabilities to the various outcomes of X, as was done in Chapter 3.

GRAPH OF THE BINOMIAL PROBABILITY FUNCTION

A graph of the binomial probabilities of Figure 5.1 is given in Figure 5.2. It shows that the most probable response is one out of the four smokers. Such a graph is called the graph (or bar graph) of the binomial probability function.

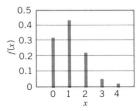

FIGURE 5.2 Bar Graph of the Binomial Probability Function for $n = 4$, $p = .25$

THE BINOMIAL DISTRIBUTION FUNCTION

The function of x that gives the sum of the probabilities for all values of X less than or equal to x is called the **cumulative distribution function** (c.d.f.), or the **population distribution function**. If there is no confusion, the c.d.f. is called simply the **distribution function**.

EXAMPLE

The probabilities given by the probability function in Figure 5.1 are accumulated in Figure 5.3 to get $P(X \leq x)$, the cumulative distribution function. The probability of getting two or fewer "yes" responses is $P(X \leq 2) = .949$, meaning that almost all of the time the number of "yes" responses will be 0, 1, or 2. The probability of three or fewer "yes" responses is .996, even closer to 1. Finally the probability of four or fewer "yes" responses is exactly 1.0. A graph of the probabilities from Figure 5.3 appears in Figure 5.4. Several points are illustrated in Figure 5.4. First note that the graph is that of a step function, illustrating the fact that the binomial distribution is discrete. Next note that the height of each step equals the probability associated with that value, as given in Figure 5.1. Finally note that this distribution function, like all distribution functions, increases from 0 on the left to 1 on the right.

x	$P(X \leq x)$	
0	.316	= .316
1	.738	= .316 + .422
2	.949	= .316 + .422 + .211
3	.996	= .316 + .422 + .211 + .047
4	1.000	= .316 + .422 + .211 + .047 + .004

FIGURE 5.3 Values of $P(X \leq x)$ in the Binomial Distribution Function for $n = 4$, $p = .25$

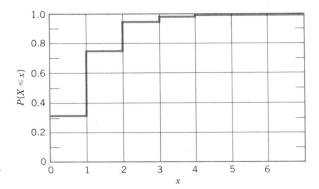

FIGURE 5.4 Graph of the Binomial Distribution Function for $n = 4, p = .25$

THE BINOMIAL DISTRIBUTION

In the **binomial distribution** the probabilities are given by the probability function

$$P(X = x) = f(x) = \binom{n}{x} p^x q^{n-x}, \qquad x = 0, 1, 2, \ldots, n,$$

where

p = probability of success on any one trial,
$q = 1 - p$ is the probability of failure on any one trial,
n = number of trials,
X = number of successes in n trials.

The distribution function
$$F(x) = P(X \le x) = \sum_{i \le x} f(i)$$

is given in Table A1 for selected values of n and p. The parameters of the binomial distribution are n and p.

TABLES FOR THE BINOMIAL DISTRIBUTION FUNCTION

The formula for computing binomial probabilities is seldom used in practice. Tables are usually used to look up the desired probabilities. One such table is given in the appendix of this book as Table A1 for values of $n \le 20$ and selected values of p. Figure 5.5 reproduces a small portion of Table A1. Table A1 is several pages long, so first the portion for the number of trials n (sample size) is found. Next, for that value of n the column corresponding to p is selected. The numbers in that column are the cumulative probabilities, $P(X \le x)$, for the binomial distribution. If the value for p falls between two of the columns given in Table A1, linear interpolation can be used to approximate the desired probabilities. (See the appendix to this chapter for an example of linear interpolation.)

EXAMPLE

Figure 5.5 shows the probabilities for $n = 4$. The column for $p = .25$ gives the probabilities that were given in Figure 5.3. Individual probabilities such as $P(X = 1)$ may be found by subtracting successive entries in the table,

$$P(X = 1) = P(X \le 1) - P(X \le 0)$$
$$= .7383 - .3164$$
$$= .4219,$$

which is in agreement with the number in Figure 5.1

n	x	p = .05	.10	.15	.20	.25	.30
4	0	.8145	.6561	.5220	.4096	.3164	.2401
	1	.9860	.9477	.8905	.8192	.7383	.6517
	2	.9995	.9963	.9880	.9728	.9492	.9163
	3	1.0000	.9999	.9995	.9984	.9961	.9919
	4	1.0000	1.0000	1.0000	1.0000	1.0000	1.0000

FIGURE 5.5 A Portion of Table A1 of the Binomial Distribution Function.

POPULATION MEAN

The **population mean** for the binomial distribution may be obtained by the direct application of the definition given by Equation (4.7). It can be shown mathematically that for binomial distributions the definition of population mean,

$$\mu = \sum_i x_i f(x_i),$$

reduces to the much simpler form

$$\mu = np. \tag{5.3}$$

For example, if $n = 4$ and $p = .25$, the mean of X is $4(.25) = 1.0$. This agrees with intuition. If four smokers are asked a certain question, and the probability is .25 of getting a "yes" from any one of them, then one would expect, on the average, about 1.0 "yes" responses from the four smokers.

POPULATION STANDARD DEVIATION

The **population standard deviation** is almost as easy to find as the population mean. The standard deviation for the binomial distribution is given by the equation

$$\sigma = \sqrt{npq}, \tag{5.4}$$

which may be obtained by direct application of Equation (4.17). Thus the variance is npq. In the example with $n = 4$ and $p = .25$ the standard deviation is

$$\sigma = \sqrt{4(.25)(.75)}$$
$$= \sqrt{.75}$$
$$= .866.$$

The actual numerical value for the standard deviation is not as easy to verify intuitively as the sample mean.

"TRYING TO IMPROVE YOUR SCORE ON THE EXAM WITH THE OLD COIN TOSS, CLAYBORNE?"

ESTIMATING P IN THE BINOMIAL DISTRIBUTION

The value of n is usually known in the binomial distribution, because one can usually count the number of trials. But the value of p is seldom known. The sample is obtained primarily to estimate p.

In Problem Setting 5.1 the lung association wants to estimate the proportion p of smokers who were influenced by the TV ads, so they ask several smokers whether they gave up smoking because of the ads. If 28 percent of the smokers they asked indicated that they gave up smoking because of the TV ads, they would estimate the unknown p to be about .28. The value .28 is called the **sample proportion** $\hat{p}$ and is found by dividing the number of favorable responses X by the number of smokers n,

$$\hat{p} = \frac{X}{n} = \text{sample proportion.} \qquad (5.5)$$

For the same reason $\hat{p}$ may be used to estimate p, it seems reasonable to use $n\hat{p}$ to estimate the mean np, and $\sqrt{n\hat{p}(1 - \hat{p})}$ to estimate the standard deviation $\sqrt{npq}$ of the number of successes X. More will be said about estimating parameters in the following chapter.

EXAMPLES OF BINOMIAL RANDOM VARIABLES

1. Accounting. A sample audit involves examining 100 checking accounts from a bank that has over 5000 checking accounts. Among other things the auditor notes whether the accounts have been overdrawn at any time in the last 30 days. In this case $n = 100$, X = the number of overdrawn accounts, and p (unknown) represents the proportion of all accounts that have been overdrawn in the last 30 days. The unknown value of p is estimated using $\hat{p}$. If the total number of overdrawn accounts is to be estimated, the value for $\hat{p}$ is simply multiplied by the total number of accounts. That is, suppose 7 of the 100 accounts have been overdrawn in the last 30 days. Then

$$\hat{p} = \frac{7}{100} = .07$$

is an estimate of the overall proportion p of overdrawn accounts. To estimate the total number of overdrawn accounts, .07 is multiplied by 5000 to get 350,

$$.07 (5000) = 350,$$

as the estimated total number of overdrawn accounts out of the total of 5000 accounts.

Note the X has the binomial distribution, because there are n trials (each account is a trial) where each trial is "overdrawn" or "not overdrawn," presumably independently of each other. If the sampling is done at random, then each account selected has the same probability p of being overdrawn.

2. Genetics. In an experiment on Mendelian inheritance, 200 plants of two particular genotypes were crossed. The number of progeny that are classified as "dwarf" is called X, and p is the probability of an individual plant being a dwarf. The number of trials n is 200, and p is unknown. The sample proportion $\hat{p}$ may be used to estimate p so that the geneticist can determine if the proportion is in agreement with theoretical predictions. If 46 out of the 200 plants were classified as dwarf, then the estimate of the probability of getting a dwarf plant is

$$\hat{p} = \frac{X}{n} = \frac{46}{200} = .23.$$

As before, the binomial distribution applies to the probability distribution of X, the number of dwarf plants.

3. Athletics. It has been stated by some individuals that the winning team in a professional football game will score more points in their best quarter than the losing team will score in the entire game. If X represents the number of pro games where this statement is true during the season and n is the total number of games played, then the unknown probability of the statement being true may be estimated by $\hat{p}$.

4. Education. Twenty beginning band students are having their instruments tuned individually by their instructor. Each instrument's pitch is either flat or sharp. The probability of an instrument having a flat pitch is estimated by $\hat{p}$, where X is the number of instruments with flat pitch and $n = 20$.

Exercises

5.1   A judge is scheduled to hear 20 appeals for traffic tickets. Each appeal has a probability of .4 of being approved, independent of each other. Find the probability that less than half (9 or fewer) of the appeals are approved.

5.2 What is the expected number of people with high blood pressure in a random sample of 80 people if approximately 10 percent of all people have high blood pressure? What is the standard deviation?

5.3 A student in chemistry who has not had time to prepare for a 20-question multiple choice test decides to use a random guess on each question. If each question on the exam has five choices, what is the student's chance of getting the correct answer on any single question? What mean score should the student expect from this type of an approach to taking an exam? What is the probability that the student will get more than half (11 or more) of the questions correct? What is the probability that the student will get more than 5 correct? What is the probability that the student will get between 2 and 6 (inclusive) correct?

5.4 If a machine has a probability .1 of manufacturing a defective part, what is the expected number of defective parts in a random sample of 20 parts manufactured by this machine? What is the probability of finding 4 or more defective parts in the sample?

5.5 In the 1972 presidential election approximately ⅓ of the voters chose McGovern and ⅔ chose Nixon. Suppose five people are selected at random from those who voted. Define the binomial random variable to be the number of people in this sample who voted for McGovern. Since the value of ⅓ is not covered in Table A1, you will have to use the binomial probability function and make a table similar to the one found in Figure 5.1 to answer the following questions.

(a) What is the probability that no one in the sample voted for McGovern?

(b) What is the probability that everyone in the sample voted for McGovern?

(c) What is the probability that a majority of those questioned in the sample voted for McGovern?

(d) What is the most probable number of people who voted for McGovern? That is, what is the mode of this probability function?

5.6 A toll-booth operator is wondering how much change (in dollar bills) she will need to begin her shift. She knows that 10 percent of all customers want change for a five dollar bill or larger. Suppose the first 20 customers are like a random sample from the population.

(a) What is the probability that 3 or fewer customers want change for a five dollar bill or larger?

(b) What is the probability that 3 or more customers want change for a five dollar bill or larger?

(c) What is the probability that exactly 3 customers want change for a five dollar bill or larger?

5.7 If 5 percent of the checks received in an all night convenience store are returned by the bank, what is the probability that the first eight checks received by the store are all cleared by the bank?

5.8 A marketing survey in Sao Paulo shows that approximately 70 percent of the upper-class shoppers purchased their TV sets in department stores. If a random sample of 15 upper-class shoppers is questioned, what is the probability that 8 or fewer of those questioned purchased their sets from stores other than department stores?

5.9 Use Table A1 to aid in making a graph of the binomial probability function for $n = 10$ and $p = .5$ as was done in Figure 5.2. Also make a similar graph for $n = 20$ and $p = .5$. Are these distributions skewed and if so in what direction? Make a graph of the binomial distribution function for both of these cases as was done in Figure 5.4.

5.2
The Normal Distribution

PROBLEM SETTING 5.2

The Environmental Protection Agency (EPA) has the responsibility of establishing estimates of miles per gallon (mpg) for both city and highway driving. These estimates are printed on the window stickers of every new car sold in this country. Television ads for new cars flash these figures on the screen usually with an asterisk referring to smaller type that cannot be readily seen or else provide a somewhat soft warning that your own mileage may vary. The truth is that your mileage will vary—downward! The reason for this is that the EPA tests are performed in the laboratory and not on the road. Therefore, without wind resistance the resulting mpg estimates are high. These estimates cause problems for both car dealers and customers because the fuel-cost conscious public would like an accurate estimate of the mpg so that it can make the best informed judgment possible with respect to mpg when purchasing an automobile.

A new dealership decides to perform a large scale test of the automobiles they sell as well as those of the competition. In this way they hope to be able to quote more realistic averages for use in comparison.

HISTOGRAMS OF THE DATA

Clearly the random variable of interest in this problem setting is the mpg, and it will have different properties from car to car as well as different properties from city to highway driving for the same car. Therefore, to reduce the magnitude of the problem, only one model of car driven under city conditions will be considered. The properties of this random variable may be considered by examining a histogram of hypothetical values of this random variable. In Figure 5.6 two histograms are presented. One shows what might be obtained from 20 mileage tests, and the other shows possible results from 2000 mileage tests. As the number of mileage tests increases, the class widths get smaller and the histogram becomes smoother. In fact, it seems reasonable to assume that as the number of observations becomes very large, the histogram might approach a graph like the one in Figure 5.7. Figure 5.7 is the graph of a normal density function.

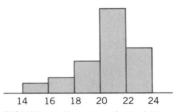 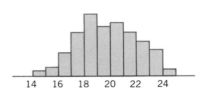

FIGURE 5.6 Two Hypothetical Histograms for City Mileage of One Model of Car

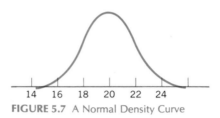

FIGURE 5.7 A Normal Density Curve

NORMAL RANDOM VARIABLES

When a histogram of the data appears to follow a symmetric, unimodal bell-shaped pattern, the normal density function may be a reasonable curve to use as the population curve. In such a case the random variable is said to have a normal distribution, and the random variable is called a **normal random variable**.

The word normal doesn't imply that other distributions are abnormal. It is simply the name customarily given to this distribution. Sometimes the word Gaussian is used instead of normal to avoid confusion. The term normal is used in this text because it is the most widely used term.

The mpg is a good example of a continuous random variable. Theoretically it can assume any number within a certain range. The normal distribution function is a continuous distribution function.

THE NORMAL DISTRIBUTION

You may have noticed that a histogram is a way of presenting the density of observations in the various intervals. More observations contained in an interval mean a higher density in that interval, and the histogram is taller in that interval. Since the histogram approaches the shape of some population curve, that curve is usually called a density curve.

The normal density function is given in Equation (5.6). Neither the density function nor the distribution function given by Equation (5.7) need to be evaluated in order to find probabilities. Both functions are very difficult to work with. Therefore, extensive tables, such as Table A2 in the appendix to this book, have been developed to enable any desired probabilities to be found quickly and easily. Before the use of Table A2 is demonstrated, you will need to learn more about the different types of normal distributions.

THE NORMAL DISTRIBUTION

The normal density function is given by

$$f(x) = \frac{1}{\sigma\sqrt{2\pi}} e^{-(x-\mu)^2/2\sigma^2}, \qquad (5.6)$$

where μ and σ are parameters equal to the population mean and standard deviation, respectively, and where $\pi = 3.14159\ldots$ and $e = 2.71828\ldots$ are well-known constants.

The normal distribution function is given by

$$F(x) = P(X \leq x) = \int_{-\infty}^{x} f(t)\, dt, \qquad (5.7)$$

where the integral represents the area under the density curve, to the left of x. Calculus is not needed here, because Table A2 may be used to find $F(x)$.

POPULATION PARAMETERS AND THEIR ESTIMATES

As you can see from Equation (5.6) the normal distribution has two parameters, μ and σ. Each different value of μ and σ represents a different normal distribution, so the normal distribution is actually a *family* of distributions indexed by μ and σ. There is a reason for choosing these particular Greek letters as the two parameters of the normal distribution. It can be shown that the mean of the normal distribution equals μ, the one parameter, and the standard deviation of the normal distribution equals σ, the other parameter. The mean and standard deviation of a continuous distribution such as the normal distribution are analogous to their counterparts for discrete distributions as defined in Sections 4.2 and 4.3. So if a particular normal distribution has $\mu = 7$ and $\sigma = 2$, as in Figure 5.8, the mean is 7 and the standard deviation is 2 for that distribution.

Suppose a population has a normal distribution with parameters μ and σ, and a random sample is drawn from that population. In practice the values of μ and σ are usually not known. The population mean equals μ, and so μ can be estimated by the sample mean $\bar{X}$. Since the population standard deviation equals the value of the second parameter, σ, the sample standard deviation s is usually used to estimate σ. More will be said about this method of estimating parameters in the next chapter.

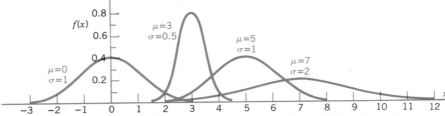

FIGURE 5.8 Graphs of Several Normal Density Functions, for Various Values of μ and σ

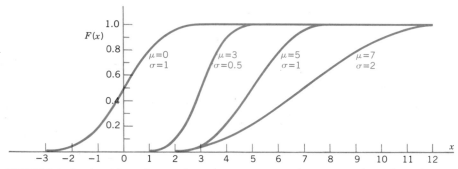

FIGURE 5.9 Graphs of Several Normal Distribution Functions, for Various Values of μ and σ

The mean μ locates the center of the normal distribution, and the standard deviation σ describes how widely the distribution is spread around the mean. Several normal density curves are given in Figure 5.8. The distribution functions corresponding to those curves appear in Figure 5.9. Note that the mean equals the median in normal distributions because the distributions are symmetric.

THE STANDARD NORMAL DISTRIBUTION

The particular normal distribution that has $\mu = 0$ and $\sigma = 1$ is called the **standard normal distribution**. Quite often the term **standard** or **standardized** is used in statistics to denote distributions or random variables that have zero mean and unit standard deviation. Values of the standard normal distribution function are tabulated in Table A2. Although this table is only for the standard normal distribution, it can be used to find probabilities or quantiles for *any* normal distribution. The letter Z is customarily used to denote a standard normal random variable, which emphasizes the fact that it has mean 0 and standard deviation 1.

CONVERTING TO A STANDARD NORMAL DISTRIBUTION

Consider the random variable X with a normal distribution and parameters μ and σ. Subtraction of μ converts X to a new random variable, Y,

$$Y = X - \mu.$$

which is also normal with standard deviation σ, but with mean 0. That is, if the mean of X is 7, each observation on X is converted to an observation on Y simply by subtracting 7 from the observation on X. The mean of Y is then 0.

Subsequent division by σ converts Y to a standard normal random variable Z:

$$Z = \frac{Y}{\sigma} = \frac{X - \mu}{\sigma}.$$

If $\sigma = 2$, then after subtracting 7 from each observation on X, the result is divided by 2 to get an observation on Z. The mean of Z is zero, and the standard deviation is one.

This is analogous to the conversion of a sample of observations to z-scores, by subtracting the sample mean $\overline{X}$ and dividing by the sample standard deviation s. The sample of z-scores always has a sample mean of zero and a sample standard deviation of one. By subtracting the **population mean** μ and dividing by the **population standard deviation** σ, the population mean and standard deviation of Z are zero and one, respectively. This does not imply that a sample of observations on Z will have a sample mean equal to zero, or a sample standard deviation equal to one. In fact, the sample mean and standard deviation most likely will not equal zero or one, respectively, because of the sampling variability always present in a sample.

HOW TO FIND QUANTILES FROM TABLE A2

Table A2 has been constructed to give quantiles for the standard normal distribution. The pth quantile z_p of a continuous distribution is the number that satisfies the equation

$$P(Z \le z_p) = p. \tag{5.8}$$

To find the pth quantile of a standard normal distribution, $0 < p < 1$, first round p off to three decimal places. Then read down the left-hand column of Table A2 to find the row that corresponds to the first two decimal places of p. Read across that row to the column whose heading (at the top of that column) is the third decimal place of p. The table entry in that row and that column is the pth quantile z_p of the standard normal distribution.

To find the pth quantile x_p for a normal distribution with mean μ and standard deviation σ, first find z_p and then use the relationship

$$x_p = \mu + \sigma z_p \tag{5.9}$$

to convert the standard normal quantiles to quantiles for the distribution with μ and σ. Quantiles from a standard normal distribution are used as intermediate values in converting to quantiles from other normal distributions. If the distribution of mpg is normal with mean $\mu = 17$ and standard deviation $\sigma = 3$, then the upper and lower quartiles might be of interest to find out where the "middle" 50 percent of car mileages is located.

TO FIND NORMAL QUANTILES

The pth quantile x_p, $0 < p < 1$, of a normal distribution with known mean μ and standard deviation σ is found from Table A2 as follows:

1. Round p to three decimal places.

2. Enter the row and column corresponding to p, and find z_p, the pth quantile of a standard normal distribution.

3. Convert to x_p using the relationship

$$x_p = \mu + \sigma z_p.$$

EXAMPLE

For example, to find the 2/7th quantile from a standard normal distribution, first convert 2/7 to a number with three decimal places; $2/7 = .286$. Read down Table A2 to the row labeled .28, then across to the column .006 to obtain the entry -0.5651,

$$z_{.286} = -0.5651,$$

which is the .286 quantile of the standard normal distribution. This means there is probability .286 of getting a value less than -0.5651 when observing a value from a standard normal distribution.

EXAMPLE

To find the 2/7th quantile from a normal distribution with $\mu = 6$ and $\sigma = 2$, first the 2/7th $= .286$ quantile for a standard normal distribution is found as described in the previous example. This is substituted into Equation (5.9) to get

$$x_{.286} = 6 + 2\,(-.5651)$$
$$= 6 - 1.1302$$
$$= 4.8698$$

as the .286 quantile of a normal distribution with $\mu = 6$ and $\sigma = 2$. This means there is probability .286 of getting a value less than 4.8698 when observing a value from a normal distribution with $\mu = 6$ and $\sigma = 2$. Figure 5.10 shows the relationship between $z_{2/7}$ and $x_{2/7}$.

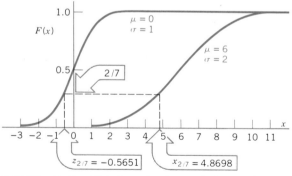

FIGURE 5.10 The 2/7th Quantile of a Standard Normal Distribution, and the 2/7th Quantile of a Normal Distribution with $\mu = 6, \sigma = 2$

HOW TO FIND PROBABILITIES FROM TABLE A2

To find probabilities from Table A2, the procedure just described is reversed. That is, to find $P(X \leq x)$, where X has a normal distribution with mean μ and standard deviation σ, two steps are involved. The first step is to **standardize** x by subtracting μ and dividing by σ,

$$z = \frac{x - \mu}{\sigma}. \tag{5.10}$$

The second step is to find the entry in Table A2 that is **closest to** z, and read the corresponding probability by noting in which row and column z lies.

To find $P(X > x)$ use the relationship

$$P(X > x) = 1 - P(X \le x)$$

and find $P(X \le x)$ in the usual way. To find $P(x_1 < X \le x_2)$ use the relationship

$$P(x_1 < X \le x_2) = P(X \le x_2) - P(X \le x_1)$$

and find $P(X \le x_2)$ and $P(X \le x_1)$ in the usual way.

EXAMPLE

Suppose it is desired to find the probability that a randomly drawn value, from a normal population with $\mu = 10$ and $\sigma = 3$, is less than or equal to 14. That is, the $P(X \le 14)$ is needed. First, 14 is standardized,

$$z = \frac{14 - \mu}{\sigma}$$

$$= \frac{14 - 10}{3}$$

$$= 1.3333.$$

The entry closest to 1.3333 is 1.3346, which is in row .90 and column .009. Therefore, the desired probability is .90 + .009 = .909. See Figure 5.11 for a geometric interpretation. The probability of a normal random variable with $\mu = 10$ and $\sigma = 3$ being less than or equal to 14 is .909.

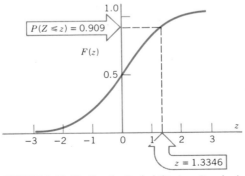

FIGURE 5.11 Finding the Probability in a Standard Normal Distribution

---

TO FIND NORMAL PROBABILITIES

To find the probability $P(X \leq x)$, where X is a normal random variable with known mean μ and standard deviation σ, and x is given:

1. Standarize x using the relationship

$$z = \frac{x - \mu}{\sigma}.$$

2. Find the entry in Table A2 that is closest to z, and note that the desired probability is the sum of the headings of the row and the column that contain z.

In addition, $P(X > x)$ and $P(x_1 < X \leq x_2)$ may be found using the above procedure and the relationships

$$P(X > x) = 1 - P(X \leq x)$$

and

$$P(x_1 < X \leq x_2) = P(X \leq x_2) - P(X \leq x_1).$$

EXAMPLE

Return now to Problem Setting 5.2 and consider 20 sample observations for a particular 4-cylinder model tested in city driving with a manual transmission.

24.21	24.35	23.82	24.21
24.14	24.60	23.75	25.01
24.66	24.72	24.47	24.38
23.08	23.88	23.09	24.57
25.16	24.62	24.62	25.14

The sample mean of these data is 24.32 with a sample standard deviation of .58. If it is safe to assume that these data come from a normal distribution (in the next section an easy way is provided for checking sample data for normality), then these sample statistics can be used as estimates of the population parameters.

Suppose you buy a car of the type being tested. What is the probability that your mileage will be less than 23 mpg? If X is a random variable representing miles per gallon, and if X is normally distributed with population mean $\mu = 24.32$ and population standard deviation $\sigma = .58$, then $P(X \leq 23)$ is needed:

$$P(X) \leq 23) = P\left(Z \leq \frac{23 - 24.32}{.58}\right)$$

$$= P(Z \leq -2.2759)$$

$$= .01 + .001$$

$$= .011.$$

What mpg would be required to be considered in the upper five percent? This question is worded such that the upper 5 percent z-value must first be found from Table A2, which is 1.6449, the .95 quantile. Next this value is substituted into Equation (5.9), which is then solved for $x_{.95}$:

$$x_{.95} = 24.32 + .58 (1.6449)$$

$$= 25.27.$$

Therefore, 25.27 is the upper 5 percent point in the distribution of gas mileage. Only 5 percent of all cars of this particular model can be expected to get better than 25.27 miles per gallon. Note that usually the population mean and standard deviation are unknown. Using sample estimates in their place is advisable only if better figures are unavailable; the resulting probabilities are only approximate when the sample mean and standard deviation are used instead of the population values.

Exercises

5.10 For a normal random variable with $\mu = 0$ and $\sigma = 1$ use Table A2 to find the following quantiles: $z_{.01}, z_{.05}, z_{.50}, z_{.95}$, and $z_{.99}$.

5.11 For a normal random variable with $\mu = 10$ and $\sigma = 2$ use Table A2 to find the quantiles $x_{.025}$ and $x_{.975}$.

5.12 If Z is a normally distributed random variable with $\mu = 0$ and $\sigma = 1$, use Table A2 to find the following probabilities.

(a) $P(Z \leq -2.3263)$ (c) $P(Z \leq 1.6449)$

(b) $P(Z \leq -1.6449)$ (d) $P(Z \leq 2.3263)$

5.13 If X is a normally distributed random variable with $\mu = 10$ and $\sigma = 2$, use Table A2 to find the following probabilities.

(a) $P(X \leq 6.0800)$

(b) $P(X \leq 13.9200)$

(c) $P(6.0800 \leq X \leq 13.9200)$

5.14 The daily output of a production line varies according to a normal distribution with a mean of 163 units and a standard deviation of 3.5 units.

(a) Find the probability that the daily output will fall below 160 units.

(b) Find the probability that the daily output will exceed 170 units.

(c) Find the probability that the daily output will be between 160 and 165 units.

(d) The production manager wants to tell his boss, "80 percent of the time our production is at least x units." What number should he use for x?

5.15 Suppose that X is a random variable representing the reaction time in seconds in a simulated driving experiment for persons who have been given a specified amount of alcohol. If X has a normal distribution with a mean of .8 seconds and a standard deviation of .2 seconds, find the following probabilities.

(a) $P(X \leq .5)$

(b) $P(X \geq 1.2)$

(c) $P(.75 \leq X \leq 1.25)$

5.16 If the errors of measurement observed on individual wristwatches attempting to record the correct time of day can be considered as normally distributed with a mean of 0 and a standard deviation of 60 seconds, what is the probability of a randomly selected watch being within 30 seconds of the correct time of day?

5.17 A student with an I.Q. of 140 claims that his I.Q. score is in the top 5 percent of students at his university. Is his claim true if I.Q. scores are normally distributed at his university with a mean of 125 and a standard deviation of 10?

5.18 Suppose that the present speeds on Michigan highways are normally distributed with a mean of 63 mph and a standard deviation of 5 mph. If the state police decide to ticket the fastest 20 percent of the motorists, how fast could you drive without risking a ticket?

5.3
The Importance of the Normal Distribution

NORMAL APPROXIMATION TO THE BINOMIAL

A question that may have occurred to the reader is how to find binomial probabilities for those values of n and p not covered in Table A1. As was shown in Exercise 5.9, the plot of the probability distribution of a binomial random variable with $p = .5$ is symmetric, and as n gets larger the graph starts to appear bell shaped. This may suggest that the normal distribution may serve as an approximation to the binomial distribution. Indeed it is true that for large sample sizes the normal distribution provides a good approximation to binomial probabilities. However, it turns out that this approximation is also fairly good for small sample sizes if p is not near 0 or 1. A formal statement of this approximation is now presented.

NORMAL APPROXIMATION TO THE BINOMIAL

If X is a binomial random variable with parameters n and p, an approximation to the probability $P(a \leq X \leq b)$, where a and b are integers, can be found using the standard normal distribution, as follows:

1. Compute $z_1 = (a - np - .5)/\sqrt{npq}$ and $z_2 = (b - np + .5)/\sqrt{npq}$.

2. Use Table A2 in the usual way to find

$$P(a \leq X \leq b) \cong P(Z \leq z_2) - P(Z \leq z_1), \tag{5.11}$$

where Z is a standard normal random variable. This is the normal approximation to the binomial distribution.

3. If $a = 0$, use the approximation

$$P(0 \leq X \leq b) \cong P(Z \leq z_2). \tag{5.12}$$

4. If $b = n$, use the approximation

$$P(a \leq X \leq n) \cong 1 - P(Z \leq z_1). \tag{5.13}$$

Note: This approximation should be used only when np and nq are greater than five.

Notice that to use the normal distribution to approximate the distribution of X, the values a and b are standardized by subtracting the mean np and dividing by the standard deviation $\sqrt{npq}$. The .5 is a **continuity correction** to adjust for the fact that a continuous distribution is being used to approximate a discrete distribution that has probabilities at both a and b.

When $p = .5$ the normal approximation to the binomial distribution will provide good answers even for small n. When $p \neq .5$ the approximation is considered adequate if both $np > 5$ and $nq > 5$. See Figure 5.12 for a graphical comparison of a binomial distribution and a normal distribution with the same mean and standard deviation.

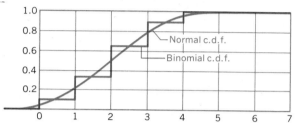

FIGURE 5.12 A Binomial Distribution Function with $n = 8$, $p = .25$, so $\mu = np = 2$ and $\sigma = \sqrt{1.5}$, Along with a Normal Distribution Function with $\mu = 2$ and $\sigma = \sqrt{1.5}$

EXAMPLE

Suppose X is a binomial random variable with $n = 3$ and $p = .5$, and it is desired to find $P(1 \leq X \leq 2)$. The exact answer is easily found from Table A1 to be .750. Even though n is small, the normal approximation will be used:

$$z_1 = (1 - 1.5 - .5)/\sqrt{.75} = -1.1547,$$
$$z_2 = (2 - 1.5 + .5)/\sqrt{.75} = 1.1547,$$

$$P(1 \leq X \leq 2) \cong P(Z \leq 1.1547) - P(Z \leq -1.1547)$$

$$= .876 - .124$$

$$= .752.$$

The approximation is good here for two reasons. First, with $p = .5$ the distribution is symmetric; and second the values of the random variable are in the center of the distribution.

EXAMPLE

What is the probability of obtaining 50 heads on the toss of 100 fair coins? Let X be a random variable representing the number of heads. Then X has a binomial distribution with $n = 100$ and $p = .5$, and

$$P(X = 50) = \binom{100}{50} (.5)^{50} (.5)^{50} = .0796.$$

This is the exact answer (to four decimal places), but don't try to check the calculations by hand. The normal approximation for the binomial is obtained as follows:

$$z_1 = (50 - 50 - .5)/\sqrt{25} = -.10,$$

$$z_2 = (50 - 50 + .5)/\sqrt{25} = .10,$$

$$P(X = 50) \cong P(Z \leq .10) - P(Z \leq -.10)$$

$$= .540 - .460$$

$$= .080.$$

The exact value, obtained from special tables, when rounded off (to three decimal places) agrees exactly with the number obtained using the normal approximation.

THE CONTINUITY CORRECTION

The previous example illustrates the principle behind the use of .5 as a continuity correction. The discrete random variable X has no probability between 49 and 50, or between 50 and 51. All of the probability is concentrated at the integers 49, 50, and 51. On the other hand, the continuous random variable has zero probability at the points 49, 50, and 51. It has nonzero probability only over intervals. So the probability for X at 50 corresponds to the interval on either side of 50 for Z. It seems reasonable to include all of the interval that is closer to 50 than to 49 or 51, so the interval extends from $50 - .5$ to $50 + .5$. This is where the .5 comes from, and this is the idea behind this continuity correction.

As another way of looking at the normal approximation and the continuity correction, refer again to Figure 5.12. The exact binomial probability equals the height of the jump at each integer, such as at $x = 4$, for example. The normal curve on the other hand rises gradually. The rise in the normal distribution function from $x = 3.5$ to 4.5 is about equal to the rise in the binomial distribution function at $x = 4$. Therefore, it could be said that the probability of the normal random variable being between 3.5 and 4.5 is about equal to the probability of the binomial random variable exactly equaling 4.

EXAMPLE

During one week when a university's computer was not operating properly due to a hardware problem, 20 percent of all jobs did not run correctly. One department of the university submitted 40 computer jobs that week. Find the probability that 5 or fewer of these computer jobs were affected by the malfunctioning computer.

Let X be the number of computer jobs affected by the malfunctioning computer. If the jobs submitted by the department resemble a random sample of all computer jobs, then X has a binomial distribution with $n = 40$ and $p = .2$. The desired probability may be approximated as follows:

$$z_2 = (5 - 8 + .5)/\sqrt{6.4} = -.9882$$
$$P(X \le 5) \cong P(Z \le z_2)$$
$$= P(Z \le -.9882)$$
$$= .162$$

from Table A2. This result compares with an exact answer of .161.

LILLIEFORS TEST FOR NORMALITY

The normal distribution is the most important and useful distribution in the study of statistics. In later chapters the normal distribution appears over and over again since many of the procedures to be presented depend on the population being at least approximately normal. Therefore, it is essential to have a procedure for checking that the sample data could have come from a normal distribution. Many procedures exist in statistics for doing this, but the one presented here is quite easy to use. It simply compares the empirical distribution function of the standardized sample values against the standard normal distribution function. This procedure is called the **Lilliefors test for normality**. In Figure 5.13 the basic diagram is presented on which the e.d.f. of the standardized sample values is to be plotted.

The e.d.f. is based on the observations within a sample, and therefore every e.d.f. shows sampling variability. For some samples the graph of an e.d.f. may tend to be above the population distribution function, while for other samples

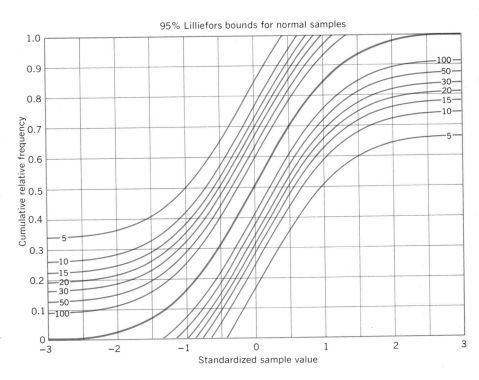

FIGURE 5.13 The Lilliefors Graph for Testing for Normality

the e.d.f. may show the opposite tendency. The probability of the graph of an e.d.f. wandering very far from the c.d.f. should tend to be small. As the sample size gets larger the e.d.f. will tend to be closer to the c.d.f. This is the basis of the diagram in Figure 5.13. For a sample of size 100, the e.d.f. of the standardized values will stay completely within the indicated bounds around the standard normal distribution function for about 95 percent of all samples, if in fact the samples are from a normal distribution. On the other hand, if the sample is taken from a nonnormal population, there is no reason for the e.d.f. to stay within those bounds. This is a convenient method for comparing an e.d.f. with a normal distribution function. Other curves in Figure 5.13 serve as boundaries for other sample sizes.

RULES FOR USING THE LILLIEFORS GRAPH AS A CHECK FOR NORMAL DATA

1. Find the sample mean $\bar{X}$ and sample standard deviation s.
2. Standardize each sample observation X_i as $Z_i = (X_i - \bar{X})/s$.
3. Plot an e.d.f. of these standardized values Z_i in the Lilliefors diagram.
4. The heavy curve in the middle of the Lilliefors diagram represents a standard normal distribution. The lighter curves immediately on either side of the heavy curve represent Lilliefors bounds for a sample of size 100 from a normal distribution. That is, if an e.d.f. plot for $n = 100$ strays outside either of these bounds, then the population should be considered nonnormal. Likewise for other sample sizes use the corresponding bounds provided for $n = 5, 10, 15, 20, 30,$ and 50.
5. For large sample sizes not covered by Figure 5.13, the normality assumption may be tested as follows. Measure the largest distance (measured in a vertical direction) between the e.d.f. and the heavy curve in Figure 5.13. If this distance is larger than $.886/\sqrt{n}$, where n is the sample size, then the population should be considered nonnormal.

    Note: Lilliefors graph paper is available in the student study guide that accompanies this text.

AN APPLICATION OF THE LILLIEFORS TEST

An assumption of normality has previously been made for the data of Problem Setting 5.2. The sample mean for the 20 observations of mpg was 24.32 and the sample standard deviation was .58. The standardized sample values found as $(X_i - 24.32)/.58$ are as follows.

$n = 20$

−.20	.04	−.87	−.20
−.32	.48	−.99	1.19
.58	.69	.25	.10
−2.15	−.77	−2.13	.43
1.45	.51	.51	1.41

The e.d.f. plot for these data is given in Figure 5.14. The check for normality is concerned with the fourth curve from both the top and bottom of the graph— that is, the curves with a label of 20. Since the e.d.f. does not cross either of

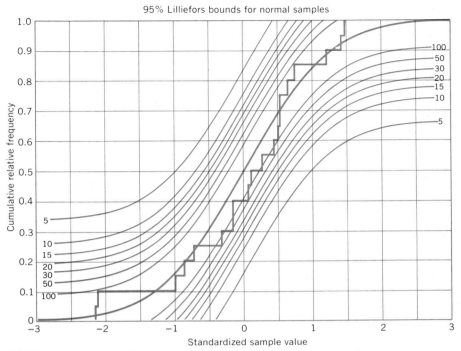

FIGURE 5.14 Using the Lilliefors Test to Check the Normality Assumption for the mpg Data

these curves at any point, it may be assumed that the sample could have come from a normal population. This does not mean that the population is actually normal. Rather, the evidence is not sufficient to conclude that the population is nonnormal. In the absence of convincing evidence that the population is non-normal, the population may be treated as if it were normal. The Lilliefors test will be used throughout the text to check for normality.

Exercises

5.19   Work Exercise 5.1 using the normal approximation to the binomial. How does your answer compare with the exact answer?

5.20   Find the required probability in Exercise 5.4 using the normal approximation to the binomial and compare your answer with the exact answer.

5.21   Suppose X is a binomial random variable with $p = .25$ and $n = 4$. Find $P(1 \leq X \leq 3)$ by exact and approximate methods. The difference between these two answers represents the error encountered by using the normal approximation to the binomial.

5.22 You are testing your roommate's ability in ESP. You draw a card from a deck of cards and concentrate on the suit (clubs, diamonds, hearts, or spades). If your roommate is guessing, he or she has probability .25 of being correct each time. Use the normal approximation to find the probability that he or she will be correct 30 or more times in 100 independent trials.

5.23 A candidate for public office is thought to have the support of 60 percent of the voters. If a random sample of 50 voters is obtained, what would be the answer to the following questions?

(a) How many supporters should the candidate expect to find in this sample?

(b) What is the probability that the number of supporters in the sample will be between 26 and 35 inclusive?

(c) What is the probability that the candidate will fail to obtain support from at least a majority of those in the sample?

5.24 The following data represent the 1980 net earnings of common stocks for 20 representative corporations. Use the Lilliefors graph to test these data for normality.

$1.68	1.72	2.50	2.90	3.11	3.35	3.80	3.85
3.89	4.36	4.64	4.76	5.35	5.81	6.11	6.35
6.69	8.41	8.83	8.97				

5.4
Sampling Distributions and the Central Limit Theorem

THE SAMPLING DISTRIBUTION OF A STATISTIC

The idea of sampling variability has been mentioned repeatedly in this text. That is, random samples from a population are by their very nature unpredictable in each individual instance, and two random samples cannot be expected to be exactly alike even if they are drawn from the same population. Any statistic computed on the basis of the observations in a sample can be expected to vary in its value from sample to sample for the same reason—sampling variability. Although the individual values of a statistic are usually unpredictable, the probabilities associated with the possible values of a statistic can be determined if the population probabilities are known.

It was mentioned in Chapter 4 that a statistic is a type of random variable, and therefore it has a cumulative distribution function, a mean, a standard deviation, and all of the other characteristics of a random variable. In the case of statistics, however, a slightly different terminology is used. The probability distribution of a statistic is often called its **sampling distribution**. The **standard error** of a statistic is another name for its standard deviation. The idea of a sampling distribution is discussed in the following example.

> **SAMPLING DISTRIBUTION** *The probability distribution of a statistic is often called the* **sampling distribution** *of that statistic.*

STANDARD ERROR	*The standard deviation of a statistic is often called the* **standard error** *of that statistic.*

EXAMPLE

As a class exercise, each member of a statistics class was asked to interview 10 randomly selected students at their university and obtain the weight of each of the 10. Each class member then reported back to the class the average weight of the 10 students interviewed; that is, the value of $\bar{X} = \Sigma X_i/10$ was reported. If the statistics class has 25 members, then the data reported would appear as

$$\bar{X}_1, \bar{X}_2, \ldots, \bar{X}_{25},$$

where $\bar{X}_i$ is the sample mean reported by the *i*th member of the class. Some relevant questions about the class exercise include:

1. Will each member of the class report the same sample mean; that is, does $\bar{X}_1 = \bar{X}_2 = \cdots = \bar{X}_{25}$ or will there be some variation in the sample means?
2. If all of the sample means are indeed not the same, is there a distribution that describes the sample means?
3. If there is a distribution describing the sample means, does it depend on the distribution of the parent population from which the sample data were obtained?

Of course the answer to these questions follow directly from the previous discussion on sampling distributions.

1. No, every member of the class will not report the same sample mean. Yes, there will be a variation in the sample means so that it would be surprising in this case to have two students report exactly the same sample mean, unless they happened to choose the same samples.
2. Yes. The distribution that describes the sample means is called its sampling distribution.
3. Yes, the sampling distribution depends on the population from which the sample was obtained. In fact, if the population distribution function is known, it is possible to find the exact distribution function of the sample statistic, although it is not always very easy to do.

THE SAMPLING DISTRIBUTION OF THE SAMPLE MEAN

The previous example uses the sample mean to illustrate the sampling distribution of a statistic. The sample mean is one of the most important statistics, and so its sampling distribution deserves further discussion. The mean and standard deviation of the sampling distribution are given by Equations (5.14) and (5.15), respectively. If the population is at least 10 times as large as the sample size, Equation (5.16) is used instead of (5.15) for the standard deviation of $\bar{X}$. This is usually the situation, in fact, so Equation (5.16) is encountered far more often than (5.15). It is also used whenever the population has a continuous distribution function or when sampling with replacement from a finite population.

THE MEAN AND STANDARD DEVIATION OF $\bar{X}$

Let $X_1, X_2, \cdots, X_n$ be a random sample from a population of size N. If μ and σ denote the population mean and population standard deviation respectively, then the mean $\mu_{\bar{x}}$ and standard error (standard deviation) $\sigma_{\bar{x}}$ of the sample mean are given by

$$\mu_{\bar{x}} = \mu \qquad \text{(the mean of } X \text{ equals the population mean)} \tag{5.14}$$

and

$$\sigma_{\bar{x}} = \frac{\sigma}{\sqrt{n}} \sqrt{\frac{N-n}{N-1}} \qquad \begin{array}{l}\text{(sampling from a finite} \\ \text{population without replacement).}\end{array} \tag{5.15}$$

If the population is large compared with the sample, say larger than 10 times the sample, or if the random variables are continuous, then

$$\sigma_{\bar{x}} = \frac{\sigma}{\sqrt{n}} \qquad \begin{array}{l}\text{(sampling with replacement, or from a continuous distribution,} \\ \text{or from a large population)}\end{array} \tag{5.16}$$

is used for the standard error of $\bar{X}$.

EXAMPLE

A state highway department has 584 tractors. A complete search of the maintenance records would show that the length of time since the most recent major repair for those 584 tractors has a population mean value $\mu = 2.37$ years with a standard deviation $\sigma = 1.58$ years. A random sample consisting of records on 15 tractors is drawn. Even before $\bar{X}$ is computed it is possible to say something about the distribution of $\bar{X}$. The mean of $\bar{X}$ is $\mu_{\bar{x}} = 2.37$ years, the same as the population mean. This gives an indication of where the middle of the distribution of $\bar{X}$ lies. The spread of the distribution of $\bar{X}$ about its mean 2.37 is measured by the standard error of $\bar{X}$, which is obtained using Equation (5.16),

$$\sigma_X = \frac{\sigma}{\sqrt{n}} = \frac{1.58}{\sqrt{15}} = .408,$$

rather than Equation (5.15), because the population size is more than 10 times as large as the sample size. If Equation (5.15) had been used instead, no harm would result; it is simply more cumbersome to use, and the result

$$\sigma_X = \frac{\sigma}{\sqrt{n}} \sqrt{\frac{N-n}{N-1}} = \frac{1.58}{\sqrt{15}} \sqrt{\frac{584-15}{584-1}} = .403$$

is not much different than before. Notice that while the mean of $\bar{X}$ is always equal to the mean of the population from which the sample was taken, the standard error of $\bar{X}$ is always smaller than the population standard deviation, and continues to get smaller as the sample size gets larger.

THE DISTRIBUTION OF $\bar{X}$ WHEN THE POPULATION IS NORMAL

The sampling distribution of $\bar{X}$ is a normal distribution when the population from which the sample was drawn has a normal distribution. This result is useful because it enables exact probability statements to be made concerning $\bar{X}$. Thus, when the population is normally distributed, with mean μ and standard deviation σ, then the mean of a sample of size n is also normally distributed, with mean μ and standard error $\sigma/\sqrt{n}$, because Equation (5.16) is always used when the population distribution function is continuous.

EXAMPLE

The exact weight of a particular type of gold coin has been found to be approximately normally distributed, with mean weight 1.000 grams and standard deviation .0038 grams. A collector buys 10 gold coins and determines their total weight to be 9.984 grams. What is the probability of getting a total weight this small or smaller if the 10 coins are truly a random sample of all coins of that type?

To answer this question note that questions about the total weight are closely related to questions about average weight, since the average is merely the total divided by the sample size. So the question translates to "What is the probability of $\bar{X}$ being .9984 or less (9.984 ÷ 10 = .9984) when the sample came from a normal population with $\mu = 1.000$ and $\sigma = .0038$?" Since $\bar{X}$ is normal, the desired probability can be found from Table A2. The mean of $\bar{X}$ is

$$\mu_{\bar{x}} = \mu = 1.000$$

and the standard error of $\bar{X}$ is given by Equation (5.16), because of the continuous distribution,

$$\sigma_{\bar{X}} = \frac{\sigma}{\sqrt{n}} = \frac{.0038}{\sqrt{10}} = .0012.$$

The value .9984 is standardized to

$$Z = \frac{\bar{X} - \mu_{\bar{X}}}{\sigma_{\bar{X}}} = \frac{.9984 - 1.000}{.0012} = -1.3333,$$

which corresponds to the probability .091 in Table A2. Thus there is probability .091, or about one chance in eleven, of a random sample of 10 gold coins having an average weight as small as the observed average weight, or equivalently having a total weight of 9.984 grams or less. Such a small probability might be the cause for suspicion that the source of gold coins for this dealer may not have a full 1.000 as a population mean weight. More discussion on this type of conclusion appears in subsequent chapters.

THE DISTRIBUTION OF $\bar{X}$ WHEN THE POPULATION IS NOT NORMAL

When the population distribution is not normal, then the distribution of $\bar{X}$ is not normal. However, one of the central results in statistics is the fact that in most situations the distribution of $\bar{X}$ becomes *approximately* normal as n gets large.

This enables Table A2 to be used to obtain approximate probabilities for $\overline{X}$ when the sample size is large. Just how large the sample size needs to be depends on the population distribution function, and also depends on how close an approximation the user requires. Therefore, it is not possible to say how large the sample size must be before the normal approximation may be used. In many cases, however, sample sizes of 20 or 30 meet many users requirements or accuracy.

CENTRAL LIMIT THEOREM

If $X_1, \ldots, X_n$ is a random sample from any infinite population having a finite variance, then the distribution of the random variable $\overline{X}$ becomes normal in form as n gets large. More precisely, the distribution of

$$\frac{\overline{X} - \mu}{\sigma/\sqrt{n}} \tag{5.17}$$

becomes the standard normal distribution as n goes to infinity, where μ and σ denote the population mean and standard deviation.

AN ILLUSTRATION OF THE CENTRAL LIMIT THEOREM IN ACTION

An order for a magazine ad is given to five advertising specialists, who work independently to draw up plans for the ad. One of the advertising specialists is Sue, who was just hired last week. The five ads are shown to a consultant, who is asked to rank the ads from 1 (best) to 5 (worst). If all five advertising specialists are equally capable, then Sue's ad should have an equal chance of being ranked first, second, third, fourth, or fifth. Let X_1 equal the rank Sue gets from the consultant. Then the probability distribution of X_1 is given by

$$f(x) = \tfrac{1}{5}, \quad x = 1, 2, 3, 4, 5$$
$$= 0, \quad \text{otherwise.} \tag{5.18}$$

A probability distribution like this, where the probabilities are equal for the various possible values of the random variable, is called a **uniform probability distribution**. A bar graph of these probabilities is given in Figure 5.15. The mean and standard deviation of this distribution are 3 and $\sqrt{2}$, respectively. It should be immediately apparent that the distribution is not a normal distribution.

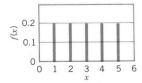

FIGURE 5.15 Graph of a Uniform Probability Distribution over the Integers from 1 to 5

Now suppose the same ads are shown to a second consultant who ranks them independently of the first consultant. Let X_2 be the rank that the second consultant gives to Sue's ad. Then X_1 and X_2 have independent outcomes, and each has the marginal probability distribution given by Equation (5.18). Let $\overline{X}$ be the sample mean

$$\overline{X} = \frac{1}{2}(X_1 + X_2).$$

It may be helpful to think of the population as consisting of five ping pong balls that have the numbers 1, 2, 3, 4, 5 painted on them and then are placed in a box. The balls are mixed, a ball is randomly drawn, and the number recorded. The ball is replaced, and after mixing again a second ball is selected with the number again recorded, thus completing one sample. A listing of every possible pair of outcomes obtained in this manner, along with the corresponding sample mean, is given in Figure 5.16.

		X_1 = Rank according to first consultant				
		1	2	3	4	5
	1	1.0	1.5	2.0	2.5	3.0
X_2 = Rank	2	1.5	2.0	2.5	3.0	3.5
according	3	2.0	2.5	3.0	3.5	4.0
to second	4	2.5	3.0	3.5	4.0	4.5
consultant	5	3.0	3.5	4.0	4.5	5.0

FIGURE 5.16 A Table of Sample Means

A list of every possible sample mean along with the probability of occurrence is given in Figure 5.17. The mean for this newly defined distribution is given by

$$\mu_{\overline{X}} = \Sigma \overline{x} P(\overline{X} = \overline{x}) = 1.0(.04) + 1.5(.08) + 2.0(.12) + \cdots + 5.0(.04) = 3.0$$

using the method described in Section 4.2.

$\overline{x}$	$P(\overline{X} = \overline{x})$
1.0	$(1/5)\,(1/5) = .04$
1.5	$2(1/5)\,(1/5) = .08$
2.0	$3(1/5)\,(1/5) = .12$
2.5	$4(1/5)\,(1/5) = .16$
3.0	$5(1/5)\,(1/5) = .20$
3.5	$4(1/5)\,(1/5) = .16$
4.0	$3(1/5)\,(1/5) = .12$
4.5	$2(1/5)\,(1/5) = .08$
5.0	$(1/5)\,(1/5) = .04$

FIGURE 5.17 The Sampling Distribution of $\overline{X}$ as given in Figure 5.16

This result is in agreement with Equation (5.14) as the mean of $\overline{X}$ always equals the population mean. The variance of this distribution of $\overline{X}$ is

$$\sigma_{\overline{X}}^2 = \Sigma \bar{x}^2 \, P(\overline{X} = \bar{x}) - \mu_{\overline{X}}^2$$

$$= (1.0)^2(.04) + (1.5)^2(.08) + (2.0)^2(.12) + \cdots + (5.0)^2(.04) - 3^2$$

$$= 10 - 9 = 1$$

using the methods described in Section 4.3.

This result is in agreement with Equation (5.16) since the standard deviation of the distribution of the sample mean is given by $\sigma/\sqrt{n}$, or $\sqrt{2}/\sqrt{2} = 1$. The graph in Figure 5.18 represents the sampling distribution of the random variable $\overline{X}$ when $n = 2$. It is roughly triangular in shape, which is closer to a normal density function in appearance, than was the original population probability distribution of Figure 5.15. The c.d.f. of $\overline{X}$ in Figure 5.20 also appears to be closer to the normal c.d.f. than is the population c.d.f. in Figure 5.19.

To proceed one more step, suppose four consultants are each asked to rank the ads, independently of each other. Let X_1, X_2, X_3, and X_4 represent the rank each of the four gives to Sue's ad, and consider the distribution of the sample mean $\overline{X}$ for this random sample of size four. The sampling distribution of $\overline{X}$ for a sample of size four is obtained in the same way as for a sample of size two, but the computations are somewhat lengthy, so they are omitted. The result is the probability distribution of $\overline{X}$ given in Figure 5.21 for a sample of size four from

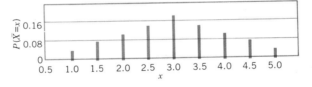

FIGURE 5.18 Sampling Distribution of $\overline{X}$ for Samples of Size Two

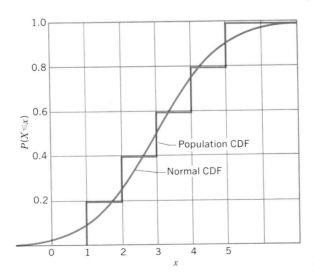

FIGURE 5.19 The Population C.D.F., Compared with a Normal C.D.F. With The Same Mean and Variance

the population given by Figure 5.15. Notice that this distribution of probabilities in Figure 5.21, and the corresponding c.d.f. given in Figure 5.22, are closer to the appearance of a normal distribution than the corresponding Figures 5.18 and 5.20 for samples of size two. As the sample size increases, the appearance will be more and more like a normal distribution, according to the Central Limit Theorem.

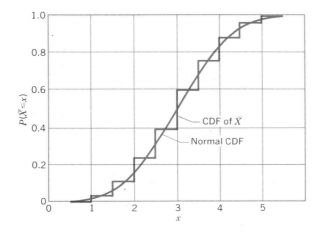

FIGURE 5.20 The C.D.F. of $\bar{X}$ $(n = 2)$, Compared With a Normal C.D.F. With the Same Mean and Variance

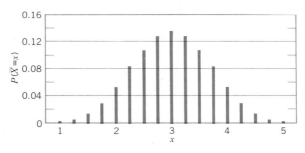

FIGURE 5.21 Sampling Distribution of $\bar{X}$ for Samples of Size Four

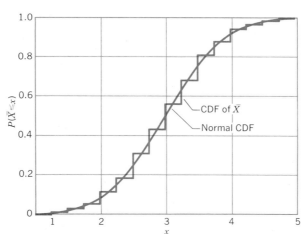

FIGURE 5.22 The C.D.F. of $\bar{X}$ $(n = 4)$ Compared With a Normal C.D.F. With the Same Mean and Variance

The following should now be noted:

1. X_1 is a discrete random variable defined for five values of x. So is X_2, X_3, and X_4.
2. $\overline{X}$ is also a discrete random variable, but defined for nine values when $n = 2$, and 17 values when $n = 4$.
3. The graph of the distribution of $\overline{X}$ is not the same as the graph of X_1, X_2, X_3, or X_4.
4. $\overline{X}$ is not normally distributed; however, the graph of the probabilities of $\overline{X}$ looks more like a bell-shaped normal graph than does the graph of the probabilities of X_1 through X_4.
5. The Central Limit Theorem does not say that the distribution of $\overline{X}$ should be normal as samples only of size $n = 2$ and $n = 4$ were used. However, it does say that as n gets large the normal approximation to the probabilities of $\overline{X}$ become more accurate.

In summary it may be said that as the sample sizes get larger, the normal distribution gets better as an approximation to the distribution of $\overline{X}$. For some population distributions the normal distribution provides a good approximation for the distribution of $\overline{X}$ even for small sample sizes. In fact, if the population has a normal distribution, then the distribution of $\overline{X}$ is *exactly* normal no matter how small n might be, as was pointed out earlier in this section. However, since exact normal distributions do not exist with real world data, this point is more of academic interest than it is of practical value.

THE APPROXIMATE NORMALITY OF THE SAMPLE PROPORTION

The Central Limit Theorem may also be used to justify the normal approximation to the distribution of the sample proportion $\hat{p}$. Recall, from Equation (5.5), that the sample proportion is X/n, where X is the number of "successes" in n trials. If the result of each trial is recorded separately by the random variable Y_i for the ith trial, then $Y_i = 1$ if the trial results in a success, which has probability p, $Y_i = 0$ if the trial does not result in a success, and X is merely the sum of the Y_i's. Since all of the Y_i's are independent and all have the same distribution, the Central Limit Theorem applies to their mean,

$$\overline{Y} = \sum_{i=1}^{n} \frac{Y_i}{n},$$

which is just another way of writing the sample proportion

$$\hat{p} = \frac{X}{n} = \sum_{i=1}^{n} \frac{Y_i}{n}$$

since X is the sum of the Y_i's.

The mean of each Y_i may be shown to be p, and the standard deviation may be shown to equal

$$\sqrt{pq}$$

since Y_i is merely a binomial random variable with $n = 1$. Therefore, the mean of $\hat{p}$ is p, its standard error is $\sqrt{pq}/\sqrt{n}$, and the statistic

$$\frac{\hat{p} - p}{\sqrt{pq}/\sqrt{n}}$$

may be approximated by the standard normal distribution if n is large. However, the above statistic may be written in a more familiar form,

$$\frac{\hat{p} - p}{\sqrt{pq}/\sqrt{n}} = \frac{\dfrac{X}{n} - p}{\sqrt{pq}/\sqrt{n}} = \frac{X - np}{\sqrt{npq}}$$

which is the standardized form of a binomial random variable. This is the justification behind the normal approximation to the binomial distribution, introduced in the previous section.

Exercises

5.25 Verify that the population mean and standard deviation of the uniform probability distribution given in Equation (5.18) are 3 and $\sqrt{2}$, respectively.

5.26 The mean and standard deviation of the sampling distribution of $\overline{X}$ for samples of size 2 represented in Figure 5.18 are three and one, respectively. What are the values of the population mean and standard deviation for the sampling distribution of $\overline{X}$ for samples of size 4 as represented in Figure 5.21?

5.27 At one time the number of shares sold per day on the New York Stock Exchange followed a normal distribution with $\mu = 16{,}000{,}000$ and $\sigma = 5{,}000{,}000$. Find the probability that the *average* daily shares sold in a one-week (five-day) period exceeds 18,000,000 shares.

5.28 Refer to Exercise 5.14. Find the probability that the average production for 10 days will be less than 160 units.

5.29 Consider a probability model that assigns the probabilities .30, .15, .10, .15, and .30 to the values of the random variable $X = 1, 2, 3, 4,$ and 5, respectively. Graph this probability function and compute the population mean and the standard deviation.

5.30 Refer to the probability model given in Exercise 5.29. Means of samples of size 2 for this model would appear as in Figure 5.16. However, the sampling distribution of $\overline{X}$ would *not* appear as in Figure 5.17. Find the correct sampling distribution of $\overline{X}$ for samples of size 2 and make a summary similar to Figure 5.17.

5.31 Make a graph of the sampling distribution of $\overline{X}$ found in Exercise 5.30 and compare it with the one found in Figure 5.18.

5.32 Compute the mean and the standard deviation of the sampling distribution found in Exercise 5.30 and compare them with population mean and variance found in Exercise 5.29. Does this comparison show agreement with Equations (5.14) and (5.16), respectively?

5.5

The Exponential Distribution

PROBLEM SETTING 5.3

The manager of a quality assurance division of an electronics firm has the responsibility of determining the length of time the firm should offer on the warranty of an amplifier manufactured by the firm. His decision will be based on laboratory tests made by the firm on the *time to failure* of the amplifier. This setting is typical of decisions or policies made on the basis of experimental results. Such experiments will usually produce some numerical outcomes (i.e., sample data) in which case the decision-making process reduces to one of applying scientific judgment to the numerical results.

DECISIONS BASED ON EXPERIMENTAL RESULTS

As a starting point on the above problem setting, it will be assumed that the manager knows the competition offers a warranty of 50 hours on their amplifier. Suppose the manager is forced to make a "seat of the pants" decision (i.e., no scientific judgment allowed) for each of the following sets of laboratory tests.

Test Set 1: 120, 115, 170, 65, 143
   Manager's Decision: It is economically sound to offer a warranty of 50 hours.
Test Set 2: 50, 30, 95, 58, 42
   Manager's Decision: A warranty of 50 may be too high.

Do you agree with the manager's decisions? At this point his decisions may appear to be correct as Test Set 1 certainly has all times well above 50 hours while Test Set 2 may be too close to 50 for comfort. What the manager really needs to know is what percent of the amplifiers his firm will have to replace because they fail in the first 50 hours of operation.

THE EXPONENTIAL RANDOM VARIABLE

An attempt is now made to associate some scientific judgment with this problem setting and to evaluate the manager's decisions. The random variable of interest is the length of life of the amplifier or, equivalently, the time to failure. Many different probability distributions are used to describe the length of life of a product; one used most often is the **exponential distribution**. It is used for products that have the same probability of failing during the next time interval when they are new as when they are used. That is, all through the life of the product, their probability of failing during the next minute (or next hour or next day) remains unchanged from when the product was new. If the probability of the amplifier failing during the next 24 hours is the same now as it was at any time in the past, including when the amplifier was new, and will be the same at any time in the future given that the amplifier is still amplifying at that future time, then the amplifier is said to have a **constant failure rate**, and the exponential distribution is appropriate to use as a model. Other examples where the exponential probability distribution is used to model random variables include the time between phone calls arriving in an office, the time between accidents at a street intersection, and the time between calls at a fire department. Because the random variable frequently involves waiting times between events, it is also called the **waiting time distribution**.

The probability distribution function and population mean and standard deviation are given as follows.

PROBABILITY DISTRIBUTION FOR AN EXPONENTIAL RANDOM VARIABLE

If a random variable X has an **exponential distribution** with parameter λ, then the density function is given by

$$f(x) = \lambda e^{-\lambda x}, \qquad x > 0, \quad \lambda > 0$$

and the distribution function is given by

$$P(X \leq x) = 0 \text{ for } x \leq 0$$
$$= 1 - e^{-\lambda x} \text{ for } x > 0. \tag{5.19}$$

Population mean $= 1/\lambda$.
Population standard deviation $= 1/\lambda$.
The parameter λ is sometimes called the **failure rate**.

It should be noted that this is another example of a continuous random variable, but unlike the normal random variable, the exponential random variable is always greater than zero. Also, the population mean equals the population standard deviation in exponential populations so the coefficient of variation always equals 1. Some typical graphs of the probability density function and distribution function for various values of λ are given in Figures 5.23 and 5.24. Note that the exponential distribution is not symmetrical.

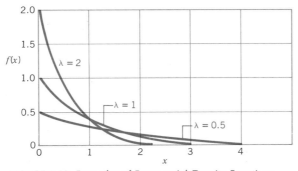

FIGURE 5.23 Examples of Exponential Density Functions

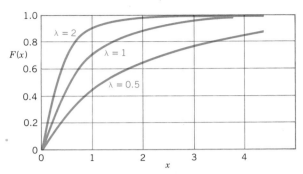

FIGURE 5.24 Examples of Exponential Distribution Functions

LILLIEFORS TEST FOR EXPONENTIAL DISTRIBUTIONS

It is very easy to compare a sample of observations with the exponential distribution function to see if the population distribution function is of the exponential type. First the observations are **standardized** by dividing each observation by the sample mean. Note that this method of standardizing exponential random variables is different from the method used for normal random variables in the previous section. Next the empirical distribution function of the standardized observations is plotted on a graph such as the one in Figure 5.25 and compared with the Lilliefors bounds. If the empirical distribution function of the standardized variables touches or crosses the Lilliefors bounds for that

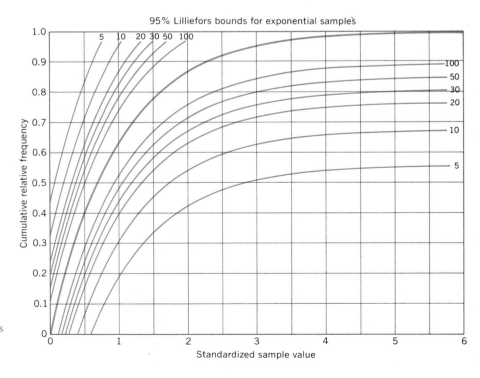

FIGURE 5.25 The Lilliefors Graph for Testing for Exponential Distributions

sample size, then the exponential distribution is not reasonable to assume for the population from which the random sample was obtained, and any procedures that use the assumption of an exponential distribution should not be used.

RULES FOR USING THE LILLIEFORS GRAPH AS A CHECK FOR EXPONENTIAL DATA

1. Find the sample mean $\overline{X}$ of the random sample.

2. Standardize each sample observation X_i as $X^*_i = X_i / \overline{X}$.

3. Plot an e.d.f. of these standardized values X^*_i in the Lilliefors diagram.

4. The heavy curve in the middle of the Lilliefors diagram represents an exponential distribution with $\lambda = 1$. The lighter curves immediately on either side of the heavy curve represent Lilliefors bounds for a sample of size 100 from that exponential distribution. That is, if the e.d.f. plot for $n = 100$ strays outside either of these bounds then the population should be considered nonexponential. Likewise for other sample sizes the corresponding bounds are provided for $n = 5, 10, 20, 30,$ and 50.

5. For large sample sizes not covered by Figure 5.25, the assumption of an exponential distribution may be tested as follows. Measure the largest distance (measured in a vertical direction) between the e.d.f. and the heavy curve in Figure 5.25. If this distance is larger than $1.0753/\sqrt{n}$, where n is the sample size, then the population should be considered as nonexponential.

Note: Lilliefors graph paper is available in the student study guide that accompanies this text.

For answering probabilistic questions about exponential random variables, Equations (5.19), (5.20), and (5.21) are used.

FINDING PROBABILITIES FOR EXPONENTIAL RANDOM VARIABLES

If X is an exponential random variable with parameter λ, then

$$P(X \geq x) = e^{-\lambda x} \text{ for } x > 0 \tag{5.20}$$

and

$$P(x_1 \leq X \leq x_2) = e^{-\lambda x_1} - e^{-\lambda x_2} \text{ for } x_1 > 0 \text{ and } x_2 > 0. \tag{5.21}$$

ESTIMATING THE PARAMETER OF AN EXPONENTIAL POPULATION

One way to estimate the parameter λ from sample data is first to find the sample mean $\overline{X}$, which is an estimate of the population mean $1/\lambda$. Then the estimator $\hat{\lambda}$ of λ is given by $1/\hat{\lambda} = \overline{X}$ or $\hat{\lambda} = 1/\overline{X}$. The manager's decisions, associated with Problem Setting 5.3, are now evaluated based on the exponential distribution.

Test Set 1: 120, 115, 170, 65, 143
$\overline{X} = 122.6,$ $\hat{\lambda} = 1/122.6 = .00816$

First the data are examined to see if it is reasonable to assume that the sample could have come from an exponential distribution. The standardized values $X^*_i = X_i/\overline{X}$ are given below.

Standardized Test Set 1:   .979, .938, 1.387, .530, 1.166

The e.d.f. of these standardized values is plotted on a Lilliefors graph in Figure 5.26. The e.d.f. is within the two curves for $n = 5$, the outermost curves in the graph, so the assumption of an exponential distribution is not unreasonable. Of course with such a small sample size almost any distribution would seem reasonable, including the normal distribution.

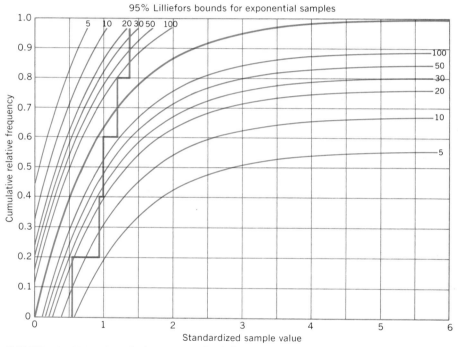

FIGURE 5.26 Using the Lilliefors Test to Check the Exponential Assumption

What proportion of the amplifiers will fail within 50 hours, the figure used on the warranty under the assumption that $\lambda = .00816$?

$$
\begin{aligned}
P(X \le 50) &= 1 - e^{-.00816(50)} \\
&= 1 - e^{-.408} \\
&= .3350
\end{aligned}
$$

The number $e^{-.408}$ may be found on most pocket calculators. As can be seen, 33.5 percent of the amplifiers could be expected to fail in the first 50 hours. Therefore, 50 appears to be too high a number to put on the warranty. The results for the second test set are even worse.

Test Set 2:   50, 30, 95, 58, 42
$\bar{X} = 55$, $\hat{\lambda} = 1/55 = .0182$
$$P(X \leq 50) = 1 - e^{-.0182(50)}$$
$$= 1 - e^{-.9091}$$
$$= .5971$$

The reader should use the Lilliefors test to verify that the exponential distribution is reasonable for the population from which this sample was obtained. This test set indicates that there is an estimated probability of 59.71 percent for each amplifier failing in the first 50 hours. Based on the first test set, what do you think would be a reasonable figure to put on the warranty?

THE POISSON DISTRIBUTION

There is a second way of looking at items failing over time, or events like telephone calls occurring over time—that is, by counting *how many* items fail, or *how many* telephone calls occur, in a fixed interval of time, such as in one hour, or one day, or two days. When the *time between failures* has an exponential distribution, then the *number of failures* in a fixed interval of time has a distribution known as the **Poisson distribution**. Notice that while the time between failures (or telephone calls) is a continuous random variable, the number of failures (or telephone calls) in a fixed time interval takes on only the integer values 0, 1, 2, 3, and so on, and is therefore a discrete random variable.

EXAMPLE

The receptionist in a bank notes the time of arrival of customers seeking special services, such as opening a new account, obtaining a loan, and so on. This is part of a study designed to determine how many employees are needed to provide sufficient service so that customers do not need to wait very long.

The arrival times are given in Figure 5.27, starting from 9:00 a.m. when the records were begun. Also given is the time between successive arrivals, or the time until the first arrival. The first arrival is at 9:03, so $X_1 = 3$ minutes. The second arrival is at 9:05, two minutes later, so $X_2 = 2$. If the probability of a customer arriving during a particular minute is independent of how long it has been since the previous customer arrived, then X_i has an exponential distribution. (Even though X_i is measured in minutes, this measurement is merely rounded off to the nearest minute for convenience. The time between arrivals is actually a continuous random variable.) The real number line plot in Figure

Time	X_i	Time	X_i	Time	X_i
9:03	3	9:29	5	10:06	7
9:05	2	9:43	14	10:13	7
9:16	11	9:54	11	10:15	2
9:23	7	9:56	2	10:27	12
9:24	1	9:59	3		

FIGURE 5.27 The Customer Arrival Times, and X_i = Time Interval Between Arrivals

5.28 displays the arrival times, graphically translated so 9:00 a.m. corresponds to time zero and time is measured in minutes.

The random variable Y_i represents the number of arrivals in the ith 10-minute interval after 9:00. There were two customers that arrived in the first 10 minutes, so $Y_1 = 2$; only one customer arrived in the next 10-minute interval, so $Y_2 = 1$, and so on. It is clear that Y_i is a discrete random variable, taking only the values 0, 1, 2, and so forth.

Y_i: 2   1   3   0   1   3   1   2   1

FIGURE 5.28 A Real Number Line Plot of Arrival Times (9:00 = Time 0) and Y_i = Number of Arrivals in Ten Minute Intervals

THE RELATIONSHIP BETWEEN POISSON AND EXPONENTIAL DISTRIBUTIONS

The probability function for the Poisson distribution is indexed by a parameter λ, which also equals the mean and the variance of the distribution. It is no coincidence that the same parameter is used for both the exponential distribution and the Poisson distribution, because of the close connection between the two distributions. If the time between arrivals X has an exponential distribution with parameter λ, then the number of arrivals in **one unit** of time, Y, has a Poisson distribution, and λ in the Poisson distribution of Y has the same numerical value as λ in the exponential distribution of X.

EXAMPLE

Suppose that the time between arrivals of airplanes at Chicago's O'Hare Airport follows an exponential distribution with $\lambda = 1.7$, where time is measured in minutes. Then the number of airplanes arriving each minute follows a Poisson distribution with parameter $\lambda = 1.7$. The mean time between arrivals is

$$\frac{1}{\lambda} = \frac{1}{1.7} = .588 \text{ minutes} \quad \text{(mean time between arrivals)}$$

with a standard deviation of .588 also. This translates to about 35 seconds between arrivals, on the average. The number of planes arriving per minute has a mean of $\lambda = 1.7$ planes, with a standard deviation of $\sqrt{\lambda} = 1.30$. This agrees with the previous calculation in that the mean number of planes arriving is somewhere between 1 and 2 per minute.

POISSON DISTRIBUTION

If a random variable Y has the **Poisson distribution**, then it takes 0, 1, 2, . . . as possible values, with probabilities given by the probability function

$$P(Y = y) = e^{-\lambda}\lambda^y/y!, \quad y = 0, 1, 2, \ldots, \tag{5.22}$$

where λ is the parameter of the Poisson distribution.
Population mean = λ.
Population standard deviation = $\sqrt{\lambda}$.

Exercises

5.33   Use the Lilliefors test for exponential distributions on the failure rates presented in Test Set 2 of this section to see if the assumption of an exponential distribution seems reasonable for these data.

5.34   Use the Lilliefors test to see if arrival times in minutes at an emergency receiving room in a large city hospital can be regarded as following an exponential distribution based on the following data:

   10   12   10   8   9   13   9   5   4   14

5.35   The time in minutes between incoming phone calls in a business office has been recorded for 10 calls.

   1.8   0.3   4.5   9.8   3.2   15.7   4.8   1.0   2.7   6.2

   (a) Estimate the mean waiting time between incoming calls.
   (b) Use the Lilliefors test for exponential distributions to see if the assumption of an exponential distribution seems reasonable for these data.
   (c) Estimate the parameter λ of an exponential distribution using these data.
   (d) Use the value of $\hat{\lambda}$ from part (c) to estimate the probability that the waiting time between incoming calls will be less than 1 minute. Also estimate the probability that the waiting time will exceed 10 minutes.

5.36   A fast food chain finds that the average waiting time required for their customers before they are served is 45 seconds. If the waiting time can be treated as an exponential random variable, find the exponential parameter λ and use this to find the probability that a customer will have to wait more than 3 minutes to be served.

5.37   At one of the busiest airports in the United States there is an average of 1 minute between arriving planes. If these arrival times can be treated as an exponential random variable, find the probability that there is more than 5 minutes between plane arrivals.

5.38   A small business uses a computer for its accounting. The computer experiences a failure of some type on the average of once every 2 hours of operation. If the time to failure follows an exponential distribution, what is the probability of the computer operating properly for an 8-hour period?

5.39   New automobiles are routinely covered by an unconditional warranty for 90 days. If the average time before a repair is required is 300 days, what proportion of cars can a new car dealer expect back within the 90-day warranty period? Assume the time to repair follows an exponential distribution.

5.40   The number of patients arriving at an emergency receiving room of a large city hospital has been recorded in 15-minute intervals during an 8-hour period. The results are as follows, where Y = number of patients arriving in 15-minute intervals.

Y:	0	1	2	3	4	5	6	7
Frequency:	5	11	11	7	4	1	0	1

Assuming that a Poisson distribution is appropriate for describing these data, estimate the parameter λ for a Poisson distribution.

5.41 Based on the estimate of λ in Exercise 5.40, estimate the probability that the number of patients arriving in a 15-minute interval will be more than 3.

5.6
Review Exercises

5.42 If 10 percent of all TV sets sold by an appliance firm require repair before the warranty expires, use the normal approximation to the binomial to find the probability that the firm will have to repair at least 20 of their last 100 sales. Find the probability that the number of sets requiring repair is between 6 and 14 inclusive.

5.43 If faculty salaries are normally distributed with a mean of $27,500 and a standard deviation of $3,000, find the probability that a faculty member chosen at random will have a salary less than $23,500; a salary greater than $30,000.

5.44 If the distribution of test scores for an examination is normal with a mean of 75 and a standard deviation of 6, what is the probability that the mean score obtained by 25 students taking this exam will be between 73 and 77?

5.45 The average monthly earnings of all employees of a large company is $1,500 with a standard deviation of $900. A random sample of 25 employees is obtained and their average salary $\overline{X}$ is computed. Find the following probabilities.

(a) $P(\overline{X} \leq \$1230)$ (b) $P(\overline{X} \leq \$1600)$ (c) $P(\$1230 \leq \overline{X} \leq 1600)$

5.46 Twenty percent of the professional employees of a large company have a doctorate degree. A random sample of 50 employees is obtained. Find the probability that the number of employees in the sample who have a doctorate is between 6 and 15 inclusive.

5.47 A machine that fills 16-ounce boxes of cereal is set to fill the boxes with a mean 16.3 ounces and a standard deviation of .15. If the weights of the cereal content are normally distributed, find the probability of a box having less than 16 ounces.

5.48 If 25 percent of the incoming freshmen at a midwestern university drop out by their sophomore year, find the probability of more than 140 students dropping out from a class of 500.

5.49 The shelf life for a brand of film has an exponential distribution with a mean of one year. If the film has a 14-month expiration date on it, what is the probability that the film will be unsatisfactory prior to its expiration date?

5.50 Twenty 10-pound bags of flour were randomly selected from the flour stock and weighed. Check these weights for normality.

9.8, 9.9, 9.9, 10.1, 9.9, 9.8, 10.1, 10.0, 10.0, 9.7, 10.5, 10.1,
9.5, 10.1, 9.9, 10.1, 10.2, 9.9, 10.0, 9.6

5.51 The time to failure in hours of 15 electronic components is given below. Check to see if an exponential distribution fits these data.

42.7, 63.1, 91.7, 55.8, 120.5, 48.6, 51.7, 36.9, 61.9, 57.8, 75.4,
81.2, 44.5, 47.2, 65.9

5.52   Graph the binomial probability function with $n = 4$ and $p = .75$. Is it skewed; and if so, in what direction?

APPENDIX

THE BINOMIAL COEFFICIENT

In this appendix the logic is presented that led to the mathematical formulation of the binomial coefficient $\binom{n}{x}$ used in Equation (5.1) and defined in Equation (5.2). The binomial coefficient is commonly referred to as a counting formula or as a **combination** of n trials selected x at a time.

COMBINATION

*The number of ways that exactly x successes can occur in n trials is referred to as a **combination** of n trials selected x at a time and is calculated as follows:*

$$\binom{n}{x} = \frac{n!}{x!\,(n-x)!}$$

where x = 0, 1, 2, . . . , n and n! = n(n − 1) (n − 2) · · · (2) (1). Note that by convention 0! is taken to be equal to 1.

The combination formula is useful because it can be used with the binomial probability function to count (compute) the number of ways that exactly x successes can occur in n trials.

EXAMPLE OF THE COMBINATION COUNTING FORMULA

Refer to Problem Setting 5.1 and do the counting first the long way (by hand) and then using the combination counting formula. For simplicity assume that the lung association in Problem Setting 5.1 is initially concerned only with the first four smokers they contact. All of the possible responses are recorded in Figure A5.1, where an N means the smoker was not influenced by the ad and Y means the smoker was influenced by the ad.

First it is noted that there are 2^4 or 16 outcomes shown in Figure A5.1, and in general there are 2^n outcomes associated with n trials of a binomial random variable. If the amount of work illustrated in Figure A5.1 is not sufficient to convince the reader of the need for a counting formula, then the reader should try working out the individual outcomes by hand for five smokers ($2^5 = 32$) or six smokers ($2^6 = 64$). Clearly listing all individual outcomes involves a lot of work and is virtually impossible as n gets to even a moderate size. Therefore,

Outcome	Response of Smoker				Number of Successes
Number	1	2	3	4	for This Outcome
1	N	N	N	N	0 } only one way to get "0" success
2	N	N	N	Y	1
3	N	N	Y	N	1
4	N	Y	N	N	1
5	Y	N	N	N	1 } four different ways to get "1" success
6	N	N	Y	Y	2
7	N	Y	N	Y	2
8	N	Y	Y	N	2
9	Y	N	N	Y	2
10	Y	N	Y	N	2
11	Y	Y	N	N	2 } six different ways to get "2" successes
12	N	Y	Y	Y	3
13	Y	N	Y	Y	3
14	Y	Y	N	Y	3
15	Y	Y	Y	N	3 } four different ways to get "3" successes
16	Y	Y	Y	Y	4 } only one way to get "4" successes

FIGURE A5.1 All Possible Outcomes for Four Trials of a Binomial Random Variable

consider how quickly the counting is done by using the combination counting formula.

Number of Yes Responses	Number of Ways This Outcome Can Occur in Four Trials
0	$\frac{4!}{0!4!} = 1$
1	$\frac{4!}{1!3!} = 4$
2	$\frac{4!}{2!2!} = 6$
3	$\frac{4!}{3!1!} = 4$
4	$\frac{4!}{4!0!} = 1$

INTERPOLATION FOR NORMAL PROBABILITIES IN TABLE A2

Since Table A2 cannot contain every possible value of z, it may be necessary to use interpolation to get a more exact answer in some situations. For example, suppose the $P(Z \leq 1.0000)$ is desired. The two values of z closest to 1.0000 in Table A2 are .9986 and 1.0027 for which

$$P(Z \leq .9986) = .84 + .001 = .841,$$
$$P(Z \leq 1.0027) = .84 + .002 = .842.$$

Clearly, the correct answer is between .841 and .842. Since three decimal place accuracy is sufficient for **most applications**, the tabled value closest to the z-value of 1.0000 would be used to obtain the probability .841. However, linear interpolation can be used here if desired.

$$P(Z \leq 1.0000) \simeq .841 + \left(\frac{1.0000 - .9986}{1.0027 - .9986}\right)(.842 - .841)$$

$$= .841 + .0003 = .8413.$$

Bibliography

Additional material on the topics presented in this chapter can be found in the following publications.

Conover, W. J. (1980). *Practical Nonparametric Statistics, 2nd ed.,* John Wiley and Sons, Inc., New York.

Durbin, J. (1975). "Kolmogorov-Smirnov Tests When Parameters are Estimated with Applications to Tests of Exponentiality and Tests on Spacings," *Biometrika,* **62,** 5–22.

Iman, R. L. (1982). "Graphs for Use with the Lilliefors Test for Normal and Exponential Distributions," *The American Statistician,* **36**(2) 109–112.

Lilliefors, H. W. (1967). "On the Kolmogorov-Smirnov Test for Normality with Mean and Variance Unknown," *Journal of the American Statistical Association,* **62,** 399–402.

Lilliefors, H. W. (1969). "On the Kolmogorov-Smirnov Test for the Exponential Distribution with Mean Unknown," *Journal of the American Statistical Association,* **64,** 387–389.

6

Estimation (One Sample)

PRELIMINARY REMARKS

In the previous chapters statistics have been presented for the summarization of sample data. These statistics include the sample mean, sample variance and standard deviation, the sample median, sample quantiles, as well as other sample statistics. Besides their usefulness in summarizing sample data, these same summary statistics often provide estimates for population parameters. For example, the sample mean $\bar{X}$ provides an estimate of the population mean μ, while the sample standard deviation s provides an estimate of the population standard deviation σ. The population proportion p is estimated by the sample proportion $\hat{p} = X/n$, where X is the number of successes observed in a random sample of size n.

TYPES OF ESTIMATES

Each of the above sample statistics provides **point estimates** of population parameters. Point estimates are quite useful but they do not always provide the type of information that is desired in a particular setting. For example, consider an engineer needing estimates of the number winter days of sunshine available for producing solar energy in different parts of the country. An example of a **point estimate** in response to this need might be that in one part of the country there will be 70 days available for producing solar energy. Another response to this need might be that with 95 percent confidence the number of solar days will be between 50 and 90. This latter estimate is an example of an **interval estimate** in that an upper and lower bound are provided along with a stated degree of confidence for the interval. In this chapter methods for finding interval estimates for the population mean, the population proportion, and the population median are explained.

6.1
General Remarks about Estimation

POINT ESTIMATES AND INTERVAL ESTIMATES

This section presents some definitions associated with estimation. Samples may be used to estimate population parameters, such as μ and σ in a normal distribution, which represent the population mean and standard deviation. Estimates may be made of other population characteristics such as the median or other quantiles not usually expressed in terms of population parameters. Estimates may take the form of a single number, called a point estimate, or an interval of values, called an interval estimate.

POINT ESTIMATE
A **point estimate** *is a single number that is used as an estimate of a population parameter or population characteristic.*

Psychology. An experiment to determine the susceptibility of individuals to hypnotism consists of having each subject clasp his hands with the fingers interlocked. If the left thumb is on top of the right thumb in the hand clasp, the subject may be a good candidate to try to hypnotize. The sample proportion of individuals with the left thumb on top is a point estimate of the population proportion.

Home Economics. A random sample consisting of 10 pieces of cloth is tested to determine the breaking strength of the warp and fill threads of each piece of cloth in the sample. The average level of breaking strength of each of these threads in the sample can be used as a point estimate of the population (all cloths of this type) mean breaking strength.

Accounting. Rather than take a complete inventory of all parts in a warehouse in an audit, a stratified sample is obtained. The stratification is based on the cost of the item. All items costing more than $50,000 are inventoried, 50 percent of the items between $10,000 and $50,000 are inventoried, and so on, until only 1 percent of items costing less than $1 are inventoried. In this case a weighted sample mean is used to estimate the total inventory value.

Banking. Banks obtain assets in the form of customer savings accounts, customer checking accounts, and money borrowed from other financial institutions. A certain percentage of this money is loaned out at a higher interest rate. This is how the bank makes money. But a certain percentage of the assets must be kept on hand to cover customer withdrawals or checks cashed, and a certain percentage must be kept in noninterest bearing accounts at the Federal Reserve. The percents vary from day to day within a bank, and from bank to bank. A sample of banks is examined to get an idea of the probability distribution of the percent assets on hand. The empirical distribution function obtained from the sample percentages can be used to estimate the population distribution function. Sample percentiles are used as point estimates of the population percentiles. The sample median provides a point estimate of the population median percentage—the value exceeded by 50 percent of all banks in the population.

INTERVAL ESTIMATE, OR CONFIDENCE INTERVAL

An **interval estimate** *is an interval that provides an upper and lower bound for a specific population parameter whose value is unknown. This interval estimate has an associated degree of confidence of containing the population parameter. Such interval estimates are also called* **confidence intervals**.

EXAMPLES OF INTERVAL ESTIMATES

In many situations where point estimates are obtained, interval estimates can also be obtained. The interval estimate is an interval that includes the point estimate. That is, in a marketing survey where the point estimate is .28 for the population proportion, an interval estimate might be of the form, "There is a 95 percent level of confidence that the true value for the population proportion lies between .25 and .31." If the point estimate for the median percent of assets held in reserve by the banks is 36 percent, an interval estimate might read "There is a 90 percent level of confidence that the median percentage of assets held in reserve by all banks is between 33 and 39 percent."

ESTIMATOR

An **estimator** *is a random variable calculated from sample data that provides either point estimates or interval estimates for some population parameter.*

EXAMPLES OF ESTIMATORS

The sample proportion $\hat{p} = X/n$, where X is the number of successes in a sample of size n, is a random variable calculated from sample data. Hence $\hat{p}$ is an estimator, while the value of $\hat{p}$ resulting from the calculation is the point estimate.

The two estimators $\overline{X} - 1.9600\,s/\sqrt{n}$ and $\overline{X} + 1.9600\,s/\sqrt{n}$, where $\overline{X}$ and s are the sample mean and standard deviation in a sample of size n, are random variables calculated from sample data. These two random variables form an approximate 95 percent interval estimator for the population mean when the sample size n is large (say, greater than 30), as will be shown in the next section. The actual numbers resulting from the calculations for a specific set of observations are the interval estimates.

PROPERTIES OF ESTIMATORS

Statisticians have developed properties to aid in evaluating an comparing the usefulness of various estimators. The two most commonly stated properties are those of **unbiasedness** and **minimum variance**. Some estimators such as $\bar{X}$ have both properties when used to estimate μ for some distributions such as the normal distribution. Other estimators, such as s, used to estimate σ, seldom satisfy either one of these properties; but s is still considered a very useful estimator. The following discussion is not intended to make the reader an expert on these properties, but it should broaden his or her perspective regarding the selection of a good estimator.

UNBIASEDNESS

An estimator is **unbiased** if its mean is the population parameter being estimated. That is, every estimator has a sampling distribution, and if the mean of the sampling distribution is the same as the parameter being estimated, then the estimator is unbiased. An unbiased estimator does not have a tendency to underestimate the population parameter, nor does it have a tendency to overestimate the population parameter in the long run. The sampling distribution of $\bar{X}$ has as its mean the same μ that is the mean of the population. Therefore $\bar{X}$ is an unbiased estimator for μ.

UNBIASED ESTIMATOR *An estimator is **unbiased** if its mean is the population parameter being estimated.*

MINIMUM VARIANCE

The best estimator comes closest to the population parameter being estimated. Since "closeness" is often measured in terms of standard deviation, or variance, the best estimator has the smallest variance associated with its sampling distribution, provided it is unbiased to begin with. Such estimators are called **minimum variance** unbiased estimators. For many distributions including the normal and the Poisson, the sample mean is a minimum variance unbiased estimator of the population mean. Such a property is usually very difficult to prove mathematically. While many estimators may be unbiased for some population parameter, usually only one will also have the minimum variance.

MINIMUM VARIANCE UNBIASED ESTIMATOR *An estimator is a **minimum variance unbiased estimator** if the variance of its sampling distribution is the smallest of all other unbiased estimators.*

The properties of unbiasedness and minimum variance can roughly be associated with targets whose bull's-eye is the parameter being estimated. If the shot pattern on the target is centered about the bull's-eye, then the bias is small. If the shot pattern is tightly clustered, then the variation is also small. A desirable situation is to have small bias with small variation. However, that is not always the case, as the following possibilities exist.

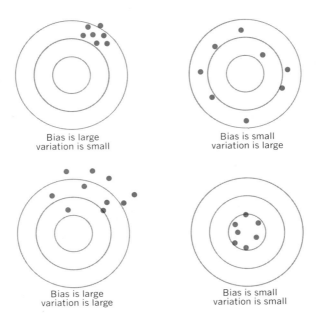

EXAMPLE TO ILLUSTRATE PROPERTIES

Assume that an air pollution index used in large cities tends to follow a normal distribution with a mean of 60 and a standard deviation of 4. Sample data for 10 days selected at random in each of five large cities are given in Figure 6.1. These observations are arranged in order, from smallest to largest within each city, for ease in reading the values.

City	Air Pollution Index										Mean
L.A.	55.9	56.3	56.8	57.2	61.2	61.9	62.5	63.8	64.4	68.2	60.82
N.Y.	55.7	55.8	57.0	57.4	59.0	59.5	59.9	60.4	64.2	67.7	59.66
K.C.	53.0	54.6	54.7	54.8	57.6	58.6	62.4	63.5	65.5	66.6	59.13
D.C.	57.3	58.1	58.6	58.7	59.0	61.9	62.6	64.4	64.9	66.7	61.22
S.F.	50.5	51.4	54.8	56.3	58.3	59.0	61.2	61.6	62.2	63.1	57.84

FIGURE 6.1 Air Pollution Index for 10 Randomly Selected Days in Five Large Cities

EXAMPLE OF UNBIASEDNESS

The sample mean $\overline{X}$ is the estimator that is best for estimating the mean of a normal population since it has both properties of unbiasedness and minimum variance. The sample means have been computed and listed with each of the samples and clearly none of the sample means is exactly equal to the population mean of 60. However, they are each quite close, as is the mean of all 50 observations, which is 59.73. The property of unbiasedness does not require that individual sample means be equal to 60, rather that the value to be expected for any one sample mean prior to taking the sample is 60 and not 60.1 or 59.99 or any other number.

EXAMPLE OF MINIMUM VARIANCE

To understand the second property, it is useful to mention that several unbiased estimators exist for providing estimates of the mean in symmetric populations. These estimators include the sample mean and the sample median as well as others. If the variability associated with some of these competing unbiased estimators were known, then the one with the minimum variance could be used. Another unbiased estimator for the mean of a normal population will now be introduced. This statistic is L = (minimum observation in the sample + maximum observation in the sample)/2, or simply $L = (X_{min} + X_{max})/2$. This is the average of the largest and smallest observations in the sample. This estimator is denoted by L (for Lazy) since it reduces the amount of work by ignoring all sample observations except for two values. The values of L are given as follows.

City:	L.A.	N.Y.	K.C.	D.C.	S.F.
$L = (X_{min} + X_{max})/2$:	62.05	61.70	59.80	62.00	56.80

Since these calculations are so easy to do, you should check them to make sure you are clear on how they were obtained. One way to get a feeling about the variability of $\bar{X}$ and L is to plot the five values of $\bar{X}$ and five values of L on the same real number line as given in Figure 6.2.

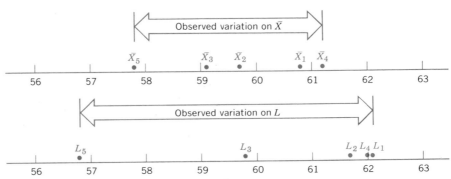

FIGURE 6.2 Variation of the Sample Statistics $\bar{X}$ and L

Both sets of points are clustered about 60, but the observed variation associated with $\bar{X}$ is less than it is with L. Although different samples could give results that contradict the above plot, the actual variance associated with $\bar{X}$ is $\sigma/\sqrt{n}$. It is shown in texts on mathematical statistics that the variance of all other estimators for μ in a normal population must be greater than $\sigma/\sqrt{n}$. The value of this property is the greater precision obtained in the estimate of μ by $\bar{X}$ rather than other estimators.

INTERVAL ESTIMATES

Point estimates have a lot of use, but it is not possible to associate a degree of confidence with the single number that is obtained. For example, the mean of sample number 1 was 60.82. It cannot be said that 90 percent, or 95 percent, or 99 percent, or in fact any percent confidence can be associated with the mean of the population. The truth is the mean of the population is either 60.82 or it isn't (in this case it is known that the true mean is 60). Hence, interval estimates that will provide the necessary tool for making related confidence statements are needed.

CONFIDENCE AND PROBABILITY

An interval estimator provides an upper and lower bound for some population parameter based on sample data. The resulting interval is referred to as a confidence interval because of the existence of an associated level of confidence that accompanies the interval. The idea of confidence associated with an interval estimate requires a bit of explanation to clarify the difference between **confidence** and **probability**. Confidence intervals are derived from probabilistic statements involving random variables.

For example, when sampling from a normal distribution the following probabilistic statement may be written:

$$P(\bar{X} - 1.6449\,\sigma/\sqrt{n} \le \mu \le \bar{X} + 1.6449\,\sigma/\sqrt{n}) = .90,$$

which is true since it involves the random variable $\bar{X}$, as will be shown in the next section. However, once a value is substituted for $\bar{X}$, the statement no longer holds. The following calculations and diagram show these intervals for the five samples of Figure 6.1.

City	$\bar{X}$	$\bar{X} - 1.6449(4)/\sqrt{10}$	$\bar{X} + 1.6447(4)/\sqrt{10}$
L.A.	60.82	58.74	62.90
N.Y.	59.66	57.58	61.74
K.C.	59.13	57.05	61.21
D.C.	61.22	59.14	63.30
S.F.	57.84	55.76	59.92

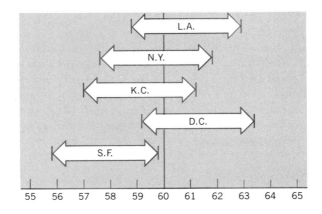

There are now five different intervals for which a basic question can be asked, "Does each of these intervals have a probability of .90 of containing the population mean μ?" The answer quite simply is "No," for any one interval either contains the population mean or it does not. In this example the true population mean is 60, and four of the five intervals provide bounds around 60 while one does not. Hence, this is where the idea of confidence comes into play. If it were stated about any one of these intervals that the confidence is 90 percent that the population mean is contained in the interval, then what is meant is: if the experiment were repeated many times, then 90 percent of the intervals would be expected to contain μ while the remaining 10 percent would not be expected to contain μ. Hence, when a 90 percent confidence interval is computed from sample data, one can never be certain that the interval is not one of the 10 percent not containing μ. Therefore, the statement is made that "There is a 90 percent level of confidence that the interval contains μ."

Although the example has dwelled on confidence intervals for means of normal populations, the remarks that were made hold for the interpretation of all confidence intervals. In the next section confidence intervals for the mean of a normal population are developed for cases where σ is known and unknown.

Exercises

6.1 In Figure 6.1 the five sample means from normal populations are given as 60.82, 59.66, 59.13, 61.22, and 57.84. The standard deviation calculated on these five sample means is 1.36. For each of these five samples find the sample median and then calculate the standard deviation for the five medians. Compare the standard deviation of the sample medians with the value of 1.36 found for the sample means. Comment on the result of this comparison for normal populations.

6.2 Refer to the table of random numbers given in Figure 1.2. Treat the last five columns as 15 samples of size 5 so that the first group of 18653 becomes the first sample of 1, 8, 6, 5, and 3. The second group of 57068 becomes the second sample of 5, 7, 0, 6, and 8. Continue on down these columns until the last group of 25013 is used to form the 15th sample of size 5.

(a) For each of these 15 samples calculate the sample mean.

(b) For each of these 15 samples calculate the sample median.

(c) Plot the 15 sample means and medians on a real number line plot similar to Figure 6.2 and indicate the observed variation for each statistic.

(d) Calculate the standard deviation of the 15 sample means.

(e) Calculate the standard deviation of the 15 sample medians.

(f) Which statistic—the sample mean or the sample median—do you believe has the smaller variance associated with it for this nonnormal distribution?

6.2
Estimating the Mean

PROBLEM SETTING 6.1

Agricultural experiment stations are set up throughout most states. These experiment stations are usually operated by a university and have as one of their goals the improvement of production by farmers. The form of the improvement has many facets, one of which is finding varieties of plants with higher yields. A basic question centers about the yield to be expected from a particular variety—that is, the mean yield. Since many factors can influence yield, it is helpful to be able to obtain confidence intervals on the mean yield.

LARGE SAMPLE CONFIDENCE INTERVAL FOR THE MEAN

The Central Limit Theorem states that for large sample sizes the distribution of $\bar{X}$ is approximately normal. This is the basis for forming a large sample confidence interval for the mean of a population, based on $\bar{X}$. For the standard normal distribution the relationship

$$P(-z_{\alpha/2} < Z < z_{\alpha/2}) = 1 - \alpha$$

holds, where $z_{\alpha/2}$ is the $1 - \alpha/2$ quantile from Table A2. The Central Limit Theorem states that for large n the distribution of $(\bar{X} - \mu)/(\sigma/\sqrt{n})$ is approximately standard normal so that

$$P\left(-z_{\alpha/2} < \frac{\bar{X} - \mu}{\sigma/\sqrt{n}} < z_{\alpha/2}\right) \cong 1 - \alpha$$

holds. By working only with the inequality within the parentheses, some elementary algebra shows that the inequality can be rewritten as

$$\bar{X} - z_{\alpha/2}\,\sigma/\sqrt{n} < \mu < \bar{X} + z_{\alpha/2}\,\sigma/\sqrt{n},$$

so the entire probability statement becomes

$$P(\bar{X} - z_{\alpha/2}\,\sigma/\sqrt{n} < \mu < \bar{X} + z_{\alpha/2}\,\sigma/\sqrt{n}) \cong 1 - \alpha.$$

This statement forms the basis for the confidence interval for the mean of a population. Such a confidence interval is a good approximation for nonnormal populations if n is large, and is exact for normal populations for all sample sizes. Note that the use of $z_{\alpha/2}$ to denote the $1 - \alpha/2$ quantile is in contradiction to prior usage of this notation, where z_q was used to denote the qth quantile. However, the usage of $z_{\alpha/2}$ to denote the $1 - \alpha/2$ quantile is so widespread that it will be used extensively throughout the remainder of this book. Each time it is used in this manner its meaning will be clearly identified.

A common question that occurs to individuals when they are first exposed to confidence intervals is "Why not always use a 99 percent or a 99.9 percent confidence interval and have the higher degree of confidence?" Examination of the above confidence intervals shows that the price paid for this greater confidence is wider intervals, and confidence intervals that are wide are often useless. For instance, it may be stated with 99.9 percent confidence that a new

---

**CONFIDENCE INTERVALS FOR THE POPULATION MEAN
(LARGE SAMPLES, i.e., $n \geq 30$**

The general confidence interval for a population mean, with an approximate level of confidence of $100(1 - \alpha)$ percent is given for large samples ($n \geq 30$) as

$$\bar{X} \pm z_{\alpha/2}\, \sigma/\sqrt{n},$$

where $\bar{X}$ is the sample mean in a simple random sample, σ is the population standard deviation (if σ is unknown, s may be used instead of σ as an approximation when n is large), and $z_{\alpha/2}$ is found in Table A2 as the $1 - \alpha/2$ quantile.

Some very commonly used confidence intervals are 90, 95, and 99 percent, which are summarized as follows.

90 percent confidence interval for μ: $\bar{X} \pm 1.6449\, \sigma/\sqrt{n}$.
95 percent confidence interval for μ: $\bar{X} \pm 1.9600\, \sigma/\sqrt{n}$.
99 percent confidence interval for μ: $\bar{X} \pm 2.5758\, \sigma/\sqrt{n}$.

If the population is normal and σ is known, this procedure provides an exact confidence interval for *all* sample sizes.

---

drug is between 8 and 92 percent effective; however, this interval is so wide that it loses any value it might otherwise have had. For a fixed level of confidence, the width of a confidence interval can be decreased by increasing the sample size, as shown following the next example.

EXAMPLE

An agricultural experiment station would like to estimate the mean yield of a new variety of wheat. Yields are observed on 25 plots of land that the experiment station operates and the average yield is found to be 32 bushels per acre. From past experience, the station knows that the distribution of wheat yields is normal and the standard deviation of these yields is four bushels. Since the population is normal and σ is known, the previous procedure may be used even though n is small. Hence, a 95 percent confidence interval for the mean yield is

$$32 \pm 1.9600(4)/\sqrt{25}$$

or

$$32 \pm 1.57$$

or

$$30.43 \text{ to } 33.57.$$

This interval is a 95 percent confidence interval for the mean yield throughout the state and the interpretation is that the experiment station can be 95 percent confident that the above interval contains the true mean yield for the variety.

FINDING THE SAMPLE SIZE FOR A CONFIDENCE INTERVAL

If a confidence interval is to be calculated for the population mean on the basis of a sample size n from a normal population for which σ is known, the width of the confidence interval can be determined as a function of the sample size. Conversely, the sample size can be determined as a function of the width. The following example illustrates how the sample size is determined.

DETERMINATION OF SAMPLE SIZE FOR A GIVEN CONFIDENCE INTERVAL WIDTH

The sample size required for a confidence interval of width w, for the mean of a normal distribution with standard deviation σ, is given by

$$n = [2\, z_{\alpha/2}\sigma/w]^2, \tag{6.1}$$

where $z_{\alpha/2}$ is the $1 - \alpha/2$ quantile from Table A2.

EXAMPLE

How large of a sample would be needed to form a 95 percent confidence interval for the mean nicotine content of a brand of cigarettes if the nicotine content has a normal distribution with $\sigma = 8.5$ milligrams and the width of the interval needs to be 6 milligrams to satisfy testing requirements? The upper and lower limits of the confidence interval are given respectively as

$$\bar{X} + 1.6449(8.5)/\sqrt{n} \quad \text{and} \quad \bar{X} - 1.6449(8.5)/\sqrt{n}.$$

Since the difference of the limits is desired to be 6 milligrams, the following equation can be solved for n:

$$(\bar{X} + 1.6449(8.5)/\sqrt{n}) - (\bar{X} - 1.6449(8.5)/\sqrt{n}) = 6$$

or

$$\sqrt{n} = 2(1.6449)(8.5)/6$$

or

$$n = 21.72,$$

which is the same as would be obtained using Equation (6.1). Since the sample size of 21.72 cannot be obtained, the required sample size is 22.

SMALL SAMPLE CONFIDENCE INTERVAL FOR THE MEAN OF NORMAL POPULATIONS

If the population is normal but the standard deviation is *unknown* and has to be estimated from a *small sample*, use of standard normal values to find confidence intervals for μ is no longer correct. Rather, the critical values are obtained from Table A3 for a distribution called **Student's *t*-distribution**. Student's *t*-distribution is the name given to a family of distributions indexed by a parameter called **degrees of freedom** (DF). As the degrees of freedom change, the distribution changes, which accounts for the different quantiles in Table A3 for the various degrees of freedom. As the degrees of freedom get large, the Student's *t*-distribution approaches the standard normal distribution. This is the reason standard normal values are appropriate for large samples. Figure 6.3 compares a Student's *t*-distribution having 5 degrees of freedom with a standard normal distribution.

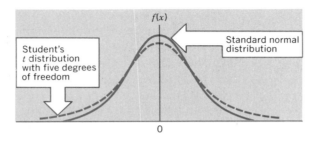

FIGURE 6.3 A Comparison of Student's t-distribution with the Standard Normal Distribution

Just as the probabilities for $(\bar{X} - \mu)/(\sigma/\sqrt{n})$ may be obtained from Table A2 when the population is normal, the probabilities for $(\bar{X} - \mu)/(s/\sqrt{n})$ may be obtained from Table A3 with DF $= n - 1$. Note that the requirement of a normal population is still present; the only change is that s is used instead of σ. The use of s to estimate σ introduces additional variability in the statistic $(\bar{X} - \mu)/(s/\sqrt{n})$ and this results in a larger variance in the distribution of the statistic, as indicated by Figure 6.3. The use of quantiles from Student's t-distribution instead of normal quantiles from Table A2 leads to the confidence interval for μ in normal populations when σ is unknown.

CONFIDENCE INTERVAL FOR THE MEAN OF A NORMAL POPULATION WHEN σ IS UNKNOWN (SMALL SAMPLES)

The confidence interval for the mean of a normal population with a level of confidence of $100(1 - \alpha)$ percent is given as

$$\bar{X} \pm t_{\alpha/2, n - 1}s/\sqrt{n},$$

where $\bar{X}$ is the sample mean that estimates the unknown value of μ, s is the sample standard deviation that estimates the unknown standard deviation σ, n is the sample size, and $t_{\alpha/2, n - 1}$ is the $1 - \alpha/2$ quantile from a Student's t-distribution with $n - 1$ degrees of freedom

---

ERROR INVOLVED WHEN USING LARGE SAMPLE CONFIDENCE INTERVALS WITH SMALL SAMPLES

Use of the Student's t-distribution will give wider confidence intervals than one would obtain by making use of the inappropriate standard normal values because of the larger variance of the Student's t-distribution. Conversely, this means that if quantiles of Z are used when values of t should be used, then the actual degree of confidence associated with the confidence interval will be less than what the individual making the confidence interval had intended. Suppose a 90 percent confidence interval is desired for μ when σ has been estimated from a sample of size 5. From Table A3 $t_{.05,4} = 2.1318$ and the proper form of the confidence interval is $\bar{X} \pm 2.1318s/\sqrt{5}$. If the standard normal distribution were used in this problem, the confidence interval would appear as $\bar{X} \pm 1.6449s/\sqrt{5}$, which is obviously not as wide as the one using Student's t-distribution. Therefore, the confidence associated with this last interval is less than 90 percent. In fact, the confidence is really only 82.4 percent, which is found by interpolation in the Student's t-table with $5 - 1 = 4$ degrees of freedom. To save the reader this work Figure 6.4 shows the relationship between the actual level of confidence and the sample size when the standard normal values are erroneously substituted for Student's t-values. Figure 6.4 illustrates that there is not much discrepancy for samples of size 50 or larger.

	Desired Level of Confidence For Intervals from Normal Populations		
n	90%	95%	99%
5	82.4%	87.8%	93.8%
10	86.5%	91.8%	97.0%
15	87.8%	92.9%	97.8%
20	88.3%	93.5%	98.1%
25	88.7%	93.8%	98.3%
30	88.9%	94.0%	98.4%
50	89.3%	94.4%	98.6%

FIGURE 6.4 Actual Level of Confidence When Student's t-Values Are Erroneously Replaced With Standard Normal Values

Two examples are now presented to demonstrate how the confidence interval for the mean of a normal population is found when the standard deviation is estimated from sample data.

EXAMPLE

For the first example of this section assume that the standard deviation was **estimated** to be 4 rather than actually known. The 95 percent confidence interval is constructed by using the .975 quantile $t_{.025,25-1} = 2.0639$ from the Student's t-distribution:

$$32 \pm 2.0639(4)/\sqrt{25}$$

or

$$32 \pm 1.65$$

or

$$30.35 \text{ to } 33.65.$$

The total width is .16 more than width of the confidence interval obtained by using the standard normal distribution. This additional width can be attributed to having to estimate the population standard deviation and the uncertainty involved in that estimate.

EXAMPLE

The number of accidents reported to a police department has been recorded for each of 20 consecutive days. Application of the Lilliefors test shows that the number of accidents could be approximated by a normal distribution. Find a 95 percent confidence interval for the daily average number of accidents based on the following data.

10, 9, 10, 11, 16, 15, 8, 6, 18, 17,
4, 12, 15, 14, 15, 9, 7, 8, 16, 14

$$\Sigma X_i = 234 \qquad \Sigma X_i^2 = 3048$$

$$\bar{X} = 234/20 = 11.7, \qquad s = \sqrt{3048/19 - 20(11.7)^2/19} = 4.041$$

From Table A3 $t_{.025,19} = 2.0930$ and the confidence interval is

$$11.7 \pm 2.0930(4.041)/\sqrt{20}$$

or 11.7 ± 1.9

or   9.8 accidents to 13.6 accidents.

Therefore, the police department can be 95 percent confident that the daily mean number of accidents reported to them is between 9.8 and 13.6.

SMALL SAMPLE CONFIDENCE INTERVAL IN NONNORMAL POPULATIONS

The previous procedures cover all cases for estimating the mean except when the samples are small and the populations are nonnormal. Although such procedures exist in books on nonparametric statistics, such a procedure has not been included in this text, because the authors feel that the procedures given in Section 6.4 can adequately cover this situation.

Exercises

6.3 A random sample of 50 checking accounts out of the 5525 accounts at the First National Bank is examined to see what the "average daily balance" was for the previous month. This is a figure that is routinely calculated once a month on each account. The mean for the 50 observations was $2,100 and the standard deviation was $400. Find a 95% confidence interval for the population mean "average daily balance" for all accounts at the First National Bank and explain the meaning of this interval.

6.4 A random sample of 9 light bulbs shows a mean burning time of 175 hours. If the population standard deviation for the burning time of the bulb under consideration is $\sigma = 16$ find 90%, 95% and 99% confidence intervals for the true mean burning time of the type of bulb being tested assuming a normal distribution. What is the meaning of the different widths associated with the confidence intervals you have found?

6.5 How many households would it be necessary to survey to form a 99% confidence interval within $200 of the mean income in a suburb of a large city if the standard deviation is $500?

6.6 If you are sampling from a population with a standard deviation of 15, how large of a sample do you need to have a 95% chance that the sample mean will fall within 2 units of the population mean?

6.7 A past analytical study has shown that the standard deviation of an achievement test is 25. How large a sample needs to be taken so that a 95 percent confidence interval for the average test score will have a width of 3?

6.8 Assume that the weekday traffic volume on a toll bridge is normally distributed. Find a 95 percent confidence interval for the population mean weekday traffic volume and interpret it if the observed volumes for five weeks were as follows.

22,130   18,465   25,616   22,440   19,869

6.9   Weekly total points scored by all professional football teams were 684, 972, 740, 868, and 777 for five successive weeks. Find a 90 percent confidence interval for the mean of the population of weekly point totals, assuming that the population is normal and that these five observations resemble a random sample. Explain the meaning of the interval you have found.

6.10   A manufacturing company claims that their new floodlight will last at least 1000 hours. A random sample of 10 lights gave an average life of 980 hours with a standard deviation of 15.8. Find a 90 percent confidence interval for the mean of the population assuming a normal distribution. Based on your confidence interval, do you think that the company's claim is justified?

6.11   A random sample of scores on an exam yielded the scores 85, 78, 52, 66, and 74. Check these data for normality using the Lilliefors test and then find a 95 percent confidence interval for the mean of all scores on the exam. Explain the meaning of the interval you have found.

6.12   A sample of size 25 from a normal population with unknown mean and variance yields $\bar{X} = 20$ and $s = 5$. An individual who claims to have some statistical training computes a 95 percent confidence interval for the population mean as $\bar{X} \pm 1.9600\, s/\sqrt{25}$ and obtains 18.04 to 21.96 as his confidence interval. He then claims that he can be 95 percent confident that this interval will contain the true value of the population mean. Approximately what is the confidence that should be associated with this interval? What is the correct 95 percent confidence interval for the population mean?

6.3
Estimating the True Proportion in a Population

ESTIMATING THE PROPORTION

Survey questionnaires are commonly used to obtain information in a variety of settings. The answers to many of these questions in the questionnaire will be tabulated and expressed as percents or proportions. For example, consider the following farm-related questions from a congressional survey.

1. Do you feel that government enforcement of agriculturally-related safety regulations is helpful in reducing the number of accidents?

   Yes _____   No _____

2. Do you feel that prices obtained on your commodities reflect a reasonable return on your investment?

   Yes _____   No _____

3. Do you believe use of a grain embargo is an effective way of dealing with the Soviet Union?

   Yes _____   No _____

4. If a grain embargo is used as an instrument of government foreign policy, do you feel that the government should offer compensation to the farmer for any resulting loss of income?

   Yes _____   No _____

All of the responses to these questions are dichotomous in nature and in each case an estimate is needed of the true proportion in the population giving a particular response to a certain question. In this section a point estimate is obtained for the true population proportion, and then two methods (one exact and one approximate) are presented for finding confidence intervals for the true population proportion.

EXACT CONFIDENCE INTERVALS

When the answers to a survey question are dichotomous, such as ''yes'' or ''no,'' and if the sample is a random sample, then the requirements for the number of favorable responses X to have a binomial distribution are satisfied. That is, each response has the same probability of being favorable if the sample is random and the responses are independent of one another. Therefore, the number of favorable responses has a binomial distribution with n (the sample size) known, and p (the probability of a favorable response) unknown. A point estimate for p is given by $\hat{p} = X/n$. This point estimate is unbiased, which as you may recall means that the mean of $\hat{p}$ is p, the population parameter being estimated.

Although large sample sizes should be obtained whenever possible, there are times when small sample sizes are unavoidable. When recording the number of ''hits'' scored by torpedos fired from a submarine, the probability of obtaining a hit is of interest, but small samples are an economic necessity. A franchising company may wish to estimate the probability of a store in its chain going out of business within the first 2 years of operation, but the number of stores available for examination is quite limited. In cases such as these there is a need

"YOUR COLLEGE SURVEY HAS MIXED RESULTS: 65% OF THE STUDENTS SAID 'GIVE UP YOUR SEAT AND RUN FOR PRESIDENT'; HOWEVER, 80% OF THOSE SAID THEY WOULDN'T VOTE FOR YOU!"

for an accurate method of obtaining a confidence interval that can be used with small samples.

The binomial tables, Table A1, may be used to find a confidence interval for p when n is 20 or less. The confidence interval found using Table A1 is exact. It is now demonstrated with an example.

POINT ESTIMATE OF THE POPULATION PROPORTION	*An unbiased* **point estimate** *of the true population proportion p is given by the sample proportion* $\hat{p} = X/n,$ *where X is the number of successes observed in a random sample of size n from the population.*

**INTERVAL ESTIMATE OF THE POPULATION PROPORTION p
(EXACT SOLUTION FOR SMALL SAMPLES, i.e., $n \leq 20$)**

Let X be the number of successes in a random sample of size n. To find an exact $100(1 - \alpha)$ percent confidence interval for p proceed as follows.

1. Let $Y = X - 1$.
2. Let $P_1 = 1 - \alpha/2$.
3. Enter Table A1 with the sample size n and find the row that starts with the value of Y. Read across the row until the entry in the table equals P_1 (approximately). The value of p found at the top of that column containing P_1 is the lower limit L for the confidence interval. It may be necessary to use interpolation due to the limited size of the table.
4. Let $P_2 = \alpha/2$.
5. Read across the row starting with the value of X until the entry in the table equals P_2 (approximately). The value of p found at the top of that column containing P_2 is the upper limit U for the confidence interval. It may again be necessary to use interpolation. The $100(1 - \alpha)$ percent confidence interval for p is from L to U.

EXAMPLE

A random survey of 20 farmers shows 12 in favor of a grain embargo. Find a 95 percent confidence interval for the proportion of all farmers in favor of a grain embargo. The solution follows the step-by-step outline given above.

1. $Y = 12 - 1 = 11$ since $X = 12$ was the observed number of successes.
2. Let $P_1 = 1 - .05/2 = .975$ since $1 - \alpha = .95$ and $\alpha = .05$.
3. Enter Table A1 with $n = 20$, on the row headed by the value of Y, which is 11. For convenience that part of the table for the rows headed by 11 and 12 is reproduced.

	$p = .30$	.35	.40	$\cdots$	.75	.80	.85
11 $\cdots$	.9949	.9804	.9435	$\cdots$	.0409	.0100	.0013
12 $\cdots$	.9987	.9940	.9790	$\cdots$	.1018	.0321	.0059

Reading across the row headed by 11 look for table entry of $P_1 = .975$. This value is found between the table entries of .9804 and .9435. The value of p at the top of each of these columns is .35 and .40, respectively. The value of L is found by interpolating between these values as follows:

$$L = .35 + (.40 - .35)\left(\frac{.975 - .9804}{.9435 - .9804}\right) = .357.$$

4. Let $P_2 = .05/2 = .025$.
5. Enter Table A1 with $n = 20$ on the row headed by the value of X, which is 12. Using the above portion of that table read across this row looking for the value of $P_2 = .025$, which is found between .0321 and .0059. The respective column headings for p are .80 and .85. Once again interpolation is used to find U:

$$U = .80 + (.85 - .80)\left(\frac{.025 - .0321}{.0059 - .0321}\right) = .814.$$

The 95 percent confidence interval for the percent of farmers favoring a grain embargo is from 35.7 to 81.4 percent.

CONFIDENCE INTERVALS BASED ON THE NORMAL APPROXIMATION

While the previous method is exact, its use requires binomial tables. If binomial tables are not available or if n is large, then an approximate method based on the normal approximation to the binomial distribution can be used. This approximation is based on the fact that, by the Central Limit Theorem, $\hat{p}$ is approximately normal with a mean p and standard deviation $\sqrt{pq/n}$. Therefore, Table A2 is used to approximate the probabilities for $(\hat{p} - p)/\sqrt{pq/n}$. This leads to the approximate probability

$$P(\hat{p} - z_{\alpha/2}\sqrt{\hat{p}\hat{q}/n} < p < \hat{p} + z_{\alpha/2}\sqrt{\hat{p}\hat{q}/n}) \cong 1 - \alpha,$$

which is the basis for the large sample confidence interval approximation.

**INTERVAL ESTIMATE OF THE POPULATION PROPORTION p
(APPROXIMATE SOLUTION APPROPRIATE FOR LARGE SAMPLES $n > 20$)**

The general confidence interval for p with a level of confidence of $100(1 - \alpha)$ percent is given approximately as

$$\hat{p} \pm z_{\alpha/2}\sqrt{\hat{p}\hat{q}/n},$$

where n is the sample size, $\hat{p}$ is the sample proportion, $\hat{q} = 1 - \hat{p}$, and $z_{\alpha/2}$ is the upper $\alpha/2$ critical value (i.e., the $(1 - \alpha/2)$ quantile) found in Table A2. Some very commonly used confidence intervals are 90, 95, and 99 percent, which are summarized as follows.

90 percent confidence interval for p: $\hat{p} \pm 1.6449\sqrt{\hat{p}\hat{q}/n}$.
95 percent confidence interval for p: $\hat{p} \pm 1.9600\sqrt{\hat{p}\hat{q}/n}$.
99 percent confidence interval for p: $\hat{p} \pm 2.5758\sqrt{\hat{p}\hat{q}/n}$.

Note: This approximate solution should be used with caution if $n\hat{p} < 5$ or $n\hat{q} < 5$.

EXAMPLE

The federal government has threatened to withdraw federal highway funds from states not complying with the 55 m.p.h. speed limit on interstate highways. Noncompliance is judged on a sliding scale and at one time was set at 60 percent. That is, if more than 60 percent of the drivers are exceeding the 55 m.p.h. speed limit, then the state is in violation of the federal requirement. A random survey was taken at different times and different locations and it was found that 6250 drivers out of 10000 were driving over 55 m.p.h. Find a 99 percent confidence interval for the true proportion of drivers exceeding 55 m.p.h. From Table A2, $z_{.005} = 2.5758$. Since $\hat{p} = .625$, the 99 percent confidence interval is

$$.625 \pm 2.5758 \sqrt{\frac{.625(.375)}{10000}},$$

or
$$.625 \pm .012,$$
or 61.3 to 63.7 percent.

Do you think there is a need for concern about compliance with the federal requirement based on these results?

Confidence intervals for p can be used to monitor the quality of a product being produced, as shown in the next example.

EXAMPLE

A medication for burns carries a warning that some burn patients may have a reaction to the medication. Further, it is thought that about 3 percent of the patients treated with this medication will have such a reaction. Fifty burn patients are treated with the medication and seven show a reaction to it. Find a 95 percent confidence interval for the proportion of patients showing a reaction to the medication based on these results.

For $n = 50$ the normal approximation is used to find the confidence interval. The estimate of the proportion of patients with reactions is given as $\hat{p} = 7/50 = .14$ and the confidence interval is given as

$$.14 \pm 1.9600 \sqrt{\frac{.14(.86)}{50}},$$

or
$$.14 \pm .10,$$
or .04 to .24.

Based on these results do you think that the probability of patients showing reaction to the medication has changed from 3 percent?

DETERMINING THE SAMPLE SIZE REQUIRED

If a confidence interval of a given width is desired for the unknown proportion p, the sample size required in order to achieve a confidence interval with a width no larger than the specified width can be obtained by noting that the total width w of the confidence interval is

$$w = 2z_{\alpha/2} \sqrt{\hat{p}\hat{q}/n}.$$

By squaring both sides and solving for n, this becomes

$$n = \hat{p}\hat{q}\, 4z_{\alpha/2}^2/w^2.$$

This still depends on the values of $\hat{p}$ and $\hat{q} = 1 - \hat{p}$, which are not known until after the sample is obtained. Therefore, the most conservative values are used; $\hat{p} = \frac{1}{2}$ gives $\hat{p}\hat{q} = \frac{1}{4}$, its largest possible value. So n must be at least equal to $z_{\alpha/2}^2/w^2$ to be sure that the interval will have width w or less.

SAMPLE SIZE REQUIRED FOR A CONFIDENCE INTERVAL FOR p OF SPECIFIED WIDTH

A sample of size n where

$$n \geq z_{\alpha/2}^2/w_2$$

will ensure that a $100(1 - \alpha)$ percent confidence interval for the binomial parameter p will have a total width of w or less, where $z_{\alpha/2}$ is the $1 - \alpha/2$ quantile from Table A2.

EXAMPLE

Suppose that in the previous example the manufacturer of the medication wants to know, within .04, the probability of the medication causing a reaction. What sample size is required for a 95 percent confidence interval? In this case the total width is .08, which allows for a .04 error on either side of $\hat{p}$. Therefore, the sample size should be at least

$$z_{\alpha/2}^2/w^2 = (1.9600)^2/(.08)^2$$
$$= 600$$

to satisfy the company. If a sample of size 600 results in 83 patients showing a reaction, the 95 percent confidence interval is

$$\hat{p} \pm z_{\alpha/2} \sqrt{\frac{\hat{p}\hat{q}}{n}} = \frac{83}{600} \pm 1.9600 \sqrt{\frac{(83/600)(517/600)}{600}}$$
$$= .14 \pm .03.$$

The point estimate, .14, is within 3 percent of the true probability p, with 95 percent confidence. Note that the actual width is less than the targeted value .08, because the actual value of $\hat{p}\hat{q}$ is .119, less than the conservative value .25 used in developing the equation.

Exercises

6.13   A random survey of 18 airline passengers indicates that 6 have never flown before. Find a 95 percent confidence interval for the true proportion of airline passengers taking their first trip. Use the exact method outlined in this section to find the desired interval.

6.14 A survey shows that 15 out of 20 students indicate that they do not smoke. Use the exact method for constructing a confidence interval for the population proportion to construct a 90 percent confidence interval for the true proportion of nonsmoking students.

6.15 Use the approximate solution appropriate for large samples to find the desired confidence interval in Exercise 6.13. Compare your answer to the exact solution obtained in Exercise 6.13.

6.16 Use the approximate solution appropriate for large samples to find the desired confidence interval in Exercise 6.14. Compare your answer to the exact solution obtained in Exercise 6.14.

6.17 A random sample of 50 checking accounts at the First National Bank is examined to see how many were overdrawn. If 6 of the 50 accounts were found to be overdrawn, find a 95 percent confidence interval for the population proportion of overdrawn accounts.

6.18 Fourteen of the 88 people randomly selected in a survey said they preferred the news on Channel 4. Find a 99 percent confidence interval for the proportion of the population that prefers the news on that channel.

6.19 A national poll in the 1972 presidential campaign based on interviews with 1200 voters gave Nixon 57 percent of the vote. Calculate a 99 percent confidence interval for the true proportion of voters supporting Nixon.

6.20 The germination rate for some flower seeds was stated as a 90 percent confidence interval from 61 to 69 percent. How large of a sample size was used to produce this interval?

6.4
Estimating the Population Median

ESTIMATING THE MEDIAN

In a previous section of this chapter a method was given for finding a confidence interval for the population mean. In some applications the population median is more appropriate than the population mean as the primary statistic for describing the population. This is especially true with highly skewed populations, such as individual incomes, where a few large incomes in the population influence the mean so that it is usually too large to reflect accurately the middle of the distribution of incomes. The population median is estimated from the sample median.

POINT ESTIMATE OF THE POPULATION MEDIAN *The sample median is used as a point estimate of the population median. The sample median is defined in Section 4.4.*

Other examples of distributions follow, where a few large observations may distort the mean so that it no longer represents the middle of the distribution as well as the median does.

1. The time to repair a boat may be very long for a few boats that require special ordering of parts.
2. The time for a letter to reach its destination may include a few letters that go astray.
3. The amount of daily receipts may be highly skewed because of a few sale days, and the Christmas rush.
4. The amount contributed by each family to a particular church may include a few families who contribute large sums of money.

In these cases the mean does not represent the middle of the distribution as accurately as the median. The first two cases may necessarily involve small sample sizes, so methods for handling small sample sizes are useful.

EXACT CONFIDENCE INTERVALS

Two methods for finding a confidence interval for the median will be presented. The first is exact and uses binomial tables. It is, therefore, of necessity restricted to those limited cases for which binomial tables are available. The second method is approximate and is appropriate for those larger sample sizes for which binomial tables are not available. It is important to realize that these procedures may be used for all types of populations, unlike the small sample confidence interval for the population mean that requires the assumption of a normal distribution.

INTERVAL ESTIMATE OF THE POPULATION MEDIAN (EXACT SOLUTION FOR SMALL SAMPLES, i.e., $n \leq 20$)

To find a confidence interval for the median, proceed as follows.

1. Order the n sample observations from smallest to largest. Use the notation $X^{(1)}$, $X^{(2)}$, $X^{(3)}$, . . . , $X^{(n)}$, where $X^{(1)}$ denotes the smallest observation, $X^{(2)}$ denotes the next smallest, and so on until $X^{(n)}$ denotes the largest observation in the sample.
2. If the approximate level of confidence desired is $100(1 - \alpha)$ percent, then let $P = \alpha/2$. That is, for a desired 95 percent confidence level, $P = .025$.
3. Enter Table A1 with the sample size n. Find the *column* headed by .5. Read down this *column* until a value is found equal to P (approximately). Call this table entry α_1.
4. Let y denote the value of the heading of the *row* in which α_1 is found.
5. Let $S_1 = y + 1$.
6. Let $S_2 = n - y$.
7. The lower bound for the confidence interval is the ordered sample observation found in position S_1—that is, $X^{(S_1)}$.
8. The upper bound for the confidence interval is the ordered sample observation found in position S_2—that is, $X^{(S_2)}$.
9. The exact level of confidence is $100(1 - 2\alpha_1)$ percent.

An easier method for finding S_1, S_2, and the exact level of confidence involves using Figure 6.5. The sample size determines which row in Figure 6.5 to use, and several useful combinations of S_1, S_2 and the exact level of confidence may be obtained.

Note that in the exact method the actual sample values are used to form the upper and lower bounds for the confidence interval. Also, the level of confidence is seldom exactly $100(1 - \alpha)$ percent due to the discrete nature of the binomial distribution. Rather, if α_1 represents the table entry in the column headed by .5 and the row headed by y, then the exact level of confidence is $100(1 - 2\alpha_1)$ percent. The following example demonstrates the procedure for finding the exact confidence interval for a population median.

			Target Confidence Level						
			90%			**95%**			**99%**
n	**S_1**	**S_2**	**Level**	**S_1**	**S_2**	**Level**	**S_1**	**S_2**	**Level**
4	1	4	87.50%						
5	1	5	93.76%						
6	2	5	78.12%	1	6	96.88%			
7	2	6	87.50%	1	7	98.44%			
8	3	6	71.10%	2	7	92.96%	1	8	99.22%
9	3	7	82.04%	2	8	96.10%	1	9	99.60%
10	3	8	89.06%	2	9	97.86%	1	10	99.80%
11	3	9	93.46%	2	10	98.82%	1	11	99.90%
12	4	9	85.40%	3	10	96.14%	2	11	99.36%
13	4	10	90.78%	3	11	97.76%	2	12	99.66%
14	5	10	82.04%	4	11	94.26%	3	12	98.70%
15	5	11	88.16%	4	12	96.48%	3	13	99.26%
16	5	12	92.32%	4	13	97.88%	3	14	99.58%
17	6	12	85.66%	5	13	95.10%	4	14	98.72%
18	6	13	90.38%	5	14	96.92%	4	15	99.24%
19	6	14	93.64%	5	15	98.08%	4	16	99.56%
20	7	14	88.46%	6	15	95.86%	5	16	98.82%

FIGURE 6.5 A Chart for Quickly Obtaining an Exact Confidence Interval for a Population Median

EXAMPLE

In order to control their quality of production, an electronics company randomly selects 16 electronic components to determine their time to failure in hours. Time to failure measurements frequently produce some observations that are much larger or much smaller than the bulk of the rest of the observations—that is, outliers. Since the mean is very sensitive to outliers, the median may provide a better point estimate of the middle of the distribution of the time to failure. For this reason it is desired to find both a point estimate and a 90 percent confidence interval for the median time to failure. Since n is an even number, the point estimate for the median is found by averaging the numbers in the $16/2 = $ 8th and $(16/2 + 1) = $ 9th positions in the ordered array of sample values listed below, or

$$X_{.5} = \text{sample median}$$
$$= (X^{(8)} + X^{(9)})/2$$
$$= (63.2 + 63.3)/2 = 63.25.$$

The step-by-step procedure for finding the confidence interval is as follows.

1. The 16 sample observations are first ordered from smallest to largest (the ordered sample observation found in the ith position is $X^{(i)}$).

$X^{(1)} = 46.7$	$X^{(5)} = 56.8$	$X^{(9)} = 63.3$	$X^{(13)} = 67.1$
$X^{(2)} = 47.2$	$X^{(6)} = 59.2$	$X^{(10)} = 63.4$	$X^{(14)} = 67.7$
$X^{(3)} = 49.1$	$X^{(7)} = 59.9$	$X^{(11)} = 63.7$	$X^{(15)} = 73.3$
$X^{(4)} = 56.5$	$X^{(8)} = 63.2$	$X^{(12)} = 64.1$	$X^{(16)} = 78.5$

2. Let $P = .10/2 = .05$ as $1 - \alpha = .90$ and $\alpha = .10$.
3. Table A1 is entered with $n = 16$ and $p = .5$. For easy reference the needed portion of that column is repeated here along with the appropriate row headings.

Row headings	$p = .5$
0	.0000
1	.0003
2	.0021
3	.0106
$y = 4$	$.0384 = \alpha_1$
5	.1051

4. The closest value to $P = .05$ is .0384; therefore $\alpha_1 = .0384$ and $y = 4$.
5. $S_1 = 4 + 1 = 5$.
6. $S_2 = 16 - 4 = 12$.
7. The lower bound for the confidence interval is found in the 5th position, or $L = X^{(5)} = 56.8$.
8. The upper bound for the confidence interval is found in the 12th position, or $U = X^{(12)} = 64.1$.

The confidence interval is from 56.8 to 64.1 hours. However, due to the discreteness of the binomial distribution and the fact that $\alpha_1 = .0384$ rather than $P = .05$, the actual level of confidence is $100(1 - 2(.0384))$ percent or 92.32 percent. Figure 6.5 could also have been used for $n = 16$ to obtain $S_1 = 5, S_2 = 12$, and the level of confidence 92.32 percent without using Table A1.

CONFIDENCE INTERVALS BASED ON THE NORMAL APPROXIMATION

If binomial tables are not available, then the values of S_1 and S_2 can be found using the normal approximation. Since the exact procedure uses $p = .5$ (i.e., a symmetric binomial distribution), the normal approximation works quite well for finding S_1 and S_2; however, without the binomial tables the exact level of confidence must be approximated. This point is illustrated in the next example.

**INTERVAL ESTIMATE OF THE POPULATION MEDIAN
(APPROXIMATE SOLUTION FOR LARGE SAMPLES)**

1. Order the n sample observations from smallest to largest. Use the following notation with the ordered sample: $X^{(1)}, X^{(2)}, \ldots, X^{(n)}$, where $X^{(1)}$ denotes the smallest observation, $X^{(2)}$ denotes the next smallest, and so on until $X^{(n)}$ denotes the largest observation in the sample.
2. Enter Table A2 to find the critical value $z_{\alpha/2}$. Some widely used values of $z_{\alpha/2}$, and their associated level of confidence, are $z_{.05} = 1.6449$ (90 percent), $z_{.025} = 1.9600$ (95 percent), and $z_{.005} = 2.5758$ (99 percent).
3. Compute $S_1^* = (n - z_{\alpha/2} \sqrt{n})/2$.
4. Let S_1 be the integer obtained by rounding S_1^* upward to the next highest integer (this is necessary since S_1^* will seldom be an integer).
5. Let $S_2 = n - S_1 + 1$.
6. The lower bound for the confidence interval is the ordered sample observation found in position S_1—that is, $X^{(S_1)}$.
7. The upper bound for the confidence interval is the ordered sample observation found in position S_2—that is, $X^{(S_2)}$.
8. To find the appropriate level of confidence, compute $z = (2S_1 - n - 1)/\sqrt{n}$. Enter Table A2 and find the entry closest to z. The value of p corresponding to this entry is the probability $\alpha_1 = P(Z \leq z)$. The approximate level of confidence is $100(1 - 2\alpha_1)$ percent.

EXAMPLE

The large sample approximation is now demonstrated by reworking the previous example.

1. The ordered sample observations appear exactly as they did in the previous example.
2. From Table A2, $z_{.05} = 1.6449$.
3. $S_1^* = (16 - 1.6449 \sqrt{16})/2$
   $= (16 - 6.58)/2$
   $= 4.71$.
4. $S_1 = 5$ (the next highest integer above $S_1^* = 4.71$).
5. $S_2 = 16 - 5 + 1 = 12$.
6. The lower bound for the confidence interval is found in the 5th position, or
   $L \quad X^{(5)} = 56.8$.
7. The upper bound for the confidence interval is found in the 12th position, or
   $U = X^{(12)} = 64.1$.
8. The confidence interval is again from 56.8 to 64.1 hours. The approximate level of confidence is obtained by entering Table A2 with

$$z = \frac{2(5) - 16 - 1}{\sqrt{16}} = -1.75$$

to find $p = .04 = \alpha_1$. The approximate level of confidence is

$100(1 - 2\alpha_1)$ percent $= 92$ percent

which agrees well with the exact value 92.32 percent found earlier.

EXAMPLE

In Section 6.2 an example was given where a 95 percent confidence interval was found for the daily **mean** number of accidents reported to the police. The same sample data are used to find a 95 percent confidence interval for the **median** number of accidents reported.

1. The ordered sample data appear as follows.

$X^{(1)} = 4$	$X^{(6)} = 9$	$X^{(11)} = 12$	$X^{(16)} = 15$
$X^{(2)} = 6$	$X^{(7)} = 9$	$X^{(12)} = 14$	$X^{(17)} = 16$
$X^{(3)} = 7$	$X^{(8)} = 10$	$X^{(13)} = 14$	$X^{(18)} = 16$
$X^{(4)} = 8$	$X^{(9)} = 10$	$X^{(14)} = 15$	$X^{(19)} = 17$
$X^{(5)} = 8$	$X^{(10)} = 11$	$X^{(15)} = 15$	$X^{(20)} = 18$

Note: The sample median of these data is $(X^{(10)} + X^{(11)})/2 = (11 + 12) = 11.5$, which compares with the sample mean, which is 11.7.

2. Since $n = 20$, Table A1 can be used to find the confidence interval; that is, let $P = .05/2 = .025$.

3. Enter Table A1 with $n = 20$ and $p = .5$. The needed portion of the table is reproduced here for ready reference.

Row headings	$p = .5$
0	.0000
1	.0000
2	.0002
3	.0013
4	.0059
$y = 5$	$.0207 = \alpha_1$
6	.0577

4. The closest value to $P = .025$ is .0207; therefore $\alpha_1 = .0207$ and $y = 5$.

5. $S_1 = 5 + 1 = 6$.

6. $S_2 = 20 - 5 = 15$.

7. The lower bound for the confidence interval is the ordered sample observation found in position 6, or $L = X^{(6)} = 9$.

8. The upper bound for the confidence interval is the ordered sample observation found in position 15, or $U = X^{(15)} = 15$.

9. Therefore, the 95.86 percent confidence interval (actual level = $100(1 - 2(.0207))$ percent = 95.86 percent) is from 9 days to 15 accidents per day. This compares with 9.8 to 13.6 accidents per day, which was the

95 percent confidence interval for the mean. The level of confidence, S_1 and S_2, could have been obtained directly from Figure 6.5 without consulting Table A1.

In this example the confidence intervals for the mean and median are in good agreement with one another. However, it is easy to change the example so that this agreement disappears. For example, if the largest four sample observations are changed to $X^{(17)} = 31$, $X^{(18)} = 33$, $X^{(19)} = 38$, and $X^{(20)} = 40$, the new sample mean and standard deviation become 15.45 and 10.87, respectively, with a corresponding 95 percent confidence interval for the daily mean accident rate from 10.4 to 20.5 accidents per day. This interval reflects the new sample mean but hardly reflects the center of the distribution of sample values as the lower end of the confidence interval (10.4 accidents) is larger than 9 of the 20 sample observations. On the other hand, since $X^{(6)}$ and $X^{(15)}$ remain unchanged, the confidence interval for the median is the same as before. This indicates the insensitivity of the sample median to the presence of outliers and shows that the confidence interval for the median still reflects the center of the distribution of sample values.

A WARNING ABOUT UNREALISTIC ASSUMPTIONS

The data in the previous example could have come from a normal distribution as determined by the Lilliefors test. This assumption was used earlier to find a confidence interval for the mean, and a confidence interval for the median was found in the preceding example. If the population is normal, the mean equals the median, and both interval values are estimates of the same quantity—namely, the middle of the population. The wider confidence interval based on the median is a result of not using or needing the normality assumption. If the procedure for finding the confidence interval for the mean is used on data from distinctly nonnormal populations, however, the resulting interval may tend to be considerably wider than the confidence interval for the median. The principle to observe is to use the procedure that is appropriate for the situation—don't make assumptions that do not conform to reality.

Exercises

6.21   For each city listed in Figure 6.1 find the sample median and use Figure 6.5 to aid in constructing a 90 percent confidence interval for the population median air pollution index. Note: The data displayed in Figure 6.1 have already been ordered from smallest to largest for each city.

6.22   Refer to the sample data displayed in Figure 6.1 and find the median air pollution index for all five cities ($n = 50$). Also find a 90 percent confidence interval for the population median based on all 50 observations.

6.23   Refer to the 20 observations on m.p.g. given in Section 5.2 for a 4-cylinder model in city driving with a manual transmission. Find the median for these data and a 95 percent confidence interval for the population median m.p.g. for this model.

6.24 Rework Exercise 6.23 using the large sample approximation and find the approximate level of confidence for this interval. Compare your answer with the exact level of significance given in Figure 6.5.

6.25 Find a 99 percent confidence interval for the median 1980 net earnings of common stocks for corporations based on the data given in Exercise 5.24.

6.26 Find a 95 percent confidence interval for the median time between incoming phone calls based on the exponential data given in Exercise 5.35.

6.27 If the scores given in Problem Setting 4.1 of Section 4.1 can be regarded as representative of all students taking that particular exam, find a 95 percent confidence interval for the median score for that exam. Note: These scores can be organized from smallest to largest by using the stem and leaf plot summary given in Figure 4.1.

6.28 Use the data given in Figure 4.12 and summarized in Figure 4.13 to find a 90 percent confidence interval for the median shoe size sold to women.

6.5
Review Exercises

6.29 Police records on 30 individuals booked on assault showed the ages recorded below. Estimate the population median age for these individuals and find a 95 percent confidence interval for the population median age.

24	20	18	32	16	25
21	18	38	22	18	16
15	14	21	17	17	17
17	23	22	16	24	20
27	21	18	26	20	15

6.30 A random sample of 40 cars from a fleet of cars driven by employees of a large company showed an average of 2870 miles driven in a month with a standard deviation of 425 miles. Find a 95 percent confidence interval for the average number of miles driven by the fleet each month.

6.31 A reading comprehension test given to 10 randomly selected fourth-grade students showed an average score of 84.2 with a standard deviation of 12.2. Assume the population is normal and find a 90 percent confidence interval for the mean reading comprehension score for fourth graders.

6.32 Find a 95 percent confidence interval for the population mean age in Exercise 6.29 and compare results with the confidence interval found for the median on that exercise.

6.33 Of 16 cars inspected during a safety campaign, 6 were found to be unsafe. Find a 95 percent confidence interval for the proportion of unsafe cars in the population.

6.34 A civic group reported to the town council that a random sample of 100 residents showed 48 in favor of an upcoming school bond issue. Find a 90 percent confidence interval for the true proportion of residents supporting the bond issue.

6.35 A random sample of tenth-grade boys resulted in the weights listed below. Estimate the median weight for the population of tenth-grade boys and find a 95 percent confidence interval for the population median.

142	134	98	119	131
103	154	122	93	137
86	119	161	144	158
165	81	117	128	103

6.36 The standard deviation of the viscosity of a brand of car oil is 0.02. How large a sample would be needed for a 95 percent confidence interval to be within 0.005 units of the population mean?

Bibliography

Additional material on the topics presented in this chapter can be found in the following publication.

Conover, W. J. (1980). *Practical Nonparametric Statistics, 2nd ed.,* John Wiley and Sons, Inc., New York.

7

Hypothesis Testing

The heart of statistics as a science is centered about gathering, displaying, analyzing, and interpreting sample data. In the previous chapters these methods have been used for estimating population quantities. In this chapter these same methods are used for the purpose of making a decision about the validity of a statement, claim, assumption, or conjecture that may have been made about a population. As an example of such a claim, consider a manufacturer of light bulbs who claims that the mean burning time of his product is 700 hours. If a random sample shows burning times of 650, 725, 675, 690, and 680, do you believe the claim of the manufacturer? The problem setting and examples in the previous chapters have demonstrated other areas where decisions need to be made based on sample data. In this chapter decision making is approached from the standpoint of **hypothesis testing** based on sample data. This approach leads to a very systematic and structured procedure for aiding the decision-making process.

7.1
Hypothesis Testing in General

PROBLEM SETTING 7.1

A lawsuit against a pharmaceutical company alleges that a drug made by the company 25 years ago, which was taken by expectant mothers, causes the children born to these mothers to be very susceptible to cancer. The lawsuit has been filed on behalf of a group of children born to these mothers and who have been stricken by cancer as young adults. Making a decision regarding the liability of the company will require analyzing sample data. Obviously, a sample of offspring of mothers who took the drug as well as a sample of offspring of mothers who did not take the drug will have to be carefully obtained and compared. That is, judgment should not be rendered on the basis of observing only those offspring involved in the lawsuit, for any such inference made to the population would be misleading.

THE UNAVOIDABILITY OF INCORRECT DECISIONS

The problem presented in the above setting is one of many that fit into the natural framework of hypothesis testing. First, the company is either guilty or innocent. Second, the jury must decide guilt or innocence. Hence, there are four distinct possibilities associated with the decision, which are summarized as follows:

	Company is Innocent	Company is Guilty
Jury Decides the Company is Innocent	Correct decision	Incorrect decision
Jury Decides the Company is Guilty	Incorrect decision	Correct decision

Of course it would be desirable to make the correct decision every time (otherwise known as Utopia). However, this is not possible. An impossible situation is created by trying to eliminate both of these incorrect decisions. This is not to say that all parties would not be doing their best to make a correct decision. Rather, even when everyone does their best the possibility of error still is present. No one should be naive enough to believe that all guilty parties are convicted or that all innocent parties are exonerated. To explore the unavoidability of making some incorrect decisions, suppose that the jury decides they never want to convict an innocent party. This can be done in only one way—declare all defendants innocent. Hardly a desirable situation! On the other hand the jury could avoid exonerating any guilty parties in only one way—declare all defendants guilty. Also not desirable! There must be some way of quantifying the trade-off between the two types of incorrect decisions. This is what is accomplished by applying the science of statistics to hypothesis testing.

THE NULL HYPOTHESIS

In hypothesis testing the formal statement or conjecture to be tested is called the **null hypothesis** and is usually denoted by H_0. The null hypothesis is often, but not always, a version of the statement "Any observed change or difference is due to chance variability," and the purpose of the hypothesis test is usually to see whether a change had indeed occurred or a real difference exists. That is why the hypothesis is called a **null** hypothesis, or hypothesis of *no* change or difference. Some examples include the following:

H_0: The defendant is not guilty.
H_0: Smoking is not harmful to your health.

H_0: The mean burning time of a new type of light bulb if not greater than the mean of a well-established type of light bulb.

H_0: Job satisfaction is independent of the level of education.

H_0: The drug is not effective in treating the disease.

NULL HYPOTHESIS	The **null hypothesis** H_0 *is the hypothesis being tested. The null hypothesis is usually, but not always, worded in a way that reflects the status quo—that is, no change or no difference.*

THE ALTERNATIVE HYPOTHESIS

For each null hypothesis to be tested there is an associated **alternative hypothesis**, which is usually denoted by H_1. The alternative hypothesis reflects the change or difference anticipated by the individuals doing the hypothesis test. That is, if the null hypothesis is not true, then what hypothesis is likely to be true? The answer to that question provides the wording used for a specific alternative hypothesis. Alternatives to the above null hypotheses include the following:

H_1: The defendant is guilty.

H_1: Smoking is harmful to your health.

H_1: The mean burning time of a new type of light bulb is greater than the mean of a well-established type of light bulb.

H_1: Job satisfaction is dependent upon the level of education.

H_1: The drug is effective in treating the disease.

ALTERNATIVE HYPOTHESIS	The **alternative hypothesis** H_1 *is the statement that reflects the situation anticipated to be true in case the null hypothesis is not true.*

HOW TO SET UP THE HYPOTHESES

The correct formulation of the null and alternative hypotheses is one of the most crucial steps in the initial stages of hypothesis testing. While hard and fast rules are not available for setting up the hypotheses correctly, some guidance will be provided by examining the previous examples. Usually it is easier to formulate the alternative hypothesis first, since the alternative hypothesis tends to reflect the effect or conditions the experimenter suspects exist, or that the experimenter hopes to show (the anticipated situation).

As an example, the statement "Smoking is harmful to your health" is the situation anticipated by the experimenter; that is, it represents what the experimenter would like to show to be true. Therefore, this statement is the alternative hypothesis. The null hypothesis is essentially the negation of this statement, or "Smoking is not harmful to your health."

Consider the statement "The mean burning time of a new type of light bulb is greater than the mean of a well-established type of light bulb." Generally a longer burning time is a desirable property of light bulbs, and a new type of light bulb should be able to exhibit this property. This is a statement that the experimenter would like to be able to show, so this is the alternative hypothesis. The null hypothesis is simply the negation of this statement.

A similar analysis may be made of the statement "The drug is effective in treating the disease," which the experimenter would like to show, or the statement "Job satisfaction is dependent on the level of education," which the experimenter suspects exists. These become the alternative hypotheses. The corresponding null hypotheses are merely the negation of the alternative hypotheses.

It should be emphasized that the hypotheses must be set up prior to an examination of the data. If hypotheses are set up on the suggestion of the data, the scientific basis behind the hypothesis test immediately disappears. The following example illustrates the proper way to set up hypotheses.

EXAMPLE

A pharmaceutical company has a new drug they would like to get approved by the Food and Drug Administration so they can market it. The company believes the drug to be at least 60 percent effective in curing a rare fatal disease.

Statement of the Alternative Hypothesis

The alternative hypothesis should be stated in such a way as to support the company's belief in the drug.

H_1: The drug is more than 60 percent effective.

Statement of the Null Hypothesis

The null hypothesis should be stated to reflect a cure rate of 60 percent or less for the drug. This is simply a negation of the alternative hypothesis.

H_0: The drug is 60 percent or less effective.

ERRORS IN HYPOTHESIS TESTING

The two types of incorrect decisions referred to previously have formal names. The incorrect decision of rejecting the null hypothesis when it is true is called a Type I (one) error, while a Type II (two) error is the name assigned to the incorrect decision of accepting the null hypothesis when it is false. If the null hypothesis is a presumption of innocence in the lawsuit of Problem Setting 7.1, then a Type I error corresponds to convicting an innocent party, while a Type II error means exonerating a guilty party.

POSSIBLE ERRORS IN HYPOTHESIS TESTING	**Type I Error**—*Rejection of the null hypothesis when it is true.* **Type II Error**—*Accepting the null hypothesis when it is false.*

Clearly not much time was spent in thinking of original names for these two types of errors. This is somewhat like naming your first two children I and II. However, be that as it may, this is the standard statistical jargon so the reader would do well to commit the terminology and definitions to memory.

"YOUR HONOR, THE JURY WANTS TO KNOW WHAT LEVEL OF SIGNIFICANCE TO USE."

LEVEL OF SIGNIFICANCE

To reflect the uncertainty associated with the decision-making process, the two types of errors in hypothesis testing are quantified by associating probabilities with them. The probability of making a Type I error is called the **level of significance** and is denoted by the Greek letter α (alpha). In practice, the individual doing the hypothesis testing chooses an appropriate value for α. A common choice for α is $\alpha = .05$, which corresponds to one chance in 20 of making a Type I error. If a Type I error is very serious, and would be a very costly mistake, then the level of significance would be set very low, say $\alpha = .01$ or smaller, thus ensuring a very low likelihood of making that type of error. Of course this smaller level of α decreases the probability of rejecting the null hypothesis and therefore increases the probability of accepting a false null hypothesis—that is, a Type II error. On the other hand, if a Type I error is not of great consequence, a level of significance of $\alpha = .10$ or so would decrease the probability of a Type II error.

LEVEL OF SIGNIFICANCE	The **level of significance**, denoted by α, is the maximum probability of making a Type I error. The value for α is selected by the individual doing the hypothesis testing.

EXAMPLE

In a continuation of the previous example, since the company wants to be very cautious about rejecting the null hypothesis when it is true, a value of $\alpha = .01$ is selected. That is, if the drug is really less than 60 percent effective (the null hypothesis is true), then the company wants to take only 1 chance in 100 of concluding that the drug is really more than 60 percent effective.

THE POWER OF THE TEST

The probability of rejecting the null hypothesis when it is false is called the **power** of the test. The opposite event, accepting the null hypothesis when it is false, is a Type II error. The probability of making a Type II error is denoted by the Greek letter β (beta), so the power equals $1 - \beta$. The numerical value for β depends on the value of α, the sample size, the type of hypothesis test being used, and the true (but unknown) population distribution function being fixed.

POWER
> The **power** of a test is the probability of rejecting a false null hypothesis, and is denoted by $1 - \beta$, where β is the probability of a Type II error. The power depends on considering one particular situation at a time that may exist when the alternative hypothesis is true.

As stated previously, an objective in hypothesis testing is to have α small (close to zero) and the power large (close to 1). There is always a trade-off between the level of significance and the power. A very small value of α (α close to zero) will result in a test with less power than a test with a larger value of α. However, an increase in the sample size usually is accompanied by an increase in power for a fixed level of significance α. Figure 7.1 summarizes the terminology and symbolism associated with hypothesis testing.

		True Situation	
		H_0 is true	H_0 is false
Decision	Accept H_0	Correct decision P(correct decision) $= 1 - \alpha$	Type II error P(Type II error) $= \beta$
	Reject H_0	Type I error P(Type I error) $= \alpha$ (level of significance)	Correct decision P(correct decision) $= 1 - \beta$ (power of the test)

FIGURE 7.1 A Summary of the Possible Decisions and Their Probabilities

SELECTING AN APPROPRIATE TEST

There are many different statistical tests available in textbooks and journals on statistics. Each statistical test has an associated **test statistic**, which is used in the decision-making process. The test procedure consists of sampling and then computing the test statistic for the observed sample. The decision to accept or reject the null hypothesis is based on how probable that value of the test statistic (or one more extreme) is, assuming the null hypothesis is true.

The selection of the best test to use in a particular situation depends, in order of importance, on the following factors:

1. The null hypothesis being tested.
2. The alternative hypothesis of interest.
3. The power of the test.

The first two factors are fairly simple to deal with, in that most statistical tests are considered appropriate for certain types of null and alternative hypotheses. The third factor is less obvious. The test with the greatest power is obviously the test to use, but the power of a test depends on several items that are usually not well understood. For example, the power of a test depends on the population distribution function of the random variable being observed. If the distribution is normal, there exists one test that has the most power; but if the distribution is not normal, this test may have less power than other tests. Therefore much attention is given to testing the assumption of normality so that an appropriate test will be selected.

Another important assumption relates to the type of sampling being used. All of the well-known tests assume that random sampling is employed in obtaining the data, so this assumption is not usually stated explicitly.

TEST STATISTIC	*A **test statistic** is a statistic computed on the sample data, and it is used in the testing of hypotheses. In this book the test statistic is frequently denoted by T, with or without subscripts.*

Once the statistical test has been selected, a **decision rule** is formulated. The decision rule is stated in terms of the test statistic. That is, the decision rule states which values of the test statistic result in the decision to reject the null hypothesis, and which values result in acceptance of the null hypothesis.

The decision rule depends on the alternative hypothesis of interest and the level of significance selected. That is, if large values of the test statistic T correspond to data sets more likely to occur when H_1 is true than when H_0 is true, then H_0 should be rejected for large values of T. Therefore, the decision rule is as follows:

Reject H_0 if $T > T_{upper}$.
Accept H_0 if $T \leq T_{upper}$.

T_{upper}, or T_U for short, is a **critical value** for T obtained from the distribution function of T when H_0 is true, as indicated in Figure 7.2a. T_U is selected so that the probability of making a Type I error is indeed equal to the preselected value α. This is called an upper-tailed test because the upper tail of T is the region that indicates H_0 should be rejected.

Similarly, a **lower-tailed test** is a test in which small values of T indicate that H_0 should be rejected. The decision rule for a lower-tailed test takes the form:

Reject H_0 if $T < T_{\text{lower}}$.
Accept H_0 if $T \geq T_{\text{lower}}$.

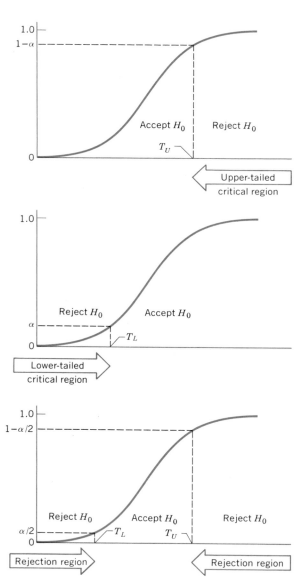

FIGURE 7.2 The Cumulative Distribution Function of the Test Statistic Assuming H_0 is True

T_{lower}, or T_L for short, is a critical value selected from the distribution of T when H_0 is true, as indicated in Figure 7.2b, such that α is the level of significance.

If H_0 is likely to be false when either T is very large or T is very small, then the decision rule takes the form:

Reject H_0 if $T > T_U$ or $T < T_L$.
Accept H_0 if $T_L \leq T \leq T_U$.

T_L and T_U are critical values obtained from the distribution function of T when H_0 is true. This is a **two-tailed test** as indicated in Figure 7.2c. The values T_L and T_U are usually selected so that there is equal probability (or nearly equal) associated with each tail of the distribution $T > T_U$ and $T < T_L$.

ONE-TAILED TEST, TWO-TAILED TEST, AND CRITICAL VALUES

A **one-tailed test** is one where the decision rule is stated as

Reject H_0 if $T > T_U$; otherwise accept H_0

for an upper-tailed test, or as

Reject H_0 if $T < T_L$; otherwise accept H_0

for a lower-tailed test. A **two-tailed test** has the decision rule:

Reject H_0 if $T > T_U$ or if $T < T_L$.
Accept H_0 if $T_L \leq T \leq T_U$.

The values T_L and T_U are **critical values** that are selected so the test will have the desired level of significance α.

EXAMPLE

Referring again to the previous example, the drug will be tested on several animals previously infected with the disease. In order to resemble more closely the human reaction to the drug, large animals are selected. The expense and the availability of the animals, and the expense of the drug being tested, dictate that a small sample size be used.

Ten animals are selected for the experiment and are infected with the disease. Each animal is given the drug after an examination shows that the animal has the disease. The animal is then observed for four weeks, which is considered a sufficient length of time for judging the effectiveness of the drug. Then each animal is examined to see if it still has the disease. Let p equal the probability that the animal is cured. Let the test statistic T be the number of animals diagnosed as being cured. Large values of the test statistic correspond to many animals being cured, and indicate that H_1 is likely to be true. This results in an upper-tailed test. Since $\alpha = .01$ has previously been determined, it is necessary to find the value of T_U such that

$$P(T > T_U) = .01$$

when H_0 is true.

The maximum value for p when H_0 is true is $p = .6$. This value of p is used in the distribution of T because $p = .6$ maximizes the probability of making a Type I error. Since the results on the animals are independent of each other, the distribution of T is binomial with $n = 10$ and $p = .6$. Entering Table A1 with these parameters gives the probability that $T > 9$, which is $1 - .9940 = .0060$. Thus $T_U = 9$ and the decision rule is as follows:

Reject H_0 if $T > 9$.
Accept H_0 if $T \leq 9$.

The exact level of significance corresponding to this critical region is $P(T > 9) = .0060$, where $P(T > 9)$ is computed assuming H_0 is true.

The power of this test depends on which value of p exists among those possible under the alternative hypothesis. For instance, if the probability of the drug curing an animal is actually .8, then the power is the probability of rejecting H_0—that is, getting $T > 9$, when $n = 10$ and $p = .8$. The statistic T still has the binomial distribution only with $p = .8$ under the new assumption regarding p. From Table A1, for $n = 10$ and $p = .8$,

$$
\begin{aligned}
P(T > 9) &= 1 - P(T \leq 9) \\
&= 1 - .8926 \\
&= .1074.
\end{aligned}
$$

The probability of making the correct decision, rejecting H_0, is only .1074 when $p = .8$, so the ability of this test to detect a false null hypothesis is not very strong under these conditions.

All of the values of power, for various values of p, are plotted as a function of p in Figure 7.3. Such a graph is called a **power curve**. The curve will be different for a different sample size n or a different level of significance α. From the power curve it can readily be seen that the power, although not large for

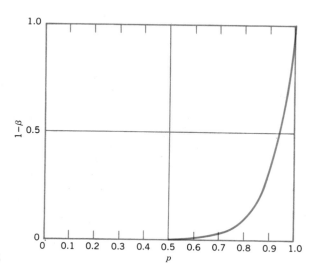

FIGURE 7.3 Power Curve for an Upper-Tailed Test when $n = 10$ and $\alpha = .006$

$p = .8$, goes to 1.0 as p gets closer to 1. This is interpreted as saying that the probability of getting more than 9 cures goes to 1.0 as the drug gets closer to 100 percent effective.

For smaller values of p, the power curve gets closer to the level of significance as p gets closer to $p = .6$. for $p < .6$ the probability of rejecting H_0 is less than the α level of .006, showing what was merely stated earlier: that the probability of making a Type I error is at its maximum value when $p = .6$.

MAKING THE DECISION

The framework for a statistical test described up to this point can be established before the sample data are collected. In fact, most well-planned experiments have all of the groundwork laid for a statistical test, including the statement of the decision rule, prior to the actual collection of the data. This assists in ensuring that the data can be analyzed once they are collected.

Only three steps remain in the hypothesis testing procedure:

1. Compute the test statistic from the sample data.
2. State the decision based on the decision rule.
3. Calculate, and state, the p-value (discussed later) associated with the observed value of the test statistic.

Computation of the test statistic is usually relatively easy, compared to the difficulties in collecting the data. Stating the decision is the easiest step of all, since it involves only a comparison of the observed value of the test statistic with one or two critical values. This decision is only an aid to the actual final decision in all cases, which must be made by someone who considers this information, as well as other information, not so easily quantified. The final step, calculation of the p-value, will now be described.

CALCULATING p-VALUES

Suppose the observed value of a test statistic greatly exceeds T_U, the upper critical value, in a test using a level of significance of $\alpha = .05$. Then there is stronger evidence supporting the decision to reject H_0 than if the test statistic had just barely exceeded T_U. Yet, in both cases the conclusion is simply stated as "The null hypothesis is rejected at $\alpha = .05$." In the former case the null hypothesis might have been rejected at $\alpha = .01$ or even $\alpha = .001$, which is a much stronger result than the one stated.

It is much more informative to report not only whether H_0 is accepted or rejected at the given α value, but also to report the **p-value**, which is the smallest level of significance that would result in the decision to reject H_0 for the observed value of the test statistic.

p-VALUE   *The **p-value** associated with an observed value of a test statistic is the smallest level of significance that would have allowed the null hypothesis to be rejected.*

For example, if T just barely exceeds T_U in an $\alpha = .05$ test, the decision is to reject H_0, and the p-value is slightly smaller than .05. A slightly smaller value of T, slightly less than T_U, results in the opposite decision—to accept H_0 at $\alpha = .05$—and the p-value is slightly larger than .05. The statements "Reject H_0" or "Accept H_0" do not tell the complete story, but a statement of the p-values—slightly less than .05 in the first case and slightly greater than .05 in the second case—conveys a lot of additional valuable information. It is always better to include some information about p-values along with the results of the analysis whenever possible. Many computer programs for statistical analyses routinely provide p-values as part of their output. If the p-value is smaller than or equal to α, the null hypothesis should be rejected.

PROCEDURE FOR TESTING HYPOTHESES

1. State the null hypothesis, H_0.
2. State the alternative hypothesis, H_1.
3. Decide on the Type 1 error rate, α.
4. Choose an appropriate testing procedure.
5. State the decision rule.
6. Compute the test statistic from the sample data.
7. Make the decision.
8. Compute the p-value and interpret it.

EXAMPLE

The previous example concludes with the following steps.

Making the decision. At the end of four weeks the number of cures observed is 8 out of the 10 animals tested. Since $T = 8$, the null hypothesis is accepted. The conclusion is that the company's claim has not been substantiated in this experiment. Even though 8 cures in 10 animals is an indication that the company's claim might be correct, such results could easily have occurred by chance even if the company's claim is not correct. The problem is that the small sample size ($n = 10$) results in very little power to detect a false null hypothesis.

Computing the p-value. The smallest critical value that permits rejection of the null hypothesis for the observed test outcome of $T = 8$ deaths is $T_U = 7$. The α level associated with $T_U = 7$ is found from Table A1 for $n = 10$ and $p = .6$:

$$P(T > T_U) = P(T > 7)$$

$$= 1 - P(T \le 7) = 1 - .8327$$

$$= .1673.$$

Therefore, the p-value associated with $T = 8$ is .1673. This is the probability of getting a value of 8 or more cures when the null hypothesis is true. Because this p-value is larger than the desired level of significance $\alpha = .01$, the null hypothesis is accepted. Another way of interpreting the p-value is that the null hypothesis could be rejected only for α levels greater than or equal to .1673.

THE LILLIEFORS TEST

At this point the reader may have realized that the Lilliefors test of Chapter 5 falls into this framework of hypothesis testing. That is, the null hypothesis is that the sample comes from a normal population while the alternative is that the population is nonnormal. The only level of significance permitted in the graph (see Figure 5.13 or Figure 5.25) is $\alpha = .05$. Graphs for additional α levels and more information about the Lilliefors test appear in the references at the end of this chapter.

The Lilliefors test as presented here is a two-tailed test. The null hypothesis is rejected if the empirical distribution function crosses the bounds corresponding to the sample size (see Figure 5.14 or Figure 5.26). It is not possible to compute p-values from the limited information presented here.

Exercises

7.1   Formulate an alternative hypothesis for each of the following null hypotheses.

(a) H_0: Employee typing speeds are not increased as a result of taking a specialized training course.

(b) H_0: Daily sales are unchanged after the start of the use of TV commercials.

(c) H_0: The average number of sickness absentees for Monday and Friday is not higher than the average number of sickness absentees for Tuesday, Wednesday, and Thursday.

(d) H_0: The proportion of viewers watching the local news on Channel 4 is less than or equal to 40 percent.

(e) H_0: The median grade point average of students at this university is 2.613.

7.2   Identify each of the hypothesis tests on Exercise 7.1 as either one-tailed or two-tailed tests.

7.3   Suppose it is desired to test the following hypothesis:

H_0: Smoking is not harmful to your health
versus H_1: Smoking is harmful to your health.

(a) In terms of the hypothesis, how would you state in words what is represented by a Type I error for this test?

(b) State in words what is represented by a Type II error for this test.

7.4   A quality control manager in a cannery will order the readjustment of all equipment involved in canning a particular food if the mean weight per can μ is less than specifications require. For cling peaches, the contents must be at least 16 ounces (the null hypothesis). A sample of cans is weighed and the sample mean $\bar{X}$ is used as the test statistic.

(a) Formulate the null and the alternative hypotheses in terms of specific values for the population mean.

(b) Is the test upper-tailed or lower-tailed?

(c) Should a small or large value of α be used with this test? (The answer to this question should include the consequences of making a Type I error.)

7.5 A bank is considering a policy of paying passbook interest on checking accounts that maintain a minimum monthly balance of at least $500. Such accounts are said to qualify for the "500 Club." The bank manager feels that more than 40 percent of the checking accounts will qualify for earned interest under this plan. The manager of a branch office of the bank wants to test the following null hypothesis:

H_0: The proportion of accounts qualifying for the "500 Club" is at most 40 percent.

The manager randomly selects 20 of the accounts at his branch and counts the number of accounts that qualify for the "500 Club."

(a) Formulate the appropriate alternative hypothesis.

(b) The test statistic has a binomial distribution with $n = 20$ and $p \leq .4$ when H_0 is true. Consult Table A1 with $p = .4$ and $n = 20$ to construct a decision rule having a level of significance of approximately .05 and state the exact level of significance associated with your decision rule.

(c) If the branch manager finds $T = 11$ of the 20 accounts qualify for the "500 Club," what decision is made regarding the null hypothesis?

(d) What is the p-value associated with the observed value of the test statistic $T = 11$? Interpret this p-value.

7.2
Testing Hypotheses about the Population Mean

PROBLEM SETTING 7.2

Producers of potatoes are well aware that as potatoes age, the sugar content increases and this in turn makes the potatoes less desirable since the eating quality declines. Therefore, the percent sugar content of stored potatoes must be continuously monitored to make sure that it stays within an acceptable range. This monitoring is done by sampling the mean percent sugar content of the potatoes.

A TEST FOR THE MEAN WITH LARGE SAMPLE SIZES

The problem setting suggests a hypothesis test regarding the mean percent sugar content of stored potatoes. Since the sample mean percent sugar content is determined by averaging many readings, the distribution of the sample mean is approximately normal by the Central Limit Theorem as discussed in Section 5.4. Because the distribution of the sample mean is approximately normal for large samples, the sample mean can be used as a test statistic for all populations when the sample sizes are large. The sample mean can also be used as a test statistic when sample sizes are small if the samples come from normal populations where σ is known, since the distribution of the sample mean is normal in this case for all sample sizes.

PROCEDURE FOR TESTING HYPOTHESES ABOUT THE POPULATION MEAN (LARGE SAMPLES)

Data. The data are represented by $X_1, X_2, \ldots, X_n$, with a sample mean $\bar{X}$ and standard deviation s.

Assumptions

1. The observations $X_1, X_2, \ldots, X_n$ represent a random sample of size n from some population.

2. This procedure is appropriate for $n \geq 30$. However, if the population is approximately normal and σ is known, this procedure can be used for $n < 30$.

Null Hypothesis. $H_0: \mu = \mu_0$, where μ_0 is some specified number. Note that this null hypothesis is equivalent to using $\mu \leq \mu_0$ or $\mu \geq \mu_0$, respectively, for the one-sided alternatives (a) or (b) below.

Test Statistic

$$T_1 = \frac{\bar{X} - \mu_0}{\sigma/\sqrt{n}}$$

For large samples, $n \geq 30$, the sample standard deviation s may be used when σ is unknown.

Decision Rules. The decision rule depends on the alternative hypothesis. Let z_α represent the $1 - \alpha$ quantile of a standard normal distribution obtained from Table A2 for a given level of significance α.

1. $H_1: \mu > \mu_0$. Reject H_0 if $T_1 > z_\alpha$; otherwise accept H_0.
2. $H_1: \mu < \mu_0$. Reject H_0 if $T_1 < -z_\alpha$; otherwise accept H_0.
3. $H_1: \mu \neq \mu_0$. Reject H_0 if $T_1 < -z_{\alpha/2}$ or $T_1 > z_{\alpha/2}$; otherwise accept H_0.

Some commonly used critical values are: $z_{.10} = 1.2816$, $z_{.05} = 1.6449$, $z_{.025} = 1.9600$, $z_{.01} = 2.3263$, and $z_{.005} = 2.5758$.

EXAMPLE

The potato producers referred to in Problem Setting 7.2 need a one-tailed test to declare stored potatoes unacceptable when the mean percent sugar content is judged to be unacceptably high. Previous experience shows 15 percent sugar content to be acceptable and that the population standard deviation of sugar content is 2.6 percent. Therefore, the test is $H_0: \mu \leq 15$ versus $H_1: \mu > 15$.

A random sample of 13 potatoes provides the following percentages of sugar content:

15.22, 12.36, 15.44, 13.53, 15.94, 17.06, 16.90,
14.31, 19.06, 18.23, 15.54, 16.28, 20.00

Since this sample is less than 30, the sample data need to be checked for normality using the Lilliefors test. These data easily pass the Lilliefors test as the reader should verify.

The mean percent sugar content for this sample is 16.14. The value of the test statistic is

$$T_1 = \frac{16.14 - 15}{2.6/\sqrt{13}} = 1.5809.$$

If the producers have decided that all such tests will be performed with $\alpha = .05$, then $z_{.05} = 1.6449$ and the decision rule is:

Reject H_0 if $T_1 > 1.6449$.

Accept H_0 if $T_1 \leq 1.6449$.

The null hypothesis is accepted and the stored potatoes are declared to be satisfactory. However, it should be kept in mind that the sample mean is 16.14 percent, above the 15 percent level of acceptability. This suggests that perhaps H_0 is accepted only because the sample size is too small to detect a difference, if there is a difference.

The largest critical value that enables the null hypothesis to be rejected for the observed value of $T_1 = 1.5809$ is obviously $T_U = 1.5809$. The level of significance associated with $T_U = 1.5809$ is obtained from Table A2 as follows:

$$P(T_1 > 1.5809) = 1 - P(T_1 \leq 1.5809)$$
$$= 1 - .943$$
$$= .057.$$

Thus the p-value is .057, which indicates that the null hypothesis was just barely accepted.

SMALL SAMPLES FROM NORMAL POPULATIONS WITH σ UNKNOWN

Frequently samples are obtained from normal populations where σ is unknown. If the samples are large, then the previous procedure can be used; but if the sample is small, a slightly different procedure is required. This procedure uses the Student's t-distribution.

PROCEDURE FOR TESTING HYPOTHESES ABOUT THE MEAN OF A NORMAL DISTRIBUTION WHEN σ IS UNKNOWN (SMALL SAMPLES, $n < 30$)

Data. The data are represented by $X_1, X_2, \ldots, X_n$.

Assumptions

1. The observations $X_1, X_2, \ldots, X_n$ represent a random sample of size n.

2. The sample comes from a normal population. (This assumption should always be verified using the Lilliefors test of Section 5.3. If this assumption is not satisfied, use the methods of Section 7.4.)

Null Hypothesis. H_0: $\mu = \mu_0$ where μ_0 is some specified number.

Test Statistic

$$T_2 = \frac{\bar{X} - \mu_0}{s/\sqrt{n}}$$

Decision Rules. Let $t_{\alpha, n-1}$ represent the $1 - \alpha$ quantile of a Student's t-distribution with $n - 1$ degrees of freedom obtained from Table A3 for a given level of significance α. The decision rule depends on the alternative hypothesis.

1. H_1: $\mu > \mu_0$. Reject H_0 if $T_2 > t_{\alpha,n-1}$; otherwise accept H_0.

2. H_1: $\mu < \mu_0$. Reject H_0 if $T_2 < -t_{\alpha,n-1}$; otherwise accept H_0.

3. H_1: $\mu \neq \mu_0$. Reject if $T_2 < -t_{\alpha/2,n-1}$ or $T_2 > t_{\alpha/2,n-1}$; otherwise accept H_0.

This set of rules is very similar to those given before for normal populations with σ known, the difference being the use of s in the calculation of the test statistic and the use of Student's t-distribution to find critical values for a pre-specified value of α. In summary, when the population is normal the statistic T_1 has a normal distribution while the statistic T_2 has a Student's t-distribution. However, as the sample sizes get large the difference between the two distributions becomes negligible. This fact can be observed by referring to the last line in Table A3 where it can be seen that the t-distribution with large degrees of freedom is virtually the same as a normal distribution. However, much of the hypothesis testing done in statistics deals with small samples and requires the use of a Student's t-distribution. That is why the distinction is made between "large" and "small" samples when stating the above rules. To reinforce this concept the previous example will now be reworked using the small sample rules.

EXAMPLE

In the previous example the population standard deviation was given as 2.6. Had this value not been known, then the sample data would be used to provide an estimate for σ. The sample standard deviation is $s = 2.15$. For the previous choice of $\alpha = .05$ the test is H_0: $\mu \le 15$ versus H_1: $\mu > 15$.

Test Statistic: $\quad T_2 = \dfrac{16.14 - 15}{2.15/\sqrt{13}} = 1.9118.$

Decision Rule: $\quad$ Reject H_0 if $T_2 \ge t_{.05,12} = 1.7823.$

Clearly the decision is to reject H_0, and the stored potatoes are declared to have too high of a sugar content. The p-value found by interpolating in Table A3 with 12 degrees of freedom is .042. This indicates that the null hypothesis was barely rejected at $\alpha = .05$.

THE EFFECT OF ESTIMATING σ

It would be very reasonable to ask why there is a difference in the conclusions on the above examples when the same sample data are used in both cases. The answer is quite simply that the statistics T_1 and T_2 have different distributions and therefore each answer is correct for its respective setting. To explore this question one step further, suppose in the second example that the sample standard deviation turned out to be 2.6; then the value of T_2 would be changed to 1.5809, which is the same value that was calculated for T_1. However, this new value of T_2 would still be compared against the Student's t-distribution value of 1.7823, while T_1 would be compared against 1.6449. Therefore, since the denominators $2.6/\sqrt{13}$ are now the same for both T_1 and T_2, the difference $\overline{X} - \mu_0$ must be larger for T_2 to detect a significant difference than it is for T_1. That is, if the standard deviation is estimated, then the sample mean $\overline{X}$ must deviate more from the hypothesized value μ_0 than is the case when σ is known in order to be significant. This phenomenon might be thought of as a penalty (or a safety measure) built into the test when the standard deviation has to be estimated, because of the additional variability introduced by estimating σ.

FINDING THE POWER OF A TEST

This section concludes with a discussion of the power associated with the previous tests for a given alternative. Recall that the power is the probability of rejecting H_0 when H_0 is false. Computing the power depends on knowing the real population mean (i.e., a given alternative mean). First consider the distribution of the percent sugar content in the first example of this section where 15 percent was regarded as an "acceptable standard" for mean percent sugar content with a standard deviation of 2.6. The distribution function of $\bar{X}$ is given in Figure 7.4.

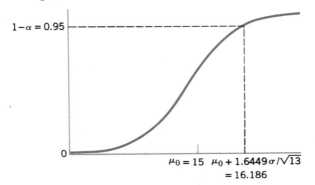

$1 - \alpha = 0.95$

$\mu_0 = 15$ $\mu_0 + 1.6449\sigma/\sqrt{13}$
$= 16.186$

FIGURE 7.4 The Distribution Function of Percent Sugar Content When H_0 is True

The decision rule says to reject the null hypothesis for sample means greater than 16.186. This value is obtained by returning to the decision rule used in the first example and noting that it can be written either as

Reject H_0 when $\dfrac{\bar{X} - \mu_0}{\sigma/\sqrt{n}} > 1.6449$, or as

Reject H_0 when $\bar{X} > \mu_0 + 1.6449\,\sigma/\sqrt{n}$

or $\bar{X} > 15 + 1.6449(2.6)/\sqrt{13} = 16.186$.

Suppose the population mean percent sugar content of a new shipment of potatoes is really 16.5 What is the probability that the decision rule will lead to rejection of H_0 and hence the shipment (i.e., what is the power)? It is helpful to put the graphs of the two distribution functions of sample means on the same axis. These functions appear in Figure 7.5.

The power $1 - \beta$ is the probability that $\bar{X}$ exceeds 16.186 as determined from the new curve. This is easily found by use of the standard normal transform and Table A2.

$$1 - \beta = P(\bar{X} \geq 16.186) = P\left(Z \geq \frac{16.186 - 16.5}{2.6/\sqrt{13}}\right)$$

$$= P(Z \geq -.4354)$$

$$= 1 - P(Z \leq -.4354)$$

$$= 1 - .332$$

$$= .668$$

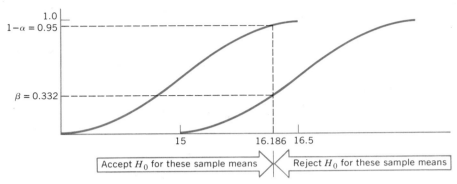

FIGURE 7.5 Distribution Functions Used for Power Calculation, $n = 13$

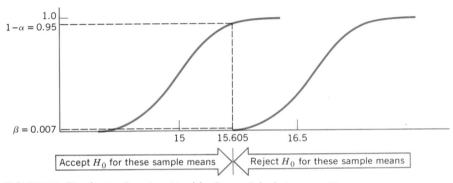

FIGURE 7.6 Distribution Functions Used for Power Calculation, $n = 50$

Therefore, if the new shipment has a population mean percent sugar content of 16.5, there is a 66.8 percent chance that the shipment will be correctly refused when a sample of size 13 and a level of significance of $\alpha = .05$ are used. Of course these calculations are valid only for $n = 13$ and $\mu = 16.5$, but clearly similar calculations could easily be done for any combination of n and μ.

To show how the power increases as the sample size increases, consider what happens as the sample size increases from 13 to 50, for example. An increase in the sample size results in a decrease in the standard error of $\overline{X}$, and therefore a clearer separation between the two distributions of $\overline{X}$, one for $\mu = 15$ and one for $\mu = 16.5$. Figure 7.6 shows the new distribution functions. The new decision rule for $n = 50$ becomes

Reject H_0 when $\dfrac{\overline{X} - 15}{2.6/\sqrt{50}} > 1.6449$, or

Reject H_0 when $\overline{X} > 15.605$,

which is a larger region of rejection than before. If μ equals 16.5, the probability of rejection becomes

$$1 - \beta = P(\bar{X} > 15.605 \mid \mu = 16.5)$$

$$= P\left(Z > \frac{15.605 - 16.5}{2.6/\sqrt{50}}\right)$$

$$= P(Z > -2.4341)$$

$$= 1 - P(Z \leq -2.4341)$$

$$= 1 - .007$$

$$= .993.$$

Thus the power goes from only .668 for a sample size of 13 to .993 for a sample size of 50. Looking at it another way, there is about one chance in three of making a Type II error when $n = 13$, but less than one chance in 100 of making a Type II error when $n = 50$.

Exercises

7.6   The quality control manager in a cannery will order the readjustment of all equipment involved in canning a particular food if the mean weight per can μ is less than specifications require. For cling peaches, the contents must be at least 16 ounces. Previous testing has established a standard deviation of 0.5 ounce per can. A sample of 100 cans is to be selected for testing.

(a) Formulate a decision rule for testing the null hypothesis H_0: $\mu \geq 16$ versus H_1: $\mu < 16$ with a level of significance of $\alpha = .05$.

(b) Based on your answer to (a), indicate whether the null hypothesis should be accepted or rejected for each of the following results. (Also, indicate for each whether or not the equipment will be readjusted.) (i) $\bar{X} = 15.5$ ounces; (ii) $\bar{X} = 15.95$ ounces; (iii) $\bar{X} = 16.1$ ounces; (iv) $\bar{X} = 15.90$ ounces.

7.7   A random sample of laboratory mice is taken to see if the average weight is 25.0 grams. The sample yields the following data.

    24.2   25.1   23.0   22.8   24.5   23.8

(a) Test the sample data for normality.

(b) Test the null hypothesis H_0: $\mu = 25$ versus H_1: $\mu \neq 25$ using a level of significance of $\alpha = .10$.

(c) What is the p-value associated with the value of the test statistic? Interpret this p-value.

7.8   Replacement of paint on highways and streets represents a large investment of funds by state and local governments each year. A new cheaper brand of paint is tested for durability after one month's time by means of reflectometer readings. In order for the new brand to be acceptable it must have a mean reflectometer reading of at least 19.6. The sample data based on 25 readings show $\bar{X} = 19.4$ and $s = 1.5$ and were accepted as being normally distributed.

(a) State the appropriate null and alternative hypotheses.

(b) Test the null hypothesis in (a) with $\alpha = .01$.

(c) State the p-value associated with the value of the test statistic, and interpret it.

7.9 Ten high school seniors took the ACT test and made the following scores.

28 26 30 24 25 29 31 26 23 27

Past ACT scores at their high school have shown the scores to be normally distributed with $\sigma = 4$ and $\mu = 25$. Test the null hypothesis $H_0: \mu = 25$ versus $H_1: \mu > 25$ with $\alpha = .05$. State the p-value associated with the value of the test statistic and interpret it.

7.10 A random sample of the recent university graduates revealed six education majors who were making $20,000, $16,000, $17,000, $16,000, $18,000, and $21,000 per year.

(a) Test these data for normality.

(b) Test the null hypothesis $H_0: \mu \geq \$20{,}000$ versus $H_1: \mu < \$20{,}000$ with $\alpha = .01$ and state and interpret the p-value.

7.11 A random sample of 200 people receiving food stamps showed their average age to be 42.8 years with $s = 6.89$ years. Test $H_0: \mu = 40$ versus $H_1: \mu > 40$ with $\alpha = .01$. State the p-value associated with the value of the test statistic and interpret it.

7.12 A survey of 40 senior citizens selected at random showed that they watched TV an average of 24 hours per week with a standard deviation of 10 hours. Test $H_0: \mu = 30$ versus $H_1: \mu < 30$ with $\alpha = .05$ and state and interpret the p-value.

7.13 Truck loads of fill dirt arriving at a construction site are contracted to carry 6.3 cubic yards of dirt per load. A sample of 25 loads showed $\overline{X} = 6.0$ yards and $s = 1.5$ yards. If these data are accepted as normal, test the null hypothesis $H_0: \mu = 6.3$ versus $H_1: \mu < 6.3$ with $\alpha = .05$. State the p-value associated with the value of the test statistic and interpret it.

7.14 The manager of a dry cleaning shop believes the average charge for cleaning is $1.80. A random sample of 100 orders shows $\overline{X} = \$1.85$ and $s = \$.025$. Test $H_0: \mu = \$1.80$ versus $H_1: \mu > \$1.80$ with $\alpha = .05$. What is the probability that the null hypothesis will be rejected if the true mean is really $1.85; that is, what is the power associated with this test?

7.3
Hypotheses about the Population Proportion p

PROBLEM SETTING 7.3

The major television networks are continuously trying to win the weekly ratings and increase their share of the viewing audience. The share is usually determined by independent groups who use statistical sampling procedures to obtain a random sample of the viewing audience. The sample must be taken in all sections of the country and must represent an attempt to reach a cross section of all viewers. A local network affiliate would like to compare its own share of the local audience with the figures claimed by its parent network.

TESTING PROCEDURE

The problem setting is really concerned with testing a hypothesis about the true population proportion p. If a large proportion of the sample obtained by the local station says they watch the channel under consideration, then the population proportion is likely to be large also. Therefore, inferences about the population proportion naturally use the number of favorable responses in the sample. About the only requirement in this procedure is that the sample be a random sample. Misleading results can be obtained from samples obtained using methods that may introduce a bias. Use of a telephone directory in obtaining the random sample may introduce a bias because the telephone directory does not list everyone in the population, although the bias may be assumed to be small.

PROCEDURE FOR TESTING HYPOTHESES ABOUT THE POPULATION PROPORTION p

Data. The sample consists of n trials, where each trial results in a success or a failure.

Assumptions.
1. The trials are independent of one another.
2. The probability of success p remains the same from trial to trial.

Null Hypothesis. $H_0: p = p_0$, where p_0 is some specified number such that $0 \leq p_0 \leq 1$.

Test Statistic. $X =$ number of successes observed in n trials.

Decision Rules. The decision rule depends on the alternative hypothesis. Let α be the appropriate level of significance desired.

1. $H_1: p > p_0$. Reject H_0 if $X > T_U$. Accept H_0 if $X \leq T_U$.

The value of T_U is found by entering Table A1 with n and p_0. T_U is the closest value such that $P(X > T_U) \cong \alpha$, or, equivalently, $P(X \leq T_U) \cong 1 - \alpha$. For large sample sizes ($n > 20$) the value of T_U is found as

$$T_U = np_0 + z_\alpha \sqrt{np_0\, q_0},$$

where z_α is the $1 - \alpha$ quantile from Table A2.

2. $H_1: p < p_0$. Reject H_0 if $X < T_L$. Accept H_0 if $X \geq T_L$.

The value of T_L is found by entering Table A1 with n and p_0. T_L is the closest value such that $P(X < T_L) \cong \alpha$, or equivalently $P(X \leq T_L - 1) \cong \alpha$. For large sample sizes ($n > 20$) the value of T_L is found as

$$T_L = np_0 - z_\alpha \sqrt{np_0\, q_0},$$

where z_α is the $1 - \alpha$ quantile from Table A2.

3. $H_1: p \neq p_0$. Reject H_0 if $X > T_U$ or if $X < T_L$. Accept H_0 if $T_L \leq X \leq T_U$.

The values of T_U and T_L are found by entering Table A1 with n and p_0. T_U is the closest value such that $P(X > T_U) \cong \alpha_1$, or, equivalently, $P(X \leq T_U) \cong 1 - \alpha_1$, and T_L is the closest value such that $P(X < T_L) \cong \alpha_2$, or equivalently $P(X \leq T_L - 1) \cong \alpha_2$, where $\alpha_1 + \alpha_2 \cong \alpha$. For large sample sizes ($n > 20$) the values of T_U and T_L are given as

$$T_U = np_0 + z_{\alpha/2} \sqrt{np_0 q_0} \quad \text{and} \quad T_L = np_0 - z_{\alpha/2} \sqrt{np_0 q_0},$$

where $z_{\alpha/2}$ is the $1 - \alpha/2$ quantile from Table A2.

Warning: If $np_0 < 5$ or $nq_0 < 5$, the large sample approximation is not very accurate.

If the sample is random, then each respondent has the same probability of saying they watched the channel in question, and the responses are independent of one another. Thus the total number of favorable responses has a binomial distribution with parameters p and n, the sample size. This suggests a natural test statistic because its distribution is well known when n is small, and can be approximated with a normal distribution when n is large.

EXAMPLE

Suppose the local TV station selects a sample of size 12 to see if the true population proportion p is different from .4 as claimed by the network. Specifically they wish to test

H_0: $p = .4$ versus H_1: $p \neq .4$.

The approximate level of significance desired is $\alpha = .10$ (a Type I error is not very serious). Therefore, the decision rule is:

Reject H_0 if $X > T_U$ or if $X < T_L$.
Accept H_0 if $T_L \leq X \leq T_U$.

T_U and T_L are obtained from Table A1 for $n = 12$ and $p = .40$.
    To find T_L enter Table A1 with $n = 12$ and $p = .40$. Read down the column of cumulative probabilities to the closest value to .05. The number is .0196 in this case, which is in the row $y = 1$, so $T_L = 2$. The exact probability associated with the lower tail is

$$P(X < T_L) = P(X < 2)$$
$$= P(X \leq 1)$$
$$= .0196.$$

The upper critical value is found by using the same column of probabilities, reading down the column to the closest value to .95. This is the cumulative probability .9427, which is in the row $y = 7$, so $T_U = 7$. The exact probability associated with the upper tail is

$$P(X > T_U) = P(X > 7)$$
$$= 1 - P(X \leq 7)$$
$$= 1 - .9427$$
$$= .0573.$$

Thus the exact level of significance is

.0196 + .0573 = .0769.

This illustrates the fact that exact levels of significance are usually not attainable when the distribution of the test statistic is discrete.
    Suppose the random sample results in $X = 2$, or 2 out of 12 responses being favorable to the channel in question Since $X = 2$ is neither less than $T_L = 2$ nor

greater than $T_U = 7$, the null hypothesis is accepted. The probability associated with a value of 2 or less, from a binomial distribution with $n = 12$ and $p = .4$, is

$$P(X \leq 2) = .0834.$$

This probability is doubled since the test is two-tailed, to give a p-value of .1668 associated with the observation obtained. That is, results as extreme as these can be expected to occur about 16.7 percent of the time when $p = .4$.

THE DISTRIBUTION OF X

In the previous example, the desired approximate level of significance $\alpha = .10$ resulted in an actual level of significance of only .0769. This difference between the target α and the actual α is due to the fact that X has a binomial distribution, which is discrete. The distribution function of X is graphed in Figure 7.7. There it is easily seen how the .05 and .95 quantiles are obtained. If the decision rule were to reject H_0 if X is less than 2 (the .05 quantile) or greater than 8 (the .95 quantile), the exact level of significance would always be less than .10. If the decision rule is changed so H_0 is rejected if $X < 3$ or $X > 7$, then the level of significance jumps to

$$P(X < 3) + P(X > 7) = P(X \leq 2) + 1 - P(X \leq 7)$$
$$= .0834 + 1 - .9427$$
$$= .1407.$$

which is too large. Therefore the compromise between these two alternatives as described in the example, rejecting H_0 if $X < 2$ or $X > 7$ with a resulting level of significance $.0196 + .0573 = .0769$, is closer to $\alpha = .10$ and is the preferred choice as a region of rejection.

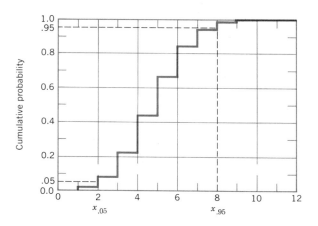

FIGURE 7.7 The Cumulative Distribution Function of the Test Statistic T when $p = .4$

THE NORMAL APPROXIMATION FOR X

The procedure for using the normal approximation is straightforward. The normal approximation is easier to use than the exact binomial distribution, but of course exact results are sometimes worth the extra effort it takes to obtain them.

The normal approximation, if used in the previous example, results in the decision rule

Reject H_0 if $X > T_U$ or $X < T_L$,

where
$$T_U = np_0 + z_{.05} \sqrt{np_0 \, q_0}$$
$$= 12(.4) + 1.6449 \sqrt{12(.4)(.6)}$$
$$= 4.80 + 2.79$$
$$= 7.59$$

and
$$T_L = 4.80 - 2.79$$
$$= 2.01$$

corresponding to the values of T_L and T_U appearing in Figure 7.2c. Thus the decision rule would be

Reject H_0 if $X > 7.59$ or $X < 2.01$,

which has an approximate level of significance of $\alpha = .10$. The decision for $X = 2$ is to reject H_0. Of course the exact level of significance associated with these numbers was found earlier, $\alpha = .1407$, from the binomial distribution tables. This illustrates the fact that exact tables should always be used when they are available. A more accurate procedure to use when the normal approximation is necessary involves using the continuity correction in finding the p-value, as illustrated below.

The p-value associated with $X = 2$ may be approximated from the normal distribution

$$P(X \le 2) \cong P\left(Z \le \frac{2 - np_0 + .05}{\sqrt{np_0 \, q_0}}\right) = P\left(Z \le \frac{2 - 4.80 + .05}{\sqrt{2.88}}\right)$$
$$= P(Z \le -1.3553)$$

and then by consulting Table A2. The tail probability associated with $z = -1.3553$ is $p = .088$, which is doubled to obtain the approximate p-value .176, because the test is two-tailed. This is reasonably close to the exact p-value of .1668 given in the example.

ACCEPTANCE SAMPLING

Attention is now turned to two techniques that are useful for monitoring the quality of a product when that quality is measured by a proportion p, such as the proportion of defective items produced. The rationale behind quality monitoring is quite simple: if the manufacturing process gets out of control and starts producing defective items, then it will be to the company's advantage to catch it as soon as possible. Otherwise, they face the prospect of replacing defective items or of losing their customers. For manufacturing processes that are geared to mass production, the quantity of the product manufactured makes inspection of every item impossible. Therefore a sampling procedure is needed that will have the capability of quickly detecting manufacturing irregularities. It is just as important for the buyer to be able to determine whether or not to accept a shipment without testing every item in the shipment. The first sampling procedure considered is known as **acceptance sampling**.

Suppose that a manufacturing process is set up such that at different time periods a random sample of size n can be obtained and each of the sample items inspected for defects. Further suppose five items are contained in the sample at any one time period. The manufacturing process is declared to be out of control if any defectives are observed. Otherwise the process is accepted as satisfactory. In this setting the probability of detecting a faulty manufacturing process as the fraction of defective products increases is of interest. The random variable of interest is the number of defective items in a sample of size 5. The manufacturing process is accepted as being in control if $X = 0$ and regarded as out of control if $X \geq 1$. From Table A1 the probabilities $P(X = 0)$ for $n = 5$ and different values of p are easily found

Fraction of Defectives	$P(X = 0)$
0.0	1.0
.05	.7738
.10	.5905
.15	.4437
.20	.3277
.25	.2373
.30	.1681
.35	.1160
.40	.0778
.45	.0503
.50	.0312
.55	.0185
.60	.0102
.65	.0053
.70	.0024
.75	.0010
.80	.0003
.85	.0001
.90	.0000

These points can be plotted on a graph referred to as an **operating characteristic** (OC) curve. The OC curve is represented in Figure 7.8 and can be used to find the probability of accepting H_0 for any value of p. Note this curve is plotted for $n = 5$ and for an acceptance rule based on $X = 0$, but the curve is easily constructed for any n and for any acceptance rule based on $X \leq a$.

Acceptance sampling is closely related to hypothesis testing. The null hypothesis is "The process is in control," which corresponds to $H_0: p \leq p_0$, where p_0 is some small probability of producing a defective item, representing the maximum permissible value. The OC curve, representing the probability of accepting the null hypothesis, provides a curve of the Type II error rate as a

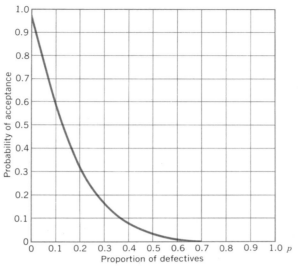

FIGURE 7.8 Operating Characteristic Curve for use with with Acceptance Sampling

function of p, for $p > p_0$. One minus the OC curve thus provides a power curve, or the probability of rejecting the null hypothesis, as a function of p.

OC CURVE	*The **operating characteristic curve** (OC Curve) is a graph of the probability of accepting a shipment as a function of the binomial parameter p, for a fixed acceptance rule and sample size n.*

QUALITY CONTROL CHARTS

The second method of monitoring quality is a **quality control chart** that consists of three horizontal lines. To use this technique the manufacturing process must have been observed when it was known to be under control. A history of observing the manufacturing process provides information about the proportion p of defective items. In the future, random samples of size n will be taken to estimate the proportion of defective items, and this sample proportion will be compared against a control chart made when the process was known to be under control. The control chart consists of three horizontal lines with the center line of the control chart located at p. The lower line is located at $p - 2.5758 \sqrt{pq/n}$ with the top line located at $p + 2.5758\sqrt{pq/n}$ if 99 percent limits are desired for the process.

Suppose a manufacturing process is known to produce $p = .08$ defective items and that repeated samples of size 30 are obtained at different points in time to monitor the manufacturing process. For $p = .08$ and $n = 30$ the 99 percent limits are located at 0 and .208. The actual calculation would put the lower line at $-.048$, but there cannot be an outcome less than 0, or for that matter greater than 1. The actual quality control chart is given in Figure 7.9.

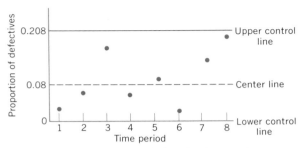

FIGURE 7.9 99 Percent Control Chart for the Sample Proportion

To use the quality control chart, random samples of size n are obtained, the number of defectives noted, and the sample proportion defective $\hat{p}$ is calculated and plotted on the chart above the point on the horizontal axis representing the corresponding time period. Whenever a point falls outside the upper or lower line, the process can be regarded as out of control. The plot should appear as a random scattering of points about the center line if the process is operating properly. Such points have been added to Figure 7.9 to give the effect of randomness. The reader should recognize that a quality control chart is merely a convenient method for testing the population proportion at successive points in time. Although the above procedure has been presented for a proportion, it can be used equally well to monitor a process where quality is measured by a mean rather than a proportion.

Exercises

7.15 A random sample of eight automobile drivers showed that only two of the drivers were wearing their seat belts. Is this sufficient evidence to reject the null hypothesis that at least 60 percent of all drivers are wearing their seat belts; that is, $H_0: p \geq .6$ is to be tested versus $H_1: p < .6$, if a level of significance of $\alpha = .05$ is used? State the p-value and interpret it.

7.16 A random sample of 20 school administrators showed that 15 favored a Kodak copy machine over a Xerox. If it is believed that there is really no difference between the machines, then each has a chance of ½ of being selected as the best when one must be selected. Let the hypothesis to be tested be expressed as follows.

$$H_0: p = .5 \quad \text{versus} \quad H_1: p \neq .5$$

If X represents the number of school administrators selecting the Kodak machine, answer the questions below based on the following decision rule: Reject H_0 if $X < 6$ or if $X > 14$.

(a) What is the exact level of significance associated with this decision rule?

(b) What decision is made based on the sample data?

(c) If the true proportion of school administrators selecting a Kodak copier is .8, find the power associated with this test.

7.17 A national magazine says that 70 percent of the owners of new Plymouths are satisfied with their purchase. In order to see if this is true for Davis, California, a new car dealer obtained a random sample of 10 recent customers to see if the population proportion is different from .7. That is, $H_0: p = .7$ is tested versus $H_1: p \neq .7$. The following decision rule is used: Reject H_0 if $X < 5$ or $X > 9$, where X is the number of successes in 10 trials.

(a) What is the exact level of significance for this test?

(b) What is the power of this test if the true value of p is .5?

7.18 Fourteen people said they preferred the news on Channel 13, out of the 88 people interviewed. Use a level of significance of .10 to test the hypothesis that exactly one-third of the population prefers the news on Channel 13. State the p-value and interpret it.

7.19 An authority of drug abuse, appearing on a local TV talk show, claims that at least 78 percent of college students are either using or have tried drugs of some kind. After the show has aired, a newspaper takes a survey at the local university and finds that 64 out of the 92 students questioned said they had either tried or were currently using drugs. Based on this evidence can the newspaper refute the authority's claim if they are willing to take a 5 percent chance of wrongly refuting his claim? State the p-value and interpret it.

7.20 Suppose that normally the percentage of persons with high blood pressure is at most 10 percent, but that for people over 40 it is suspected to be higher. A random sample of 50 individuals over 40 indicates that 10 have high blood pressure. Test the appropriate hypothesis using a level of significance of $\alpha = .05$. State the p-value and interpret it.

7.21 A manufacturing process is monitored by periodically taking samples of size 10 and observing the number of defective items. The manufacturing process is said to be in control if $X \leq 1$, where X is the number of defective items observed in the sample. Construct an operating characteristic curve for this process similar to the one given in Figure 7.8.

7.22 A machine producing typing elements for electric typewriters has been observed over a period of time to produce 3 percent defective parts. Samples of size 50 are obtained periodically and inspected for defective items. Construct a 99 percent quality control chart for this process. If a sample of size 50 shows five defective items, would the manufacturing process be considered to be out of control?

7.23 Suppose it is desired to test $H_0: p = .7$ versus $H_1: p = .8$ for the population proportion. A sample of size 30 is taken from the population and the null hypothesis is to be rejected if the number of successes is greater than 23.

(a) What is the significance level for this test?

(b) What is the power of this test?

7.4
Testing Hypotheses about the Population Median

THE SAMPLE MEDIAN

Statistics such as the sample mean are not always a meaningful way to summarize a set of sample data. For example, a few years ago a small town in Kansas reported an average family income of $50,000. The truth was that a rich oilman lived in the town and his income was in the millions while the other families in the town made about $10,000. This situation is typical of cases where the median should be used as a summary statistic rather than the mean.

In this section a procedure is given for testing hypotheses about the median of a population. This procedure can be used with all populations and all sample sizes, unlike the procedure for the population mean presented earlier in this chapter, which is valid only for large samples or for small samples from normal distributions.

AN INTUITIVE APPROACH

Prior to the presentation of the test procedure consider the real number line plots of four different samples in Figure 7.10. In each case the null hypothesis is that the median of the population is 15.

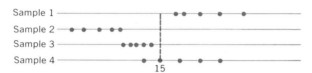

FIGURE 7.10 Real Number Line Plots for Hypothetical Samples

The hypothesized median value of 15 is marked on each of the real number line plots. What do you think about 15 as the population median for each one of these samples? Each of these samples will be examined one at a time in order to help answer this question.

Sample 1 / All five sample observations are above the hypothesized median; therefore, it would seem likely that the true median is *larger than 15*.

Sample 2 / All five sample observations are below the hypothesized median; therefore, it would seem likely that the true median is *smaller than 15*.

Sample 3 / Same as sample #2 except that all five observations are closer to the hypothesized median. Should this make a difference? Not really, since *all* sample observations are still below 15. Being closer to the median could merely reflect a scale change, such as pennies to dollars, which shouldn't affect the decision.

Sample 4 / The decision on this sample becomes more difficult to make since some sample observations are above the hypothesized median and some are below. Samples like this one require a more formal statistical analysis in order to aid the decision-making process.

These four examples suggest that a test statistic could be based on counting the number of sample observations on either side of the hypothesized median. That is precisely the way the test statistic is obtained. All that is missing for the hypothesis test is a probability distribution for the test statistic that will allow the computation of the probability of Type I and Type II errors. The binomial distribution adapts easily to this situation.

Suppose the median grade point average (GPA) of students at your school is 2.613. If you were to obtain a random sample of 20 students, how would you expect these points to appear on a real number line plot with respect to 2.613? Would you be surprised if all 20 were either above or below 2.613? Of course, all 20 on one side or the other of 2.613 is a possibility if 2.613 is the true median, but this is not a very likely outcome (in fact, it is like tossing 20 coins in the air and having them show all heads when they land, or all tails). What you should expect is 10 on either side of 2.613 or some kind of split not too far away from this. This model with its independent observations or trials naturally leads to a binomial random variable with parameter $p = .5$. That is, if the random variable X is the number of observations less than or equal to the median, then the binomial distribution in Table A1 for $n = 20$ and $p = .5$ can be used to find critical values for the test statistic.

PROCEDURE FOR TESTING HYPOTHESES ABOUT THE MEDIAN OF A POPULATION

Data. The data are represented by $X_1, X_2, \ldots, X_n$.

Assumptions. The observations $X_1, X_2, \ldots, X_n$ represent a random sample of size n from some population.

Null Hypothesis. H_0: The population median is M, where M is a number specified in the context of the experiment.

Test Statistic. Let T_1 = the number of observations in the sample that are *less than or equal* to M.

Let T_2 = the number of observations in the sample that are *less than* M. Note that if no sample observations are equal to M, then $T_1 = T_2$.

Decision Rules. The decision rule depends on the alternative hypothesis. Let α be the approximate level of significance desired.

1. H_1: The population median is greater than M.

   Reject H_0 if $T_1 \leq T_L$.
   Accept H_0 if $T_1 > T_L$.

The value of T_L is found by entering Table A1 with n and $p = .5$. T_L is the closest value such that $P(T_1 \leq T_L) \cong \alpha$. For large sample sizes $(n > 20)$ the value of T_L is found as $T_L = (n - z_\alpha\sqrt{n})/2$, where z_α is the $1 - \alpha$ quantile from Table A2.

2. H_1: The population median is less than M.

   Reject H_0 if $T_2 > T_U$.
   Accept H_0 if $T_2 \leq T_U$.

The value of T_U is found by entering Table A1 with n and $p = .5$. T_U is the closest value such that $P(T_2 > T_U) \cong \alpha$, or, equivalently, $P(T_2 \leq T_U) \cong 1 - \alpha$. For large sample sizes $(n > 20)$ the value of T_U is found as $T_U = (n + z_\alpha\sqrt{n})/2$, where z_α is the $1 - \alpha$ quantile from Table A2.

3. H_1: The population median is not equal to M.

   Reject H_0 if $T_1 \leq T_L$ or if $T_2 > T_U$.
   Otherwise accept H_0.

The values of T_U and T_L are found by entering Table A1 with n and $p = .5$. T_L is the closest value such that $P(T_1 \leq T_L) \cong \alpha/2$ and T_U is the closest value such that $P(T_2 > T_U) \cong \alpha/2$, or, equivalently, $P(T_2 \leq T_U) \cong 1 - \alpha/2$.

For large sample sizes $(n > 20)$ the values of T_L and T_U are given as $T_L = (n - z_{\alpha/2}\sqrt{n})/2$ and $T_U = (n + z_{\alpha/2}\sqrt{n})/2$, where $z_{\alpha/2}$ is the $1 - \alpha/2$ quantile from Table A2.

Note: Critical values may be obtained more easily from the summary given in Figure 7.11 instead of Table A1.

THE MEDIAN TEST

The test for the population median uses the number of observations below the hypothesized median as the test statistic. The exact distribution of this statistic when H_0 is true is the binomial distribution, which can be obtained from Table A1 for $n \leq 20$. The normal approximation is used for larger sample sizes.

If the population distribution is normal, then the tests of Section 7.2 are more powerful and should be preferred. If the population is not approximately normal, which can be examined by making a Lilliefors test on the sample data, then this median test may have more power than the tests of Section 7.2, especially if the sample tends to have outliers. For symmetric probability distributions the mean equals the median, so a test for the median also tests for the mean, and vice versa. In many cases the population probability distribution is not symmetric, so a test for the median and a test for the mean are actually testing slightly different parameters, although both the mean and the median are measures of central tendency in a population.

FIGURE 7.11 Critical Values for the Median Test for $n \leq 20$

Sample Size n	T_L	T_U	α Level	
			For One-Tailed Tests	For Two-Tailed Tests
4	0	3	.0625	.1250
5	0	4	.0312	.0624
6	0	5	.0156	.0312
	1	4	.1094	.2188
7	0	6	.0078	.0156
	1	5	.0625	.1250
8	0	7	.0039	.0078
	1	6	.0352	.0704
9	0	8	.0020	.0040
	1	7	.0195	.0390
	2	6	.0898	.1796
10	0	9	.0010	.0020
	1	8	.0107	.0214
	2	7	.0547	.1094
11	0	10	.0005	.0010
	1	9	.0059	.0118
	2	8	.0327	.0654
12	0	11	.0002	.0004
	1	10	.0032	.0064
	2	9	.0193	.0386
	3	8	.0730	.1460

Sample Size *n*	T_L	T_U	α Level For One-Tailed Tests	α Level For Two-Tailed Tests
13	0	12	.0001	.0002
	1	11	.0017	.0034
	2	10	.0112	.0224
	3	9	.0461	.0922
14	1	12	.0009	.0018
	2	11	.0065	.0130
	3	10	.0287	.0574
	4	9	.0898	.1796
15	1	13	.0005	.0010
	2	12	.0037	.0074
	3	12	.0176	.0352
	4	10	.0952	.1184
16	1	14	.0003	.0006
	2	13	.0021	.0042
	3	12	.0106	.0212
	4	11	.0384	.0768
	5	10	.1051	.2102
17	1	15	.0001	.0012
	2	14	.0064	.0128
	3	13	.0245	.0490
	4	12	.0717	.1434
18	1	16	.0001	.0002
	2	15	.0007	.0014
	3	14	.0038	.0076
	4	13	.0154	.0308
	5	12	.0481	.0962
19	2	16	.0004	.0008
	3	15	.0022	.0044
	4	14	.0096	.0192
	5	13	.0318	.0636
	6	12	.0835	.1670
20	2	17	.0002	.0004
	3	16	.0013	.0026
	4	15	.0059	.0118
	5	14	.0207	.0414
	6	13	.0577	.1154

EXAMPLE

Consider an example with several parts to illustrate these rules on a sample of size 10 with α as close as possible to .05. All parts are concerned with testing the following null hypothesis:

H_0: The population median is 15.

In calculating the test statistic in the following example real number line plots of the sample values are used without listing the actual sample values, since all that is needed for the calculation of T_1 and T_2 is the position of the points relative to the hypothesized median.

    Part 1. H_1: The population median is greater than 15.

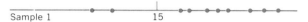

From Table A1 for $n = 10$, $p = .5$, $P(T_1 \leq 2) = .0547$; therefore, $T_L = 2$. Note that $P(T_1 \leq 1) = .0107$, which is not used because it is farther from .05. The exact level of significance is $\alpha = .0547$. As an alternative procedure Figure 7.11 could be used instead of Table A1. The sample size $n = 10$ leads to $T_L = 2$ and $\alpha = .0547$ in Figure 7.11.

    Decision Rule: Reject H_0 if $T_1 \leq 2$. Otherwise accept H_0.
    Test Statistic: $T_1 = 2$, $T_2 = 2$.

H_0 is rejected in favor of the alternative. Notice that almost all of the sample points are greater than 15. The p-value is .0547, since this is the smallest level of significance for which H_0 can be rejected.

    Part 2. H_1: The population median is less than 15.

From Table A1, $P(T_2 \leq 7) = .9453$ or $P(T_2 > 7) = 1 - .9453 = .0547$; therefore, $T_U = 7$ and $\alpha = .0547$. Again direct use of Figure 7.11 gives $T_U = 7$ and $\alpha = .0547$.

    Decision Rule: Reject H_0 if $T_2 > 7$. Otherwise accept H_0.
    Test Statistic: $T_1 = 9$, $T_2 = 9$.

H_0 is rejected in favor of the alternative. Almost all of the sample points are less than 15, thus it would appear that the population median is also less than 15. The p-value is $P(T_2 \geq 9) = .0107$.

    Part 3. H_1: The population median is less than 15.

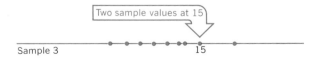

    Decision Rule: Reject H_0 if $T_2 > 7$. (Same as in Part 2.)
    Test Statistic: $T_1 = 9$, $T_2 = 7$.

H_0 is *not* rejected as in Part 2. Note that T_1 and T_2 are not equal this time and that a test statistic with a value of 9 would have rejected H_0 while one with a value of 7 does not. That is why it is important to account for the sample values equal to the hypothesized median. The p-value is $P(T_2 \geq 7) = 1 - P(T_2 \leq 6) = 1 - .8281 = .1719$.

Part 4. H_1: The population median does not equal 15.

Sample 4

From Table A1 $P(T_1 \leq 1) = .0107$ and $P(T_2 > 8) = 1 - .9893 = .0107$. Therefore, $T_L = 1$ and $T_U = 8$, with $\alpha/2 = .0107$ and $\alpha = .0214$. (This is the closest two-tailed test to .05.) Again these numbers may read directly from Figure 7.11 using $n = 10$ and the α level for the two-tailed test.

Decision Rule: Reject H_0 if $T_1 \leq 1$ or $T_2 > 8$; otherwise accept H_0.

Test Statistic: $T_1 = 9$, $T_2 = 9$.

H_0 is rejected. The p-value is $2 \cdot P(T_2 \geq 9) = .0214$. The normal approximation is now demonstrated on this last part of the example. It gives $T_L = (10 - 1.9600 \sqrt{10})/2 = 1.9$ and $T_U = (10 + 1.9600 \sqrt{10})/2 = 8.1$. Therefore, the decision rule is to reject H_0 if $T_1 \leq 1.9$ or $T_2 > 8.1$, which is equivalent to the previous test.

Exercises

7.24 Shoe sizes for the 20 most recent sales made to women were given in Problem Setting 4.2 and are repeated below. Use a level of significance of $\alpha = .05$ to test the following hypothesis.

H_0: the population median shoe size sold to women is 7½, versus
H_1: the population median shoe size sold to women is less than 7½.

9½	7	5½	7½	9
5½	5	7	5½	8
7	6½	6½	6	6½
6½	9	7½	6½	6

Also find the p-value and interpret it.

7.25 The exam scores of 40 students in a statistics class were given in Problem Setting 4.1 and are summarized below. Use a level of significance of $\alpha = .01$ to test the following hypothesis.

H_0: the population median score for this exam is 70, versus
H_1: the population median score for this exam is greater than 70.

77	68	86	84	95	98	87	71
84	92	96	83	62	83	81	85
91	74	61	52	83	73	85	78
50	81	37	60	85	100	79	81
75	92	80	75	78	71	64	65

7.26 The m.p.g. recorded on 20 models of a 4-cylinder automobile in city driving are given below. Use a level of significance of $\alpha = .05$ to test the following hypothesis. State the p-value and interpret it.

H_0: the population median m.p.g. for this model is 24, versus
H_1: the population median m.p.g. for this model is greater than 24.

24.21	24.14	24.66	23.08	25.16
24.35	24.60	24.72	23.88	24.62
23.82	23.75	24.47	23.09	24.62
24.21	25.01	24.38	24.57	25.14

7.27 The 1980 net earnings of common stocks for 10 randomly selected corporations are reported below. Use a significance level of $\alpha = .01$ to test the following hypothesis. State the p-value and interpret it.

H_0: the population median net earnings on common stocks for corporations is $2.00, versus
H_1: the population median net earnings on common stocks for corporations is greater than $2.00.

$1.71	2.17	2.25	2.43	2.32
3.15	3.30	5.52	3.32	3.76

7.28 The time in minutes between incoming phone calls in a business office has been recorded for 10 calls.

1.8   0.3   4.5   9.8   3.2   15.7   4.8   1.0   2.7   6.2

Use these data to test the following hypothesis with $\alpha = .10$.

H_0: the population median time between calls is 7 minutes, versus
H_1: the population median time between calls is not 7 minutes.

Find the p-value and interpret it.

7.29 The number of overtime hours worked in one week by 13 randomly selected employees in a large department store during the Christmas season is as follows:

19.5	16.6	16.7	17.8	20.2	23.3	21.2
18.6	23.3	22.5	19.8	20.5	29.3	

Use these data to test the following hypothesis with $\alpha = .10$.

H_0: the population median number of overtime hours for workers in this store is 20, versus
H_1: the population median number of overtime hours for workers in this store is greater than 20.

Find the p-value and interpret it.

7.30 Data are given below that represent the survival time in days for 13 patients with cancer of the bronchus. These patients were diagnosed as terminally ill and were then given a treatment of 10 grams of vitamin C daily. Use a level of significance of $\alpha = .01$ to test the following hypothesis. State the p-value and interpret it.

H_0: the population median survival time is 1 month (30 days), versus
H_1: the population median survival time is greater than 1 month.

39	427	17	460	90	187	58	52
100	200	42	167	33			

7.5
Review Exercises

7.31 Refer to the sample data given in Exercise 6.29. Test the hypothesis that the median age for these individuals is 22 versus the alternative that the median age is less than 22. Let $\alpha = .05$ and state and interpret the p-value.

7.32 Some samples of pure iron are obtained and their melting point is determined in a laboratory with the results given below. Test these data for normality.

1486, 1502, 1478, 1507, 1489, 1504, 1489, 1503
1503, 1484, 1509, 1507, 1507, 1513, 1514, 1488

7.33 For the sample data in Exercise 7.32 test H_0: $\mu = 1492$, versus H_1: $\mu > 1492$. Let $\alpha = .05$ and state and interpret the p-value.

7.34 A standard insulin (known potency) is in ected into 30 laboratory test animals and the sample mean percentage decrease in blood sugar four hours after injection is found to be 35.3 percent. The standard deviation of the dosage is known to be $\sigma = 15.7$ percent. Use these sample results to test H_0: $\mu = 30\%$, versus H_1: $\mu > 30\%$ with $\alpha = .01$.

7.35 If the true mean percentage decrease in blood sugar in Exercise 7.34 is 35 percent, find the power of the test used in that exercise.

7.36 State officials have expressed the opinion that 15 percent of the cars on the road would not pass a safety inspection. A random sample of 20 cars shows 6 to be unsafe. Use these results to test H_0: $p = .15$, versus H_1: $p \neq .15$. Let $\alpha = .05$ and state and interpret the p-value.

7.37 The results of a voter preference poll involving 200 randomly selected voters shows 53 percent favor the Republican candidate for U.S. Senator. Use these sample results to test H_0: $p = .5$, versus H_1: $p > .5$. Let $\alpha = .05$ and state the p-value.

7.38 A machine shop wants to institute a control chart for the length of bolts it is producing. The job order specifies that the mean length of the bolts is to be 2.7 cm. Past experience indicates that the standard deviation of the bolts manufactured by this shop is 0.2 cm. The quality control engineer suggests that a sample of size 15 be examined every hour to make sure the manufacturing process stays in control. Construct a 95 percent quality control chart for this manufacturing process.

7.39 For the situation described in Exercise 7.38 sample means are obtained for eight successive hours with the following results: 2.72, 2.65, 2.68, 2.71, 2.73, 2.78, 2.90, and 2.95. Plot these points on the quality control chart of Exercise 7.38 to determine if the process is out of control.

Bibliography

Additional material on the topics presented in this chapter can be found in the following publications.

Conover, W. J. (1980). *Practical Nonparametric Statistics, 2nd ed.*, John Wiley and Sons, Inc., New York.

Iman, R. L. (1982). "Graphs for Use with the Lilliefors Test for Normal and Exponential Distributions," *The American Statistician, 36*(2), 109–112.

8

Two Related Samples (Matched Pairs)

PRELIMINARY REMARKS

At this point in the text the reader should have developed an understanding of procedures for gathering, displaying, summarizing, and analyzing sample data. It has been pointed out that some procedures are appropriate for normal populations while others are better in nonnormal situations. Additionally, estimation and hypothesis testing provide a structured framework as an aid in the decision-making process.

With this background the reader should now be in a position to branch out and apply the techniques learned to a wide variety of everyday problem settings. The problem settings encountered in the rest of this text should account for many of the data-based decision-making situations that are likely to be encountered in actual applications. It is not the authors' intent in the upcoming chapters to make the reader a statistical analyst; but, rather, to provide a clean, concise, and readable explanation of each of the procedures, and to do this in a manner that will allow this text to serve as a ready reference that can be read at some future date to aid in understanding some particular data analysis problem related to decision making. Therefore, the format followed in this text involves first presenting the need for a procedure to handle a particular situation, then explaining how the procedure is to be performed and analyzed along with examples. More importantly, the format includes a clear statement of the assumptions necessary for a correct analysis and interpretation of sample data. Techniques are provided for checking these assumptions and alternative procedures are provided that are appropriate when assumptions are not satisfied (which occurs frequently).

8.1

The Paired *t*-Test

PROBLEM SETTING 8.1

New teaching methods and teaching aids are in a continual state of development in virtually all subject areas. For example, phonetic spelling is frequently recommended by some individuals as an aid to learning how to spell better. In another area, the early 1960s found new math in vogue for teaching mathematical concepts. The development of techniques and aids that enable the subject matter to be effectively communicated or the skill acquired is certainly an admirable goal and should be encouraged. However, a carefully planned evaluation is needed to establish the validity of the procedure under consideration. For example, different techniques and aids have been used over the years to increase such a basic skill as reading. If a new technique designed to increase reading speed is made available for use by a school system, then those individuals in charge of the program should be in a position to determine accurately the effectiveness of the technique on reading skills for which a quantitative measurement of improvement exists.

CONTROLLING UNWANTED VARIATION

The evaluation of increased or decreased reading skills is only one of a multitude of problems that can be solved using the techniques presented in this chapter. However, before these techniques are presented some background is needed on a very fundamental concept in statistics. That concept is **the reduction or controlling of unwanted or extraneous variation** in the gathering of the sample data.

Consider a study designed to determine the effectiveness of a medication used to control high blood pressure. An experiment could be set up by randomly selecting two groups of individuals. The first group is given a placebo (an inert or innocuous medication) and the other group is given the medication of interest. After some period of time the average blood pressures for each group are recorded to aid in making a decision regarding the medication. Do you see any flaws in this experimental setup? One obvious flaw is that the variation due to medication is obscured by variability from person to person. That is, suppose that the group selected for the placebo has initially a higher average blood pressure than the treatment group, or that many of the individuals in the groups don't even suffer from high blood pressure. The original objective was to determine the effectiveness of the medication in reducing an individual's blood pressure. Some of the variation in the data could have been reduced by designing the experiment differently.

This is best done by taking measurements only on individuals with high blood pressure and then following up by taking a second set of measurements on these same individuals after they have received the medication for a period of time. By approaching the problem in this manner some unwanted variation in the observed measurements is controlled, since the difference observed in individual readings will be due primarily to the medication if it is effective. If it is not effective the only variation observed will be random variation that would normally be observed on any one individual's blood pressure when taken at different time periods. This method of pairing observations is usually very effective in reducing unwanted variability in the data. These paired observations are called **matched pairs**. A procedure for analyzing such matched pairs is called the **paired t-test**, and is the subject of this section.

The use of matched pairs in an experiment is an important and useful method for reducing extraneous variation in the outcomes. Some examples where matched pairs are useful follow:

1. Twin calves are used to control variation due to genetic differences in experiments to determine the effectiveness of feed additives, the usefulness of preventive medication, or the effect of environmental factors on meat texture. In each case one twin receives one treatment while the other receives a different treatment.

2. The same individual may be observed twice—once before a treatment and once after the treatment—to measure the effectiveness of the treatment. The pre- and post-treatment measurements are the two observations that constitute the paired observations. The "treatment" may be a training session, where the observations are measurements of teaching effectiveness before and after the training session. Or the "treatment" may be an experimental medication for treatment of allergy problems, while the observations are levels of severity of the allergy symptoms before and after the medication is applied. The "treatment" may be the installation of a new computer system and the measurements may be user error rates before the installation and after a suitable training period following the installation.

THE PAIRED t-TEST

The paired t-test is used to analyze pairs of observations (X_i, Y_i), $i = 1, \ldots, n$, to see if the mean of the X's equals the mean of the Y's. Confidence intervals for the differences in means may also be found. These procedures resemble the procedures in the two previous chapters, which were intended for a single sample of observations. The matched pairs are reduced to a single sample by taking differences between the two observations in the matched pair, such as by subtracting the blood pressure reading after the medication from the reading for that person before the medication is administered. These **differences** in readings, denoted by $D_i = X_i - Y_i$, form a single sample of observations, as in the previous two chapters, so the methods of those chapters are applied to the differences. For convenience those methods are given in this section again, and restated to apply to the matched pairs problem.

Care should be taken to keep the notation consistent in each application, especially with the one-tailed tests. That is, initially either measurement may be designated X or Y, but once a variable is identified as X, it needs to be identified as X consistently throughout the entire analysis, and the mean for that variable needs to be called μ_X. Otherwise, the rejection region used may inadvertently consist of values of the test statistic that tend to support H_0 rather than discredit H_0. Figure 8.1 may be helpful in preventing errors of this type from occurring.

PROCEDURE FOR TESTING HYPOTHESES ABOUT DIFFERENCES IN RELATED SAMPLES (PAIRED t-TEST)

Data. The data $D_1, D_2, \ldots, D_n$ (where $D_i = X_i - Y_i$) are computed from pairs of measurements (X_i, Y_i) on each of the n elements in the sample. Note that the two related sample values X_i and Y_i have all been reduced to a single sample of D_i's. Let

$$\overline{D} = \frac{\Sigma D_i}{n} \quad \text{and} \quad s_D = \sqrt{\frac{1}{n-1}\left[\Sigma D_i^2 - \frac{(\Sigma D_i)^2}{n}\right]}.$$

Assumptions

1. The random variables $D_1, D_2, \ldots, D_n$ are independent of one another and are identically distributed with mean $\mu_x - \mu_y$.
2. If $n < 30$, the D_i's should be normally distributed. If $n \geq 30$, the normality assumption is not necessary because the Central Limit Theorem applies to $\overline{D}$.

Estimation. A confidence interval for $\mu_X - \mu_Y$ with a level of confidence of $100(1-\alpha)$ percent is given as $\overline{D} \pm t_{\alpha/2,n-1} \, s_D/\sqrt{n}$, where $t_{\alpha/2,n-1}$ is the $1 - \alpha/2$ quantile of the Student's t-distribution with $n-1$ degrees of freedom as given in Table A3.

Null Hypothesis. H_0: $\mu_X = \mu_Y$ (The mean of the X measurement is the same as the mean of the Y measurement)

Test Statistic

$$T = \frac{\overline{D}}{s_D/\sqrt{n}}$$

Decision Rule. The decision rule is based on the alternative hypothesis of interest.

1. H_1: $\mu_X > \mu_Y$ (the X measurement tends to be higher than the Y measurement). Reject H_0 if $T < t_{\alpha,n-1}$; otherwise accept H_0. See Figure 8.1a.
2. H_1: $\mu_X < \mu_Y$ (the X measurement tends to be lower than the Y measurement). Reject H_0 if $T < -t_{\alpha,n-1}$; otherwise accept H_0. See Figure 8.1b.
3. H_1: $\mu_X \neq \mu_Y$ (the X measurement tends to be either larger or smaller than the Y measurement). Reject H_0 if $T > t_{\alpha/2,n-1}$ or if $T < t_{\alpha/2,n-1}$; otherwise accept H_0. See Figure 8.1c.

EXAMPLE

Speed reading courses have been a popular fad over the last several years. Suppose a school decides to do some independent evaluation before enrolling a large number of their students to take a speed reading course. Ten students are randomly selected to take the course. The reading score of each student, combining measurements of speed and comprehension, is recorded before and after the course and the differences in reading scores are calculated for each student. A pair of measurements (X_i, Y_i) is available on each student. Clearly it doesn't matter if the reading score before the course is denoted by X or Y as long as the notation is consistent throughout. Therefore, let X represent the score after completing the course and let Y represent the reading score for the same person before taking the course. The data for the two related samples are given in Figure 8.2.

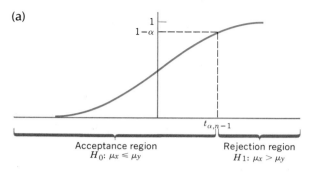

(a)

Acceptance region
$H_0: \mu_x \leq \mu_y$

Rejection region
$H_1: \mu_x > \mu_y$

$t_{\alpha, n-1}$

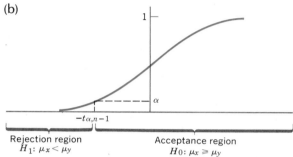

(b)

Rejection region
$H_1: \mu_x < \mu_y$

Acceptance region
$H_0: \mu_x \geq \mu_y$

$-t_{\alpha, n-1}$

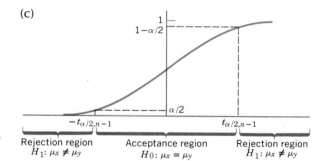

(c)

Rejection region
$H_1: \mu_x \neq \mu_y$

Acceptance region
$H_0: \mu_x = \mu_y$

Rejection region
$H_1: \mu_x \neq \mu_y$

$-t_{\alpha/2, n-1}$ $t_{\alpha/2, n-1}$

FIGURE 8.1 The Cumulative Distribution Function for Student's t-Distribution Showing the Rejection Regions

Person	X_i Reading Score After Course	Y_i Reading Score Before Course	$D_i = X_i - Y_i$ Difference
1	221	211	10
2	231	216	15
3	203	191	12
4	216	224	−8
5	207	201	6
6	203	178	25
7	201	188	13
8	179	159	20
9	179	177	2
10	211	197	14

FIGURE 8.2 Paired Data Before and After a Speed Reading Course

The hypothesis to be tested is

$H_0: \mu_X = \mu_Y$ (there is no change in the mean reading score), versus
$H_1: \mu_X > \mu_Y$ (the mean reading score is higher after the course).

For a choice of $\alpha = .01$, Table A3 is entered with $10 - 1 = 9$ degrees of freedom to complete the following decision rule:

Reject H_0 if $T > t_{.01,9} = 2.8214$.

Sample calculations give

$$\bar{D} = 10.9 \quad \text{and} \quad s_D = \sqrt{\tfrac{1}{9}[1963 - (109)^2/10]} = 9.28 \text{ since } \Sigma D_i = 109 \text{ and}$$

$\Sigma D_i^2 = 1963$. The test statistic T is found as

$$T = \frac{10.9}{9.28/\sqrt{10}} = 3.715.$$

H_0 is rejected and it is concluded that the mean reading score has increased. The *p*-value is approximately .003. Because the *p*-value is so small, there is little chance of making a Type I error in this case.

Since it is easy to make an error in the calculations, it is advisable to plot the paired data to see if the decision is reasonable. A scatterplot is useful for analyzing this type of data; one such plot appears in Figure 8.3. The 45 degree line where $X = Y$ has been added to the plot to aid in examining the results. If the speed reading course is not effective, then the pairs of points should be randomly scattered about the 45 degree line. If, however, the course is effective, then most of the points should be below the 45 degree line in the region labeled $X > Y$. Nine of the ten pairs are in this region; hence the decision seems justified.

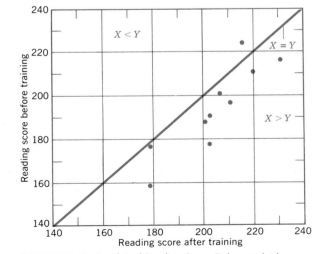

FIGURE 8.3 Scatterplot of Reading Score Before and After Training

"WE NEED A QUARTERBACK WHO ISN'T AFRAID OF BEING INUNDATED BY NUMBERS."

A 95 percent confidence interval for the mean change in reading score is easily found as follows:

$$\overline{D} \pm t_{\alpha/2, n-1} \, s_D/\sqrt{n} = 10.9 \pm 2.2622(9.28)/\sqrt{10}$$

$$= 10.9 \pm 6.6$$

$$= 4.3 \text{ to } 17.5.$$

The mean change in reading score is believed to be between 4.3 and 17.5 with 95 percent confidence.

CHECKING THE ASSUMPTIONS

In the previous example two important assumptions for the paired t-test have been neglected. Do you know what they are? The assumptions of independence and normality have been ignored. The assumption of independence was satisfied by randomly selecting 10 individuals for the comparison; however, the normality assumption must be checked by plotting the standardized sample values $D_i^* = (D_i - \overline{D})/s_D$ as an empirical distribution function on the Lilliefors graph. The standardized values are as follows.

D_i:	10	15	12	-8	6	25	13	20	2	14
D_i^*:	$-.10$	.44	.12	-2.04	$-.53$	1.52	.23	.98	$-.96$	.33

The Lilliefors graph is given in Figure 8.4 and the plotted e.d.f. of the 10 observations is well within the bounds for a sample of size 10. Therefore, it is reasonable to conclude that the results and interpretation of the above test are correct.

It is impossible to overemphasize the importance of checking the D_i's for normality when the sample size is small, and that is why the Lilliefors bounds have been provided as a very easy way to do this checking. The reason this check is necessary is that the test statistic T cannot be correctly compared with a Student's t random variable unless the normality assumption is satisfied. Hence, the results of the test are correct for small sample sizes only if the normality assumption is satisfied.

Of course, the Lilliefors test does not prove that the population is normally distributed, but it indicates whether the data could reasonably be assumed to have come from a normal distribution. The importance of checking for normality is partially offset by the low power of any test for normality when the sample sizes are small. For this reason some statisticians prefer to omit the Lilliefors test and use the procedures in the next section whenever sample sizes are small. This is a perfectly acceptable alternative.

Many well-intentioned individuals have been led to erroneous conclusions by ignoring test assumptions. Use of a test when its assumptions are not satisfied may result in a procedure with very little power to detect the differences present in the population, or a test with an actual Type I error rate much larger

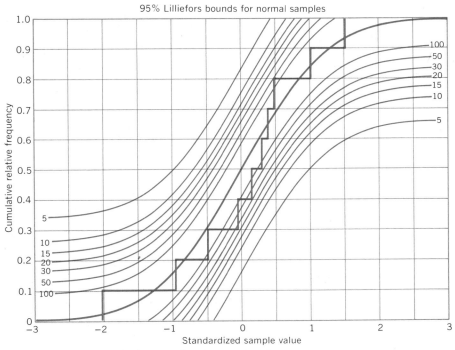

FIGURE 8.4 Normality Check on Reading Scores Example

or smaller than the person doing the testing desired. The direct result of using the wrong test may be quite costly, as an incorrect decision could be made regarding the merits of a new product, or the results of a marketing survey could be interpreted incorrectly. The next section considers related samples when the normality assumption is not satisfied (which frequently seems to be the case when dealing with real world data).

Exercises

8.1 A teacher selects 10 typing students to participate in a week of training designed to improve their typing speed. The teacher times their speed before and after the course to see if the course is effective. State the appropriate null and alternative hypotheses and test the null hypothesis with $\alpha = .025$. (The Lilliefors test should be used to check the normality assumption because of the small sample size.) State the p-value. Find a 95 percent confidence interval for the mean increase in typing speed.

Speed After Course	Speed Before Course
55	50
46	42
78	70
61	63
52	58
45	35
47	46
57	52
71	60
58	49

8.2 Twelve students in a statistics course recorded the scores listed below on their first and fourth exams in the course.

Score on Test #1	Score on Test #4
64	80
28	87
90	90
30	57
97	89
20	51
100	81
67	82
54	89
44	78
100	100
71	81

Test the hypothesis that there is no difference in the scores for the two exams versus the alternative that the scores were lower on the first test. Let $\alpha = .05$. (The Lilliefors test should be used to check the normality assumption because of the small sample size.) State the *p*-value and interpret it.

8.3 Twenty patients on a diet to lose weight had their weight recorded before starting the diet and after one month's time on the diet. The weight change (before minus after) was recorded for each patient as follows.

7	−6	3	1	6	4	9	−5	9	7
−3	7	−9	8	6	−4	4	9	−6	1

State the appropriate null and alternative hypothesis to test the effectiveness of this diet. Use a significance level of $\alpha = .05$. (The Lilliefors test should be used to check the normality assumption because of the small sample size.)

8.4 The management of a factory with large assembly lines wants to try out a new assembly technique for one of its lines. Fifteen employees are randomly selected and the number of units each assembled in one week is noted. These 15 employees are then trained on the new assembly technique and the number of units assembled by each is recorded for one week using the new technique. The data are given below. State the appropriate null and alternative hypotheses to test to see if the new technique is effective in increasing worker productivity. Use a level of significance of $\alpha = .01$. (The Lilliefors test should be used to check the normality assumption because of the small sample size.) State the *p*-value. Find a 90 percent confidence interval for the mean increase in productivity.

Worker	Number of Units Assembled Using Present Method	Number of Units Assembled Using New Technique
1	34	33
2	28	36
3	29	50
4	45	41
5	26	37
6	27	41
7	24	39
8	15	21
9	15	20
10	27	37
11	23	21
12	31	18
13	20	29
14	35	38
15	20	27

8.5   Prices for 15 randomly selected food items have been taken from advertisements placed in the local newspaper by two competing supermarkets and are listed below. Use a level of significance of $\alpha = .05$ for a test of a hypothesis to see if there is any difference in the prices offered by the two supermarkets. (The Lilliefors test should be used to check the normality assumption because of the small sample size.) State the p-value.

Food Item	Prices Offered by Supermarket #1	Prices Offered by Supermarket #2
Chile	$0.98	$0.89
Saltines	.69	.79
Cake Mix	.69	.87
Eggs	.83	.92
Catsup	.79	.82
Bologna	1.39	1.65
Orange Drink	1.39	1.59
Macaroni	.30	.33
Avocados	.25	.39
Orange Juice	1.45	1.39
Noodles	.59	.63
Tomato Sauce	.17	.14
Margarine	.79	.89
Grapefruit	.17	.13
Crisco	2.39	1.99

8.6   An insurance adjuster has received estimates from two different repair garages for minor repairs to 20 automobiles. State the appropriate null and alternative hypotheses to see if there is any difference in the estimates of the two garages. Let $\alpha = .05$. (The Lilliefors test should be used to check the normality assumption because of the small sample size.) State the p-value.

Claim Number	Estimate by Garage #1	Estimate by Garage #2
1	$48	$46
2	56	49
3	87	71
4	88	56
5	86	62
6	64	54
7	80	52
8	78	88
9	72	82
10	70	80
11	80	64
12	58	78
13	72	42
14	60	56
15	64	72

Claim Number	Estimate by Garage #1	Estimate by Garage #2
16	46	64
17	56	54
18	59	50
19	73	61
20	78	70

8.7 Twenty employees are asked to rate each of their fellow employees on a scale from 0 to 100 with at least one employee getting a 0 and at least one employee getting a 100. The average of these ratings is then compared with the supervisor's rating of the employees. Use a level of significance of $\alpha = .05$ to test the appropriate hypothesis to see if the two sets of ratings differ. (The Lilliefors test should be used to check the normality assumption because of the small sample size.) State the *p*-value.

Employee Number	Average Peer Rating	Supervisor Rating
1	1	7
2	3	0
3	10	3
4	7	11
5	7	14
6	9	14
7	9	24
8	19	7
9	27	26
10	25	54
11	36	47
12	41	29
13	46	12
14	44	72
15	57	42
16	48	96
17	72	90
18	80	84
19	83	100
20	100	75

8.8 Two individual stores of a large chain are thought to have equally attractive locations in a large city. The daily receipts for 60 business days show store number 1 to have an average daily difference of $58 higher than store number 2 with a standard deviation of $200. Use these summary statistics with a level of significance of $\alpha = .05$ to see if there is any difference in the daily receipts of the two stores. State the *p*-value.

8.9 One hundred automobiles have their m.p.g. checked with a tank of unleaded gasoline and with a tank of gasohol. A toss of a coin is used to determine whether the automobile receives the unleaded gasoline or the gasohol first. The drive is not told which type of fuel is in the tank, but does receive instructions to keep the driving conditions as nearly constant as possible on the two tanks. Why is this procedure used? At the end of the test the gasohol fuel shows an average m.p.g. of 0.43 higher than the unleaded fuel with a standard deviation of 1.8 m.p.g. for the differences.

Use a level of significance of $\alpha = .01$ to test the hypothesis that there is no difference in the m.p.g. of the two fuels. State the p-value.

8.2
The Wilcoxon Signed Ranks Test

CHOOSING THE BEST TEST

The assumption of normality is necessary for the paired t-test when the sample sizes are small. If the normality assumption is not satisfied, then the procedure presented in this section should be used to test the same hypothesis $\mu_X = \mu_Y$ for matched pairs. The paired t-test is an acceptable method to use for all populations when the sample sizes are large. However, if the populations are not normal, the procedures in this section often have more power than the paired t-test even though the sample sizes are large. A careful analysis of the data should include a Lilliefors test for normality in order to select the most appropriate test.

Tests that require the assumption of normality, or any other specific distribution, are called **parametric** tests, while tests that are valid for all populations are called **nonparametric** tests. The t-test is a parametric test, and the test presented in this section is a nonparametric test.

A REVISION OF THE READING SPEED EXAMPLE

Consider the reading speed example of the previous section where a new sample produces the following differences in scores.

$$D_i: \quad 10 \quad 45 \quad 9 \quad -5 \quad 4 \quad 49 \quad 8 \quad 52 \quad -2 \quad 6$$

Sample calculations give $\overline{D} = 17.6$ and $s_D = 22.01$. The corresponding standardized values $D_i^* = (D_i - 17.6)/22.01$ are given as follows.

$$D_i^*: \quad -.35 \quad 1.25 \quad -.39 \quad -1.03 \quad -.62 \quad 1.43 \quad -.44 \quad 1.56 \quad -.89 \quad -.53$$

The test for normality consists of plotting the e.d.f. of these standardized values on the Lilliefors graph. Such a plot is given in Figure 8.5. This graph shows that the e.d.f. falls outside of the upper curve labeled with a 10; thus the assumption of normality is rejected for these sample data. Since a basic assumption for the paired t-test has not been satisfied, the paired t-test should not be used on these data. That is, suppose the test statistic for the paired t-test is computed as was done in the example in the previous section, $T = 17.6/(22.01/\sqrt{10}) = 2.53$. If this value were compared against the tabled t-value of that example, namely, $t_{.01,9} = 2.8214$, the null hypothesis would be accepted. However, it is not correct to compare T against 2.8214 as T no longer has a Student's t-distribution. Hence, the test is without meaning and the decision to accept H_0 may be correct or incorrect. One never knows for sure.

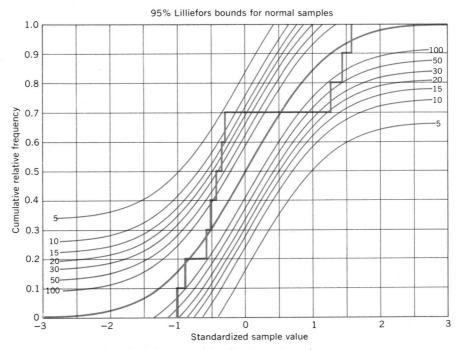

FIGURE 8.5 Normality Check for Revised Reading Score Example

THE WILCOXON SIGNED RANKS TEST

A procedure for testing H_0: $\mu_X = \mu_Y$ with matched pairs in the data does not necessarily need to be based on the assumption of a normal distribution. Many statistical procedures for such a situation are valid for all kinds of populations. They will not be as powerful as the paired t-test when the population is normal, but they are often more powerful than the paired t-test when the populations are not normal. Recall that tests should be selected on the basis of their power, when all other factors are equal. The Wilcoxon signed ranks test, presented in this section, is well known to have good power for a wide variety of distributions. Although the test does not require normality of the D_i's, it is assumed that the distribution of D_i's is symmetric. A symmetric distribution is one where the right half of the graph of the density or probability function is the mirror image of the left half. The form of the Wilcoxon signed ranks test statistic presented here is simpler than some of the more standard forms for presentation, but it is equivalent to the other forms that appear elsewhere.

Note that the critical values for the Wilcoxon signed ranks tests are the same as for the paired t-test even though the test statistic is computed on ranks. The reason for this is that the t-distribution provides an excellent approximation to the exact distribution of the Wilcoxon signed ranks test statistic T_R even for small values of n. In addition, this makes the Wilcoxon signed ranks test very easy to use, as the only difference in computing the Wilcoxon signed ranks test statistic and the paired t-test statistic is the additional step of ranking the $|D_i|$. Therefore, the authors refer to the Wilcoxon signed ranks test as a **rank transformation test** since it is the result of computing the paired t-test on rank transformed data. The Wilcoxon signed ranks test will now be demonstrated by completing the above example where the data are nonnormal.

PROCEDURE FOR TESTING HYPOTHESES ABOUT DIFFERENCES IN RELATED SAMPLES (WILCOXON SIGNED RANKS TEST)

Data. The data $D_1, D_2, \ldots, D_n$ (where $D_i = X_i - Y_i$) are computed from pairs of measurements (X_i, Y_i) on each of the n elements in the sample.

Assumptions. The random variables $D_1, D_2, \ldots, D_n$ are independent and identically distributed, and their distribution is symmetric.

Null Hypothesis. $H_0: \mu_X = \mu_Y$ (the mean of the X measurement is the same as the mean of the Y measurement).

Test Statistic. Compute the absolute value of the D_i. Assign ranks 1 to n to the $|D_i|$. Use average ranks in case of ties. Denote the rank of $|D_i|$ by R_i. If the original value of D_i was negative, then give a negative sign to the corresponding rank R_i. Otherwise R_i will remain positive. Let

$$\bar{R} = \sum \frac{R_i}{n} \quad \text{and} \quad s_R = \sqrt{\frac{1}{n-1}\left[\sum R_i^2 - \frac{(\sum R_i)^2}{n}\right]}.$$

The test statistic T_R is given as $T_R = \bar{R}/(s_R/\sqrt{n})$.

Decision Rule. The decision rule is based on the alternative of interest.

1. $H_1: \mu_X > \mu_Y$ (the X measurement tends to be higher than the Y measurement). Reject H_0 if $T_R > t_{\alpha, n-1}$; otherwise accept H_0.
2. $H_1: \mu_X < \mu_Y$ (the X measurement tends to be lower than the Y measurement). Reject H_0 if $T_R < -t_{\alpha, n-1}$; otherwise accept H_0.
3. $H_1: \mu_X = \mu_Y$ (the X measurement tends to be either higher or lower than the Y measurement).
   Reject H_0 if $T_R > t_{\alpha/2, n-1}$ or if $T_R < -t_{\alpha/2, n-1}$; otherwise accept H_0.

The value of $t_{\alpha, n-1}$ is the $1 - \alpha$ quantile found in Table A3 for a Student's t random variable with $n - 1$ degrees of freedom. This provides a good approximation to the exact distribution of T_R for all sample sizes.

EXAMPLE

Recall that the revised reading scores in this section represent differences for 10 randomly selected individuals who were measured before (Y) and after (X) taking a speed reading course. These differences were determined by the Lilliefors test to be not normally distributed. Therefore, the Wilcoxon signed ranks test is used to test H_0 with $\alpha = .01$.

$H_0: \mu_X = \mu_Y$ (the mean reading scores are not changed by the course),
versus
$H_1: \mu_X > \mu_Y$ (the mean reading scores are higher after the course).

Person	Score After Course X_i	Score Before Course Y_i	Difference in Scores $D_i = X_i - Y_i$	Absolute Value of Difference $|D_i|$	Rank of $|D_i|$	Signed Rank R_i
1	261	251	10	10	7	7
2	292	247	45	45	8	8
3	317	308	9	9	6	6
4	253	258	−5	5	3	−3
5	271	267	4	4	2	2
6	305	256	49	49	9	9
7	238	230	8	8	5	5
8	320	268	52	52	10	10
9	267	269	−2	2	1	−1
10	281	275	6	6	4	4
						$\Sigma R = 47$

$$\bar{R} = \Sigma R_i/10 = 47/10 = 4.7 \qquad \Sigma R^2 = 385$$

$$s_R = \sqrt{\frac{1}{9}[385 - (47)^2/10]} = 4.27$$

Test Statistic:

$$T_R = 4.7/(4.27/\sqrt{10}) = 3.48$$

Decision Rule

Reject H_0 if $T_R \geq t_{.01,9} = 2.8214$; otherwise accept H_0. H_0 is easily rejected. The p-value for $T_R = 3.48$ is estimated by interpolation in Table A3 to be

$$P(T_R > 3.48) = .004$$

from the Student's t-distribution with 9 degrees of freedom.

Recall that earlier in this section the paired t-test was applied to these same nonnormal sample data and the value of the resulting test statistic was $T = 2.53$. When T was compared against the critical value of 2.8214, the null hypothesis was accepted. This example serves to point out that no meaning should be attached to a paired t-test when the normality assumption is not satisfied.

CHECKING THE REASONABLENESS OF THE DECISION

As in the last section it is worthwhile to make a scatterplot of the scores before and after the speed reading course to see if the conclusion of the previous example appears reasonable. Such a plot is given in Figure 8.6. An examination of Figure 8.6 shows that 8 of the 10 sample pairs are plotted in the region where scores after the course are higher than those before the course (i.e., $X > Y$). With the remaining two scores showing almost no change.

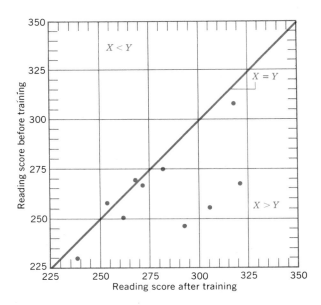

FIGURE 8.6 Scatterplot of Reading Scores Before and After Training

WHICH MATCHED PAIR TEST TO USE

A natural question to consider at this point is "Why not always use the Wilcoxon signed ranks test?" This is not a bad policy to follow. However, the choice of which test to use should be based on the following considerations.

1. If the normality assumption is satisfied, then the paired t-test has slightly more power than the Wilcoxon signed ranks test. On the other hand, if the normality assumption is not satisfied, the paired t-test may have considerably less power than the Wilcoxon signed ranks test. The test that tends to have more power should be used.

2. The paired t-test is more commonly used, appearing in virtually all books for a first course in statistics. Therefore, someone reading your analysis is likely to feel more comfortable with a paired t-test. However, this is not a justification for ignoring the assumptions behind the t-test.

Exercises

8.10 Rework Exercise 8.1 using the Wilcoxon signed ranks test. Compare results of using the Wilcoxon signed ranks test with those obtained from the paired t-test.

8.11 Rework Exercise 8.2 using the Wilcoxon signed ranks test. Compare results of using the Wilcoxon signed ranks test with those obtained from the paired t-test.

8.12 Rework Exercise 8.3 using the Wilcoxon signed ranks test. Compare results of using the Wilcoxon signed ranks test with those obtained from the paired t-test.

8.13 Rework Exercise 8.4 using the Wilcoxon signed ranks test. Compare results of using the Wilcoxon signed ranks test with those obtained from the paired t-test.

8.14 Rework Exercise 8.5 using the Wilcoxon signed ranks test. Compare results of using the Wilcoxon signed ranks test with those obtained from the paired t-test.

8.15 Rework Exercise 8.6 using the Wilcoxon signed ranks test. Compare results of using the Wilcoxon signed ranks test with those obtained from the paired t-test.

8.16 Rework Exercise 8.7 using the Wilcoxon signed ranks test. Compare results of using the Wilcoxon signed ranks test with those obtained from the paired t-test.

8.17 An insurance adjuster wants to compare estimates from two different repair garages for minor repairs on automobiles. Thirteen pairs of estimates are available.

Claim Number	Estimate by Garage #1	Estimate by Garage #2
1	$165	$139
2	156	132
3	165	134
4	135	133
5	134	130
6	131	133
7	130	130
8	126	125
9	120	122
10	120	119
11	118	114
12	115	116
13	108	105

(a) State the appropriate null and alternative hypotheses to see if there is any difference in the mean estimates of the two garages. Let $\alpha = .05$ and test the null hypothesis with the Wilcoxon signed ranks test. State the p-value.

(b) Check the differences in estimates from the two garages for normality using the Lilliefors test.

(c) Based on the results of part (b) the paired t-test should not be applied to these data; however, compute the paired t-test to test the null hypothesis on part (a) and compare with the results of the Wilcoxon signed ranks test.

8.3
Review Exercises

8.18 Twelve patients have their systolic blood pressure checked before and after receiving a new medication to reduce blood pressure. The changes in blood pressure readings (before minus after) were as follows: 11, 7, 2, 9, −7, −5, 3, 4, 13, 8, 5, and −6. Use these data to test H_0: $\mu_B = \mu_A$ versus H_1: $\mu_B > \mu_A$. Let $\alpha = .05$. (The Lilliefors test should be used to check these differences for normality due to the small sample size.)

8.19 Use the Wilcoxon signed ranks test to test the hypothesis in Exercise 8.18.

8.20 A shoe manufacturer is field testing the durability of a leather sole made using a new process, and comparing it with the type of leather sole presently in use. Fifteen pairs of shoes are constructed, where one shoe in each pair has the new type of leather sole. After six months of daily usage by 15 supermarket employees, the shoes were examined and measured with respect to percent of wear still remaining in the sole. The results are as follows.

Employee	New Sole	Present Sole
1	73	64
2	43	41
3	47	43
4	53	41
5	58	47
6	47	32
7	52	24
8	38	43
9	61	53
10	56	52
11	56	57
12	34	44
13	55	57
14	65	40
15	75	68

Check the assumption of normality as required prior to a test for equality of means.

8.21 Use the appropriate test for equality of means in Exercise 8.20 to assist in deciding whether to replace the old process for making leather soles with the new process. Use $\alpha = .05$.

8.22 Make a scatterplot of the data in Exercise 8.20 to see if the decision reached in the solution to Exercise 8.21 appears to be justified.

8.23 In international gymnastics competition a panel of judges rates each gymnast's performance on a scale of 0 to 10, with 10 being the best. Twelve performances were observed where one of the judges was from the contestant's home country.

Contestant	Native Judge	Average of Foreign Judges	Contestant	Native Judge	Average of Foreign Judges
1	6.8	6.7	7	6.6	5.4
2	4.5	4.3	8	5.8	5.9
3	8.0	8.1	9	6.0	6.1
4	7.2	7.2	10	8.8	9.1
5	8.7	8.3	11	8.7	8.7
6	4.5	4.6	12	4.4	4.3

Check the assumption of normality, as required prior to a test of equality of means.

8.24 Use the appropriate test of equality of means in Exercise 8.23 to see if there is any support for the belief that judges from the contestant's home countries tend to be biased in one direction or the other.

8.25 Make a scatterplot of the data in Exercise 8.23 to see if it supports the conclusion reached in Exercise 8.24.

8.26 Find a 95 percent confidence interval for the bias involved in the decision of the judge from the contestant's home country. Refer to Exercise 8.23.

8.27 A company with a fleet of cars contracts out for body repair work. In order to see if there is a tendency for the bid from one body shop to be higher or lower than a bid on the same job from another body shop, eight cars needing repairs were sent to both body shops for bids with the following results.

Car:	1	2	3	4	5	6	7	8
Shop A:	364	112	840	610	172	83	165	216
Shop B:	412	110	960	640	163	75	160	274

Use a test for equality of mean bids that does not require a preliminary test of normality, at $\alpha = .05$.

8.28 Use a scatterplot on the data in Exercise 8.27 to see if they tend to support the conclusion of Exercise 8.27.

Bibliography

Additional material on the topics presented in this chapter can be found in the following publications.

Conover, W. J. (1980). *Practical Nonparametric Statistics, 2nd ed.*, John Wiley and Sons, Inc., New York.

Conover, W. J. and Iman, R. L. (1981). "Rank Transformations as a Bridge between Parametric and Nonparametric Statistics," *The American Statistician,* **35**(3), 124–129.

Iman, R. L. (1974). "Use of a *t*-Statistic as an Approximation of the Exact Distribution of the Wilcoxon Signed Ranks Test Statistic," *Communications in Statistics,* **3**(8), 795–806.

9

Estimation and Hypothesis Testing with Two Independent Samples

PRELIMINARY REMARKS

In the previous chapter the value of pairing observations was discussed. This technique of matching observations should be used whenever possible to control extraneous variation in the experimental results. However, much sample data involves two sets of measurements that do not occur in pairs; in fact, the samples are not related at all. Perhaps they were obtained from separate populations. Such samples are **independent** of one another, whereas the related samples of the previous chapter are called **dependent** samples. Independent samples are usually denoted by $X_1, X_2, \ldots, X_{n_x}$ for the n_x items in one sample and by $Y_1, Y_2, \ldots, Y_{n_y}$ for the n_y items in the second sample. Further, the sample sizes are not necessarily equal. Some examples of settings where independent samples are used include the following.

1. X's: the distance recorded in yards for Brand A golf balls when tested by a mechanical driver;
   Y's: the distance recorded in yards for Brand B golf balls when tested by a mechanical driver.

2. X's: daily sales by a company for one month prior to the start of their use of TV commercials;
   Y's: daily sales by the company for one month after the start of their use of TV commercials.

3. X's: the mileage observed on several Goodyear tires;
   Y's: the mileage observed on several Firestone tires.

4. X's: the number of sickness absentees on Monday and Friday for a period of one year;
   Y's: the number of sickness absentees on Tuesday, Wednesday, and Thursday for a period of one year.

5. X's: achievement scores in arithmetic recorded by students taught by a teacher in an ordinary classroom setting;

Y's: achievement scores in arithmetic recorded by students using a programmed text where the students consulted with the teacher only when they had questions.

For each of these samples the focal point of interest is a comparison of the means of their respective populations to see if they are the same or different. For the above examples interest would center on the following comparisons.

1. Is the mean driving distance the same for both brands of golf balls?
2. Is the mean of sales before the TV commercials the same as the mean of sales after the start of the use of TV commercials?
3. Is the mean number of miles driven the same for both Goodyear and Firestone tires?
4. Is the mean number of sickness absentees observed on Monday and Friday the same as that observed on Tuesday, Wednesday, and Thursday?
5. Is the mean achievement score the same for both methods of instruction?

The investigation of these questions requires a formal statement of the hypothesis of interest as well as the calculation of a test statistic. The first section of this chapter is devoted to methods that can be used with most types of populations if the sample sizes are large (≥ 30). These methods may be used when the sample sizes are small if the populations are normal and variances are known. If the variances are unknown and the sample sizes are small, the methods of the second section may be used, provided both populations are

"OUR TESTS SHOW BRAND 'X' TO BE SUPERIOR
TO THE NUMBER 1 SELLING BRAND."

normal. The method of the third section is appropriate for all populations and all sample sizes, but especially for small samples from nonnormal populations because the methods of the first two sections are not valid in that situation. However, population variances should be approximately equal for this latter method.

9.1
Large Samples: Inferences about the Difference Between Two Means

PROBLEM SETTING 9.1

Employee absenteeism due to sickness is a continuing source of concern to employers due to the number of employees who take advantage of the system and tend to regard sick leave as "vacation to be taken." Employers have developed several methods to handle these problems such as charging the first day of each occurrence of reported sick leave to vacation, requiring official letters from doctors for sick leave coinciding with a scheduled company holiday, and calling in employees for a conference when the number of sick days is excessive compared to other employees. One problem group is those employees who call in sick only on Monday or Friday in order to have a three-day weekend.

SHOWING DIFFERENCES GRAPHICALLY

In such a problem setting management is interested in knowing if the mean number of sickness absences reported for Monday and Friday is higher than the mean number of sickness absences during the remainder of the work week. Suppose that during a recent five-week period the following number of absences were reported.

Mon. and Fri.: 81, 86, 73, 77, 90, 91, 75, 62, 98, 74
Tues.-Wed.- Thur.: 89, 55, 59, 64, 37, 58, 35, 57, 65, 68, 42, 71, 69, 49, 67

As a first step on a problem on this nature it is useful to plot the data. This is a place where the e.d.f. plot comes in handy as both e.d.f.'s can be put in the same figure, as in Figure 9.1.

What type of information is contained in Figure 9.1? First of all, if there were no difference in the absentee rates for the grouping of days, then the two e.d.f.'s would probably cross each other several times. As it is , the Mon.–Fri. group is clearly shifted to the right of the Tues.-Wed.-Thur. group. The only way such a shift like this can occur is for one group to have consistently higher values than the other. A comparison of the medians of the two groups, obtained easily from Figure 9.1, shows the Mon.–Fri. group has a median of 79 and the Tues.–Wed.–Thur. group has a median of 64. Hence, this lends support to the interpretation of the e.d.f. plot that one sample is associated with a higher degree of absenteeism than is the other.

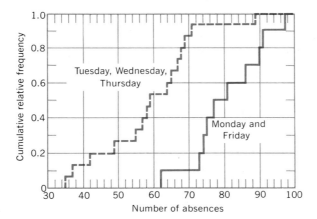

FIGURE 9.1 Empirical Distribution Functions of the Number of Sick Leave Absences for Two Independent Samples

THE TEST FOR $\mu_x = \mu_y$ WHEN $n_x \geq 30$ AND $n_y \geq 30$

If both samples come from normal populations, then a **two-sample t-test** can be used to compare the means μ_x and μ_y of the respective parent populations as discussed in the next section. However, there is no need to check the normality assumption if the samples are large, say ≥ 30 each, since the Central Limit Theorem may usually be used to show that $\bar{X}$, $\bar{Y}$, and $\bar{X} - \bar{Y}$ are approximately normally distributed in that case. It is the fact that $\bar{X} - \bar{Y}$ is approximately normal that furnishes the foundation for the validity of the test of this section.

The mean of the sampling distribution of $(\bar{X} - \bar{Y})$ is $(\mu_x - \mu_y)$ and the standard deviation is $\sqrt{\sigma_x^2/n_x + \sigma_y^2/n_y}$, where σ_x and σ_y are the respective population standard deviations. Therefore, for large samples the approximate probability

$$P\left(-z_{\alpha/2} < \frac{\bar{X} - \bar{Y} - (\mu_x - \mu_y)}{\sqrt{\sigma_x^2/n_x + \sigma_y^2/n_y}} < z_{\alpha/2}\right) \cong 1 - \alpha$$

obtained from Table A2 leads to the probability statement

$$P(\bar{X} - \bar{Y} - z_{\alpha/2}\sqrt{\sigma_x^2/n_x + \sigma_y^2/n_y} < \mu_x - \mu_y$$
$$< \bar{X} - \bar{Y} + z_{\alpha/2}\sqrt{\sigma_x^2/n_x + \sigma_y^2/n_y}) \cong 1 - \alpha,$$

which is the basis for the confidence interval for $(\mu_x - \mu_y)$.

EXAMPLE

A school board would like to compare the mean income in two different school districts in a large city. Fifty households are randomly selected in each district with the following results.

	District 1	District 2
Sample mean	$32,070	$30,750
Sample std. dev.	$2,500	$3,000

PROCEDURES FOR INFERENCES ABOUT THE DIFFERENCE IN MEANS OF TWO POPULATIONS (LARGE SAMPLES, BOTH SAMPLE SIZES $\geq$ 30)

Data. The data $X_1, X_2, \ldots, X_{n_x}$ and $Y_1, Y_2, \ldots, Y_{n_y}$ represent two samples of sizes n_x and n_y, respectively, taken from two populations. Denote the respective population means and variances by $\mu_x, \sigma_x^2, \mu_y,$ and σ_y^2. Compute the sample means $\bar{X}$ and $\bar{Y}$, and if σ_x^2 and σ_y^2 are unknown, compute the sample variances s_x^2 and s_y^2.

Assumptions
1. Both samples are random samples from their respective populations.
2. The two samples are independent of one another.
3. The samples are large enough so that $\bar{X}$ and $\bar{Y}$ are approximately normally distributed by the Central Limit Theorem. Usually $n_x \geq 30$ and $n_y \geq 30$ is sufficient for this assumption to hold.

Confidence Interval. A $100(1 - \alpha)$ percent confidence interval for the difference $\mu_x - \mu_y$ is given approximately by

$$\bar{X} - \bar{Y} \pm z_{\alpha/2} \sqrt{\sigma_x^2/n_x + \sigma_y^2/n_y},$$

where $z_{\alpha/2}$ is the $1 - \alpha/2$ quantile from Table A2. If σ_x and σ_y are unknown, s_x and s_y may be used instead.

Null Hypothesis. $H_0: \mu_x = \mu_y$

Test Statistic

$$Z = \frac{\bar{X} - \bar{Y}}{\sqrt{\sigma_x^2/n_x + \sigma_y^2/n_y}} \tag{9.1}$$

If σ_x and σ_y are unknown, s_x and s_y may be used since the sample sizes are large.

Decision Rule. The decision rule is based on the alternative hypothesis of interest.
1. $H_1: \mu_x > \mu_y$. Reject H_0 if $Z > z_\alpha$; otherwise accept H_0. See Figure 9.2a.
2. $H_1: \mu_x < \mu_y$. Reject H_0 if $Z < -z_\alpha$; otherwise accept H_0. See Figure 9.2b.
3. $H_1: \mu_x = \mu_y$. Reject H_0 if $Z > z_{\alpha/2}$ or if $Z < -z_{\alpha/2}$; otherwise accept H_0. See Figure 9.2c.

The value of z_α is the $1 - \alpha$ quantile from Table A2.

Note: This procedure is exact for *all* sample sizes if the populations are normal and σ_x and σ_y are known, conditions that rarely occur in actual applications.

Use these data to test $H_0: \mu_x = \mu_y$ versus $H_1: \mu_x \neq \mu_y$ with $\alpha = .05$. The test statistic is given by

$$Z = \frac{\bar{X} - \bar{Y}}{\sqrt{s_x^2/n_x + s_y^2/n_x}}$$

$$= \frac{32,070 - 30,750}{\sqrt{(2500)^2/50 + (3000)^2/50}} = 2.3901$$

Decision Rule
Reject H_0 if $Z \geq z_{.025} = 1.9600$ or if $Z \leq -z_{.025} = -1.9600$. H_0 is rejected and it is concluded that the mean income in the two school districts is not the same.

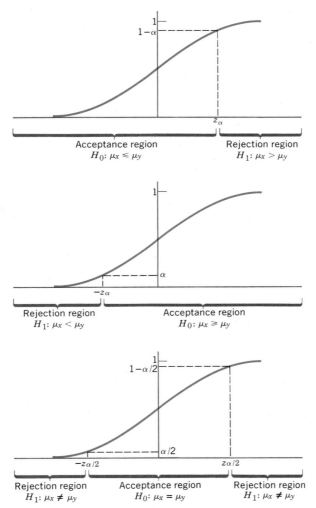

FIGURE 9.2 The Normal Cumulative Distribution Function Showing the Rejection Regions for Large Samples

The p-value associated with $Z = 2.3901$ is found from Table A2 to be twice .008, because this is a two-tailed test and $P(Z > 2.3901) = .008$. Thus the p-value is .016 and H_0 is rejected since this value is less than the α-level of .05. A 95 percent confidence interval for the difference in means $\mu_x - \mu_y$ uses many of the same calculations.

$$\bar{X} - \bar{Y} \pm z_{.025} \sqrt{s_x^2/n_x + s_y^2/n_y} = 1320 \pm 1.9600(552.27)$$

$$= 1320 \pm 1082$$

Thus a 95 percent confidence interval for the difference $\mu_x - \mu_y$ is the interval from \$238 to \$2402. Since this interval does not contain the value of zero this indicates that the difference $\mu_x - \mu_y$ is larger than zero, or, equivalently, $\mu_x > \mu_y$.

NORMAL POPULATIONS AND KNOWN VARIANCES

Suppose samples of less than 30 are available. Then the Central Limit Theorem cannot be used to justify the approximate normality of $\overline{X} - \overline{Y}$ and the Lilliefors test should be applied to each sample. If the Lilliefors test does not indicate lack of normality in the populations, then the procedures of this section are still valid provided the population variances are known.

EXAMPLE

Two brands of golf balls are to be compared with respect to driving distance. The balls are tested on a mechanical driver known to give normally distributed distances with a standard deviation of 15 yards. Hence, the driving distance observed for Brand X and Brand Y can be thought of as normally distributed random variables with unknown means μ_x and μ_y, each with a population variance of $15^2 = 225$. Each ball is tested 25 times with $\overline{X} = 275$ and $\overline{Y} = 290$. Test the hypothesis H_0: $\mu_x = \mu_y$ versus H_1: $\mu_x \neq \mu_y$ with $\alpha = .01$.

Test Statistic

$$Z = \frac{\overline{X} - \overline{Y}}{\sqrt{\sigma_x^2/n_x + \sigma_y^2/n_y}}$$

$$= \frac{275 - 290}{\sqrt{225/25 + 225/25}} = -3.5355$$

Decision Rule

Reject H_0 if $Z \geq z_{.005} = 2.5758$ or if $Z \leq -z_{.005} = -2.5758$. Therefore, H_0 is rejected and it is concluded that the mean driving distance is not the same for the two golf balls. The p-value associated with $Z = -3.5355$ is $2P(Z < -3.5355)$ because this is a two-tailed test. Use of Table A2 gives the p-value of less than $2(.0005) = .001$. This small p-value indicates that the evidence is strong for rejecting H_0. A 99 percent confidence interval for the difference $\mu_x - \mu_y$ in mean driving distance of the two balls is

$$\overline{X} - \overline{Y} \pm z_{.005} \sqrt{\sigma_x^2/n_x + \sigma_y^2/n_y} = -15 \pm 2.5758 \sqrt{225/25 + 225/25}$$

$$= -15 \pm 10.93$$

or from -25.93 yards to -4.07 yards. Since this interval does not contain zero, there is no reason to believe that the driving distance is the same for the two brands of golf balls.

Exercises

9.1 A random sample of truck drivers for one nationwide trucking firm revealed 30 drivers with an average of 18,000 miles during a six-week period with a standard deviation of 2,100 miles while 30 drivers from another trucking firm showed an average of 12,000 miles with a standard deviation of 820 miles for the same six-week period. Find a 95 percent confidence interval for the difference in the mean mileage for the two trucking firms and interpret the results.

9.2 Two laboratories have tested one brand of tobacco for nicotine content. Past experience shows the standard deviation of the nicotine content to be 12. The results in grams are recorded as follows for the standardized samples given to the laboratories for testing.

Laboratory 1:   56   37   61   56   65   41   63   50   42
Laboratory 2:   46   25   46   64   34   56

State and test the appropriate null and alternative hypotheses to see if the mean nicotine content is the same as measured by the two laboratories. Use $\alpha = .10$ and state the p-value. (The Lilliefors test should be used to check each sample for normality because of the small sample sizes used.)

9.3 A home economics lab class compared gas to electric ovens by baking one type of bread in 30 ovens of each type. The gas ovens had an average baking time of 0.90 hours with a standard deviation of 0.09 hours and the electric ovens had an average baking time of 0.70 hours with a standard deviation of 0.16 hours. At the 5 percent level of significance, should the null hypothesis of identical mean baking times for the two kinds of ovens be accepted or rejected? State the p-value. Find a 95 percent confidence interval for the difference in mean baking times.

9.4 Pharmaceutical companies desiring to get a share of the mild tranquilizer (like valium and librium) market are continuously testing new compounds in their laboratories. In one of the experiments used for screening new compounds laboratory mice are placed in individual sealed beakers and the asphyxiation times in minutes are recorded when testing for drugs that affect the central nervous system. The reason for this test is that an experimental drug may either speed up the metabolism rate, in which case the air supply is quickly diminished, or it may have the opposite effect, or no effect at all. A placebo was tested on 40 mice as a control group and showed a mean asphyxiation time of 16.4 minutes with a standard deviation of 1.55 minutes. An experimental group of size 40 was treated with a new compound and had a mean asphyxiation time of 16.9 minutes with a standard deviation of 1.67 minutes. Test the null hypothesis that mean asphyxiation times are the same for both groups versus an alternative that the mean times are different with $\alpha = .01$. State the p-value.

9.5 A laboratory starts their experiments with mice on Monday and Wednesday of each week. In order that the experimental results be reliable, the weights of the mice are required to be the same on both starting days. A lab technician suspects that the mice starting on Wednesday tend to be lighter than those starting on Monday. Random samples of weights obtained on 45 mice on each day gave the following results.

Monday: $\bar{X} = 15.88$ grams, $s_x = 1.25$ grams
Wednesday: $\bar{Y} = 15.58$ grams, $s_y = 1.21$ grams

Test the appropriate hypothesis with $\alpha = .10$. State the p-value.

9.6 In Exercise 8.9 a comparison using matched pairs was made on m.p.g. obtained with unleaded gasoline and gasohol. The experiment could also be performed by using two groups of drivers under similar driving conditions, one of which uses only unleaded gasoline and one uses only gasohol. If 50 drivers of each type got 23.8 m.p.g. on gasohol with a standard deviation of 1.7, and 23.4 m.p.g. on unleaded gasoline with a standard deviation of 1.9, test the hypothesis that there is no difference in the mean m.p.g. of the two fuels with $\alpha = .01$. Comment on whether you think the matched pairs approach or the two independent samples approach is the best way to conduct this experiment.

9.7   Two plastics produced by different processes were given breaking strength tests. Thirty measurements were made on plastic A and showed a mean breaking strength of 28.3 with a standard deviation of 3.3, while 40 tests on plastic B showed a sample mean of 26.7 and a standard deviation of 4.9. Do these plastics differ significantly in breaking strengths? Use $\alpha = .05$ and state the p-value.

9.8   A measure of productivity has been devised for workers on an assembly line. A random sample of 40 workers in 1980 is compared with the results using a random sample of 40 workers in 1975. The sample statistics are $\overline{X} = 115$ and $s_x = 12$ for 1980 and $\overline{Y} = 108$ and $s_y = 22$ for 1975. Find a 95 percent confidence interval for the mean increase in productivity and interpret the results.

9.2
Small Samples from Normal Distributions: Inferences about the Difference Between Two Means

If the sample sizes are large enough, the test in the previous section can be used for samples from almost any population. That test is also the best test to use if the populations are normal and the variances are known, for all sample sizes. But what can be used with smaller samples when the variances are unknown? The tests in this section satisfy a part of this gap, and the procedure in the next section completes the job.

In this section methods are given for handling small samples when the populations are normal and the variances are unknown. The reasonableness of the normality assumption should be checked with a goodness-of-fit test, such as the Lilliefors test, in each case. If the normality assumption is not satisfied for either sample, the procedure in the next section should be used because it usually has more power in such cases than the procedures given in this section.

THE TWO-SAMPLE t-TEST

Two-sample problems are most commonly associated with small sample sizes and with equal but unknown variances. For this situation the small samples must be used to provide an estimate of the common but unknown variance σ^2. The sample variances s_x^2 and s_y^2 each provide an estimate of σ^2. However, a better estimate of σ^2 can be obtained by combining the two estimates in a weighted average as follows:

$$s_p^2 = \frac{(n_x - 1)s_x^2 + (n_y - 1)s_y^2}{n_x + n_y - 2}. \tag{9.2}$$

For equal sample sizes this formula reduces to $s_p^2 = (s_x^2 + s_y^2)/2$, which is the simple average of the two estimates. Otherwise, each of the estimates receives weights proportional to its respective degrees of freedom. As in the previous chapters, when σ^2 is estimated from small samples the resulting test and confidence interval involves a Student's t random variable when the populations are normal. A frequently used two-sample technique—commonly called the **two-sample t-test**—is now presented and illustrated.

PROCEDURES FOR INFERENCES ABOUT THE DIFFERENCE IN MEANS OF TWO NORMAL POPULATIONS WITH EQUAL VARIANCES (TWO-SAMPLE T-TEST)

Data. The data consist of two samples $X_1, X_2, \ldots, X_{n_x}$ of size n_x and $Y_1, Y_2, \ldots, Y_{n_y}$ of size n_y from two populations. Denote the respective population means by μ_x and μ_y. Compute the sample means $\overline{X}$ and $\overline{Y}$ and the sample variances s_x^2 and s_y^2. (If the population variances σ_x^2 and σ_y^2 are known, use the methods outlined in Section 9.1.) Also compute the pooled sample standard deviation:

$$s_p = \sqrt{\frac{(n_x - 1)\, s_x^2 + (n_y - 1)\, s_y^2}{n_x + n_y - 2}}. \tag{9.3}$$

Assumptions
1. Both samples are random samples from their respective populations.
2. The two samples are independent of one another.
3. Both populations are normal with equal variances ($\sigma_x^2 = \sigma_y^2$).

Confidence Interval. A $100(1 - \alpha)$ percent confidence interval for the difference $\mu_x - \mu_y$ is given by

$$(\overline{X} - \overline{Y}) \pm t_{\alpha/2, n_x + n_y - 2}\, s_p\, \sqrt{1/n_x + 1/n_y}$$

where $t_{\alpha/2, n_x + n_y - 2}$ is the $(1 - \alpha/2)$ quantile of the Student's t-distribution with $n_x + n_y - 2$ degrees of freedom, obtained from Table A3.

Null Hypothesis. $H_0: \mu_x = \mu_y$

Test Statistic

$$T = \frac{\overline{X} - \overline{Y}}{s_p\, \sqrt{1/n_x + 1/n_y}} \tag{9.4}$$

Decision Rule. The decision rule depends on the alternative hypothesis of interest.
1. $H_1: \mu_x > \mu_y$. Reject H_0 if $T > t_{\alpha, n_x + n_y - 2}$; otherwise accept H_0.
2. $H_1: \mu_x > \mu_y$. Reject H_0 if $T < -t_{\alpha, n_x + n_y - 2}$; otherwise accept H_0.
3. $H_1: \mu_x \ne \mu_y$. Reject H_0 if $T > t_{\alpha/2, n_x + n_y - 2}$ or $T < -t_{\alpha/2, n_x + n_y - 2}$; otherwise accept H_0.

The value of $t_{\alpha, n_x + n_y - 2}$ is the $(1 - \alpha)$ quantile of the t-distribution with $n_x + n_y - 2$ degrees of freedom, obtained from Table A3.

Note: This procedure is appropriate to use for all sample sizes when both populations are normal with equal population variances.

EXAMPLE

The sickness absenteeism data associated with Problem Setting 9.1 are used to test $H_0: \mu_x = \mu_y$ versus $H_1: \mu_x > \mu_y$ with $\alpha = .01$, where μ_x is the population mean for absenteeism on Monday and Friday and μ_y is the population mean for absenteeism on Tuesday–Wednesday–Thursday. The Lilliefors test on each sample shows that the normality assumption is reasonable. The assumption of equal variances will be addressed later in this section. Because of the small sample sizes and the reasonableness of the assumptions the test of this section is used.

The sample data of Section 9.1 give the following results.

M–F: $n_x = 10$, $\bar{X} = 80.70$, $s_x^2 = 113.34$
T–W–T: $n_y = 15$, $\bar{Y} = 59.00$, $s_y^2 = 201.43$

$$s_p = \sqrt{\frac{(10-1)113.34 + (15-1)201.43}{10 + 15 - 2}}$$

$$= \sqrt{166.96} = 12.92$$

Test Statistic

$$T = \frac{80.70 - 59.00}{12.92 \sqrt{\dfrac{1}{10} + \dfrac{1}{15}}} = \frac{21.70}{5.28} = 4.1141$$

Decision Rule

From Table A3, $t_{.01,10+15-2} = 2.4999$ and H_0 is rejected if $T \geq 2.4999$. Therefore, H_0 is soundly rejected, and it is concluded that the absenteeism rate is significantly higher on Monday and Friday than it is during the rest of the week. The p-value is found by interpolation in Table A3 to be about .0003, which indicates that these results could hardly have occurred by chance. A 95 percent confidence interval for the difference in absenteeism rates is given by

$$\bar{X} - \bar{Y} \pm t_{.025,23}\, s_p \sqrt{1/n_x + 1/n_y} = 21.70 \pm 2.0687(5.28),$$

or from 10.78 to 32.62 workers per day, which again confirms the fact that absenteeism rates are higher on Monday and Friday than during the rest of the week, because the confidence interval does not include 0.

THE ASSUMPTION OF EQUAL VARIANCES

In the presentation of the two-sample t-test the estimates s_x^2 and s_y^2 were pooled together to form s_p^2 as an estimate of the common variance σ^2. However, this pooling was done under the *assumption* that the populations had equal variance; otherwise the pooling is not justified and the t-test cannot be used. A procedure is now presented for testing the assumption of equal variances when two samples come from *normal populations*. This test **should be used only after the normality assumption has been checked and satisfied.** Tests are valid only when their assumptions are satisfied; the distribution of the F statistic (below) is affected by nonnormalities of the population.

The test for equality of variances involves a distribution not yet encountered in this book, known as an F distribution. A random variable that consists of the *ratio* of two sample variances, s_1^2/s_2^2, has an F distribution if the two samples are independent and from normal populations with equal population variances $\sigma_1^2 = \sigma_2^2$. Table A5 gives selected quantiles for the family of F distributions. The family of F distributions is indexed by two parameters:

$k_1 = $ degrees of freedom in the numerator,
$k_2 = $ degrees of freedom in the denominator.

The correct F distribution is selected from Table A5 by noting that the degrees of freedom parameters may be found easily from the two sample sizes:

$k_1 = n_1 - 1$, where n_1 is the sample size for s_1^2,
$k_2 = n_2 - 1$, where n_2 is the sample size for s_2^2.

PROCEDURE FOR TESTING THE EQUALITY OF VARIANCES FOR TWO NORMAL POPULATIONS

Data. Same as for the two-sample t-test.

Assumptions. The random variables $X_1, X_2, \ldots, X_{n_x}$ are independent of one another and are **normally distributed** with unknown variance σ_x^2. The random variables $Y_1, Y_2, \ldots, Y_{n_y}$ are independent of one another and are **normally distributed** with unknown variance σ_y^2.

Estimation. The unknown population variances σ_x^2 and σ_y^2 are estimated by the sample variances s_x^2 and s_y^2, respectively.

Null Hypothesis. H_0: $\sigma_x^2 = \sigma_y^2$

Alternative Hypothesis. H_1: $\sigma_x^2 \neq \sigma_y^2$

Test Statistic. $F = \max(s_x^2, s_y^2)/\min(s_x^2, s_y^2)$; that is, divide the larger of the two sample variances by the smaller of the two sample variances.

Decision Rule. If $s_x^2 > s_y^2$, let $k_1 = n_x - 1$ and $k_2 = n_y - 1$. If $s_x^2 < s_y^2$, let $k_1 = n_y - 1$ and $k_2 = n_x - 1$. Reject H_0 at the level of significance α if $F \geq F_{\alpha/2, k_1, k_2}$.

The value F_{α, k_1, k_2} is the upper α-level critical value of the F distribution with k_1 and k_2 degrees of freedom from Table A5.

Note: If the desired degrees of freedom for the F distribution cannot be found in Table A5, simply use the next higher degrees of freedom given in the table.

If the null hypothesis of equal variances is accepted, then the two-sample t-test can be used given that the populations are normal. If the null hypothesis of equal variances is rejected and the populations are normal, then the Satterthwaite's approximation (to follow) should be used. Note: There is no satisfactory procedure to use if the populations are nonnormal, the variances are unequal, and the sample sizes are small.

EXAMPLE

The sickness absentee data presented in Section 9.1 are now used to test H_0: $\sigma_x^2 = \sigma_y^2$ versus H_1: $\sigma_x^2 \neq \sigma_y^2$ to see if the assumption of equality of variance for the two-sample t-test was justified. Let $\alpha = .10$.

Test Statistic

The sample variances s_x^2 and s_y^2 were previously given as 113.34 and 201.43, respectively. Since s_y^2 is the maximum of these two values, the F ratio is found as

$F = s_y^2/s_x^2 = 201.43/113.43 = 1.78.$

Decision Rule

Since $s_x^2 < s_y^2$, let $k_1 = n_y - 1 = 15 - 1 = 14$ and $k_2 = n_x - 1 = 10 - 1 = 9$. The F distribution for $k_1 = 14$, $k_2 = 9$ is not given in Table A5, so the F distribution with the next higher degrees of freedom is used, which in this case is 15 and 9. From Table A5, $F_{.05,15,9} = 3.006$ so the decision rule is to reject H_0 if $F \geq 3.006$. Since the observed F is less than 3.006, H_0 is *not* rejected. Keep in mind that this solution depends on the normality assumption being satisfied, for if it is not satisfied, this test is without meaning.

SMALL SAMPLES AND UNEQUAL VARIANCES

An approximate procedure, called Satterthwaite's approximation, has been devised for the case where the populations are normal but the variances are unknown and unequal. If the sample sizes are large, the method of the previous section can be used; but if the sample sizes are small, some modification is required. The change suggested by Satterthwaite is to use the t-distribution instead of the normal distribution, where the degrees of freedom in the t-distribution is given by the equation

$$f = \frac{(s_x^2/n_x + s_y^2/n_y)^2}{\dfrac{(s_x^2/n_x)^2}{n_x - 1} + \dfrac{(s_y^2/n_y)^2}{n_y - 1}}. \tag{9.5}$$

The number f will always be less than or equal to $n_x + n_y - 2$, the degrees of freedom used in the t-test. The test statistic is not the same as in the t-test either, because a pooled estimate of the variance is no longer justified; so the test statistic in Equation (9.1) is used.

As the sample sizes get large, f becomes large also, and the t-distribution with f degrees of freedom approaches the normal distribution. This procedure then becomes the procedure of the previous section, as one would expect.

EXAMPLE

Suppose that the absentee data in Problem Setting 9.1 had been collected during a five-week period in January and February, and the company is interested in conducting a similar study during the summer months. Data for a five-week period in the summer were collected and summarized as follows:

$$\text{M–F:} \quad n_x = 10, \quad \overline{X} = 112.62, \quad s_x^2 = 868.11$$
$$\text{T–W–T:} \quad n_y = 15, \quad \overline{Y} = 69.72, \quad s_y^2 = 206.65$$

Both samples pass the Lilliefors test for normality. The null hypothesis of interest is $H_0: \mu_x = \mu_y$ versus the alternative $H_1: \mu_x > \mu_y$, as before. However, the assumption of equal variances should be tested first. The test for equality of variances uses the F statistic

$$F = \frac{\max(s_x^2, s_y^2)}{\min(s_x^2, s_y^2)} = \frac{s_x^2}{s_y^2} = \frac{868.11}{206.65} = 4.201.$$

The critical value $F_{.05,9,14} = 2.646$ is used for an $\alpha = .10$ test, because the degrees of freedom are 9 and 14. Since the observed F exceeds 2.646, the null hypothesis of equal variances is rejected. Therefore in the test of H_0: $\mu_x = \mu_y$, Satterthwaite's approximation is used.

The approximate degrees of freedom from Equation (9.6) becomes

$$f = \frac{(868.11/10 + 206.65/15)^2}{\dfrac{(868.11/10)^2}{9} + \dfrac{(206.65/15)^2}{14}} = 11.9.$$

PROCEDURES FOR INFERENCES ABOUT THE DIFFERENCE IN MEANS OF TWO NORMAL POPULATIONS WITH UNEQUAL VARIANCES

Data. The data consist of two samples $X_1, X_2, \ldots, X_{n_x}$ of size n_x and $Y_1, Y_2, \ldots, Y_{n_y}$ of size n_y from two populations. Denote the respective population means by μ_x and μ_y. Compute the sample means $\bar{X}$ and $\bar{Y}$ and sample variances s_x^2 and s_y^2. (If the population variances σ_x^2 and σ_y^2 are known, use the methods outlined in Section 9.1.) Also compute the approximate degrees of freedom f given by

$$f = \frac{(s_x^2/n_x + s_y^2/n_y)^2}{\dfrac{(s_x^2/n_x)^2}{n_x - 1} + \dfrac{(s_y^2/n_y)^2}{n_y - 1}}. \tag{9.6}$$

Assumptions
1. Both samples are random samples from their respective populations.
2. The two samples are independent of one another.
3. Both populations are normal with unequal variances ($\sigma_x^2 \neq \sigma_y^2$).

Confidence Interval. An approximate $100(1 - \alpha)$ percent confidence interval for the difference $\mu_x - \mu_y$ is given by

$$\bar{X} - \bar{Y} \pm t_{\alpha/2,f} \sqrt{s_x^2/n_x + s_y^2/n_y},$$

where $t_{\alpha/2,f}$ is the $1 - \alpha/2$ quantile of the Student's t-distribution with f degrees of freedom, obtained from Table A3.

Null Hypothesis. H_0: $\mu_x = \mu_y$

Test Statistic

$$T = \frac{\bar{X} - \bar{Y}}{\sqrt{s_x^2/n_x + s_y^2/n_y}} \tag{9.7}$$

Decision Rule. The decision rule depends on the alternative hypothesis of interest.
1. H_1: $\mu_x > \mu_y$. Reject H_0 if $T > t_{\alpha,f}$; otherwise accept H_0.
2. H_1: $\mu_x < \mu_y$. Reject H_0 if $T < -t_{\alpha,f}$.
3. H_1: $\mu_x \neq \mu_y$. Reject H_0 if $T > t_{\alpha/2,f}$ or $T < -t_{\alpha/2,f}$; otherwise accept H_0.

The value $t_{\alpha,f}$ is the $(1 - \alpha)$ quantile of the t-distribution with f degrees of freedom. Linear interpolation may be used to find $t_{\alpha,f}$ in Table A3 if f is not an integer.

Note: This test is only approximate. It may be used for all sample sizes but is most appropriate when either $n_x < 30$ or $n_y < 30$.

The test statistic, using Equation (9.7), becomes

$$T = \frac{112.62 - 69.72}{\sqrt{\dfrac{868.11}{10} + \dfrac{206.65}{15}}} = 4.277$$

For an $\alpha = .05$ test, the null hypothesis is rejected in favor of the alternative $H_1: \mu_x > \mu_y$ if T exceeds $t_{.05,f}$ for $f = 11.9$ degrees of freedom. From Table A3, $t_{.05,11.9} = 1.7837$ is used as the critical value, so the null hypothesis is rejected. The p-value is less than .001, indicating a very clear increase in the absentee rate for Monday and Friday as compared with the days in the middle of the week.

Exercises

9.9 A college professor teaches two sections of the same subject. One section meets at 8 a.m. and the other meets at 10 a.m. with one hour separating the two classes. At examination time she gives identical exams to both classes but is concerned that the 10 a.m. section might have higher scores based on possible information received about the test from individuals in the 8 a.m. section. Test the null hypothesis $H_0: \mu_8 = \mu_{10}$ versus $H_1: \mu_8 < \mu_{10}$ with $\alpha = .05$ and state the p-value. (Each sample should be tested for normality using the Lilliefors test.)

8 a.m. Scores			10 a.m. Scores		
98	95	94	100	98	98
91	90	89	97	96	94
87	77	76	92	92	89
75	73	70	86	85	85
67	67	66	85	82	81
65	57	56	79	79	78
55	53	45	78	77	77
43	39		74	72	67
			65	62	56
			56	49	47

9.10 Find a 95 percent confidence interval for the difference in the means of the two classes in Exercise 9.9.

9.11 Rework Exercise 9.9 using the Satterthwaite approximation and compare results with those obtained in Exercise 9.9. (You should include a test of equality of variance with your analysis to see if the use of the Satterthwaite approximation is justified.)

9.12 Use the two-sample t-test on the data presented in Exercise 9.2 and compare results with those obtained using the large sample test of Section 9.1. The comparison should include the p-values for both tests.

9.13 The burning time in minutes is recorded below for each of two types of smudge pots used to protect orchards from frost. Is there a significant difference in the burning times of the two types of smudge pots? State and test the appropriate hypotheses with $\alpha = .05$. State the p-value. (Each sample should be tested for normality using the Lilliefors test.)

Burning Time for Brand 1	Burning Time for Brand 2
564	617
560	681
634	628
641	644
597	591
621	645
562	610
597	678
664	755
565	597

9.14 Find a 90 percent confidence interval for the mean difference in burning times based on the data presented in Exercise 9.13.

9.15 A college senior is considering job offers of the same amount of money in two different cities and is concerned that housing costs might not be the same in each city. He obtains a copy of the Sunday paper from each city and uses a random selection procedure to choose 20 classified ads from the "Apartments for Rent" section. He finds the average rental cost in one city to be $355 with a standard deviation of $25 and the average cost in the other city to be $375 with a standard deviation of $18. Assuming that the rental costs are normally distributed, test the appropriate hypothesis with $= .05$ to see if the average rental cost differs in the two cities. State the p-value.

9.16 Two grade school classes were taught arithmetic by two different methods: (a) the students were taught by the teacher in an ordinary classroom situation and (b) the students taught themselves using a programmed text and consulted the teacher only when they had questions. At the end of the year each group was given a standard exam with the results given below. Can it be concluded that the experimental group has a higher mean level of achievement if it is assumed that these classes were selected randomly? Let $\alpha = .01$ and state the p-value. (Each sample should be tested for normality using the Lilliefors test.)

Scores for Standard Group	Scores for Experimental Group
72	111
75	118
77	128
80	138
104	140
110	150
125	163
	164
	169

9.3

General Populations: Inferences about the Differences Between Two Means

THE WILCOXON–MANN–WHITNEY RANK SUM TEST

The two-sample techniques of the previous section have the underlying assumption that the samples come from normal populations. If a Lilliefors test shows this assumption to be unwarranted, then the procedure presented in this section should be used. Some statisticians prefer to use this test in all situations because the Lilliefors test does not always detect nonnormal populations, especially when sample sizes are small. In addition, this technique is not nearly as sensitive to the assumption of equal variances as is the two-sample t-test, so it may be used as an approximate test when that assumption is not valid. The name of this procedure is the **Wilcoxon–Mann–Whitney rank sum test**. It is calculated by performing the two-sample t-test on rank transformed data. Hence, this is also a **rank transformation test**, as was the Wilcoxon signed ranks test. With this test the data are ranked from 1 to $n_x + n_y$. Average ranks are used in case of ties. For example, consider the following set of 10 observations to see how the ranking is done.

$$\text{Data} \quad \begin{cases} X: & 4.8,\ 9.2,\ 3.6,\ 6.3 \\ Y: & 7.1,\ 6.3,\ 11.8,\ 10.5,\ 8.7,\ 6.5 \end{cases}$$

$$\begin{matrix} \text{Joint} \\ \text{Ranking} \end{matrix} \quad \begin{cases} R_x: & 2,\ 8,\ 1,\ 3.5 \\ R_y: & 6,\ 3.5,\ 10,\ 9,\ 7,\ 5 \end{cases}$$

The smallest observation is 3.6, so it gets rank 1. The largest observation is 11.8, so it gets the largest rank 10. The two observations tied at 6.3 each get the average rank 3.5, which represents the average of the ranks 3 and 4, which they would have received if they were not tied. The two-sample t-test is calculated on the ranks assigned to the X and Y samples rather than on the sample observations.

Of course this procedure is exactly the same as the two-sample t-test except that it is applied to the ranks of the data rather than to the data themselves. This makes the Wilcoxon–Mann–Whitney rank sum test very easy to use. This nonparametric procedure is now illustrated in a rather long example, which helps to point out some of the advantages of this test.

EXAMPLE

Two homes for the elderly are selected for a test to see if the physically able residents would use an exercise room if it were available in lieu of spending so much time watching television. One of the homes (Home 2) is then equipped with an exercise room and 10 residents are randomly selected from each home for this experiment; however, the residents are unaware that the experiment is taking place so as not to introduce bias into the results. The number of hours each resident watches television in a given week is monitored in each home with the results recorded as follows.

PROCEDURE FOR TESTING HYPOTHESES ABOUT THE DIFFERENCE IN MEANS OF TWO GENERAL POPULATIONS (WILCOXON–MANN–WHITNEY RANK SUM TEST)

Data. The data consist of two samples $X_1, X_2, \ldots, X_{n_x}$ of size n_x and $Y_1, Y_2, \ldots, Y_{n_y}$ of size n_y from two populations. Denote the respective population means by μ_x and μ_y. Assign ranks 1 to $(n_x + n_y)$ to the two samples jointly, using average ranks in case of ties. Compute the sample means of the ranks $\bar{R}_x$ and $\bar{R}_y$, and the sample variances $s_{R_x}^2$ and $s_{R_y}^2$ of the ranks for the two samples, respectively. Also compute the pooled standard deviation of the ranks

$$s_p = \sqrt{\frac{(n_x - 1)s_{R_x}^2 + (n_y - 1)s_{R_y}^2}{n_x + n_y - 2}}, \tag{9.8}$$

as was done with the two-sample t-test.

Assumptions
1. Both samples are random samples from their respective populations.
2. The two samples are independent of one another.
3. The two population distribution functions are identical except for possibly different means.

Null Hypothesis. $H_0: \mu_x = \mu_y$

Test Statistic

$$T_R = \frac{\bar{R}_x - \bar{R}_y}{s_p \sqrt{1/n_x + 1/n_y}} \tag{9.9}$$

Decision Rule. The decision rule depends on the alternative hypothesis of interest.
1. $H_1: \mu_x > \mu_y$. Reject H_0 if $T_R > t_{\alpha, n_x + n_y - 2}$; otherwise accept H_0.
2. $H_1: \mu_x < \mu_y$. Reject H_0 if $T_R < -t_{\alpha, n_x + n_y - 2}$; otherwise accept H_0.
3. $H_1: \mu_x \neq \mu_y$. Reject H_0 if $T_R > t_{\alpha/2, n_x + n_y - 2}$ or if $T_R < -t_{\alpha/2, n_x + n_y - 2}$; otherwise accept H_0.

The value $t_{\alpha, n_x + n_y - 2}$ is the $1 - \alpha$ quantile of the Student's t-distribution with $n_x + n_y - 2$ degrees of freedom obtained from Table A3.

Note: This procedure is valid for all sample sizes and all distributions, although the t-distribution is merely a good approximation to the exact distribution of T_R. The approximation works well even for small sample sizes.

Home 1 X	Home 2 Y
7.05	9.25
14.25	2.05
8.75	2.75
10.50	2.50
11.90	6.40
4.50	6.90
37.60	10.00
9.40	8.00
8.10	33.00
45.20	5.20

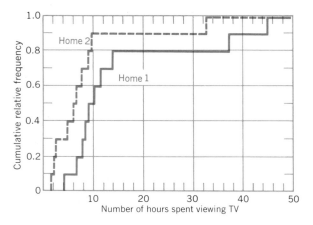

FIGURE 9.3 Empirical Distribution Functions of Hours Spent Viewing TV by Residents of Two Rest Homes

To get a "visual feel" for the data, an e.d.f. is plotted in Figure 9.3 for the hours of TV viewing reported by each of the homes. Since the e.d.f. for Home 1 is always to the right (indicating larger numbers) than the e.d.f. for Home 2, it would seem that the residents of Home 1 spend more hours watching television than those in Home 2. For the purposes of illustration only, the **two-sample t-test on the raw data** with $\alpha = .05$ will be used to test the hypothesis

$$H_0: \mu_x = \mu_y \quad \text{versus} \quad H_1: \mu_x > \mu_y.$$

The sample data give the following statistics:

$$\bar{X} = 15.73 \text{ hours}, \qquad \bar{Y} = 8.61 \text{ hours},$$
$$s_x = 13.90 \text{ hours}, \qquad s_y = 9.02 \text{ hours},$$

$$s_p = \sqrt{\frac{9(13.90)^2 + 9(9.02)^2}{18}} = \sqrt{137.39} = 11.72,$$

$$T = \frac{15.73 - 8.61}{11.72\sqrt{\frac{1}{10} + \frac{1}{10}}} = \frac{7.12}{5.24} = 1.36.$$

The decision rule is given as: Reject H_0 if $T > t_{.05,18} = 1.7341$. For these data H_0 is not rejected in spite of the strong evidence suggested by the joint plot of the e.d.f.'s. The assumption of normality should have been verified before the two-sample t-test by using a Lilliefors plot. The standardized sample values are as follows.

Home 1 X*	Home 2 Y*	Home 1 X*	Home 2 Y*
−.62	.07	−.81	−.19
−.11	−.73	1.57	.15
−.50	−.65	−.46	−.07
−.38	−.68	−.55	2.70
−.28	−.25	2.12	−.38

The Lilliefors graphs in Figure 9.4 and 9.5 make it clear that neither of the samples can be regarded as coming from normal populations. Hence, the above t-test on the raw data has no meaning. Therefore, the Wilcoxon–Mann-Whitney rank sum test—that is, **the two-sample t-test on ranks**—is used to test

   H_0: the mean number of hours spent viewing TV is the same for the residents of both homes, versus

   H_1: the mean number of hours spent viewing TV is higher for the home without an exercise room than for the home with an exercise room

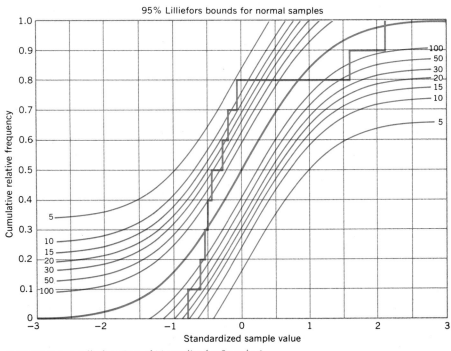

FIGURE 9.4 A Lilliefors Test of Normality for Sample 1

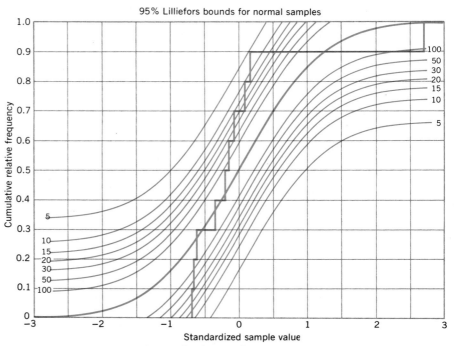

FIGURE 9.5 A Lilliefors Test of Normality for Sample 2

The first step is to replace the sample data with a joint ranking from 1 to 20.

Home 1 R_X	Home 2 R_Y
8	12
17	1
11	3
15	2
16	6
4	7
19	14
13	9
10	18
20	5

Sample statistics calculated on these ranks yield the following:

$$\bar{R}_x = 13.30, \qquad \bar{R}_y = 7.70,$$
$$s_{R_x} = 5.08, \qquad s_{R_y} = 5.54,$$

$$s_p = \sqrt{\frac{9(5.08)^2 + 9(5.54)^2}{18}} = \sqrt{28.25} = 5.315.$$

Test Statistic

$$T_R = \frac{13.30 - 7.70}{5.315 \sqrt{\frac{1}{10} + \frac{1}{10}}} = \frac{5.60}{2.38} = 2.36$$

Decision Rule

Reject H_0 if $T_R > t_{.05,18} = 1.7341$. Using this test H_0 is easily rejected, which agrees with the obvious interpretation of the joint plot of the e.d.f.'s in Figure 9.3 for the two homes. Hence, it is safe to conclude that the number of hours of TV viewing is higher for the residents of Home 1 than for Home 2. The p-value for these data is found from Table A3 to be about .015. Note that the test on ranks does better than the usual t-test in this situation because of the presence of outliers.

MANY TIES OR ORDERED CATEGORIES

The Wilcoxon–Mann–Whitney test is especially appropriate when the data for two samples have been collected into groups, and only the frequencies are given for the various group intervals. This resembles a situation where there are many ties, because all of the observations in one group may be regarded as tied. Average ranks are used to handle ties as usual, even though the number of ties is quite extensive. This application of the Wilcoxon–Mann–Whitney test is illustrated in the following example.

EXAMPLE

A random sample of faculty salaries from Big-8 schools is compared with a random sample from Big-10 schools. However, only a grouped summary is available for analysis.

Faculty Salaries	Frequencies	
	Big-8	Big-10
Less than $15,000	8	6
$15–$20,000	14	13
$20–$25,000	12	16
$25–$35,000	6	12
Over $35,000	2	7

The null hypothesis of equal mean salaries is to be tested against the alternative $\mu_x \neq \mu_y$ at $\alpha = .05$. Note that the interval sizes are not equal, and that the end interval "over $35,000" normally would cause a problem in calculating sample means or standard deviations. The use of ranks eliminates this problem.

The ranks and average ranks are given below for these data.

Faculty Salaries	Frequencies f_i		Ranks That Would Be Assigned	Average Rank R_i
	Big-8	Big-10		
Less than $15,000	8	6	1–14	$\dfrac{1 + 14}{2} = 7.5$
$15–20,000	14	13	15–41	$\dfrac{15 + 41}{2} = 28$
$20–25,000	12	16	42–69	$\dfrac{42 + 69}{2} = 55.5$
$25–35,000	6	12	70–87	$\dfrac{70 + 87}{2} = 78.5$
Over $35,000	2	7	88–96	$\dfrac{88 + 96}{2} = 92$
Totals	42	54		

The **average ranks** are used in computing $\bar{R}$ and s in the same way **class marks** were used in computing the mean and standard deviation from grouped data:

$$\bar{R}_x = \frac{1}{n_x}\sum_{i=1}^{k} f_i R_{x_i}$$

$$= \frac{1}{42}\,[8(7.5) + 14(28) + 12(55.5) + 6(78.5) + 2(92)]$$

$$= \frac{1}{42}\,(1773) = 42.214,$$

$$\bar{R}_y = \frac{1}{n_y}\sum_{i=1}^{k} f_i R_{y_i}$$

$$= \frac{1}{54}\,[6(7.5) + 13(28) + 16(55.5) + 12(78.5) + 7(92)]$$

$$= \frac{1}{54}\,(2883) = 53.389,$$

$$s_{R_x}^2 = \frac{1}{n_x - 1}\left[\sum_{i=1}^{k} f_i R_{x_i}^2 - \frac{(\Sigma f_i R_{x_i})^2}{n_x}\right]$$

$$= \frac{1}{41}\,[8(7.5)^2 + 14(28)^2 + 12(55.5)^2 + 6(78.5)^2 + 2(92)^2 - (1773)^2/42]$$

$$= 699.3798,$$

$$s_{R_y}^2 = \frac{1}{n_y - 1}\left[\sum_{i=1}^{k} f_i R_{y_i}^2 - \frac{(\Sigma f_i R_{y_i})^2}{n_y}\right]$$

$$= \frac{1}{53} [6(7.5)^2 + 13(28)^2 + 16(55.5)^2 + 12(78.5)^2 + 7(92)^2 - (2883)^2/54]$$

$$= 737.5157.$$

The pooled standard deviation of the ranks is

$$s_p = \sqrt{\frac{41(669.3798) + 53(737.5157)}{42 + 54 - 2}}$$

$$= 26.60$$

and the test statistic becomes

$$T_R = \frac{\bar{R}_x - \bar{R}_y}{s_p\sqrt{1/n_x + 1/n_y}} = \frac{42.214 - 53.389}{26.60\sqrt{\frac{1}{42} + \frac{1}{54}}}$$

$$= -2.042.$$

The decision rule is to reject H_0: $\mu_x = \mu_y$ if $T_R > 1.9855$ or if $T_R < -1.9855$ where 1.9855 is the .975 quantile from Table A3 for 94 degrees of freedom. T_R is less than -1.9855. Therefore, the null hypothesis is rejected. The p-value is between .02 and .05. That is, if there is no difference in mean salaries the chances of observing a sample difference at least as extreme as the one actually observed is between .02 and .05.

A SYNOPSIS OF THIS CHAPTER

In this chapter four methods have been presented for comparing two population means on the basis of two independent samples. In some situations several of these methods may be appropriate to use, although one may be more appropriate than the others. For this reason a brief chart is given in Figure 9.6 to summarize just where to use each test.

FIGURE 9.6 Chart for Locating the Appropriate Method for Testing the Equality of Two Means Using Two Independent Samples

Situation	Preferred Method	Also Acceptable
I. Distributions are Normal		
A. Variances are Unknown and Equal		
1. Small Samples	Section 9.2	Section 9.3
2. Large Samples	Sections 9.1 or 9.2	Section 9.3
B. Variances are unknown		
1. Small Samples	Section 9.2	
2. Large Samples	Section 9.1	
C. Variances are Known	Section 9.1	Section 9.2
II. Distributions are Nonnormal		
A. Variances are Equal		
1. Small Samples	Section 9.3	
2. Large Samples	Sections 9.3 or 9.1	
B. Variances are Unequal		
1. Small Samples	None	
2. Large Samples	Section 9.1	

Exercises

9.17 Two intersections in a city seem to have an unusually large number of automobile accidents. The city decides to spot station a police officer at the intersection with the highest traffic count. An individual is hired to make the traffic counts for one-hour intervals at randomly selected times during the week. The counts are as follows.

Traffic Count at Location #1	Traffic Count at Location #2
592	622
625	644
777	664
613	853
587	608
637	635
629	885
843	668
544	649
	714
	668

(a) Use a significance level of $\alpha = .05$ with the Wilcoxon–Mann–Whitney rank sum test to test the hypothesis that there is no difference in the mean traffic counts for the two locations.

(b) Use the Lilliefors test to check each of these samples for normality to see if the methods of Section 9.2 would be appropriate, and then compare the results of using the two-sample t-test with the Wilcoxon–Mann–Whitney rank sum test.

(c) Compare plots of the e.d.f.'s for the two samples to see which of the conclusions in (a) or (b) seems to be correct.

(d) This data gathering experiment could also have been done by having individuals at each location count at the same one-hour periods during the day. Comment on what might have been the advantage of using matched pairs.

9.18 The number of visitors to Carlsbad Caverns was counted for a one-week period that included the fourth of July in 1979 and 1980. Treat these data as random samples and use the Wilcoxon–Mann–Whitney rank sum test to see if the mean number of visitors is the same for both years. Use $\alpha = .10$ and state the p-value.

Number of Visitors for the Week of the Fourth of July 1979	Number of Visitors for the Week of the Fourth of July 1980
397	314
286	257
268	278
254	252
571	613
604	646
384	253

9.19 Given below are population projections to the nearest million for the U.S.A. for the years 1985 and 2000. Treat these numbers as if they represented random samples from the two populations ($n_x = 233$ and $n_y = 260$) and use the Wilcoxon–Mann–Whitney rank sum test to see if the mean age of the population is the same for both years. Let $\alpha = .10$ and state the p-value.

Age Group	1985 Projection	2000 Projection
Under 5 years	19	18
5 to 13 years	29	35
14 to 17 years	14	16
18 to 24 years	28	25
25 to 34 years	40	34
35 to 44 years	31	41
45 to 54 years	23	36
55 to 64 years	22	23
65 and over	27	32

9.20 Use the Wilcoxon–Mann–Whitney rank sum test on the employee absenteeism data of Problem Setting 9.1 to test for equal means. Compare results with the example of Section 9.2, which used the two-sample t-test on these data.

9.21 Use the Wilcoxon–Mann–Whitney rank sum test on the data given in Exercise 9.2 and compare results with the solution in that section.

9.22 Use the Wilcoxon–Mann–Whitney rank sum test on the data given in Exercise 9.10 and compare results with the solution found by using the two-sample t-test.

9.23 Use the Wilcoxon–Mann–Whitney rank sum test on the data given in Exercise 9.13 and compare results with the solution found by using the two-sample t-test.

9.24 Use the Wilcoxon–Mann–Whitney rank sum test on the data given in Exercise 9.16 and compare results with the solution found by using the two-sample t-test.

9.4
Review Exercises

9.25 Two emission control devices were tested on 20 cars of the same model with 10 cars using device A and the other 10 cars using device B. The emission levels were checked and showed $\bar{X}_A = 1.24$ with $s_A = 0.06$ while $\bar{X}_B = 1.00$ with $s_B = 0.03$. Assuming that the emission levels are normally distributed, test $H_0: \mu_A = \mu_B$ versus $H_1: \mu_A > \mu_B$ with $\alpha = .01$. Find a 95 percent confidence interval for the difference in mean emission levels.

9.26 Two random samples of sizes 40 and 60 are selected to test the effectiveness of two sleeping pills. The number of hours of sleep is recorded for each individual following the administration of the sleeping pill. The results are summarized below. Find a 90 percent confidence interval for the true mean difference in sleeping time for the two pills.

Pill 1	Pill 2
$\bar{X} = 8.65$	$\bar{Y} = 7.15$
$s_x^2 = 9.00$	$s_y^2 = 5.00$
$n_x = 40$	$n_y = 60$

9.27 Test for equality of variance based on the sample results in Exercise 9.26. Let $\alpha = .05$.

9.28   Octane determinations have been made on a brand of gasoline at two different geographic locations. The first location was at an elevation of 5600 feet and the second location was at an elevation of 1200 feet. Test each of these two samples for normality.

Elevation 5600:   82.1, 82.1, 83.1, 83.0, 82.8, 83.0, 82.1, 83.0, 82.3, 81.7, 82.9, 82.8, 82.2

Elevation 1200:   84.0, 83.5, 84.0, 85.0, 83.1, 83.5, 81.7, 85.4, 84.1, 83.0, 85.8, 84.0, 84.2, 82.2, 83.6, 84.9

9.29   Test for equality of variance based on the sample results in Exercise 9.28. Let $\alpha = .05$.

9.30   Use the appropriate test for equality of means in Exercise 9.28 to determine if the octane rating is the same for this brand of gasoline at the two elevations. Let $\alpha = .05$.

9.31   Make a graph of the two empirical distribution functions for the sample data in Exercise 9.28 to see if the decision in Exercise 9.30 appears to be justified.

9.32   Use the Wilcoxon–Mann–Whitney rank sum test to see if the mean high temperature in Des Moines is greater than the mean high temperature in Spokane. Let $\alpha = .05$ and state the p-value.

Des Moines: 83, 91, 94, 89, 89, 96, 91, 92, 90
Spokane: 78, 82, 81, 77, 79, 81, 80, 81

9.33   Make a graph of the two empirical distribution functions for the sample data in Exercise 9.32 to see if the decision made in that exercise seems to be jusitified.

Bibliography

Additional material on the topics presented in this chapter can be found in the following publications.

Conover, W. J. (1973). "Rank Tests for One Sample, Two Samples and k Samples without the Assumption of a Continuous Distribution Function," *The Annals of Statistics, 1,* 1105–1125.

Conover, W. J. (1980). *Practical Nonparametric Statistics, 2nd ed.,* John Wiley and Sons, Inc., New York.

Conover, W. J. and Iman, R. L. (1981). "Rank Transformations as a Bridge between Parametric and Nonparametric Statistics," *The American Statistician,* **35**(3), 124–129.

Iman, R. L. (1976). "An Approximation to the Exact Distribution of the Wilcoxon–Mann–Whitney Rank Sum Test Statistic," *Communications In Statistics,* **A5**(7), 587–596.

10
Contingency Tables

PRELIMINARY REMARKS

A contingency table is an array of counting numbers (i.e., 0, 1, 2, etc.) in matrix form, as specified below, where those numbers represent frequencies. For example, at the end of the semester a representative of the university's registrar's office may say he examined the grade reports of 72 students, or he may wish to be more specific and say he examined 18 grade reports of engineering students, 16 grade reports of education majors, and 38 grade reports of agriculture students. These grade reports can be expressed in the form of a 1 × 3 ("one by three") contingency table.

ENGR	EDUC	AGRL	Total
18	16	38	72

The contingency table has one row and three columns.

The representative may wish to be more specific and state whether these grade reports indicated semester honors for each student. This leads to a 2 × 3 contingency table.

	ENGR	EDUC	AGRL	Total
No Honors	10	10	21	41
Honors	8	6	17	31
Totals	18	16	38	72

The totals, including two **row totals**, three **column totals**, and one **grand total**, are optional but are usually included for the reader's convenience.

In general an $r \times c$ contingency table is an array of counts, or frequencies, in r rows and c columns. Such tables provide a convenient method of displaying data, especially when the data may be classified by two criteria: one represented in the rows and the other represented in the columns. The original data will usually have to be summarized so it can be put in the form of a contingency table. For example, the original data for the above contingency tables may have looked like this.

Student Name	Major	GPA	Honors
Murray, J. L.	Education	3.25	Yes
Pierce, H. J.	Agriculture	3.60	Yes
Klatt, D. L.	Engineering	2.83	No
etc.			

One qualitative variable, "major," is used to classify each grade report into different columns. Another qualitative variable, "honors," is used to classify each grade report further into different rows. In this case both variables are qualitative variables, but quantitative variables may be used as well if the range for the variable is divided into intervals. For example, the row variable could have been GPA instead of "honors," where row 1 is "less than 3.25," and row 2 is "3.25 or more." The numbers in the **cells** of the table (i.e., the intersection of a row and column) are the frequencies with which the respective row and column classifications occurred together in the original data.

10.1
2 × 2 Contingency Tables

PROBLEM SETTING 10.1

A production supervisor is concerned about a possible difference in the quality of production for the two shifts operating in a factory that manufactures shirts. Fifty shirts are randomly sampled from the output of the day shift, and another 50 shirts are randomly sampled from the night shift. Each shirt is carefully inspected to see if it is defective. Defective shirts are either sold as "factory seconds" at a lower price, repaired and sold at the higher price, or discarded if the defect is serious. The production supervisor summarizes the sample results in a contingency table having 2 rows and 2 columns as follows.

Made by	Classification of Shirt		Totals
	Defective	Nondefective	
Day Shift	3	47	50
Night Shift	5	45	50
Totals	8	92	100

PROBLEM SETTING 10.2

A university admissions officer checks to see if the graduation rate is the same for male and female students admitted as beginning freshmen students. A random sample of students admitted as beginning freshmen in September, 1975 are classified as male or female, and as to whether or not they graduated from that school within 5 years. Again the data may be presented in a contingency table with 2 rows and 2 columns.

	Entering Freshmen		
Sex of Student	Graduated	Did Not Graduate	Totals
Male	16	28	44
Female	18	19	37
Totals	34	47	81

THE HYPOTHESIS OF INTEREST

Although there are wide variations in the types of data presented in the two problem settings, the question of interest is basically the same in both settings: Is the row classification independent of the column classification? In Problem Setting 10.1 this translates as "Is the defective proportion the same for both shifts?" In Problem Setting 10.2 the question becomes "Of the entering freshmen students, is the proportion graduating within 5 years the same for male students as it is for female students?"

Problem Settings 10.1 and 10.2 are different in several basic respects. In Problem Setting 10.1 row totals are not random, since they represent sample sizes that are determined prior to sampling. In contrast, the row totals in Problem Setting 10.2 are not known prior to drawing and examining the sample of students, so they are random. The exact distributions of the respective test statistics, used for analyzing the questions posed above, are different from one another because of this difference in the nature of the row totals. However, the exact distributions of the test statistics are almost never used because they are difficult to derive. The approximate distribution for both test statistics is a distribution known as the **chi-square distribution.**

The chi-square distribution involves one parameter, called **degrees of freedom**. Table A4 gives selected quantiles for the family of chi-square distributions.

INDEPENDENCE

In Chapter 3 independence between rows and columns of a frequency table was established by seeing if the joint probabilities equaled the product of the marginal probabilities. The frequency tables in Chapter 3 represented the entire population, so the frequency table was converted easily into probabilities by dividing the frequencies by the total population size.

In the problem settings of this section the frequency tables represent random samples from populations, so the probabilities are not known exactly but may be estimated from the sample data. Then a measure of the degree of independence exhibited by the data is given by a comparison of the estimated joint probabilities with the product of the estimated marginal probabilities.

EXAMPLE

In Problem Setting 10.2 the sample data may be converted to estimated probabilities by dividing each number by the total sample size 81.

	Graduated	Did Not Graduate	Totals
Male	16	28	44
Female	18	19	37
Totals	34	47	81

Sample Frequencies

	Graduated	Did Not Graduate	Totals
Male	.1975	.3457	.5432
Female	.2222	.2346	.4568
Totals	.4198	.5802	1.0000

Estimated Probabilities

The estimated joint probabilities are given in the cells, such as the joint probability for a student being "Female" and "Did Not Graduate," which is estimated as .2346. The marginal probabilities are estimated by the numbers in the margins. The marginal probability of a student being "Female" is estimated as .4568, and the marginal probability of a student being in the "Did Not Graduate" classification is estimated as .5802. The product of the two estimated marginal probabilities,

$$(.4568)(.5802) = .2650,$$

is not exactly equal to the estimated joint probability .2346.

Hence, one might conclude that these two variables are not independent of one another. However, one should keep in mind that this is a sample from the population and some differences should be expected due to sampling error, or variability. Just how much sampling variability can be tolerated before reaching the conclusion that these variables are in fact not independent is determined by the methods covered in this chapter.

SOME NOTATION

The test statistic measures the variability between the estimated joint probability, and the product of the estimated marginal probabilities, for all four cells in the contingency table. The following notation will be used.

cell (i, j) = the location in row i and column j

p_{ij} = the probability of a sample observation being classified in cell (i, j).

$\hat{p}_{ij}$ = the relative frequency in cell (i, j), from an observed sample.

$p_{i.}$ = the marginal probability for row i, equals $\Sigma_j p_{ij}$

$\hat{p}_{i.}$ = the relative frequency for row i, from an observed sample, equals $\Sigma_j \hat{p}_{ij}$

$p_{.j}$ = the marginal probability for column j, equals $\Sigma_i \hat{p}_{ij}$

$\hat{p}_{.j}$ = the relative frequency for column j, from an observed sample, equals $\Sigma_i \hat{p}_{ij}$

n = total number of observations

EXAMPLE

In the previous example the joint probabilities p_{11}, p_{12}, p_{21}, and p_{22} were unknown, as were the marginal probabilities $p_{1.}$, $p_{2.}$, $p_{.1}$, and $p_{.2}$. Their estimates are given in the table of estimated probabilities as $\hat{p}_{11} = .1975$, $\hat{p}_{12} = .3457$, $\hat{p}_{21} = .2222$, and $\hat{p}_{22} = .2346$. The estimated marginal probabilities are $\hat{p}_{1.} = .5432$, $\hat{p}_{2.} = .4568$, $\hat{p}_{.1} = .4198$, and $\hat{p}_{.2} = .5802$.

THE RATIONALE BEHIND THE TEST STATISTIC

If row classification is independent of column classification, each joint probability p_{ij} equals the product of the respective marginal probabilities $p_{i.}$ and $p_{.j}$. Since the probabilities are unknown, the independence property is judged on the basis of the estimates $\hat{p}_{ij}$ to see if they are approximately equal to the products $\hat{p}_{i.}$ times $\hat{p}_{.j}$.

A statistic used to measure the degree of independence exhibited by a table of sample frequencies is

$$T = \sum_{i=1}^{2} \sum_{j=1}^{2} \frac{(\hat{p}_{ij} - \hat{p}_{i.}\hat{p}_{.j})^2}{\hat{p}_{i.}\hat{p}_{.j}/n},$$

which is a function of the difference between $\hat{p}_{ij}$ and the product $\hat{p}_{i.}\hat{p}_{.j}$, squared to make all of the terms positive, standardized by dividing by $\hat{p}_{i.}\hat{p}_{.j}/n$ to account for unequal cell sizes, and then summed over all of the cells. This statistic T simplifies considerably in Equation (10.1) to make the computations easier.

A TEST FOR INDEPENDENCE IN A 2 × 2 CONTINGENCY TABLE

Data. The data may be obtained two ways.
1. A random sample of size n is drawn from a population, and each item is classified into one of two columns of a 2 × 2 contingency table according to one criterion, and into one of two rows according to another criterion. (See Problem Setting 10.2.) Or:

2. The first row of a 2 × 2 contingency table represents a random sample of size R_1 from one population, and the second row represents a random sample of size R_2 from another population. Each observation is classified into one of the two columns. (See Problem Setting 10.1).

In either case the data are represented in a 2 × 2 contingency table,

	Column 1	Column 2	Row Totals
Row 1	a	b	R_1
Row 2	c	d	R_2
Column Totals	C_1	C_2	$n = $ Grand Total

where a, b, c, and d represent the number of observations classified in each of the four cells.

Assumptions. Each observation is classified by row and/or column independently of every other observation. This assumption is satisfied if the sample is random.

Null Hypothesis. The classification by rows is independent of the classification by columns. In symbols,

$$H_0: p_{ij} \neq p_i.p_{.j} \text{ for all four cells } (i, j).$$

Test Statistic

$$T = \frac{n(ad - bc)^2}{R_1 R_2 C_1 C_2} \tag{10.1}$$

Decision Rule. The alternative hypothesis is that the classification by rows of the sample observations is dependent on the column classification. In symbols this becomes

$$H_1: \quad p_{ij} \neq p_i.p_{.j} \text{ for some cell } (i, j).$$

Reject H_0 at the α level of significance if T exceeds the $1 - \alpha$ quantile from a chi-square distribution with 1 degree of freedom, as given in the first row of Table A4.

EXAMPLE

Recall the case of the university admissions officer in Problem Setting 10.2 who was concerned whether the graduation rate was the same for male and female students who were admitted as incoming freshmen. A random sample of 81 students who were incoming freshmen in September 1975 were classified according to whether they graduated within 5 years or not (columns 1 and 2) and according to male or female (rows 1 and 2). This is an example of the single random sample, classified two ways, described in part 1 of **Data**. The results are presented in a 2 × 2 contingency table.

	Graduated	Did Not Graduate	Totals
Male	16	28	44
Female	18	19	37
Totals	34	47	81

The test statistic is

$$T = \frac{n(ad - bc)^2}{R_1 R_2 C_1 C_2} = \frac{81[16(19) - 28(18)]^2}{44(37)(34)(47)} = 1.245.$$

The critical value for a 5 percent level test is 3.841. Since $T = 1.245$ is less than 3.841, the null hypothesis is accepted. The p-value is found from Table A4 by interpolation to be approximately .27. The graduation rate could be the same for male and female students. The observed apparent inequities could be due to chance fluctuation in the sample.

THE PHI-COEFFICIENT

There is a popular measure of association that is used with 2 × 2 contingency tables, called the **phi-coefficient**, or **ϕ-coefficient**. It is used as a measure of the row-column dependence in 2 × 2 contingency tables. If all of the items in row 1 are in one column, and all of the items in row 2 are in the other column, the phi-coefficient will equal $+1$ or -1, representing the most extreme case of row-column dependence. If the proportions classified in each column are the same for both rows, then there is apparent independence between the rows and the columns and the phi-coefficient is close to zero.

The phi-coefficient can also be used in testing the hypothesis of row-column independence. The principal advantage in using ϕ instead of T, presented earlier in this section for testing the same hypothesis, is that one-tailed alternative hypotheses may be considered when ϕ is used as a test statistic. The two-tailed version of the test using ϕ is equivalent to the test using T as a test statistic.

This latter use of ϕ in a hypothesis test is to see if there is a significant association between row and column classifications. This is a different concept than the former use of ϕ, as a measure of the strength of that association.

THE PHI-COEFFICIENT AS A MEASURE OF ASSOCIATION IN 2 × 2 CONTINGENCY TABLES

Data. The data may be presented as counts in a 2 × 2 contingency table as follows:

	Column 1	Column 2	Row Totals
Row 1	a	b	R_1
Row 2	c	d	R_2
Column Totals	C_1	C_2	n

Phi-Coefficient

$$\phi = \frac{ad - bc}{\sqrt{R_1 R_2 C_1 C_2}}$$

Properties

1. ϕ is a measure of the amount of dependence between the row classification and the column classification.
2. ϕ will be close to zero if there is independence between the row classification and the column classification.
3. ϕ will be close to its maximum value of $+1.0$ or its minimum value of -1.0 if the observations in row 1 tend to be classified in an opposite column from those in row 2.

THE PHI-COEFFICIENT AS A STATISTIC TO TEST FOR INDEPENDENCE

The **Data, Assumptions,** and **Null Hypothesis** are the same as in "A Test for Independence in a 2 × 2 Contingency Table," presented earlier in this section.

Test Statistic

$$\phi = \frac{ad - bc}{\sqrt{R_1 R_2 C_1 C_2}}$$

Decision Rule. Use of the phi-coefficient as a test statistic permits one-sided alternative hypotheses to be considered. Let z_α be the $1 - \alpha$ quantile from Table A2.

1. $H_1: p_{11} > p_{1.}p_{.1}$ and $p_{22} > p_{2.}p_{.2}$. Reject H_0 if $\phi > z_\alpha/\sqrt{n}$.
2. $H_1: p_{12} > p_{1.}p_{.2}$ and $p_{21} > p_{2.}p_{.2}$. Reject H_0 if $\phi < -z_\alpha/\sqrt{n}$.
3. $H_1: p_{ij} \neq p_{i.}p_{.j}$ for some cell (i, j). Reject H_0 if $|\phi| > z_{\alpha/2}/\sqrt{n}$.

This is equivalent to the previous test based on T.)

EXAMPLE

In Problem Setting 10.1 the number of defective items in random samples from the production of the day shift and the night shift were summarized in a 2 × 2 contingency table:

	Defective	Nondefective	Totals
Day Shift	3	47	50
Night Shift	5	45	50
Totals	8	92	100

A measure of dependence is desired between the proportion of defectives and the shift responsible for the production. The phi-coefficient is

$$\phi = \frac{ad - bc}{\sqrt{R_1 R_2 C_1 C_2}} = \frac{3(45) - 47(5)}{\sqrt{50(50)(8)(92)}} = -.074,$$

which shows a slight, probably negligible, dependence between the proportion defective and the shift responsible for the production.

If management suspected, prior to sampling, that the night shift tended to produce a higher proportion of defective items, because of fatigue and other factors, a one-tailed test would be set up to test the null hypothesis of independence against the alternative that p_{21} (representing the probability of the night shift producing a defective item) tends to be too large, or

$$H_1 = p_{21} > p_{2.}p_{.1}.$$

The decision rule is to reject H_0 if $\phi < -z_\alpha/\sqrt{n}$, which is $\phi < -1.6449/\sqrt{100} = -.164$ for a 5 percent level of significance. Since ϕ equals $-.074$, the null hypothesis of independence is accepted.

COMMENTS

1. The phi-coefficient and the test statistic T are directly related through the relationship

$$T = n \phi^2.$$

2. The phi-coefficient is simply the correlation coefficient calculated on the observations in the contingency table. That is, if each observation is given numerical values like 1 or 2 depending on whether it is in column 1 or 2 (called the "value of X"), and values 1 or 2 depending on whether it is in row 1 or 2 (called the Y variable), then the value of the correlation coefficient r computed on the X's and Y's is exactly the same as ϕ.

3. The chi-square distribution and normal distribution used in the significance tests involving T and ϕ are only approximations to the exact distribution, which is usually not practical to work with. The approximations work satisfactorily if the row and column totals are not too small. Current theory on this subject indicates that this approximation is satisfactory if each product $R_1 C_1$, $R_1 C_2$, $R_2 C_1$, and $R_2 C_2$ is greater than n.

4. The test presented in this section can be used for testing the hypothesis that two populations have the same proportion p of successes. Let a be the number of successes in a random sample of size R_1 from population 1 and let c be the number of successes in a random sample of size R_2 from population 2. If p_1 and

p_2 represent the true proportions in the two populations, then the null hypothesis is $p_1 = p_2$. To test $H_0: p_1 = p_2$ simply put the data in the form of a 2 × 2 contingency table:

	Number of Successes	Number of Failures	Totals
1st Population	a	$R_1 - a = b$	R_1
2nd Population	c	$R_2 - c = d$	R_2
Totals	C_1	C_2	n

Then use the test of this section.

Exercises

10.1 A random sample of 100 employees in an automobile assembly plant contains 63 men, 48 who belong to the union and 15 who do not, and 37 women, of whom 22 belong to the union and 15 do not. Test the null hypothesis that the proportion of all employees belonging to the union is the same for men and women, against the one-sided alternative that the proportion is higher among male employees. Use a level of significance of 5 percent and state the p-value.

10.2 A company that manufactures aftershave lotion wants to compare two different packaging techniques. One technique uses a picture of a woman and the other does not. One hundred people were polled using the lotion with the picture of the woman and 85 of them rated the lotion as favorable and 15 said it was unfavorable. The package without a picture was shown to 300 people, with 210 favorable responses and 90 unfavorable. Test the hypothesis that the package does not affect opinion—that is, that the proportion of people rating the lotion favorable is the same for both packaging techniques against the one-sided alternative that the proportion is higher for the package with the picture. Use a level of significance of $\alpha = .05$ and find the p-value.

10.3 It has been hypothesized that being left-handed is independent of eye color. Test this hypothesis using the results of a study of 100 individuals with blue eyes and 100 individuals with brown eyes, the results of which are summarized below. Use $\alpha = .05$ and state the p-value.

	Handedness	
Color of Eyes	Left	Right
Blue	5	95
Brown	15	85

10.4 A study by five American and Danish psychiatrists concerned adopted children who became alcoholics as adults. The results of this study, released in 1973, reported on Danish men who had been separated from their biological parents during early infancy.

Fifty-five of these men had one parent who was diagnosed as alcoholic. Ten of these 55 children of an alcoholic parent were found to be alcoholic. These were compared with 78 other adopted men whose biological parents had no known history of alcoholism. Four of the offspring of this nonalcoholic group were found to be alcoholic. Test the hypothesis that the proportion of alcoholic children is the same for both alcoholic and nonalcoholic parents. Let $\alpha = .01$ and find the p-value.

10.5   The Dow Jones industrial average was examined in a random sample of 40 days. Each day's record noted whether the D-J was "up" compared with the previous day or not, and whether the following trading day resulted in an "up" day for the D-J or not. The results are summarized in the 2 × 2 contingency table.

| First Day | Dow Jones Performance on the Following Day | |
	Up	Down
Up	8	14
Down	9	9

Test the hypothesis that the first and second-day performances of the Dow Jones Industrial average are independent of one another. Let $\alpha = .10$ and find the p-value.

10.6   "Free Pizza" (two for the price of one) coupons were distributed by direct mail and by handbill distributed from a local pizzaria. Of the 250 coupons sent by direct mail, 43 were redeemed. Of the 200 handbills distributed, 28 coupons were redeemed. Test the hypothesis that the proportion of persons redeeming coupons is the same, whether the coupon is received by direct mail or by handbill. Let $\alpha = .05$ and state the p-value.

10.7   Compute the phi-coefficient for the data in Exercise 10.5.

10.8   A random sample of grade point averages was obtained for 20 graduate students with U.S. citizenship.

3.28	3.08	2.70	3.76	2.90
2.90	3.58	3.30	3.75	3.21
3.41	3.55	3.20	3.63	4.00
3.85	2.87	3.69	3.80	3.21

A random sample of grade point averages was also obtained for 20 graduate students without U.S. citizenship.

3.71	3.36	4.00	2.92	3.85
3.70	3.90	3.75	3.82	3.85
4.00	3.20	3.50	3.21	4.00
3.05	3.91	3.46	3.92	3.59

Use a level of significance of $\alpha = .05$ to test the hypothesis that the proportion of graduate students with a grade point average above 3.50 is the same for both groups of students.

10.2

The $r \times c$ Contingency Table

PROBLEM SETTING 10.3

In a study published in the *Journal of the American Medical Association* (Nov. 1980) investigators studied the effect of smoking abstinence and length of survival for patients diagnosed with small cell lung cancer. A total of 112 cigarette smoking patients with small cell lung cancer were investigated. The patients were classified according to their smoking habits at the time of diagnosis. Twenty patients had stopped smoking prior to diagnosis (NS-Prior), 35 had stopped smoking at diagnosis (NS-Dx), and 57 patients continued smoking. The study was aimed at determining if discontinuation of smoking, even at diagnosis, would have beneficial effects on survival. As part of such a study the characteristics of the patients must be examined to make sure there is no built-in bias in the experiment. That is, the investigators checked the three categories of patients to make sure that they were not different with respect to such items as extent of disease, pretreatment performance, age and sex distribution, and pack-years smoked. Data such as these are conveniently listed as counts in a **contingency table**. For example, the counts for the characteristic of pack-years smoked are summarized in a **3 × 6** (3 rows, 6 columns) **contingency table.**

	Pack-Years Smoked						
	0–19	20–39	40–59	60–79	80–99	100+	Totals
NS-Prior	1	16	15	12	9	4	57
NS-Dx	0	13	11	8	2	1	35
S	1	7	6	3	2	1	20
Totals	2	36	32	23	13	6	112

THE CHI-SQUARE TEST FOR CONTINGENCY TABLES

The above problem setting is an example of the more general $r \times c$ contingency table that has r rows and c columns. The hypothesis of interest is that the row and column classifications are independent of one another. As in the previous section, row and column independence implies that the joint probability p_{ij} associated with an observation being in the cell in row i and column j must equal the product of the marginal probabilities for row i, $p_{i.}$, and column j, $p_{.j}$,

$$p_{ij} = p_{i.}p_{.j} \quad \text{(independence)},$$

for all of the cells. If O_{ij} represents the actual cell count, then p_{ij} is estimated from the observed relative frequency

$$\hat{p}_{ij} = O_{ij}/n$$

for each cell. The marginal probabilities are estimated from the observed row totals R_i and column totals C_j, after division by the total sample size n:

$$\hat{p}_{i.} = R_i/n, \qquad \hat{p}_{.j} = C_j/n.$$

The **observed** cell count is O_{ij}. The **expected** cell count E_{ij} under the assumption of independence is obtained by multiplying the estimated probability $\hat{p}_{ij}$ of each observation being in cell (i, j) times the total number of observations n,

$$
\begin{aligned}
E_{ij} &= n\hat{p}_{i.}\hat{p}_{.j} \quad \text{(by independence)} \\
&= n \cdot \frac{R_i}{n} \cdot \frac{C_j}{n} \\
&= \frac{R_i C_j}{n}.
\end{aligned}
$$

A statistic that is used to measure the degree of independence exhibited by a sample frequency table involves computing the difference between the observed and expected cell counts, $O_{ij} - E_{ij}$, squaring to make all terms positive, dividing by E_{ij} to standardize for the different cell frequencies, and summing over all cells:

$$T = \sum_i \sum_j \frac{(O_{ij} - E_{ij})^2}{E_{ij}}.$$

This statistic T is equal to the statistic given in Equation (10.1) for 2×2 contingency tables. If the null hypothesis is true, T may be compared with quantiles from a chi-square distribution (Table A4) with $(r - 1)(c - 1)$ degrees of freedom as an approximate distribution for T. This approximation appears to be satisfactory if most of the E_{ij}'s are 1 or greater, and all E_{ij}'s are at least 0.5 in size.

EXAMPLE

The random sample of 112 patients in Problem Setting 10.3 was classified by smoking status and pack-years smoked. The expected values E_{ij} are listed in the same cell with the observed values O_{ij} for convenience.

	Pack-Years Smoked						
	0–19	20–39	40–59	60–79	80–99	100+	Totals
NS-Prior	1 / 1.02	16 / 18.32	15 / 16.29	12 / 11.71	9 / 6.62	4 / 3.05	57
NS-Dx	0 / .63	13 / 11.25	11 / 10.00	8 / 7.19	2 / 4.06	1 / 1.88	35
S	1 / .36	7 / 6.43	6 / 5.71	3 / 4.11	2 / 2.32	1 / 1.07	20
Totals	2	36	32	23	13	6	112

CHI-SQUARE TEST OF INDEPENDENCE FOR $r \times c$ CONTINGENCY TABLES

Data. The data are classified into rows by one criterion (or variable) and into columns by another criterion (or variable). They are represented as frequency counts O_{ij} in an $r \times c$ contingency table (one with r rows and c columns), where O_{ij} equals the number of observations in cell (i, j)—that is, the intersection of row i with column j.

	Column				Totals
	1	2	$\cdots$	c	
Row 1	O_{11}	O_{12}	$\cdots$	O_{1c}	R_1
2	O_{21}	O_{22}	$\cdots$	O_{2c}	R_2
$\cdots$	$\cdots$	$\cdots$	$\cdots$	$\cdots$	$\cdots$
r	O_{r1}	O_{r2}	$\cdots$	O_{rc}	R_r
Totals	C_1	C_2	$\cdots$	C_c	n

Assumptions. Each observation is classified independently of every other observation. This assumption is satisfied if there is one random sample being classified by two criteria (rows and columns), or if each row represents a random sample of observations being classified into columns.

Null Hypothesis. The classification by rows is independent of the classification by columns. This hypothesis requires individual interpretation for each application. In symbols,

H_0: $p_{ij} = p_{i.}p_{.j}$ for all cells (i, j).

Test Statistic. Let

O_{ij} = the **observed** count in row i, column j,

$E_{ij} = \dfrac{R_i C_j}{n}$

= the **expected** count in row i, column j, if the null hypothesis is true.

Then one form for the test statistic is

$$T = \sum_{\substack{\text{all} \\ \text{cells}}} \frac{(O_{ij} - E_{ij})^2}{E_{ij}} \qquad (10.2)$$

where $(O_{ij} - E_{ij})^2/E_{ij}$ is computed separately for each cell, and those quantities are summed over all rc cells. An alternative form for the test statistic, which involves fewer computations, is

$$T = \sum_{\substack{\text{all} \\ \text{cells}}} \frac{O_{ij}^2}{E_{ij}} - n. \qquad (10.3)$$

Decision Rule. Reject the null hypothesis at the level of significance α if T exceeds the $1 - \alpha$ quantile of the chi-square distribution with $(r - 1)(c - 1)$ degrees of freedom. These critical values may be obtained from Table A4.

Comment. Note that the chi-square distribution is only an approximation to the true distribution of T, which is too difficult to obtain in most cases. The approximation is considered quite good unless there are many very small expected values E_{ij}. If most E_{ij}'s are greater than 1.0, the approximation may still be very good, especially if the number of degrees of freedom is large. If some of the E_{ij} are less than 0.5, however, it may be advisable to combine several similar rows or columns, if it is convenient, to raise those low values. This is a judgment decision that requires consideration of individual circumstances in each case.

As a check on the computations the row and column totals for the E_{ij} should be the same as listed for the O_{ij}. In general, any row or column with no observations should be dropped from the analysis. Column 1 has low expected values, hence it could be combined with the adjacent column. Combining those two columns makes sense in this case, since they have similar interpretations. Combining columns means adding the corresponding observed values and expected values. The revised contingency table is given below.

Pack-Years Smoked

	0–39	40–49	60–79	80–99	100+	Totals
NS-Prior	17 / 19.34	15 / 16.29	12 / 11.71	9 / 6.62	4 / 3.05	57
NS-Dx	13 / 11.88	11 / 10.00	8 / 7.19	2 / 4.06	1 / 1.88	35
S	8 / 6.79	6 / 5.71	3 / 4.11	2 / 2.32	1 / 1.07	20
Totals	38	32	23	13	6	112

With the aid of Equation (10.2) the computations proceed as follows.

Cell	O_{ij}	E_{ij}	$\dfrac{(O_{ij} - E_{ij})^2}{E_{ij}}$
Row 1, Col. 1	17	$\dfrac{57(38)}{112} = 19.339$	$\dfrac{(17 - 19.339)^2}{19.339} = .028$
Row 1, Col. 2	15	$\dfrac{57(32)}{112} = 16.286$	$\dfrac{(15 - 16.286)^2}{16.286} = 0.10$
. . .	. . .	. . .	. . .
Row 3, Col. 5	1	$\dfrac{20(6)}{112} = 1.071$	$\dfrac{(1 - 1.071)^2}{1.071} = .00$
Totals	112	112	$T = 3.88$

If Equation (10.3) had been used instead of Equation (10.2), the calculations would look like this:

Cell	O_{ij}	E_{ij}	$\dfrac{O_{ij}^2}{E_{ij}}$
Row 1, Col. 1	17	$\dfrac{57(38)}{112} = 19.339$	14.94
Row 1, Col. 2	15	$\dfrac{57(32)}{112} = 16.286$	13.82
. . .	. . .	. . .	. . .
Row 3, Col. 5	1	$\dfrac{20(6)}{112} = 1.071$	0.93
Totals	112	112	115.88

$$T = \sum_{\substack{\text{all} \\ \text{cells}}} \frac{O_{ij}^2}{E_{ij}} - n = 115.88 - 112 = 3.88$$

The result is the same as before, but with fewer computations. The null hypothesis is that the characteristic of pack-years smoked is the same for all patient groups. To test this hypothesis at the 5 percent level of significance the .95 quantile of a chi-square distribution with

$$(r - 1)(c - 1) = (3 - 1)(5 - 1) = 2(4) = 8$$

degrees of freedom is obtained from Table A4. This value is 15.51. Since T does not exceed 15.51, the null hypothesis is accepted. The number of pack-years smoked does not tend to be significantly different for the three patient groups. The p-value associated with $T = 3.88$ is seen from Table A4 to be much more than .4.

CLASSIFICATION USING QUANTITATIVE VARIABLES

The "natural" application of the contingency table analysis is when each observation is measured by two **qualitative** variables, as in the above example. However, **quantitative** variables may also be used to classify the observations into rows, or columns, or both. The range of each quantitative variable may have to be arbitrarily divided into intervals so that each interval represents one row or one column. If the quantitative variable is "years of formal education," the intervals may be "did not graduate from high school" "high school graduate," "holds bachelor's degree," "holds master's degree," and "holds doctoral degree." If the quantitative variable is salary, the intervals may be "< \$5000," "≥ \$5000 but < \$10,000," and so on, with a final category including all salaries above a certain value.

EXAMPLE

When a company receives the customer registration cards from customers who have purchased its products, it may want to do some analysis on the data received. For example, does advertising effectiveness vary according to the age of the customer? Here it is necessary to find out if there is a dependence between type of advertising that led to the sale, and the age of the customer. A random sample of 200 cards from the thousands of cards in the files provides the counts in the following contingency table.

Reason for Selecting Product	\multicolumn{5}{c}{Age of Customer}					Totals
	0–20	21–30	31–40	41–50	Over 50	
Previous Ownership	10	21	28	8	6	73
Store Display	4	8	2	0	0	14
Catalog	3	4	5	0	0	12
Magazine	4	2	23	8	0	36
Newspaper	12	18	14	2	4	50
Other	5	8	2	0	0	15
Totals	38	60	74	18	10	200

The calculations proceed as follows.

Cell	O_{ij}	E_{ij}	O_{ij}^2/E_{ij}
Row 1, Col. 1	10	13.87	7.21
Row 1, Col. 2	21	21.90	20.14
. . .	. . .	. . .	. . .
Row 6, Col. 5	0	0.75	0.00
		Total	252.28

$$T = \sum_{\substack{all \\ cells}} O_{ij}^2/E_{ij} - n = 252.28 - 200 = 52.28$$

Since there are

$$(r - 1)(c - 1) = 5(4) = 20$$

degrees of freedom, the critical value for a 5 percent test is the .95 quantile from the chi-square distribution with 20 degrees of freedom, which is given in Table A4 as 31.41. Since T exceeds this critical value, the null hypothesis of independence between the type of advertising and the age of the customer is rejected, and the conclusion may be made that different types of advertising influence the various age groups differently. The p-value is approximately .0001.

CRAMER'S CONTINGENCY COEFFICIENT

Sometimes a **measure of association** in a contingency table is a useful descriptive statistic to obtain. There are many different measures that are used with contingency tables. The one presented here is widely used.

CRAMER'S CONTINGENCY COEFFICIENT

Data. The data are counts or frequencies, in an $r \times c$ contingency table. Let q equal the smaller of r and c.

Cramer's Contingency Coefficient

$$\Phi = \frac{T}{n(q-1)} \tag{10.4}$$

T is the statistic defined earlier in this section by Equation (10.2) or Equation (10.3).

Properties
1. Φ (capital "phi") is a measure of the association between the row classification and column classification in a contingency table.
2. Φ will tend to be close to zero, its minimum value, if there is independence between the row and column variables.
3. If all of the observations in each row tend to collect in one column, but in a different column for each row, Φ will tend to be close to 1.0, its maximum value.
4. Φ is the $r \times c$ analog of the phi-coefficient for 2×2 contingency tables. In fact, Φ equals ϕ squared if $r = 2$ and $c = 2$.

Hypothesis Test. If a hypothesis test is desired using Φ, the test should be conducted using T as a test statistic, in the manner described earlier in this section. Since Φ is a function of T, any conclusions or p-values obtained using T may be applied to Φ just as well.

EXAMPLE

A measure of association may be expressed for the relationship between advertising effectiveness and age of the customer, using the data given in the previous example. Since $r = 6$ and $c = 5$, q equals the smaller of the two, $q = 5$, and

$$\Phi = \frac{T}{n(q-1)} = \frac{52.28}{200(4)} = .065,$$

which indicates that the relationship is slight, even though it is statistically significant as shown by the test using T in the previous example.

COMMENTS

1. If the different rows represent different *populations*, then significance in the overall contingency table may be interpreted as "Some of the populations

have different *distributions* than others." In cases like this a followup question "Which pairs of populations have different distributions?" is often important to answer. This question may be answered by forming smaller contingency tables with only 2 rows each, and analyzing each of these tables separately using the test in this section to see which pairs of populations are significantly different. This pairwise comparison of rows should be undertaken only if the original contingency table shows a significant row-column dependence.

2. If one of the variables (rows, say) is quantitative, and the other variable represents a qualitative variable, it often happens that the alternative hypothesis of interest is whether some of the columns tend to yield *larger* observations than other columns. In this case the rows represent *ordered* categories, and a multi-sample rank test is more appropriate to use because of its greater power to detect such differences if they exist. A rank test for two samples (2 columns) and ordered row categories was given in Section 9.3.

3. If both variables, represented by rows and columns, are quantitative in nature, then there is a choice between the methods of this section to test for independence, and the methods of the next chapter, which are oriented toward detecting particular types of dependence, as will be explained in the next chapter.

Exercises

10.9 The value of the test statistic T was found to be 52.28 in the second example of this section by use of Equation (10.3). Compute the value of T using Equation (10.2). Which equation for computing T do you think is the easier to use?

10.10 Compute Cramer's contingency coefficient for the first example in this section. How does this value compare with the value given in the final example of this section? What interpretation is associated with the value you have computed?

10.11 The placement service at a university selects the records of 200 applicants in three major areas, classified according to major and degree. A summary table is given as follows.

Major	Degree		
	B.S.	M.S.	Ph.D.
Accounting	78	17	5
Elec. Engr.	33	12	3
Biology	13	31	8

Test the hypothesis that degree type is independent of major. Let $\alpha = .05$ and state the *p*-value.

10.12 A personnel manager of a large company wishes to know if employee satisfaction is independent of their job category. Test the hypothesis of independence based on the following data with $\alpha = .05$. State the p-value.

Satisfaction	Categories			
	I	II	III	IV
High	40	60	52	48
Medium	103	87	82	88
Low	57	53	66	64

10.13 A political survey has been taken to compare opinion of the administration's foreign policy against the political party of the respondent. Test the null hypothesis of independence of opinion on foreign policy and political party. Use $\alpha = .01$ and state the p-value.

Party	Opinion	
	Approve	Disapprove
Republican	114	53
Democrat	87	27
Other	17	8

10.14 At the end of a course a college professor makes the following tally based on grade received and class attendance. Test the hypothesis that grade received in the course is independent of the number of days absent. Use $\alpha = .05$ and state the p-value.

Number of Days Absent	Grade Received	
	Pass	Fail
0–3	24	0
4–6	18	2
More than 7	3	13

10.15 A random sample of freshmen entering Texas Tech University in 1972 was examined again 5 years later to see how many had graduated. The students were divided into three categories: those from Lubbock County, those from other counties in Texas, and those from out of state. Test the hypothesis that graduation status is independent of where the student's home is located. Let $\alpha = .10$ and state the p-value.

| | Graduation Status | |
Student's Home	Did Graduate	Did Not Graduate
Lubbock County	78	15
Other Texas Counties	92	24
Outside of Texas	38	13

10.16   Three supermarkets have a policy of advertising specials on certain days of the week in order to attract customers. A customer count is maintained at these supermarkets from the hours of 11 a.m. to 2 p.m. for the five weekdays. Use the summary of these counts to test the hypothesis that choice of supermarket is independent of the day of the week. Let $\alpha = .01$ and state the p-value.

| | Day of the Week | | | | | |
Supermarket	Mon.	Tues.	Wed.	Thur.	Fri.	Totals
Furrs	605	639	790	617	573	3224
Safeway	674	657	723	937	686	3677
A & P	564	790	477	529	501	2861
Totals	1843	2086	1990	2083	1760	9762

10.3
The Chi-Square Goodness-of-Fit Test

PROBLEM SETTING 10.4

The personnel supervisor in a plant employing over 500 people is interested in the absentee pattern of the employees. She decides to keep a record of all sick leave taken in a two-week period, with the following results.

| | Day on Which Sick Leave Was Taken | | | | |
Mon.	Tues.	Wed.	Thurs.	Fri.	Total
19	24	8	14	31	96

She is interested in testing the hypothesis that there is an equal chance of sick leave being taken on any of the five work days of the week.

GOODNESS-OF-FIT TEST

This is a situation that often arises, where the experimenter has in mind certain probabilities, and wants to see if the observations confirm or deny the preconceived probabilities. Any test that compares the observations with the probabilities, to see how well the probabilities "fit" the observed frequencies, is called a **goodness-of-fit test.** One of the most widely used goodness-of-fit tests is the **chi-square goodness-of-fit test.** It is especially appropriate if the data naturally fall into a single-row contingency table; however, it may also be used with data that are grouped into intervals, and the count in each interval is used to form a single-row contingency table.

The concept underlying this goodness-of-fit test is the same concept described in earlier sections of this chapter. The observed counts O_{ij} are compared with the expected counts E_{ij} using a statistic that sums $(O_{ij} - E_{ij})^2/E_{ij}$ over

---

THE CHI-SQUARE GOODNESS-OF-FIT TEST

Data. The data appear as counts or frequencies in a contingency table with one row. In some cases the observations may be on a continuous-valued random variable, but are grouped into intervals and presented in a contingency table:

			Cells		
1	2	3	$\cdots$	k	Total
O_1	O_2	O_3	$\cdots$	O_k	n

Assumptions. The data represent observations from a random sample.

Null Hypothesis. The probability of an observation falling into cell 1 is p_1, cell 2 is $p_2, \ldots$, cell k is p_k, for some specified values of p_1 to p_k.

Test Statistic. The test statistic is

$$T = \sum_{i=1}^{k} \frac{(O_i - np_i)^2}{np_i}, \tag{10.5}$$

which may be written also as

$$T = \sum_{i=1}^{k} \frac{O_i^2}{np_i} - n \tag{10.6}$$

for easier computations.

Decision Rule. Reject the null hypothesis at the α level of significance if T exceeds the $1 - \alpha$ quantile from the chi-square distribution with $k - 1$ degrees of freedom, where k is the number of cells. These critical values may be found in Table A4.

Comment. The chi-square distribution is only an approximation to the true distribution, which is usually too difficult to obtain. The chi-square approximation is usually fairly good, however, unless some of the values of np_i are quite small. For example, values of np_i as small as 1.0 seem to cause no difficulty. However, values less than 0.5 should be avoided by combining adjacent cells, if this is reasonable to do in the experiment.

all of the cells. In this application the expected counts E_{ij} are not estimated from the marginal probabilities as before, but they are given directly from the individual cell probabilities p_i specified in the null hypothesis. Also only one subscript is needed because of the single-row nature of the contingency table. Thus the statistic looks like

$$T = \sum \frac{(O_i - E_i)^2}{E_i},$$

where $E_i = np_i$ gives the expected cell count.

EXAMPLE

If the personnel supervisor in Problem Setting 10.4 feels that "bonafide" sick leave should occur with equal probability over all five work days, then she is interested in testing the null hypothesis

$$H_0: \quad p_1 = \tfrac{1}{5}, \quad p_2 = \tfrac{1}{5}, \quad p_3 = \tfrac{1}{5}, \quad p_4 = \tfrac{1}{5}, \quad p_5 = \tfrac{1}{5}.$$

She is interested in seeing how well these probabilities fit the observations.

	Mon.	Tues.	Wed.	Thurs.	Fri.	Totals
Observed	19	24	8	14	31	96
Expected (np_i)	19.2	19.2	19.2	19.2	19.2	96

The expected values are found using $np_i = 96(\tfrac{1}{5}) = 19.2$. The test statistic for the chi-square goodness-of-fit test is

$$T = \sum_{i=1}^{5} \frac{O_i^2}{np_i} - n$$

$$= \frac{(19)^2}{19.2} + \frac{(24)^2}{19.2} + \cdots + \frac{(31)^2}{19.2} - 96$$

$$= \frac{2158}{19.2} - 96 = 16.40.$$

For a 5 percent test the .95 quantile from a chi-square distribution with $k - 1 = 4$ degrees of freedom yields a critical value of 9.488. Therefore, the null hypothesis is easily rejected, and the conclusion is that sick leave is not equally likely to be taken over all five working days. The p-value associated with $T = 16.40$ is found from Table A4 by interpolation to be about .003.

COMMENT

If some of the parameters in the hypothesized probability distribution are not specified, then they must be estimated from the data. Each such parameter estimated from the data results in a decrease of one unit in the number of degrees of freedom used in the chi-square goodness-of-fit test. See the following example for an illustration.

EXAMPLE

An operations manager is interested in building a mathematical model that will accurately describe the number of breakdowns in his production machines, and will imitate the variability of breakdowns as well. Certain assumptions of independence and equal probability of breakdown among the machines leads to a binomial distribution, where n is the number of machines (10 in this case), and p is the probability of a machine breaking down sometime during the week. The value of p is unknown and must be estimated from the data.

The data are collected for 26 weeks with the following results.

	Number of Machines Breaking Down Each Week							Total
	0	1	2	3	4	5	6 or more	
Number of Weeks Observed	5	9	10	1	0	1	0	26

The parameter p represents the probability of a machine breaking down in one week, so it may be estimated by dividing the total number of machines breaking down by the total number of machines times the 26 weeks:

$$\text{estimate for } p = \frac{\text{total number of machines breaking down}}{\text{(total number of machines)} \times \text{(total number of weeks)}}$$

$$= \frac{0(5) + 1(9) + 2(10) + 3(1) + 5(1)}{10(26)}$$

$$= .142.$$

The binomial probabilities for $n = 10$ machines and $p = .142$ may be estimated from Table A1, by interpolation between $p = .10$ and $p = .15$. These probabilities are multiplied by the total number of weeks to get the expected number of weeks with $X = 0, 1, 2, \ldots$ breakdowns.

From Table A1	p_i By subtraction	$26\,p_i$
$P(X \leq 0) = .2212$	$P(X = 0) = .2212$	5.75
$P(X \leq 1) = .5750$	$P(X = 1) = .3538$	9.20
$P(X \leq 2) = .8377$	$P(X = 2) = .2627$	6.83
$P(X \leq 3) = .9560$	$P(X = 3) = .1183$	3.08
$P(X \leq 4) = .9914$	$P(X = 4) = .0354$	0.92
$P(X \leq 5) = .9988$	$P(X = 5) = .0074$	0.19
$P(X \leq 6) = .9999$	$P(X = 6) = .0011$	0.03
$P(X \leq 7) = 1.0000$		

Because of the small expected values in the last two cells, the last 3 cells are combined.

	Number of Machines Breaking Down Each Week					
	0	1	2	3	4 or more	Totals
Observed	5	9	10	1	1	26
Expected $(26\,p_i)$	5.75	9.20	6.83	3.08	1.14	26

The use of Equation (10.6) gives

$$T = \frac{25}{5.75} + \frac{81}{9.20} + \frac{100}{6.83} + \frac{1}{3.08} + \frac{1}{1.14} - 26$$

$$= 29.00 - 26$$

$$= 3.00.$$

The number of degrees of freedom is 3, one less than $k - 1 = 4$, because the parameter p in the binomial distribution is estimated from the data. The null hypothesis is that the number of machines breaking down each week follows a binomial distribution with $n = 10$ and unknown p. The test statistic $T = 3.00$ is compared with the critical value of 5 percent test, 7.815, which is the .95 quantile from a chi-square distribution with 3 degrees of freedom. The null hypothesis is accepted. The binomial distribution appears to provide a reasonable model for the operations manager to work with. The p-value is about .40.

An example showing how the chi-square goodness-of-fit test may be used with continuous random variables will now be given.

EXAMPLE

If accidents occur at random in a factory, it is sometimes assumed that the time between successive accidents follows an exponential distribution. As explained in Section 5.5 the exponential distribution is a continuous distribution, so the time intervals between successive accidents need to be collected into intervals in order to use the chi-square goodness-of-fit test. Observations over a month resulted in the following times between accidents.

	Times Between Accidents (hours)						Total
	0–1	1–2	2–3	3–4	4–5	> 5	
Frequency	108	62	21	18	11	9	229

The exponential distribution has one parameter that needs to be estimated from the data using the reciprocal of the sample mean. The sample mean may be estimated by assuming that the observations in each interval all fall at the midpoint of the interval; .5 for the first interval, 1.5 for the second interval, and so on. The last interval includes all numbers greater than 5, so 6.0 will be used as a "midpoint," under the belief that most of the observations in that interval should be between 5 and 7. Thus

$$\bar{X} = \frac{1}{229}[0.5(108) + 1.5(62) + 2.5(21)$$

$$+ 3.5(18) + 4.5(11) + 6.0(9)]$$

$$= 1.60.$$

The probability that an exponential random variable will fall between t_0 and t_1 is

$$p_1 = e^{-\lambda t_0} - e^{-\lambda t_1},$$

where λ is estimated by $1/\bar{X}$, and where e is a well-known constant equal to 2.718, approximately. Thus:

Interval	Probability p_i		np_i
0 to 1	$e^0 - e^{-1/1.6}$	= .4647	106.4
1 to 2	$e^{-1/1.6} - e^{-2/1.6}$	= .2488	57.0
2 to 3	$e^{-2/1.6} - e^{-3/1.6}$	= .1331	30.5
3 to 4	$e^{-3/1.6} - e^{-4/1.6}$	= .0713	16.3
4 to 5	$e^{-4/1.6} - e^{-5/1.6}$	= .0381	8.7
> 5	$e^{-5/1.6}$	= .0440	10.1
Totals		1.0000	229.0

The test statistic is

$$T = \sum_{i=1}^{6} \frac{O_i^2}{E_i} - n$$

$$= \frac{(108)^2}{106.4} + \frac{(62)^2}{57.0} + \frac{(21)^2}{30.5} + \frac{(18)^2}{16.3} + \frac{(11)^2}{8.7} + \frac{(9)^2}{10.1} - 299$$

$$= 4.33.$$

Since this value of the test statistic T is less than 9.488, the .95 quantile of a chi-square distribution with 4 degrees of freedom, the null hypothesis is accepted at the 5 percent level. The test utilizes 4 degrees of freedom rather than 5 since λ was estimated from the sample data. The exponential distribution with $\lambda = 1/1.6 = 0.625$ seems to fit well the observed time intervals between accidents. The p-value associated with $T = 4.33$ is approximately .37, as shown by Table A4.

A COMPARISON WITH THE LILLIEFORS TEST

The test presented in this section is a goodness-of-fit test. The Lilliefors test of Chapter 5 is also a goodness-of-fit test, so it is appropriate at this time to compare the two tests.

Recall that the Lilliefors test is presented for only two situations. One situation is to see if the sample data could reasonably be regarded as having come from a normal population, where neither parameter μ nor σ is specified. The other situation is to see if the sample data could reasonably be regarded as

"GOOD NEWS! CLAYBORNE'S LOW SCORE WAS JUST THE OBSERVATION WE NEEDED TO FIT A NORMAL DISTRIBUTION TO THE TEST SCORES! CLAYBORNE?"

having come from an exponential population, where the parameter λ is again unspecified. The Lilliefors goodness-of-fit test, as presented, is limited to those two situations. The chi-square goodness-of-fit test, on the other hand, is more versatile. It may be applied to test the goodness of fit for any distribution— discrete or continuous. It is more natural to use the Lilliefors test to test for normal or exponential distributions, however, because the data may be used directly in the Lilliefors test. The chi-square test requires grouping the data into arbitrary intervals, and in so doing it loses some "information" contained in the data.

The authors recommend the Lilliefors test whenever possible because it tends to have more power and it is easier to use. The chi-square goodness-of-fit test may be used when the other, more appropriate, tests are not available. In the previous example the data were already grouped, so the chi-square test is the only one that could be used. With large data sets both procedures are tedious and so a computer should be used.

Exercises

10.17 The brief table of random numbers given in Figure 1.2 was generated by a computer program designed to produce the digits 0, 1, 2, 3, 4, 5, 6, 7, 8, and 9 in a random fashion and with equal frequencies. Use a chi-square goodness-of-fit test to see if the digits in the first 10 columns of Figure 1.2 occur with equal frequencies. Let $\alpha = .05$. (*Hint:* See Exercise 1.9 for a summary of these data.)

10.18 The genotypes of a cross in a genetics experiment should occur in the ratio of 9:3:3:1. Test the hypothesis that this ratio fits the following experimental results 150, 46, 40, and 20. Let $\alpha = .05$.

10.19 Midway through the greyhound racing season at Denver, Colorado, the tabulation below was made on the number of times that each starting position finished in the money (i.e., 1st, 2nd, or 3rd). Test the null hypothesis that each position is equally likely to finish in the money; that is, test that a uniform distribution is the correct probability model for these counts. Let $\alpha = .01$.

Starting Position:	1	2	3	4	5	6	7	8	Total
Finish (1st, 2nd, or 3rd):	189	177	146	139	131	145	122	143	1192

10.20 A classical set of data in statistical texts concerns the number of deaths (per year per corps) of Prussian cavalry after being kicked by a horse. A Poisson probability distribution (see Section 5.5) given by the function $f(x) = e^{-\lambda}\lambda^x/x!$, $x = 0, 1, 2, \ldots$, where λ is the mean of the population, has been proposed to fit these horsekick fatalities. Use the chi-square goodness-of-fit test to see if the Poisson model fits these data at a level of significance of $\alpha = .01$. The chi-square distribution with $5\text{-}1\text{-}1 = 3$ degrees of freedom should be used to find critical values for this test since the mean had to be estimated from the sample data ($\hat{\lambda} = .61$).

Number of Deaths:	0	1	2	3	4	Total
Poisson Probability:	.54	.33	.10	.02	.01	
Observed Frequency:	109	65	22	3	1	200

10.21 In Figure 4.8 a probability model was given for the number of fire calls in one day. Use a chi-square goodness-of-fit test to see if the model probabilities fit the number of calls at a fire station for a 100-day period. Let $\alpha = .05$ and state the p-value.

Number of Fire Calls:	0	1	2	3	4	5	6
Probability:	.10	.15	.15	.20	.15	.15	.10
Observed Frequency:	15	20	17	21	9	10	8

10.22 An insurance company wants to compare their dollar claim settlement against the summary given by *Consumer Report* in Exercise 4.10. Use a chi-square goodness-of-fit test to compare the insurance company claim settlements against the percentages given by *Consumer Report* based on 300 randomly selected claims. Let $\alpha = .10$ and state the p-value.

Dollar Amount of Claim Settlement	Percent of Claims	Claims for Insurance Company
1–99	13.8	50
100–199	15.8	40
200–299	10.9	30
300–399	7.9	28
400–499	5.0	20
500–999	16.8	36
1000–1499	6.9	15
1500–1999	4.0	17
2000–2499	3.0	5
2500–4999	5.9	14
5000–9999	4.0	12
10,000–4999	5.0	16
50,000 or more	1.0	7
		300

10.23 In Figure 1.1 a summary of order of draft selection and date of birth was given for the 1970 draft lottery. A summary of these data is given below to show the number of selections in each month that were among the first half of potential draftees (i.e., draft number less than or equal to 183). If the draft were truly random, then each month should be approximately evenly split above and below draft order number 183.5. Use a chi-square goodness-of-fit test to check the randomness of the draft based on the following summaries. Let $\alpha = .01$ and state the p-value.

Month	Number of Selections in the First 183	Number Expected
Jan.	13	15.5
Feb.	12	14.5
March	9	15.5
April	11	15
May	14	15.5
June	14	15
July	14	15.5
Aug.	18	15.5
Sept.	19	15
Oct.	13	15.5
Nov.	21	15
Dec.	25	15.5

10.24 A class of 43 students took an exam having 10 questions. The results are summarized below. Use a chi-square goodness-of-fit test to see if a binomial distribution can be used as a mathematical model to describe the test results as was done in the second example of this section. Use $\alpha = .05$ and state the p-value.

Number of Questions Missed	Number of Students
0	2
1	8
2	16
3	12
4	5
5 or more	0

10.4

Review Exercises

10.25 A medication for treating the common cold is tested by giving the medication to half of a randomly selected group of people with colds. The other half receives a placebo. The summary of each patient's opinion of the treatment is given in the following table. Test the hypothesis that patient opinion is independent of treatment. Let $\alpha = .05$ and state the p-value.

Treatment	Patient's Opinion		
	Harmful	No Effect	Helpful
Medication	16	30	104
Placebo	20	42	88

10.26   The number of magazines sold each hour at a corner newsstand is recorded below for several hours of business. Use a chi-square goodness-of-fit test to see how well a Poisson distribution (see Section 5.5) with $\lambda = 2$ fits these data. Let $\alpha = .01$.

Number of Magazines Sold	Frequency
0	10
1	25
2	33
3	20
4	7
5	3
6	2
	100

10.27   In order to see if customers might be influenced by different colors used on cards good for a discount at a retail store, the store mailed out 1000 cards to its customers using 250 cards for each of four color types. The assignment of color to a customer was made at random. The store then recorded the number of cards of each type that were returned, and this summary is given below. Test the hypothesis that the return of a discount card is independent of the color used on the card. Let $\alpha = .05$ and state the p-value.

	Color Used on Card				Totals
	Red	White	Blue	Green	
Returned	107	106	115	127	455
Not Returned	143	144	135	123	545
Totals	250	250	250	250	1000

10.28   A production supervisor is interested in knowing if the number of breakdowns on four machines is independent of the shift using the machines. Test this hypothesis based on the following sample counts. Let $\alpha = .05$ and state the p-value.

	Number of Breakdowns for Machine:				Totals
Shift	A	B	C	D	
7–3	14	9	17	19	59
3–11	14	17	27	30	88
11–7	19	14	19	26	78
Totals	47	40	63	75	225

10.29   The number of automobile fatalities for 1 year in a large city has been grouped by the day of the week. Use a chi-square goodness-of-fit test to see how well a uniform distribution fits these data. Let $\alpha = .01$ and state the p-value.

Sun.	Mon.	Tue.	Wed.	Thur.	Fri.	Sat.	Total
27	20	17	17	21	28	18	148

10.30 New employees at an army base are given a brief instruction on security regulations. The instructions can be given with a videotape, which the employees watch, or a brochure, which the employees read. New employees are assigned randomly to one or the other, and they are given a test over the security regulations at the completion of their indoctrination. Results for one set of employees show that of the 48 employees who watched the videotape, 13 failed to pass the test the first time. Of the 52 employees who read the brochure, 9 failed to pass the test. If this represents a random sample of all new employees, what do these results say about the effectiveness of the two instruction programs compared with each other?

10.31 The U.S. Army wants to know if results on a mechanical aptitude test can be used to predict success in a machinery repair course. Sixty-five applicants for the course are all accepted into the course. Forty-six applicants had scores above 110 on their mechanical aptitude tests, and 38 of them successfully completed the course. The remaining 19 applicants scored below 110, and only 6 of them successfully completed the course. Test the hypothesis that the mechanical aptitude test results are independent of the course results, against the one-sided alternative that high test scores tend to predict successful course performance.

10.32 A consumer group is examining the contents of breakfast cereal boxes to see if the weight of the cereal is at least as much as advertised on the box. Random samples of four different brands of cereal were examined.

	Cereal Brand			
	A	B	C	D
Total Number of Boxes Examined	40	50	40	45
Number of Boxes Judged Too Light	6	3	8	2

Test the hypothesis that each of the four cereal manufacturers has the same probability of producing a box of cereal that contains less than the specified weight of contents.

10.33 A manufacturer of farm machinery receives bolts from six different suppliers. Random samples of 100 bolts were drawn from shipments from each of the six suppliers, and the number of defective bolts from each was noted. These numbers of defective bolts were 6, 7, 17, 8, 10, and 4. Test the null hypothesis that the probability of receiving a defective bolt is the same from all six suppliers.

Bibliography

Additional material on the topics presented in this chapter can be found in the following publications.

Conover, W. J. (1974). "Some Reasons for Not Using the Yates Continuity Correction on 2×2 Contingency Tables," *Journal of the American Statistical Association,* **69**, 374–376.

Conover, W. J. (1980). *Practical Nonparametric Statistics,* 2nd ed., John Wiley and Sons, Inc., New York.

Shapiro, S. S., Wilk, M. B., and Chen, Mrs. H. J. (1968). "A Comparative Study of Various Tests for Normality," *Journal of the American Statistical Association,* **63,** 1343–1372.

11

Correlation

In Chapter 10 the phi-coefficient resulting from the calculation of the correlation coefficient on the observations in a contingency table is used as a measure of association between the row classification and the column classification of the contingency table. In this chapter the correlation coefficient of Section 4.5 is used as a measure of association between observations on a pair of variables X and Y. When such variables occur in pairs a scatterplot is useful for providing information about possible relationships between X and Y that might not be easily discernible from observing the data. That is, it may be of interest to know if the data exhibit a straight line relationship (or nearly so), or a curvilinear relationship, or perhaps no relationship at all. The correlation coefficient calculated on raw data is quite useful in determining the strength of **linear** relationships between paired observations. Since linear relationships play an important role in statistics, particularly in the regression techniques of the next chapter, the first section of this chapter is devoted to studying the correlation coefficient computed on raw data.

Relationships that are not linear may exist between a pair of variables. In such cases the value of Y may increase as X increases, or the value of Y may decrease as X increases. Relationships such as these are more generally described as **monotonic**. The strength of monotonic relationships is also of interest in statistics and such relationships are more accurately measured by the correlation coefficient calculated on rank transformed data. The second section of this chapter considers this calculation, which is referred to as rank correlation and which was introduced in Section 4.6 under the name of Spearman's rho.

323

11.1

Introduction to Correlation

PROBLEM SETTING 11.1

School systems give standardized tests each year, such as the Comprehensive Tests of Basic Skills (CTBS), in order to evaluate the performance of students. Areas tested by the CTBS include reading, language (spelling, mechanics, and expression), math, reference skills, science, and social studies. The CTBS scores can be used in a number of ways, such as comparing schools within a school system, evaluating the school's performance during the current year compared to previous years, and for comparing the school system's performance against the national average. The CTBS scores can also be used to provide a standard against which the results of experimental programs can be evaluated. For example, has a new method of teaching science caused a significant change in the science scores of students taking the CTBS?

Additionally, since so many areas are tested, scatterplots of pairs of scores can be made to see if the scores tend to follow a straight line. For example, do reading scores and math scores exhibit a straight line relationship? A plot of these scores for 20 randomly selected middle schools is given in Figure 11.1. Notice the tendency of the points to follow a straight line. Such a tendency indicates a linear relationship exists between reading scores and math scores. However, the relationship shows some scattering around the straight line. There is a need to measure the **strength of the linear relationship** between the two variables. The usual statistic used for the measure is the **sample correlation coefficient**, which was introduced in Section 4.5.

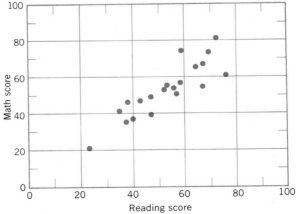

FIGURE 11.1 A Scatterplot of Reading Score versus Math Score for 20 Middle Schools

THE SAMPLE CORRELATION COEFFICIENT

The sample correlation coefficient results from a calculation made on a bivariate sample (X_i, Y_i), $i = 1, \ldots, n$, using Equation (4.19) given in Section 4.5, which is repeated here for easy reference.

$$r = \frac{\Sigma X_i Y_i - n\bar{X}\bar{Y}}{[(\Sigma X_i^2 - n\bar{X}^2)(\Sigma Y_i^2 - n\bar{Y}^2)]^{1/2}}$$

This calculation produces a number between -1 and $+1$, which indicates how closely a scatterplot of the sample pairs (X_i, Y_i) resembles a straight line. That is, r measures the strength of a linear relationship between X and Y. For example, in Figure 11.2 a straight line can be drawn through all pairs of points in each scatterplot.

SCATTERPLOTS AND CORRELATION COEFFICIENTS

In Figure 11.2 X and Y increase together in (a) and (b) (positive correlation), while in (c) and (d) Y decreases as X increases (negative correlation). In fact, since all pairs of points are on a straight line, (a) and (b) represent the extreme in positive correlation of $+1$ while (c) and (d) represent the extreme in negative correlation of -1. Of course, when dealing with real world problems, nature

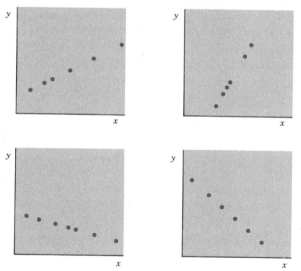

FIGURE 11.2 Sample Pairs (X_i, Y_i) Which Have a Perfect Positive or Perfect Negative Correlation

seldom responds with all sample pairs (X_i, Y_i) on a straight line. Rather, some scatterplots will appear close to a linear relationship while others will not. Examples associated with different values of r are shown in Figure 11.3.

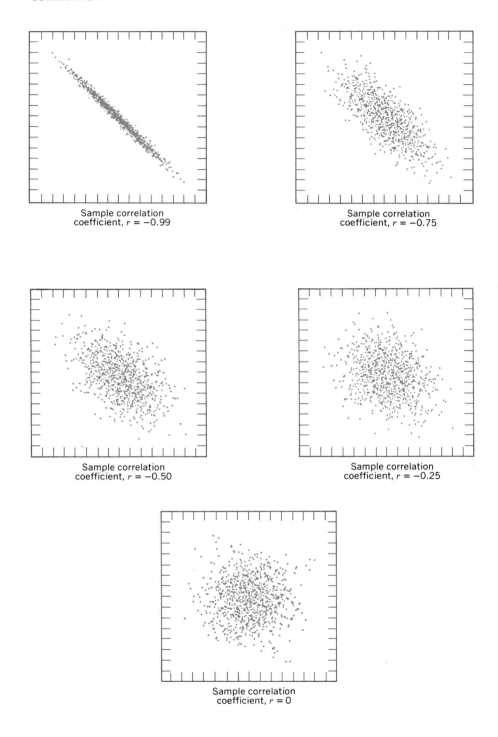

FIGURE 11.3 Scatterplots for Different Values of the Sample Correlation Coefficient

(Figure 11.3 continued)

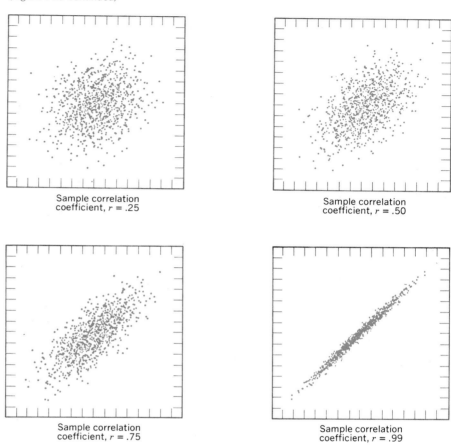

Sample correlation
coefficient, $r = .25$

Sample correlation
coefficient, $r = .50$

Sample correlation
coefficient, $r = .75$

Sample correlation
coefficient, $r = .99$

BIVARIATE DISTRIBUTIONS

The scatterplots in Figure 11.3 illustrate the concept that two random variables X and Y behave jointly according to some **bivariate probability distribution**. That is to say, X has a probability distribution and Y also has a probability distribution; however, their joint behavior is not completely determined by their marginal behavior. When the two random variables are correlated, they are no longer independent of one another. Rather the random variable Y is influenced by the random variable X and vice versa.

If each of the marginal distributions for the random variables X and Y is normal, then the joint probability distribution of X and Y is usually given by a particular distribution known as the **bivariate normal distribution**. A graph to represent the three-dimensional nature of the bivariate normal density function is given in Figure 11.4. In Figure 11.4 lines have been added that are parallel to each of the X and Y axes such that they each intersect the elliptical mound-shaped portion of the graph. These lines illustrate the fact pointed out previously that the distributions for the random variables X and Y are each normal.

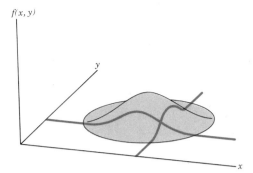

$f(x, y)$

y

x

FIGURE 11.4 Plot of a Bivariate Normal Density

One of the parameters of a bivariate normal distribution is the population correlation coefficient ρ, which is estimated by r. If ρ equals zero in a bivariate normal distribution, then the two variables are independent. In this section a procedure is given for testing whether an observed sample correlation coefficient is sufficiently large in absolute value to indicate that the population

---

PROCEDURES FOR INFERENCES REGARDING CORRELATION (BIVARIATE NORMAL DISTRIBUTION)

Data. The data consist of n pairs (X_i, Y_i), $i = 1, \ldots, n$ from a bivariate population. Compute the sample correlation coefficient

$$r = \frac{\Sigma X_i Y_i - n\,\overline{X}\,\overline{Y}}{[(\Sigma X_i^2 - n\overline{X}^2)(\Sigma Y_i^2 - n\overline{Y}^2)]^{1/2}}.$$

Assumptions
1. The sample of pairs (X_i, Y_i), $i = 1, \ldots, n$ is a random sample.
2. The joint distribution of (X, Y) is a bivariate normal distribution.

Estimation. The sample correlation coefficient provides a point estimate of the population correlation coefficient ρ. An approximate $100(1 - \alpha)$ percent confidence interval may be formed as follows.

1. Convert r to a new variable w using the table in Figure 11.5.
2. Find

$$w_L = w - \frac{z_{\alpha/2}}{\sqrt{n-3}} \quad \text{and} \quad w_U = w + \frac{z_{\alpha/2}}{\sqrt{n-3}}$$

where $z_{\alpha/2}$ is the $1 - \alpha/2$ quantile of a normal distribution given in Table A2.
3. Use the table in Figure 11.5 to convert w_L to r_L and w_U to r_U. The $100(1 - \alpha)$ percent confidence interval for ρ is from r_L to r_U.

Null Hypothesis. H_0: $\rho = 0$ (i.e., X and Y are independent).

Test Statistic. The test statistic is r.

Decision Rule. The decision rule depends on the alternative hypothesis of interest. Critical values are given in Figure 11.6.

1. H_1: $\rho > 0$. Reject H_0 if $r > T_{\alpha,n}$; otherwise accept H_0.
2. H_1: $\rho < 0$. Reject H_0 if $r < -T_{\alpha,n}$; otherwise accept H_0.
3. H_1: $\rho \neq 0$. Reject H_0 if $|r| > T_{\alpha/2,n}$; otherwise accept H_0.

correlation coefficient is not equal to 0. Also a method for finding a confidence interval for ρ is presented. Both procedures rely heavily upon the assumption of a bivariate normal distribution for X and Y. This assumption represents the primary weakness of these procedures, as the bivariate normal distribution assumption is difficult to check. One way to check for bivariate normality is to test for normality of the marginal distributions of X and Y by themselves, using the Lilliefors test. If the X's are regarded as possibly having a normal distribution, then the Y's are tested for normality using the same test. Acceptance of normality for both variables *indicates the possibility* of a bivariate normal distribution, but *does not assure* a bivariate normal distribution. However, rejection of normality by either test indicates the assumption of bivariate normality is not justified. Thus it is possible to reject bivariate normality, but not prove it. If the assumption is reasonable, these methods are the best available in terms of power. If the assumption is not valid, the methods in the next section should be used. Two examples will now be given to illustrate these procedures.

FIGURE 11.5 Table for Finding a Confidence Interval for ρ in Bivariate Normal Populations. (If r or w is negative, use this Table, adding a negative sign to both.)

r	w	r	w	r	w	r	w
0.00	0.0000	0.25	0.2554	0.50	0.5493	0.75	0.9730
0.01	0.0100	0.26	0.2661	0.51	0.5627	0.76	0.9962
0.02	0.0200	0.27	0.2769	0.52	0.5763	0.77	1.0203
0.03	0.0300	0.28	0.2877	0.53	0.5901	0.78	1.0454
0.04	0.0400	0.29	0.2986	0.54	0.6042	0.79	1.0714
0.05	0.0500	0.30	0.3095	0.55	0.6184	0.80	1.0986
0.06	0.0601	0.31	0.3205	0.56	0.6328	0.81	1.1270
0.07	0.0701	0.32	0.3316	0.57	0.6475	0.82	1.1568
0.08	0.0802	0.33	0.3428	0.58	0.6625	0.83	1.1881
0.09	0.0902	0.34	0.3541	0.59	0.6777	0.84	1.2212
0.10	0.1003	0.35	0.3654	0.60	0.6931	0.85	1.2562
0.11	0.1104	0.36	0.3769	0.61	0.7089	0.86	1.2933
0.12	0.1206	0.37	0.3884	0.62	0.7250	0.87	1.3331
0.13	0.1307	0.38	0.4001	0.63	0.7414	0.88	1.3758
0.14	0.1409	0.39	0.4118	0.64	0.7582	0.89	1.4219
0.15	0.1511	0.40	0.4236	0.65	0.7753	0.90	1.4722
0.16	0.1614	0.41	0.4356	0.66	0.7928	0.91	1.5275
0.17	0.1717	0.42	0.4477	0.67	0.8107	0.92	1.5890
0.18	0.1820	0.43	0.4599	0.68	0.8291	0.93	1.6584
0.19	0.1923	0.44	0.4722	0.69	0.8480	0.94	1.7380
0.20	0.2027	0.45	0.4847	0.70	0.8673	0.95	1.8318
0.21	0.2132	0.46	0.4973	0.71	0.8871	0.96	1.9459
0.22	0.2237	0.47	0.5101	0.72	0.9076	0.97	2.0923
0.23	0.2342	0.48	0.5230	0.73	0.9287	0.98	2.2976
0.24	0.2448	0.49	0.5361	0.74	0.9505	0.99	2.6467

Note: These values of w can be found from r using the relationship $w = \frac{1}{2} \ln [(1 + r)/(1 - r)]$. Likewise values of r can be found from w using the relationship $r = (e^{2w} - 1)/(e^{2w} + 1)$.

n	$\alpha = 0.05$	$\alpha = 0.025$	$\alpha = 0.005$
4	0.900	0.950	0.990
5	0.805	0.878	0.959
6	0.729	0.811	0.917
7	0.669	0.754	0.875
8	0.621	0.707	0.834
9	0.582	0.666	0.798
10	0.549	0.632	0.765
11	0.521	0.602	0.735
12	0.497	0.576	0.708
13	0.476	0.553	0.684
14	0.458	0.532	0.661
15	0.441	0.514	0.641
16	0.426	0.497	0.623
17	0.412	0.482	0.606
18	0.400	0.468	0.590
19	0.389	0.456	0.575
20	0.378	0.444	0.561
21	0.369	0.433	0.549
22	0.360	0.423	0.537
23	0.352	0.413	0.526
24	0.344	0.404	0.515
25	0.337	0.396	0.505
26	0.330	0.388	0.496
27	0.323	0.381	0.487
28	0.317	0.374	0.479
29	0.311	0.367	0.471
30	0.306	0.361	0.463
31	0.301	0.355	0.456
32	0.296	0.349	0.449
33	0.291	0.344	0.442
34	0.287	0.339	0.436
35	0.283	0.334	0.430
36	0.279	0.329	0.424
37	0.275	0.325	0.418
38	0.271	0.320	0.413
39	0.267	0.316	0.408
40	0.264	0.312	0.403
41	0.260	0.308	0.398
42	0.257	0.304	0.393
43	0.254	0.301	0.389
44	0.251	0.297	0.384
45	0.248	0.294	0.380
46	0.246	0.291	0.376
47	0.243	0.288	0.372
48	0.240	0.285	0.368
49	0.238	0.282	0.365
50	0.235	0.279	0.361

FIGURE 11.6 Values of $T_{\alpha,n}$ Used for Tests of Significance on ρ in Bivariate Normal Populations.

Note: Values of $T_{\alpha,n}$ not given can be found as $t_{\alpha,n-2}/(n - 2 - t_{\alpha,n-2}^2)^{1/2}$ where $t_{\alpha,n-2}$ i the $1 - \alpha$ quantile of the Student's t-distribution obtained from Table A3 with $n - 2$ degrees of freedom.

EXAMPLE

The measurements of reading and math scores for the random sample of 20 middle schools in Problem Setting 11.1 are represented in a scatterplot in Figure 11.1. A worksheet for these data is given in Figure 11.7. Use these data to test $H_0: \rho = 0$ versus $H_1: \rho \neq 0$ with $\alpha = .01$.

Substitution of the sample calculations in Figure 11.7 into Equation (11.1) gives

$$r = \frac{62375 - 20(53.2)(55.4)}{[(60352 - 20(53.2)^2)\ (65344 - 20(55.4)^2)]^{1/2}} = .890.$$

since $\overline{X} = 53.2$ and $\overline{Y} = 55.4$.

Middle School	Reading Score X	Math Score Y	XY	X^2	Y^2
1	47	42	1974	2209	1764
2	71	81	5751	5041	6561
3	64	68	4352	4096	4624
4	35	43	1505	1225	1849
5	43	50	2150	1849	2500
6	60	75	4500	3600	5625
7	38	47	1786	1444	2209
8	59	59	3481	3481	3481
9	67	69	4623	4489	4761
10	56	57	3192	3136	3249
11	67	57	3819	4489	3249
12	57	54	3078	3249	2916
13	69	75	5175	4761	5625
14	38	38	1444	1444	1444
15	55	59	3186	2916	3481
16	76	63	4788	5776	3969
17	53	57	3021	2809	3249
18	40	40	1600	1600	1600
19	47	52	2444	2209	2704
20	23	22	506	529	484
Totals	1064	1108	62375	60352	65344

FIGURE 11.7 Worksheet for the Calculation of the Sample Correlation Coefficient

Decision Rule

Reject H_0 if $|r| > T_{.005,20} = 0.561$, where $T_{.005,20}$ is obtained from Figure 11.6. Since $|.890|$ is greater than 0.561, H_0 is easily rejected at $\alpha = .01$. The p-value is much smaller than .01. Since the value of the sample correlation is positive, it is concluded that greater math scores are associated with higher reading scores, and vice versa.

Confidence Interval

A 95 percent confidence interval is found by converting $r = .890$ to $w = 1.4219$ from Figure 11.5. For $n = 20$ and $z_{.025} = 1.9600$, the values of w_L and w_U become

$$w_L = 1.4219 - \frac{1.9600}{\sqrt{17}} = .9465$$

and

$$w_U = 1.4219 + \frac{1.9600}{\sqrt{17}} = 1.8973,$$

which convert to $r_L = .74$ and $r_U = .96$ respectively, using the relationship $r = (e^{2w} - 1)/(e^{2w} + 1)$ from Figure 11.5. The approximate 95 percent confidence interval for ρ is from .74 to .96.

EXAMPLE

The state racing commission is interested in knowing if a significant correlation exists between attendance and the amount of money wagered. Use the data below for 10 racing days to test H_0: $\rho = 0$ versus H_1: $\rho > 0$ with $\alpha = .05$.

	Attendance (hundreds) X	Amount Wagered (millions) Y	XY	X²	Y²
	117	2.07	242.19	13689	4.2849
	128	2.19	280.32	16384	4.7961
	122	3.14	383.08	14884	9.8596
	119	2.26	268.94	14161	5.1076
	131	3.40	445.40	17161	11.5600
	135	2.89	390.15	18225	8.3521
	125	2.93	366.25	15625	8.5849
	120	2.66	319.20	14400	7.0756
	130	3.33	432.90	16900	11.0889
	127	3.53	448.31	16129	12.4609
Totals	1254	28.40	3576.74	157558	83.1706

$$\bar{X} = 125.4 \quad \bar{Y} = 2.84$$

$$r = \frac{3576.74 - 10(125.4)(2.84)}{[(157558 - 10(125.4)^2)(83.1706 - 10(2.84)^2)]^{1/2}} = .554$$

Decision Rule

From Figure 11.6, $T_{.05,10} = 0.549$ so H_0 will be rejected if $r > 0.549$. Since $r = .554$, H_0 is rejected. The p-value is slightly smaller than .05. Of course each

*"COACH, HAVE YOU NOTICED A CORRELATION BETWEEN
THE NUMBER OF INJURIES AND THE NUMBER OF TIMES
THEY RUN TO HUXLEY'S SIDE OF THE FIELD?"*

of these sample sets should be checked for normality using the Lilliefors test. Also, a scatterplot of these data would be helpful in determining if a straight line relationship exists.

COMMON MISINTERPRETATIONS OF CORRELATION

This section is closed with a warning about the misinterpretation of the word correlation as it is used in statistics. Statisticians usually use the term correlation as a measure of the strength of a linear relationship between two variables. The problem arises when the word correlation is used by nonstatisticians to imply what is commonly referred to as a "cause and effect" relationship. That is to say, the occurrence of one event causes another event to occur, hence the events are correlated. However, "cause and effect" is never implied in the statistical use of the word correlation. For example, during the 1950s it was discovered that during months when consumption of soft drinks was high, there were also large numbers of cases of polio; that is, a statistical correlation existed between these two sets of numbers. Was consumption of soft drinks causing polio? No! In truth, in the summer months when the weather was hot the number of polio cases and the consumption of soft drinks both naturally increased. As the weather cooled off during the rest of the year, both variables decreased in value. While there was a strong correlation between polio cases and soft drink sales, there was no reason to believe that one was causing the other.

Another common misinterpretation is that $r = 0$ indicates the lack of a relationship between X and Y. Values of r close to zero merely indicate the lack of a *linear* relationship between X and Y; other forms of relationships may still exist.

Exercises

Note: The Lilliefors test should be used to test the marginal distributions for normality in each of the following exercises. Remember that these tests will not ensure a joint bivariate normal distribution, but may rule it out.

11.1 A summary of financial statistics of institutions of higher education compiled by the National Center for Education Statistics for the fiscal year 1977 gives a breakdown of current revenues. Given below is a summary by different regions of the country of the percentage of revenue obtained from tuition and fees and government appropriations (mostly state government). Compute the sample correlation coefficient for these data (see Exercise 4.42) and test H_0: $\rho = 0$ versus H_1: $\rho < 0$ with $\alpha = .05$. What meaning is associated with your decision?

	Percentage of Total Revenue	
Region	From Tuition and Fees	From Government Appropriations
New England	33.0	17.1
Mideast	29.1	27.2
Great Lakes	22.8	33.3
Plains	19.7	35.4
Southeast	18.0	37.9
Southwest	13.6	45.8
Rocky Mountains	17.0	36.3
Far West	11.9	41.9

11.2 The National Center for Education Statistics has compiled data on the type of degree granted to males and females for 11 consecutive academic years. Data regarding the awarding of first-professional degrees is given below. Compute the sample correlation coefficient for these data and test H_0: $\rho = 0$ versus H_1: $\rho > 0$ with $\alpha = .05$. What meaning is associated with your decision?

Academic Year	Number of First-Professional Degrees	
	To Men	To Women
1966–67	31,064	1,429
1967–68	33,083	1,645
1968–69	34,069	1,612
1969–70	33,344	1,908
1970–71	35,797	2,479
1971–72	41,021	2,753
1972–73	46,827	3,608
1973–74	48,904	5,374
1974–75	49,230	7,029
1975–76	53,210	9,851
1976–77	52,668	12,112

11.3 The Bureau of Labor Statistics reported expenditures on clothing and personal care for low-, intermediate-, and high-budget four-person families in the northcentral portion of the United States for the autumn of 1979 as follows.

	Type of Budget					
	Low		Intermediate		High	
	Cloth.	Per. Care	Cloth.	Per. Care	Cloth.	Per. Care
Chicago	797	326	1153	426	1684	586
Cincinnati	994	305	1430	401	2083	550
Cleveland	920	409	1323	544	1930	756
Detroit	830	339	1195	452	1750	619
Kansas City	927	391	1329	516	1940	724
Milwaukee	973	357	1390	465	2034	645
Minneapolis	872	357	1251	470	1820	651
St. Louis	820	334	1184	426	1747	573

Compute the sample correlation coefficient for each budget type—low, intermediate, and high—and for all data pooled together (see Exercise 4.48). Test the hypothesis $H_0: \rho = 0$ versus $H_1: \rho \neq 0$ for each budget type. Also, test the same hypothesis for all data pooled together. Let $\alpha = .05$ for all tests. How do you explain the difference in the decision made on each of the tests on budget type and the decision made for all data pooled together?

11.4 The monetarist school of economic thought states that all monetary expansion in excess of real economic growth is eventually transmitted into inflation. In other words, money supply (cash and bank deposits) should only be permitted to grow as fast as the output of goods and services grows. Any more rapid growth will merely create excess dollars to chase after the same supply of goods and services—and send prices soaring. Given below are 20-year summary data for 14 countries. Compute the sample correlation coefficient for these data (see Exercise 4.46) and test H_0: $\rho = 0$ versus H_1: $\rho > 0$ with $\alpha = .005$. What does your decision indicate about the monetarist hypothesis?

Country	Annual Rates of Change 1960–1979			
	(1) Money Supply Growth	(2) Real Economic Growth	(1) − (2) Anticipated Inflation	Actual Inflation
Brazil	44.7%	9.1%	35.6%	32.1%
Japan	16.8	8.4	8.4	6.1
Italy	16.5	4.3	12.2	9.1
Spain	16.2	6.3	9.9	10.5
Mexico	15.8	6.1	9.7	8.4
France	10.4	4.1	6.3	6.4
Netherlands	9.5	4.2	5.3	6.4
Sweden	9.1	3.1	6.0	6.6
Germany	8.7	3.8	4.9	4.4
U.K.	8.1	2.5	5.6	8.3
Canada	8.1	4.8	3.3	5.3
Switzerland	7.8	2.8	5.0	5.1
Belgium	7.1	4.2	2.9	5.2
U.S.A.	5.3	3.6	1.7	4.7

11.5 One way to evaluate a stock is to see how many shares are held by financial institutions—that is banks, investment and insurance companies. Given below is such a listing for 10 discount variety chain stores along with their recent earnings per share as obtained from a recent issue of *Standard and Poor's Stock Guide*. Compute the sample correlation coefficient for these data (See Exercise 4.47) and test H_0: $\rho = 0$ versus H_1: $\rho \neq 0$ with $\alpha = .10$. What meaning is associated with your decision?

Company	Number of Institutional Shares Held (thousands)	Earnings per Share
Danners Inc.	197	$1.28
Duckwall-Alco	24	2.20
Fed-Mart	99	4.80
Grand Central	2	1.22
Hartfield-Zodys	377	1.97
K-Mart	53,170	2.96
Thrifty Corp.	544	1.46
Wal-Mart	3,452	2.34
Woolworth	6,587	4.84
Zayre	632	2.64

11.6 Market share trends of the liquor industry are given below for the years 1959 and 1974. Compute the sample correlation coefficient for these data (see Exercise 4.43) and then find a 95 percent confidence interval for the population correlation coefficient. What interpretation do you give to this interval?

| | Percent of Market | |
	1959	1974
Whiskey types:		
Straights	25.9	16.3
Spirit blends	32.1	13.8
Scotch	7.7	13.7
Canadian	5.0	11.4
Bonds	4.6	1.0
Other	0.3	0.3
Vodka	7.3	16.9
Gin	9.2	10.0
Cordials	3.5	6.2
Brandy	2.4	4.0
Rum	1.5	3.7
Other	0.5	2.7

11.7 The data given in Figure 2.19 for typing speeds and typing error rates for 20 students are repeated below. The sample correlation coefficient for these data was found in Exercise 4.45. Find a 95 percent confidence interval for the population correlation coefficient. What interpretation do you give to this interval?

Student Number	Typing Score	Number of Errors
1	68	8
2	72	2
3	35	9
4	91	14
5	47	9
6	52	13
7	75	12
8	63	3
9	55	0
10	65	14
11	84	0
12	45	14
13	58	14
14	61	12
15	69	2
16	22	2
17	46	5
18	55	5
19	66	13
20	71	2

11.8   The data given in Figure 2.21 for bank deposits and deposit growth for 12 Albuquerque, New Mexico banks are repeated below. The sample correlation coefficient was found for these data in Exercise 4.44. Find a 90 percent confidence interval for the population correlation coefficient. What interpretation do you give to this interval?

	1979 Deposits ($ Thousands)	1978–1979 Percent of Deposit Growth
Albuquerque National Bank	675,709	5.1
First National Bank	457,085	12.0
Bank of New Mexico	242,682	7.1
American Bank of Commerce	99,615	−2.9
Rio Grande Valley Bank	78,947	22.3
Citizens Bank	39,603	−7.5
Fidelity National Bank	38,213	6.9
Southwest National Bank	42,813	13.5
Republic Bank	33,445	−2.3
Western Bank	30,271	5.1
Plaza del Sol National Bank	10,370	13.9
El Valle State Bank	5,635	−6.0

11.2
The Rank Correlation Coefficient

PROBLEM SETTING 11.2

In peer evaluation, subordinates are asked to rank themselves and to rank each other. This can be quite enlightening for the subordinates, who are encouraged to think seriously about their own job performance relative to the job performance of their peers. It can also boost morale if the results of the rankings are used in the evaluations by supervisors. The primary purpose of such a procedure is to provide an objective independent standard against which the supervisor can check his or her own rankings. Any obvious discrepancies deserve further consideration and investigation into the cause of the discrepancy.

Peer rankings may be based on some kind of scaling, such as a scale from 0 to 100. It is difficult to standardize such a scale in the minds of the subordinates, so a rating of 50 may represent a poor performance in one subordinate's mind, but may be interpreted as "average" by another. However, the relative ordering, "Joe, then Frank, then either Mary or Charlie," and so on, is easily understood by all parties involved. So the rating scale may be used primarily as a vehicle for ranking the subordinates from "worst" to "best." In some cases the subordinates may bypass the rating scale entirely and simply provide an ordering of their peers.

The degree of agreement or disagreement between the peer rankings and the supervisor's rankings may be summarized by a **rank correlation coefficient**. The rank correlation coefficient presented here is called **Spearman's rho**. It was introduced earlier, in Section 4.6.

MONOTONIC RELATIONSHIPS

There are many situations where the bivariate data (X_i, Y_i) do not come from a bivariate normal population and where a definite relationship exists between X_i and Y_i, but the relationship is not necessarily linear. The following plots of bivariate sample data illustrate nonlinear relationships.

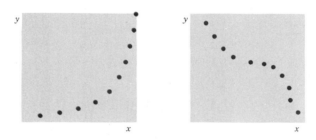

It is clear that a strong relationship exists between X and Y in each of the plots; however, it is equally clear that a straight line will not provide an adequate description of the data. Hence, the sample correlation, which provides a measure of the strength of the **linear** relationship between two variables, would not correctly describe the relationship between X and Y. Rather, such relationships as these are more aptly described as **monotonic**. That is to say, a strictly **monotonic relationship** is one where X and Y strictly increase together, or one strictly increases as the other decreases. If these graphs are converted to the ranks of X versus the ranks of Y, then the relationship between the ranks is a straight line.

THE RANK CORRELATION COEFFICIENT

Spearman's rho is a convenient way to test the strength of the monotonic relationship between X and Y without being concerned whether or not the relationship is linear. The formula for calculating the rank correlation has previously been given in Equation (4.25) of Section 4.6 as

$$r_s = \frac{\Sigma R_{x_i} R_{y_i} - C}{[(\Sigma R_{x_i}^2 - C)(\Sigma R_{y_i}^2 - C)]^{1/2}} \tag{11.2}$$

where R_{x_i} represents the ranks $1, 2, \ldots, n$, which are assigned to the X_i, R_{y_i} has the same meaning only for the Y_i, and $C = n(n+1)^2/4$. As with the sample correlation coefficient, r_s will yield values between $+1$ and -1. Further, the value of r_s can be $+1$ or -1, without r being $+1$ or -1, but the converse is not true.

A NUMERICAL ILLUSTRATION

The different values of r_s and r are now illustrated for a simple example using points from the deterministic relationship $Y = e^X$.

X	Y
0	1
1	2.72
2	7.39
3	20.09
4	54.60

First r is calculated.

	X	Y	XY	X^2	Y^2
	0	1	0	0	1
	1	2.72	2.72	1	7.39
	2	7.39	14.78	4	54.60
	3	20.09	60.26	9	403.43
	4	54.60	218.39	16	2980.96
Totals	10	85.79	296.15	30	3447.37

$\bar{X} = 2 \quad \bar{Y} = 17.16$

$$r = \frac{296.15 - 5(2)(17.16)}{[(30 - 5(2)^2)(3447.37 - 5(17.16)^2)]^{1/2}} = .886$$

This value of r yields a test statistic of $r = .886$, which would just be significant at $\alpha = .05$ when testing $H_0: \rho = 0$ versus $H_1: \rho \neq 0$. Now the correlation coefficient on ranks r_s is computed.

	R_x	R_y	$R_x R_y$	R_x^2	R_y^2
	1	1	1	1	1
	2	2	4	4	4
	3	3	9	9	9
	4	4	16	16	16
	5	5	25	25	25
Totals	15	15	55	55	55

$$R_s = \frac{55 - 45}{[(55 - 45)(55 - 45)]^{1/2}} = 1.000$$

Hence, r_s has achieved its maximum value of $+1$ indicating the strength of the **monotonic** relationship between X and Y is much stronger than the strength of the **linear** relationship indicated by r. A plot of the five pairs of data points in Figure 11.8 shows this statement to be well founded since the points lie on a curve and not a straight line. Hence, r_s can measure the strength of a monotonic nonlinear relationship more accurately than r.

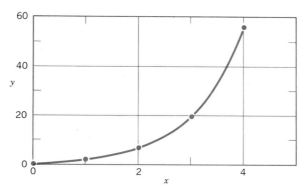

FIGURE 11.8 A Scatterplot of Five Values from $Y = e^x$

HYPOTHESIS TESTING

Tests of significance can be performed for the significance of Spearman's rho much as was done with the sample correlation coefficient. Essentially the data are replaced by ranks, and the test described in the previous section is used on the ranks. If the null hypothesis is rejected, then the conclusion is that there is a tendency toward a monotonic relationship between X and Y, and therefore X and Y are not independent. For this reason the null hypothesis is usually stated in terms of independence between X and Y, although there are some types of dependence—nonmonotonic dependence—that will not be detected by this test. A brief table of more exact critical values is provided in Figure 11.9 for the convenience of the reader.

A TEST OF INDEPENDENCE USING SPEARMAN'S RHO

Data. The data consist of n observations (X_i, Y_i), $i = 1, \ldots, n$, on the bivariate random variable (X, Y). Let R_{x_i} represent the ranks of the X's from 1 to n, and let R_{y_i} represent the ranks of the Y's. Note that the X's are ranked only among themselves, and the same is true for the Y's. Average ranks are used in case of ties.

Assumption. The sample (X_i, Y_i), $i = 1, \ldots, n$, is a random sample from some population.

Null Hypothesis. X and Y are independent.

Test Statistic. The test statistic is r_s as given by Equation (11.2).

Decision Rule. The decision rule depends on the alternative hypothesis of interest. Critical values for the test statistic are given in Figure 11.9.

1. H_1: X and Y tend to increase or decrease together (i.e., the relationship tends to be monotonically increasing). Reject H_0 if $r_s > T^*_{\alpha,n}$: otherwise accept H_0.

2. H_1: X tends to decrease as Y increases and vice versa (i.e., the relationship tends to be monotonically decreasing). Reject H_0 if $r_s < - T^*_{\alpha,n}$; otherwise accept H_0.

3. H_1: X and Y are not independent (i.e., there is a tendency toward a monotonic relationship between them). Reject H_0 if $| r_s | > T^*_{\alpha/2,n}$; otherwise accept H_0.

n	$\alpha = 0.05$	$\alpha = 0.025$	$\alpha = 0.005$
4	0.800		
5	0.800	0.900	
6	0.771	0.829	0.943
7	0.679	0.750	0.893
8	0.619	0.714	0.857
9	0.583	0.683	0.817
10	0.552	0.636	0.782
11	0.527	0.609	0.746
12	0.497	0.580	0.727
13	0.478	0.555	0.698
14	0.459	0.534	0.675
15	0.443	0.518	0.654
16	0.427	0.500	0.632
17	0.412	0.485	0.615
18	0.399	0.472	0.598
19	0.390	0.458	0.583
20	0.379	0.445	0.568
21	0.369	0.435	0.555
22	0.360	0.424	0.543
23	0.352	0.415	0.531
24	0.344	0.406	0.520
25	0.336	0.398	0.510
26	0.330	0.389	0.500
27	0.324	0.382	0.492
28	0.318	0.375	0.483
29	0.311	0.369	0.474
30	0.306	0.362	0.467
31	0.301	0.355	0.456
32	0.296	0.349	0.449
33	0.291	0.344	0.442
34	0.287	0.339	0.436
35	0.283	0.334	0.430
36	0.279	0.329	0.424
37	0.275	0.325	0.418
38	0.271	0.320	0.413
39	0.267	0.316	0.408
40	0.264	0.312	0.403
41	0.260	0.308	0.398
42	0.257	0.304	0.393
43	0.254	0.301	0.389
44	0.251	0.297	0.384
45	0.248	0.294	0.380
46	0.246	0.291	0.376
47	0.243	0.288	0.372
48	0.240	0.285	0.368
49	0.238	0.282	0.365
50	0.235	0.279	0.361

FIGURE 11.9 Values of $T^*_{\alpha,n}$ used with the Spearman's Rank Correlation Coefficient for Tests of Independence of the X_i and Y_i

Note: Values of $T^*_{\alpha,n}$ not given in this table can be found approximately as

$$T^*_{\alpha,n} \cong \frac{t_{\alpha,n-2}}{\sqrt{n-2+t^2_{\alpha,n-2}}},$$

where $t_{\alpha,n-2}$ is the $1-\alpha$ quantile of the Student's t-distribution with $n-2$ degrees of freedom given in Table A3. This approximation is fairly good even for small values of n.

EXAMPLE

Motivation for this example is provided in the problem setting of this section. Twenty subordinates are asked to rate their peers on a scale of 0 to 100. They are given some standard guidelines, such as to be sure to rate at least one person as 0 and at least one person as 100, with the other ratings spread out between those two extremes. The average rating is computed for each of the subordinates. This represents the rating that person has been given, averaged over all 20 evaluations, including his or her own rating. These are plotted in Figure 11.10 along with the supervisor's rating for that person, to give some idea of the pattern of the relationship.

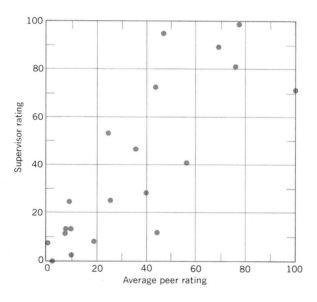

FIGURE 11.10

Of primary interest however is the agreement between the **rankings** of the peer group and the supervisor, rather than the agreement between ratings. These rankings are given in Figure 11.11 along with some preliminary calculations. The calculation of Spearman's rho proceeds as follows:

$$r_s = \frac{2751 - 2205}{[(2869 - 2205)(2869 - 2205)]^{1/2}} = .822.$$

To test the null hypothesis of independence between the peer rankings and the supervisor rankings at $\alpha = .05$, $|r_s|$ is compared with $T^*_{.025,20} = 0.445$. Thus the null hypothesis of independence is easily rejected, and it is concluded that supervisor rankings and peer rankings are not independent of one another. In fact, the positive correlation indicates that they tend to agree. The p-value is seen from Figure 11.9 to be less than .01. A more accurate approximation for the p-value, if desired, could be found from the approximation provided at the bottom of Figure 11.9 based on the t-distribution of Table A3.

Average Subordinate Rating X	Supervisor Rating Y	R_x	R_y	R_x^2	R_y^2	R_xR_y
1	7	1	3.5	1.00	12.25	3.50
2	0	2	1	4.00	1.00	2.00
7	11	3.5	5	12.25	25.00	17.50
7	13	3.5	7.5	12.25	56.25	26.25
9	13	5.5	7.5	30.25	56.25	41.25
9	24	5.5	9	30.25	81.00	49.50
10	2	7	2	49.00	4.00	14.00
19	7	8	3.5	64.00	12.25	28.00
24	53	9	14	81.00	196.00	126.00
26	26	10	10	100.00	100.00	100.00
36	47	11	13	121.00	169.00	143.00
40	29	12	11	144.00	121.00	132.00
43	72	13	15	169.00	225.00	195.00
45	12	14	6	196.00	36.00	84.00
47	95	15	19	225.00	361.00	285.00
56	42	16	12	256.00	144.00	192.00
70	89	17	18	289.00	324.00	306.00
78	83	18	17	324.00	289.00	306.00
80	100	19	20	361.00	400.00	380.00
99	74	20	16	400.00	256.00	320.00
		210	210	2869.00	2869.00	2751.00

$\bar{R}_x = 10.5$ $\bar{R}_y = 10.5$

FIGURE 11.11 Worksheet for Calculating the Sample Rank Correlation Coefficient

EXAMPLE

The example of the last section using the Attendance-Amount Wagered data is reworked here as an analysis on the ranks of the data. For a value of $\alpha = .05$ the null hypothesis

H_0: daily racing attendance and the amount of money wagered are independent of one another

is tested versus the alternative

H_1: daily racing attendance and the amount of money wagered increase together monotonically.

	R_x	R_y	R_xR_y	R_x^2	R_y^2
	1	1	1	1	1
	7	2	14	49	4
	4	7	28	16	49
	2	3	6	4	9
	9	9	81	81	81
	10	5	50	100	25
	5	6	30	25	36
	3	4	12	9	16
	8	8	64	64	64
	6	10	60	36	100
Totals	55	55	346	385	385

$$\bar{R}_x = 5.5 \qquad \bar{R}_y = 5.5$$

$$r_s = \frac{346 - 302.5}{[(385 - 302.5)(385 - 302.5)]^{1/2}} = .527$$

Decision Rule

From Figure 11.9, $T^*_{.05,10} = 0.552$. Reject H_0 if $r_s > 0.552$. The null hypothesis is accepted using a one-tailed test. That is, the data are not sufficient to detect a significant monotonic relationship between daily attendance and amount of money wagered. In this case the lack of significance may be due to the small sample size used.

Exercises

11.9   Compute Spearman's rank correlation coefficient on the data given in Exercise 11.5 (see Exercise 4.53) and compare your result with the correlation coefficient calculated in Exercise 11.5. How do you explain the sizable difference in the two correlation coefficients for these data? Use the value of the Spearman's rank correlation coefficient to test the hypothesis that the earnings per share is independent of the number of institutional shares held. Let $\alpha = .10$.

11.10   Compute Spearman's rank correlation coefficient on the data given in Exercise 11.6 (see Exercise 4.50). Test the hypothesis that the percent of the market in 1959 is independent of the percent of the market in 1974, versus the alternative that the percents tend to agree for the 2 years. Let $\alpha = .005$.

11.11   Nine students made the scores listed below on their first two tests in a statistics class. Compute Spearman's rank correlation coefficient for these scores and test the hypothesis that the scores are independent of one another versus an alternative that the scores tend to agree (increase together). Let $\alpha = .005$.

Student Number	First Test	Second Test
1	90	84
2	91	83
3	82	80
4	94	90
5	92	91
6	88	89
7	89	88
8	63	62
9	86	92

11.12 Two coaches review last week's game film and independently grade each of the player's performances on a scale from 0 to 100. Compute Spearman's rank correlation coefficient for these scores and test the following hypothesis with $\alpha = .05$.

H_0: there is no association between the scores given the players by the two coaches, versus

H_1: there is an association between the two scores (either the coaches tend to agree on the best and poorest performances or they tend toward opposite ratings on the players, i.e., what one coach calls good the other calls bad and vice versa)

Player Number	Scores from Coach 1	Scores from Coach 2
1	64	80
2	28	87
3	90	90
4	30	57
5	97	89
6	20	51
7	100	81
8	67	82
9	54	69
10	44	78
11	100	100
12	71	81

11.13 Given below are the results of a City Commission race in Manhattan, Kansas. Is there a significant correlation between the order (rank) of finish in the primary and the general elections? State and test the appropriate hypothesis. Let $\alpha = .05$.

Candidate	Votes in Primary	Votes in General
R	2251	3929
B	2073	3578
A	1993	4041
S	1489	2941
Y	1332	1802
K	1253	2525

11.14   Two noted football authorities gave the following preseason rankings to the Big-8 football conference race. Is there any agreement in their rankings? State and test the appropriate hypothesis with $\alpha = .05$.

Team	Street and Smith	Minneapolis Line
Nebraska	1	2
Oklahoma	2	1
Kansas State	3	8
Colorado	4	3
Missouri	5	6
Iowa State	6	7
Kansas	7	4
Oklahoma State	8	5

11.15   Two customers are randomly selected in a supermarket and asked to smell five aftershave lotions and rank them in order of preference from 1 (best) to 5 (least desirable). Based on the results given below, test the hypothesis that the two sets of rankings are independent of one another versus the alternative that they tend to agree (increase together). Let $\alpha = .05$.

	Brand of After-Shave				
	A	B	C	D	E
Customer Number 1 Ranking:	1	4	3	2	5
Customer Number 2 Ranking:	3	5	2	1	4

11.16   Consider the following set of ranks for the nation's top 20 football teams as given by the two leading wire services. Use Spearman's rank correlation coefficient to test the hypothesis that the two sets of rankings are independent of one another, versus the alternative that the two sets of rankings tend to agree. Let $\alpha = .005$.

Football Team	AP Ranking	UPI Ranking
Georgia	1	1
Florida State	4	2
Pittsburgh	2	3
Oklahoma	5	4
Michigan	3	5
Alabama	6	6
Baylor	9	7
Notre Dame	10	8
Nebraska	7	9
Penn State	8	10
North Carolina	11	11
UCLA	14	12
Southern Cal	13	13
Ohio State	15	14
Brigham Young	12	15

Continues

Football Team	AP Ranking	UPI Ranking
Washington	16	16
Mississippi St.	18	17
South Carolina	17	18
SMU	19	19
Maryland	20	20

11.3
Review Exercises

11.17   Use the sample data given in Exercise 2.39 to test $H_0: \rho = 0$ versus $H_1: \rho > 0$. Let $\alpha = .05$ and state the p-value. (See Exercise 4.55 for the necessary calculations.)

11.18   Use the sample data given in Exercise 2.39 to test the hypothesis that height and weight are independent of one another versus the alternative that they tend to increase together. Let $\alpha = .05$ and state the p-value. (See Exercise 4.56 for the necessary calculations.)

11.19   Given below are the percent of registered voters and the percent turnout for several cities in a nationwide election. Use these data to test $H_0: \rho = 0$ versus $H_1: \rho > 0$. Let $\alpha = .05$ and state the p-value.

City	Percent Registered	Percent Voting
Detroit	92.0	70.0
Topeka	81.9	69.3
Philadelphia	77.6	69.8
Los Angeles	77.0	64.2
Boston	74.0	63.3
Tampa	68.8	63.6
San Francisco	68.0	64.4
Honolulu	60.0	54.7
Birmingham	39.1	13.8
Atlanta	33.8	25.6

11.20   Refer to Exercise 11.19 and test the hypothesis that the percent of voters registered and the percent of voters actually voting are independent of one another, versus the alternative that they tend to increase together. Let $\alpha = .05$ and state the p-value.

11.21   Two judges rate 10 contestants in a beauty contest. Test the hypothesis that the ratings are independent of one another versus the alternative that the judges tend to agree in the judgment. Let $\alpha = .05$ and state the p-value.

					Contestant					
	A	B	C	D	E	F	G	H	I	J
Judge 1:	1	2	3	4	5	6	7	8	9	10
Judge 2:	2	3	1	4	6	5	9	10	8	7

11.22   Use the data in Exercise 1.22 to examine the relationship between the age of inauguration and the age of death for past U.S. presidents. Make a scatterplot of the data.

Although this is not a random sample since all past presidents who are no longer living are considered, examination of these data may convey some insight into data that may occur in the future.

11.23 Use the Lilliefors test to check the normality assumption for the data in Exercise 1.22. Let Z_I represent the standardized values for the age of first inauguration, and let Z_D be the corresponding values for the age of death. A convenient formula for r when Z_I and Z_D are available is

$$r = (\Sigma Z_I \cdot Z_D)/(n - 1)$$

where the sum is over all pairs of observations. Use this formula to compute r for the data in Exercise 1.22.

11.24 The percent change in college football season ticket sales may be related to the percent change in the proportion of games won during the previous season. A random sample of 12 colleges showed the following percent changes.

Percent Change in Season Ticket Sales	Percent Change in Proportion of Games Won (Prev. Season)
+4.3	+11
+7.0	+22
+3.0	0
−6.2	−9
−0.1	+10
−1.7	−18
+4.2	0
+6.9	+33
+2.3	+27
−10.2	−44
−8.1	−30
−1.7	−11

Display these data in a scatterplot and discuss the results. Use these data to see if there is a tendency for sales to increase after an improvement in the won-loss record of a college team. Use $\alpha = .05$. Discuss other variables that may have some influence on the percent change in season ticket sales.

Bibliography

Additional material on the topics presented in this chapter can be found in the following publications.

Conover, W. J. (1980). *Practical Nonparametric Statistics, 2nd ed.,* John Wiley and Sons, Inc., New York.

Conover, W. J. and Iman, R. L. (1981). "Rank Transformations as a Bridge between Parametric and Nonparametric Statistics," *The American Statistician,* **35**(3), 124–129.

Iman, R. L. and Conover, W. J. (1978). "Approximations of the Critical Region for Spearman's Rho with and without Ties Present," *Communications In Statistics,* **B7**(3), 269–282.

Iman, R.L. and Davenport, J. M. (1982). "Rank Correlation Plots for Use with Correlated Input Variables," *Communications In Statistics,* **B11**(3).

12

Regression

PRELIMINARY REMARKS

When a farmer prepares to plant a crop there are many factors that he could consider with respect to the final yield. These factors might include what type of rainfall is likely before the crop is harvested; what daily temperatures are to be expected; if it freezes, what will be the date of the first killing frost; should he fertilize the crop, and if so how much; what variety of plant should he use; what types of farming techniques are likely to produce the best results, and so on. Most of these factors the farmer won't have any control over. Although the farmer has no way of predicting what will happen in the future, he can observe what has happened in the past, and in the long run the information should be helpful in predicting the yield.

In a different area, before you enrolled in college you may have wondered if you would be successful in pursuing a college degree. Other than the degree itself, the measure of success in college most frequently used is cumulative grade point average (G.P.A.). Many factors influence your G.P.A. Quantitative measures exist for some of the factors such as high school G.P.A., rank in class, ACT score, and I.Q. score. Other factors are difficult to obtain a quantitative estimate of, such as parent's attitude toward your university plans, curriculum in which you are enrolled, the university you choose, and where you reside at college. Information such as this will not always enable your G.P.A. to be anticipated correctly, but on the average this additional information should prove useful in predicting your G.P.A. success.

Both of these examples have much in common. They involve complex relationships between one variable, such as the yield of a crop in one case or the G.P.A. of a student in the other case, and a collection of other variables. They involve financial considerations. And they have been the subject of extensive research into the best ways of expressing a mathematical relationship among the variables.

Finding a suitable relationship between the one unknown variable, called the **dependent variable**, and a group of other known quantities, called **independent variables**, is a challenging problem. One methodology for handling this type of problem is called **regression analysis**. This chapter provides a simple introduction to regression analysis. Regression analysis *defines the mathematical relationship* between two (or more) variables, in contrast with correlation analysis, which merely *measures the strength* of the relationship between two variables.

12.1
An Introduction to Regression

PROBLEM SETTING 12.1

The calculation of insurance rates is a complex process that requires highly skilled professionals. The people who do this type of analysis are called actuaries, and the body of knowledge dealing with the subject is called actuarial science. As a part of this training there are 10 challenging examinations that a trainee can take. The first exam covers undergraduate mathematics and calculus. The second exam includes topics in probability and statistics. The third, fourth, and fifth exams are concerned with various topics in mathematics, finance, and insurance. Passage of the first five exams qualifies one to be called an *Actuarial Associate*. Passage of all 10 exams results in the coveted title *Fellow of the Society of Actuaries*.

An undergraduate program in actuarial science is designed to help the student pass as many of the first five exams as possible before graduation. An undergraduate who is just completing the actuarial program inquires of the university placement center about the salary she might expect to earn. A placement center counselor obtains a listing of recent graduates of the university who became employed as actuaries by insurance companies. The starting salaries and number of exams passed by the graduates are given in Figure 12.1.

Graduate	Starting Salary	No. of Exams Passed	Graduate	Starting Salary	No. of Exams Passed
1	$11,900	1	15	$12,000	1
2	18,000	3	16	15,900	2
3	17,300	2	17	13,000	1
4	18,000	2	18	8,900	0
5	21,300	5	19	17,000	3
6	10,200	0	20	20,500	4
7	11,100	0	21	14,800	2
8	15,900	3	22	10,200	0
9	18,000	4	23	13,300	1
10	13,100	2	24	12,500	1
11	9,400	0	25	17,800	4
12	16,500	2	26	17,000	3
13	22,500	5	27	11,200	1
14	11,100	1	28	16,600	4

FIGURE 12.1 Starting Salaries and Number of Exams Passed for 28 Recent Graduates

It seems reasonable to assume that these 28 observations behave as a random sample of all graduates in the recent past or near future, who are graduating in actuarial science from that same university.

In order to answer the student's inquiry regarding the starting salary she can expect, the counselor plots an empirical distribution function of starting salaries, as in Figure 12.2. From this e.d.f. the sample median is easily seen to be $15,350, and the lower and upper sample quartiles are $11,200 and $17,300, respectively. The counselor explains to the student that she can obtain some idea of what she might expect to earn, knowing that the middle 50 percent of recent graduates are in the $11,200 to $17,300 range.

At this point the student mentions the fact that she has passed all of the first five actuarial exams, and asks the counselor if this will help him in predicting her future starting salary. How can the counselor use this additional information to provide the student with a more precise estimate of the starting salary she might expect?

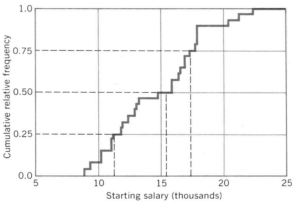

FIGURE 12.2 The Empirical Distribution Function of Starting Salaries

REGRESSION

The question posed by the student in the problem setting is similar to questions posed in many other real situations. That is, when the sample consists of bivariate data (X, Y), can inferences regarding the variable Y be improved by knowing something about the variable X? It seems reasonable to expect that the additional information furnished by X should improve knowledge about the population of Y if X and Y are related in some way.

For example, the bivariate data in Figure 12.1 can be summarized by the number of exams passed. Such a summary is given in Figure 12.3. It is fairly obvious from Figure 12.3 that the average salary $\bar{Y}$ tends to increase as the number of exams passed X increases. A graphical representation of this point is given by the scatterplot in Figure 12.4.

0 Exams	1 Exam	2 Exams	3 Exams	4 Exams	5 Exams
$ 8,900	$11,100	$13,100	$15,900	$16,600	$21,300
9,400	11,200	14,800	17,000	17,800	22,500
10,200	11,900	15,900	17,000	18,000	
10,200	12,000	16,500	18,000	20,500	
11,100	12,500	17,300			
	13,000	18,000			
	13,300				
$\bar{Y}_0 = \$9,960$	$\bar{Y}_1 = \$12,143$	$\bar{Y}_2 = \$15,933$	$\bar{Y}_3 = \$16,975$	$\bar{Y}_4 = \$18,225$	$\bar{Y}_5 = \$21,900$

FIGURE 12.3 Group Means for Salaries Grouped by Number of Exams Passed

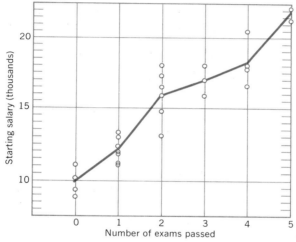

FIGURE 12.4 A Scatterplot of Starting Salaries and Number of Exams Passed, With the Sample Means Connected

If the counselor in the problem setting examines Figures 12.3 and 12.4, he can note that only two of the 28 graduates has passed all five exams as the student has done, and their starting salaries were $21,300 and $22,500, both in the upper range of starting salaries. Clearly this additional information regarding the number of exams passed is useful for predicting the student's starting salary. The topic of **regression** is concerned with methods for utilizing this additional information.

POPULATION REGRESSION CURVE

New graduates would get a better idea of starting salaries if the number of exams passed is taken into consideration, because salaries appear to be different for the different numbers of exams passed. A graduate who has passed all five exams should expect a salary closer to the average for other graduates who passed all five exams. In the scatterplot given in Figure 12.4, lines connecting the average values of Y for each individual value of X are provided to give a clearer indication of the tendency for the mean of Y to depend on the observed value of X. The population mean of Y for each individual value of X is called the **conditional mean of Y given $X = x$**, and is usually denoted by $\mu_{y|x}$.

A graph of $\mu_{y|x}$ as a function of x is called the **population regression curve**. It may be used to make predictions on individual values of Y given that X has a value x. Suppose that the population mean starting salaries for all actuarial graduates are given as a function of the number of exams passed as follows.

$$X = 0: \quad \mu_{y|0} = \$11,000 \qquad X = 3: \quad \mu_{y|3} = \$16,000$$
$$X = 1: \quad \mu_{y|1} = \$13,000 \qquad X = 4: \quad \mu_{y|4} = \$19,500$$
$$X = 2: \quad \mu_{y|2} = \$14,500 \qquad X = 5: \quad \mu_{y|5} = \$23,000$$

A graph of these values is given in Figure 12.5, as the population regression curve. In practice the population regression curve is seldom known and must be estimated from a sample.

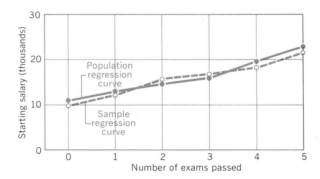

FIGURE 12.5 A Graph of the Population Regression Curve And the Sample Regression Curve

The **conditional mean of Y given $X = x$**, denoted by $\mu_{y|x}$, represents the population mean of all values of Y that share that same value for X.

The **population regression curve** for Y is the curve that represents the mean of Y given $X = x$, $\mu_{y|x}$, plotted as a function of x.

SAMPLE REGRESSION CURVE

There are many different ways of estimating the population regression curve. Any estimate based on a sample of observations is called a **sample regression curve**. One way of obtaining a sample regression curve is to graph the sample means of Y for each value of X, as was done in Figures 12.4 and 12.5. Another way is to assume that the population means all lie on a straight line and then fit a straight line to the scatterplot of bivariate data. The straight line method is easy to handle mathematically and is the method developed in subsequent sections of this chapter.

The y-value for each value x, obtained from a sample regression curve, is usually denoted by $\hat{Y}$, and is called **the predicted value for Y**. Each possible value of X may be used to obtain a corresponding value $\hat{Y}$ from the sample regression curve, and $\hat{Y}$ is then used as a point estimate of $\mu_{y|x}$ for that value of X. The same $\hat{Y}$ may be used as a point prediction value for any individual value of Y, when the only information available is X. Every observed value of Y is associated with its **predicted value** $\hat{Y}$ through the X value paired with Y. For example, the first graduate listed in Figure 12.1 has a salary Y = $11,900 and number of exams passed X = 1. The sample regression curve in Figure 12.4 gives $\hat{Y}$ = $12,143 (see $\overline{Y}_1$ from Figure 12.3) as the predicted salary for X = 1. The predicted value for an observation denoted by (X_i, Y_i) is denoted with the same subscript $\hat{Y}_i$.

HOW MUCH OF THE VARIATION IN Y IS EXPLAINED BY KNOWLEDGE OF X

If nothing were known about the number of exams passed by the undergraduate student in Problem Setting 12.1, the sample mean of starting salaries in Figure 12.1, Y = $14,821, could be used to provide the student with an estimate of starting salaries. A measure of variation in starting salaries is given, of course, by the sample variance

$$s_y^2 = \frac{1}{n-1} \sum_{i=1}^{n} (Y_i - \overline{Y})^2 = \frac{1}{27} (374{,}767{,}143)$$

$$= 13{,}880{,}265.$$

The term $\Sigma(Y_i - \overline{Y})^2$ is called the **total sum of squares**. In regression this is frequently used instead of s_y^2 to measure the total variability in Y. For the salary data the total sum of squares is given above as 374,767,143.

TOTAL SUM OF SQUARES (TOTAL VARIATION IN Y)	*The total of the squared deviations between each observation* Y_i *and the sample mean* $\overline{Y}$ *is the* **total sum of squares**, $$\text{Total SS} = \Sigma(Y_i - \overline{Y})^2.$$

Since the sample data in Figure 12.1 records both starting salary and number of exams passed for each graduate, it is natural to ask how much of the total variation in starting salaries can be explained by knowledge of the number of exams passed. Or, in general, for a set of bivariate observations on (X, Y), how much of the variation in Y can be explained by knowledge of X?

First the amount of variation that is *not* explained by knowledge of X is discussed. The prediction $\hat{Y}$ for Y, given a value of X, is obtained from the sample regression curve. The unexplained variation is measured by the difference between Y and $\hat{Y}$. The sum of the squares of these differences, $\Sigma(Y_i - \hat{Y}_i)^2$, is called the **error sum of squares** and represents the variation in Y remaining after regression on X.

ERROR SUM OF SQUARES (VARIATION IN Y REMAINING AFTER REGRESSION ON X)	*The total of the squared deviations between each observation Y_i and its point estimate $\hat{Y}_i$ from the sample regression curve is the* **error sum of squares**, $$Error\ SS = \Sigma(Y_i - \hat{Y}_i)^2.$$

The ratio of the *unexplained* variation to the *total* variation represents the proportion of variation in Y that is not explained by regression on X. Subtraction from 1.0 gives the proportion of variation in Y that is explained by **regression on X**. The statistic used to express this proportion is called the **coefficient of determination** and is denoted by R^2. It may be written in various equivalent ways, as follows.

$$R^2 = 1 - \frac{\text{Variation in } Y \text{ Remaining After Regression on } X}{\text{Total Variation in } Y}$$

$$R^2 = 1 - \frac{\text{Error Sum of Squares}}{\text{Total Sum of Squares}}$$

$$R^2 = 1 - \frac{\Sigma(Y_i - \hat{Y}_i)^2}{\Sigma(Y_i - \bar{Y})^2}$$

The value of R^2 is the proportion of the variation in the dependent variable Y explained by regression on the independent variable X. The calculation of the coefficient of determination is illustrated in the next example.

COEFFICIENT OF DETERMINATION	*The* **coefficient of determination** R^2 *is the proportion of variation in Y explained by a sample regression curve. The equation for R^2 is* $$R^2 = 1 - \frac{\Sigma(Y_i - \hat{Y}_i)^2}{\Sigma(Y_i - \bar{Y})^2}.$$ (12.1)

EXAMPLE

For the data given in Problem Setting 12.1 the total sum of squares is

$$\Sigma(Y_i - \bar{Y})^2 = 374,767,143.$$

The sample regression curve for these data is given in Figure 12.4. The error sum of squares associated with the sample regression curve can be computed by referring to Figure 12.3. The computations are given in Figure 12.6.

The coefficient of determination is obtained using Equation (12.1).

$$R^2 = 1 - \frac{\Sigma(Y_i - \hat{Y}_i)^2}{\Sigma(Y_i - \bar{Y})^2}$$

$$= 1 - \frac{33,857,476}{374,767,143}$$

Group	Y_i	$\hat{Y}_i$	$(Y_i - \hat{Y}_i)^2$
1	$ 8,900	$ 9,960	1,123,600
	9,400	9,960	313,600
	10,200	9,960	57,600
	10,200	9,960	57,600
	11,100	9,960	1,299,600
. . .	. . .	. . .	. . .
6	$21,300	$21,900	360,000
	22,500	21,900	360,000
$\Sigma(Y_i - \hat{Y}_i)^2$ = Error SS			33,857,476

FIGURE 12.6 Worksheet for Calculating Error Sum of Squares

$$= 1 - .090$$

$$= .910$$

Thus 91 percent of the total variation in starting salaries is accounted for by regression on the number of exams passed. In most applications 91 percent would probably be regarded as a relatively large value for R^2.

COMMENT

Where there are many observations, the regression curve for Y may be estimated using the sample means of Y's for the various values of X, or by averaging the Y values whose X values are *close to* each other in case each X is observed only a few times. When there are not many observations, some assumptions are often made regarding the form of the regression curve. In the next section the regression curve is assumed to be a straight line.

Exercises

12.1 A random sample of recent university graduates was classified according to grade point average and starting salary. The grade point averages were grouped into intervals and were summarized as follows.

Starting Salaries (nearest thousand) Classified by G.P.A.				
2.0–2.4	2.4–2.8	2.8–3.2	3.2–3.6	3.6–4.0
$11.4	$12.7	$13.4	$16.0	$15.3
10.7	11.5	12.3	13.3	15.1
10.1	10.9	12.1	12.8	14.3
10.3	10.2	13.3	12.5	14.7
	12.2	12.6	12.9	14.8
	13.0	13.5	13.7	
		14.8	15.7	
			15.5	

Graph the G.P.A. values versus starting salaries in a scatterplot. (Use the median of each G.P.A. interval to make the scatterplot—i.e., 2.2, 2.6, 3.0, 3.4, and 3.8.) Compute the total sum of squares $= \Sigma(Y_i - \bar{Y})^2$ for the starting salaries.

12.2   For the sample data in Exercise 12.1 find the sample conditional mean of Y for each G.P.A. classification. Add these points to the scatterplot of Exercise 12.1 and then connect them to form a sample regression curve.

12.3   Find the error sum of squares for the sample data in Exercise 12.1.

12.4   Find the coefficient of determination for the sample data in Exercise 12.1. What is the interpretation of this value?

12.5   The number of automobile accidents reported to the police over a four-week period during the winter in a large city is given below. Make a scatterplot of day of the week versus number of accidents reported for these data and compute the total sum of squares $= \Sigma(Y_i - \bar{Y})^2$ for the number of accidents.

Sun.	Mon.	Tue.	Wed.	Thur.	Fri.	Sat.
145	137	120	133	165	181	201
139	129	118	129	170	173	198
138	136	123	138	158	185	214
140	131	129	131	162	187	207

12.6   For the sample data in Exercise 12.5 find the sample conditional mean of number of accidents reported for each day of the week. Add these points to the scatterplot of Exercise 12.5 and then connect them to form a sample regression curve.

12.7   Find the error sum of squares for the sample data in Exercise 12.5.

12.8   Find the coefficient of determination for the sample data in Exercise 12.5. What is the interpretation of this value?

12.2
Least Squares Computations for Linear Regression

PROBLEM SETTING 12.2

As mentioned at the beginning of this chapter the prediction of college G.P.A. prior to enrollment depends on many factors. Certainly some factors are more instrumental in influencing G.P.A. than others and the determination of the most important of these factors has been the subject of much research over the years. Of course, the reason for wanting to identify these factors is to aid in the counseling of high school students. Of the factors that are available for which a quantitative measure exists, the most reliable tend to be the standardized test scores such as ACT and SAT. The reason for this is that these tests compare students on a nationwide basis as opposed to high school G.P.A.'s, which only reflect one school's standard.

The same type of problem occurs at a higher level when medical and law schools evaluate a large number of applicants and select only a few for admittance. Certainly a fundamental requirement of this process is selecting candidates whose chances of completing the curriculum appear excellent.

A university admissions officer compares the ACT scores of 12 students with their G.P.A. after 1 year of college. The results are given in Figure 12.7. Since the purpose of studying data of this type is to try to predict a student's success in college, the relationship between ACT scores and G.P.A. is interesting to examine. Is it possible to obtain a mathematical relationship between the two that will enable G.P.A. to be predicted with some degree of confidence, when only the ACT score is known? This question is examined in this section.

Student	ACT X	G.P.A. Y
1	24	2.66
2	20	2.25
3	23	2.82
4	27	2.91
5	31	3.45
6	29	3.47
7	28	3.75
8	26	3.33
9	31	4.00
10	26	3.49
11	20	2.61
12	18	1.80

FIGURE 12.7 Freshmen ACT Scores and G.P.A. After 1 Year for 12 Randomly Selected Students

LINEAR REGRESSION

One way of providing an answer for the university admissions officer in the problem setting is to fit a straight line to the observations in Figure 12.7. The formulas for fitting a straight line to a set of sample data are given in this section.

The population regression curve was defined in the previous section as the curve that represents the mean of Y given $X = x$, which is usually denoted $\mu_{y|x}$, plotted as a function of x. When the population regression curve is a straight line, the regression is said to be linear. The **linear regression** concept is important because scatterplots of data frequently indicate that a straight line relationship may exist between $\mu_{y|x}$ and x. In reality the regression may be curved, but a straight line relationship may be helpful as an approximation to the true but unknown regression curve. The first step in describing the relationship between two variables usually consists of fitting a straight line to the observed points plotted in the scatterplot.

If the regression curve between two variables is a straight line

$$\mu_{y|x} = \alpha + \beta x,$$ (12.2)

LINEAR REGRESSION *then the* **regression is linear**. *(Although it is customary to use α and β in linear regression, these symbols are not to be confused with probabilities of Type I and Type II errors, which is another way these same symbols are sometimes used.)*

METHOD OF LEAST SQUARES

The general form for the equation of a straight line may be recalled from algebra as $y = a + bx$, where a is defined as the y intercept of the line and b is the slope of the line, as illustrated in Figure 12.8. That is, a is the number of units either above or below the origin $(0, 0)$ where the straight line crosses the y-axis and b is the ratio of the change in vertical distance $(y_2 - y_1)$ between any two points on the line divided by the horizontal change $(x_2 - x_1)$ between the same two points. The problem in estimating a linear regression curve is how to find a and b for a set of sample values (X_i, Y_i) when all points do not naturally fall on a straight line. That is, what should the criterion be for computing a and b so the line $y = a + bx$ appears to fit the data well? Since the line fitted to the sample values is generally used for predicting values of Y, simple logic would imply that there should be some agreement between values observed for Y and values predicted for Y. The criterion that has been traditionally used for this problem is to find a and b such that the following sum of squares is minimized:

$$\sum_{i=1}^{n} (\text{observed } Y_i - \text{predicted } Y_i)^2 = \sum_{i=1}^{n} (Y_i - (a + bX_i))^2.$$

CONNORS 3-81

"NO CLAYBORNE! SLOPE AND INTERCEPT ARE NOT RELATED TO SKIING AND FOOTBALL!"

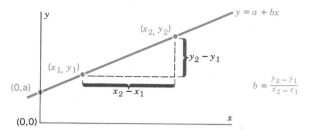

FIGURE 12.8 A Review of the Equation for a Straight Line

This approach is known as **the principle of least squares**. The minimization of this sum of squares follows from elementary calculus, and yields the solution given by Equations (12.3) and (12.4).

LEAST SQUARES SOLUTION FOR LINEAR REGRESSION

The slope and intercept needed to fit a straight line, $y = a + bx$, to a set of bivariate sample data (X_i, Y_i), $i = 1, \ldots, n$, by the principle of least squares are computed as follows.

$$\text{Slope:} \quad b = \frac{\Sigma X_i Y_i - n\bar{X}\bar{Y}}{\Sigma X_i^2 - n\bar{X}^2} \tag{12.3}$$

$$\text{Intercept:} \quad a = \bar{Y} - b\bar{X} \tag{12.4}$$

The regression line $\mu_{y|x} = \alpha + \beta x$ is estimated using $\hat{\mu}_{y|x} = a + bx$.

A NUMERICAL ILLUSTRATION

The least squares calculations will now be illustrated on a simple set of four bivariate observations.

	X	Y	XY	X²
	1	2	2	1
	2	1	2	2
	3	3	9	9
	4	3	12	16
Totals	10	9	25	30

$$\bar{X} = \frac{10}{4} = 2.5 \qquad \bar{Y} = \frac{9}{4} = 2.25$$

$$b = \frac{25 - 4(2.5)(2.25)}{30 - 4(2.5)^2} = .5$$

$$a = 2.25 - .5(2.5) = 1$$

The sample regression equation is $\hat{\mu}_{y|x} = 1 + .5x$. A scatterplot of the four points along with the least squares regression line appears in Figure 12.9. The

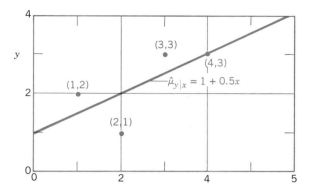

FIGURE 12.9 A Scatterplot with the Sample Regression Line

predicted values $\hat{Y}_i$ of Y for each observed X are calculated by substituting each value of X into the sample regression equation as illustrated in Figure 12.10. The principle of least squares guarantees that no other choice of values for a and b can be made that will make the error sum of squares smaller than 1.50 for these four data points. However, if a different set of sample observations were used where all pairs of points naturally fall on a straight line, then it would be possible for the error sum of squares to be equal to zero. The reason for this is that the observed and predicted values would always agree and hence their difference would be zero.

X	Observed Y	Prediction $\hat{Y} = 1 + .5X$	Error $(Y_i - \hat{Y}_i)^2$	Total $(Y_i - \bar{Y})^2$
1	2	1.5	.25	.0625
2	1	2.0	1.00	1.5625
3	3	2.5	.25	.5625
4	3	3.0	0.00	.5625
	9		Error SS = 1.50	Total SS = 2.7500

FIGURE 12.10 A Numerical Illustration of the Sums of Squares

THE COEFFICIENT OF DETERMINATION

The **coefficient of determination** for the numerical illustration may be computed from Equation (12.1) as

$$R^2 = 1 - \frac{\text{Error SS}}{\text{Total SS}} = 1 - \frac{1.50}{2.75} = .455,$$

which means 45.5 percent of the variation in Y is accounted for by the simple linear regression on X. In the case of simple linear regression R^2 always equals the square of the sample correlation coefficient, r; therefore the notation r^2 is customarily used to represent the coefficient of determination in simple linear regression.

EXAMPLE

Grade point averages for 12 randomly selected students who have just completed their freshmen year are paired with their ACT scores as part of a study to examine the predictability of success in college. Freshmen are selected since a sample beyond that class runs the risk of missing students who may have been dismissed from school for academic reasons. The results, and a worksheet for calculations, are given in Figure 12.11.

$$b = \frac{\Sigma XY - n\overline{X}\,\overline{Y}}{\Sigma X^2 - n\overline{X}^2}$$

$$= \frac{950.77 - 12(303/12)(36.54/12)}{7857 - 12(303/12)^2}$$

$$= \frac{28.14}{206.25} = .1364$$

This indicates that an increase of 1 point in the ACT score tends to be associated with an increase of .136 in the freshman year G.P.A. The intercept for the least squares regression line is

$$a = \overline{Y} - b\overline{X} = (36.54/12) - .1364(303/12) = -.399.$$

Thus the least squares regression line becomes

$$\hat{\mu}_{y|x} = -.399 + .1364x.$$

Student	ACT X	G.P.A. Y	X^2	XY
1	24	2.66	576	63.84
2	20	2.25	400	45.00
3	23	2.82	529	64.86
4	27	2.91	729	78.57
5	31	3.45	961	106.95
6	29	3.47	841	100.63
7	28	3.75	784	105.00
8	26	3.33	676	86.58
9	31	4.00	961	124.00
10	26	3.49	676	90.74
11	20	2.61	400	52.20
12	18	1.80	324	32.40
Total	303	36.54	7857	950.77

FIGURE 12.11 Freshmen G.P.A.'s and ACT scores for 12 Randomly Selected Students

EXAMPLE

A manufacturing company would like to know if storage life (in months) of their batteries can be predicted based on the number of days of conditioning. A random sample of batteries is taken for 10 batteries having from 1 to 10 days conditioning. For the sample data in Figure 12.12 do the following.

1. Plot the data in a scatterplot to see if a linear relationship looks reasonable for the data.
2. Find the sample regression equation by the method of least squares and include it in the scatterplot.
3. Find the sample correlation coefficient.
4. Find the percent of variation in storage life accounted for by regression on the number of days of conditioning.
5. Predict the storage life for a battery with six days conditioning.

	Days Conditioning X	Storage Life Y	XY	X^2	Y^2
	1	3	3	1	9
	2	2	4	4	4
	3	5	15	9	25
	4	4	16	16	16
	5	6	30	25	36
	6	8	48	36	64
	7	9	63	49	81
	8	9	72	64	81
	9	12	108	81	144
	10	10	100	100	100
Totals	55	68	459	385	560

$\bar{X} = 55/10 = 5.5$ $\bar{Y} = 68/10 = 6.8$

FIGURE 12.12 Storage Life Noted From a Random Sample of Batteries with Specified Days of Conditioning

1. See Figure 12.13 for a scatterplot of the data. Note how the relationship between X and Y appears to be close to a straight line.

2. $b = \dfrac{459 - 10(5.5)(6.8)}{385 - 10(5.5)^2} = 1.0303$

   $a = 6.8 - 1.0303(5.5) = 1.133$

   The sample regression equation is

   $\hat{\mu}_{y|x} = 1.133 + 1.03x,$

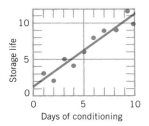

FIGURE 12.13 A Scatterplot of the Sample Data and the Least Squares Regression Line

and the sample regression line has been plotted in the scatterplot in Figure 12.13. The estimated slope of 1.03 means that the storage life will tend to increase by 1.03 for each additional day of conditioning.

3. $r = \dfrac{459 - 10(5.5)(6.8)}{\sqrt{[385 - 10(5.5)^2][560 - 10(6.8)^2]}} = .947$

indicating a strong linear relationship between X and Y.

4. $r^2 = (.947)^2 = .897$, so 89.7 percent of the variation in storage life can be explained by the regression on days of conditioning.

5. From the sample regression equation the prediction for six days is found as

$$\hat{\mu}_{y|x} = 1.133 + 1.03(6) = 7.3.$$

LACK OF ASSUMPTIONS

The formulas given in this section for fitting a straight line to a set of bivariate data are derived using calculus and the least squares principle. They are valid for fitting a straight line to any set of bivariate data without regard to how the data were obtained or the type of population. In particular no distributional assumptions were made and no assumptions were made about the sample being random. Those assumptions are used when testing hypotheses or when forming confidence intervals, as in the following sections. But the methods introduced so far in this chapter, including finding the least squares regression line, computing r^2 or sums of squares, and making point estimates of the true slope and intercept, do not require any assumptions and may be used wherever and whenever they are appropriate.

Exercises

12.9 Find the value of r^2 for the G.P.A.–ACT score example of this section and interpret its meaning.

12.10 Make a scatterplot for the G.P.A.–ACT score example of this section and include a graph of the least squares line in the scatterplot.

12.11   Annual rainfall in inches and wheat yield in bushels per acre have been recorded for a farm in northwest Kansas. These data are reported below. Find the value of r^2 for these data.

Year	Annual Rainfall	Wheat Yield
1	23.5	33.8
2	20.4	23.0
3	22.8	33.2
4	25.9	35.5
5	28.9	44.5
6	27.1	40.8
7	26.8	45.3
8	25.2	39.0
9	29.0	50.0
10	25.1	40.4
11	20.3	24.0
12	18.5	24.8

12.12   Make a scatterplot for the data given in Exercise 12.11 and include a graph of the least squares line in the scatterplot.

12.13   Find the least squares regression equation for the sample data given in Exercise 12.1. Use the median of each G.P.A. interval as the X value—that is, 2.2, 2.6, 3.0, 3.4, and 3.8, respectively. Find the value of r^2 for these data.

12.14   Predict the starting salary for a graduate with a G.P.A. of 3.45 using the least squares regression equation found in Exercise 12.13.

12.15   Find the least squares regression equation for the sample data given in Exercise 12.5. Since the day of the week is a qualitative variable it will have to be replaced by a quantitative variable by letting $X = 1$ for Sunday, $X = 2$ for Monday, and so on until $X = 7$ for Saturday. Find the value of r^2 for these data.

12.16   Make a scatterplot for the data of Exercise 12.5 and add the graph of the least squares regression line found in Exercise 12.15 to the scatterplot. Do you think a straight line gives an adequate representation of these data?

12.3
Assumptions for Linear Regression

PROBLEM SETTING 12.3

Workers on a production line assemble small calculators. A production engineer is interested in studying the production rate on one part of the assembly line, to see how the individual worker's production varies as a function of the time spent on the production line. Twenty-four workers were randomly selected for observation, and the number of pieces completed Y was noted for each after a length of time X had elapsed. The data are given in Figure 12.14.

What sort of model may be used to describe the regression of Y on X? How can the parameters of the model be estimated? What assumptions are needed

X Time Spent On Assembly Line	Y Number of Units Assembled
.5 hours	36
.5	41
.5	33
1.0	83
1.0	61
1.0	68
1.5	116
1.5	99
1.5	91
2.0	120
2.0	135
2.0	125
2.5	170
2.5	143
2.5	151
3.0	180
3.0	205
3.0	188
3.5	212
3.5	241
3.5	208
4.0	236
4.0	261
4.0	253

FIGURE 12.14 The Number of Units Assembled and the Time Spent on the Assembly Line for 24 Workers

in order to place confidence intervals on the true model parameters? How can the estimated regression curve be used to predict values for the dependent variable when the independent variable is known? These and other questions will be addressed in this section and the following sections.

A MODEL FOR LINEAR REGRESSION

While the previous section provides the mechanics of fitting a straight line to bivariate sample data, it does not address the problem of inference with respect to linear regression. In order to make any inferences about the population regression curve, some assumptions need to be satisfied regarding the population. In linear regression the following assumptions are usually made.

1. The random variables $Y_1, Y_2, \ldots, Y_n$ satisfy the linear relationship

$$Y_i = \alpha + \beta X_i + \epsilon_i, \quad i = 1, \ldots, n, \tag{12.5}$$

where α and β are population parameters that represent respectively the y intercept and slope of the population regression curve, ϵ_1 through ϵ_n are random variables, and the X's may be random variables or may be fixed values.

2. The mean of ϵ_i is zero for all i.

3. The variance of ϵ_i is constant, independent of the values for X, and is denoted by σ^2 for all i.

4. The distribution of ϵ_i is normal for all i.

5. The ϵ_i's are independent of each other.

Assumptions 2 through 5 are equivalent to saying that $\epsilon_1, \ldots, \epsilon_n$ is a random sample from a normal distribution with mean zero and variance σ^2. Each of these assumptions is discussed separately in this section.

ASSUMPTION 1: $Y = \alpha + \beta X + \epsilon$

The method of least squares is used to estimate α and β, with a and b defined in the previous section. Under the assumptions of this model, a and b are statistics and have known sampling distributions. In fact, the mean of a is α, and the mean of b is β, so both a and b are unbiased estimators of their respective population counterparts. Further, both a and b have normal distributions. This fact is useful when testing hypotheses or forming confidence intervals.

The values of X may be observations on a random variable, such as when X and Y are measurements on randomly sampled individuals from a population. For example, in a marketing survey X may be the amount spent on groceries and Y may be the corresponding amount spent on meat, and both are random variables. Or the X's may be fixed constants, such as in cooking experiments where X may represent the predetermined oven temperature settings and Y measures the time until a 1 lb. roast is finished at that temperature setting.

The ϵ's are called **population residuals** (or population errors), and represent the difference between Y and the population regression curve. This fact is more apparent if the model is rewritten as

$$\epsilon_i = Y_i - (\alpha + \beta X_i). \tag{12.6}$$

Population residuals ϵ_i are the differences between the observed values Y_i and the conditional means $\mu_{y|x}$,

$$\epsilon_i = Y_i - \mu_{y|x}.$$

RESIDUALS   **Sample residuals** e_i are the differences between the observed values Y_i and the predicted values $\hat{Y}_i = a + bX_i$,

$$e_i = Y_i - \hat{Y}_i,$$

from a sample regression curve.

After a least squares line is fit to the data, the difference $Y_i - \hat{Y}_i$ is called a **sample residual** and is denoted by e_i. This may be written as

$$e = Y_i - \hat{Y}_i = Y_i - (a + bX_i). \tag{12.7}$$

The sample residuals $e_i, \ldots, e_n$, although not independent, are used to study the validity of the Assumptions 3, 4, and 5 concerning the population residuals.

ASSUMPTION 2: $E(\epsilon) = 0$

This assumption usually causes no problem in linear regression. The sample residuals always have a sample mean equal to 0,

$$\bar{e} = \frac{1}{n} \sum_{i=1}^{n} e_i = 0,$$

as a byproduct of the least squares method.

ASSUMPTION 3: EQUAL VARIANCES

The assumption of equal variances is often violated in applications of regression. Violations of this assumption should not be taken lightly, since the validity of the following statistical methods may be seriously impaired. The assumption of equal variances is also called the assumption of **homogeneity of variances**, or **homoscedasticity**. Lack of equal variances is also called **nonhomogeneity of variances**, or **heteroscedasticity**.

When variances are nonhomogeneous, the spread of Y often tends to increase with larger values of X. Therefore, one way of testing the validity of this assumption is to divide the e_i's into two groups: one group associated with small values of X and the other group associated with large values of X. The sample variances s_1^2 and s_2^2 are computed on the two groups of e's separately. Then the F test for equal variances given in Section 9.2 is used to see if the two groups of residuals have equal population variances. Rejection of the null hypothesis using this test is strong evidence that this assumption is not satisfied. Recall, however, that this F test is very sensitive to the assumption of normality; so if the residuals are not normally distributed, the results of this F test are not valid. Even if the residuals are normal, the effect of using the least squares regression line is to make the F test an approximate procedure.

The **residual variance** σ^2 is estimated from the sample residual variance, with $n - 2$ as a divisor instead of $n - 1$, as

$$\hat{\sigma}^2 = \frac{\text{Error Sum of Squares}}{n - 2} = \frac{\sum_{i=1}^{n} (Y_i - \hat{Y}_i)^2}{n - 2}.$$

A form of $\hat{\sigma}^2$ easier to use in computing is given by

$$\hat{\sigma}^2 = \frac{n - 1}{n - 2}(s_y^2 - b^2 s_x^2),$$

where s_x^2 and s_y^2 are the sample variances of the X's and the Y's respectively.

RESIDUAL VARIANCE

The **population residual variance** *is the variance of* ϵ_i *and is denoted by* σ^2. *The* **sample residual variance** *is denoted by* $\hat{\sigma}^2$ *and denotes the variation of* Y_i *from the sample regression curve. It may be computed using either*

$$\hat{\sigma}^2 = \frac{1}{n-2} \sum_{i=1}^{n} (Y_i - \hat{Y}_i)^2 = \frac{1}{n-2} \sum_{i=1}^{n} e_i^2, \tag{12.8}$$

or the more convenient

$$\hat{\sigma}^2 = \frac{n-1}{n-2} (s_y^2 - b^2 s_x^2), \tag{12.9}$$

where s_y^2 *and* s_x^2 *are the sample variances of* Y_i *and* X_i *respectively, and* b *is the sample slope.*

ASSUMPTION 4: NORMALITY

The sample residuals $e_1, \ldots, e_n$ may be used to check the validity of the normality assumption on the ϵ_i's. The Lilliefors test may be used, but it is an approximate procedure in this case since the e_i's are not independent of each other. The approximate Lilliefors test is performed by plotting the empirical distribution function of the residuals standardized as follows:

$$e_i^* = e_i / \hat{\sigma}. \tag{12.10}$$

Note that $\bar{e} = 0$, so the only change in e_i^* from the usual z-scores is that $n - 2$ is used as a divisor for $\hat{\sigma}^2$ instead of $n - 1$. If the residuals are not normal, the statistical methods in Section 12.5 should be used rather than those in Section 12.4.

ASSUMPTION 5: INDEPENDENCE

The independence of the ϵ_i's may be tested using methods of **serial correlation** explained later in this chapter.

EXAMPLE

The production engineer in Problem Setting 12.3 looks at a scatterplot of the data given in Figure 12.14 and decides to try a linear regression model, because of the nearly straight line relationship between X and the conditional means of Y given X (see Figure 12.15). The least squares method yields the unbiased estimate of β,

$$b = \frac{\Sigma XY - n\bar{X}\bar{Y}}{\Sigma X^2 - n\bar{X}^2} = \frac{9678.5 - 24(2.25)(144.00)}{153 - 24(2.25)^2} = 60.4,$$

and α,

$$a = \bar{Y} - b\bar{X} = 144 - 60.4(2.25) = 8.1,$$

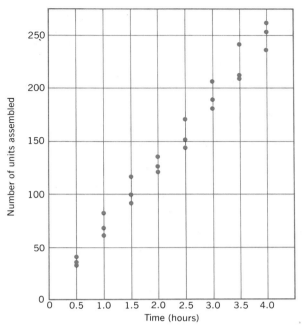

FIGURE 12.15 A Scatterplot of Units Assembled (Y) versus Time (X)

from the calculations given in the worksheet in Figure 12.16. Note that the X values are not observations on a random variable in this case because the production engineer is free to inspect each individual worker's production total at any time X he chooses. This fact does not alter the analysis in any way. The sample residuals $e_i = Y_i - \hat{Y}_i$ are given in the worksheet also. A plot of the residuals against X, such as in Figure 12.17, is useful to detect any obvious violations of the assumptions of homogeneity of variance, normality, or independence. In this case the residuals appear to be well behaved because of their apparent random scattering, but tests of the assumptions will be made to provide some confidence.

The homogeneity of variance assumption is tested by computing the sample variance for the 12 e_i's associated with the smallest X_i's (two hours or less), which is $s_1^2 = 72.357$, and comparing it with $s_2^2 = 159.404$, which is the sample variance for the 12 e_i's associated with X_i's greater than two hours. As an approximate procedure the resulting F statistic

$$F = \frac{\text{larger } s_i^2}{\text{smaller } s_i^2} = \frac{159.404}{72.357} = 2.203$$

is compared with the .975 quantile from the F distribution with 11 degrees of freedom in both the numerator and denominator, which is 3.478 from Table A5 using interpolation. Therefore, the null hypothesis of equal variances is accepted at $\alpha = .05$. This is an indication of, but not a proof of, the validity of Assumption 3. It is never possible to completely verify the validity of any of the

	X	Y	X^2	Y^2	XY	$\hat{Y}$	e_i	e_i^*
	0.5	36	0.25	1296	18.0	38.31	−2.31	−0.21
	0.5	41	0.25	1681	20.5	38.31	2.69	0.25
	0.5	33	0.25	1089	16.5	38.31	−5.31	−0.49
	1.0	83	1.00	6889	83.0	68.50	14.50	1.35
	1.0	61	1.00	3721	61.0	68.50	−7.50	−0.70
	1.0	68	1.00	4624	68.0	68.50	−0.50	−0.05
	1.5	116	2.25	13456	174.0	98.70	17.30	1.61
	1.5	99	2.25	9801	148.5	98.70	0.30	0.03
	1.5	91	2.25	8281	136.5	98.70	−7.70	−0.71
	2.0	120	4.00	14400	240.0	128.90	−8.90	−0.83
	2.0	135	4.00	18225	270.0	128.90	6.10	0.57
	2.0	125	4.00	15625	250.0	128.90	−3.90	−0.36
	2.5	170	6.25	28900	425.0	159.10	−10.90	1.01
	2.5	143	6.25	20449	357.5	159.10	−16.10	−1.49
	2.5	151	6.25	22801	377.5	159.10	−8.10	−0.75
	3.0	180	9.00	32400	540.0	189.30	−9.30	−0.86
	3.0	205	9.00	42025	615.0	189.30	15.70	1.46
	3.0	188	9.00	35344	564.0	189.30	−1.30	−0.12
	3.5	212	12.25	44944	742.0	219.50	−7.50	−0.70
	3.5	241	12.25	58081	843.5	219.50	21.50	2.00
	3.5	208	12.25	43264	728.0	219.50	−11.50	−1.07
	4.0	236	16.00	55696	944.0	249.69	−13.69	−.127
	4.0	261	16.00	68121	1044.0	249.69	11.31	1.05
	4.0	253	16.00	64009	1012.0	249.69	3.31	0.31
Totals	54.0	3456	153.00	615122	9678.5		0.00	0.03

$$\overline{X} = 2.25 \quad \overline{Y} = 144.00 \quad s_x^2 = 1.3696 \quad s_y^2 = 5106.87$$

FIGURE 12.16 A Worksheet for Checking Assumptions in the Linear Regression Model

assumptions, but a check such as this is useful to detect obvious violations of the assumptions. To check the assumption of normality, the sample residual variance is calculated as

$$\hat{\sigma}^2 = \frac{n-1}{n-2}(s_y^2 - b^2 s_x^2)$$

$$= \frac{23}{22}[5106.87 - 3647.7(1.3696)]$$

$$= 116.02.$$

The residuals are divided by $\hat{\sigma} = 10.77$ to get the standardized residuals e_i^*. These standardized residuals are given in the worksheet in Figure 12.16. A plot of the empirical distribution function of the e_i^*'s on the Lilliefors test chart (not shown here) shows that the e_i's may reasonably be regarded as having come from a normal distribution, as the reader can easily verify. Thus the data in Problem Setting 12.3 appear to satisfy all of the assumptions of the linear model. This enables the methods of the following section to be applied to make inferences in this situation.

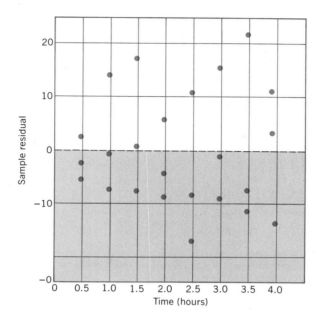

FIGURE 12.17 A Scatterplot of the Sample Residuals (e_i) Versus Time (X).

Exercises

12.17 Use the methods of this section to test the normality assumption for the least squares residuals for the sample data given in Exercise 12.1 (see Exercise 12.13 for the least squares fit to these data).

12.18 Use the methods of this section to test the normality assumption for the least squares residuals for the sample data given in Exercise 12.5 (see Exercise 12.15 for the least squares fit to these data).

12.19 Test the assumption of homogeneity of variance for the sample regression data given in Exercise 12.1.

12.20 Test the assumption of homogeneity of variance for the sample regression data given in Exercise 12.5

12.4

Confidence Intervals and Hypothesis Tests When the Residuals are Normal

REVIEW

In the previous section a method is presented for determining whether the residuals

$$e_i = Y_i - \hat{Y}_i = Y_i - a - bX_i, \quad i = 1, \ldots, n$$

should be considered normally distributed or not. If the residuals are not normally distributed, the method of the next section should be used. Although that method is also valid for normally distributed residuals, the methods presented in this section should be preferred in such a case, because they are easier to use and there is a wider variety of methods available for normally distributed residuals.

Of lesser importance is the slight gain in power associated with the use of these methods when they are appropriate. From a practical standpoint this slight gain in power is offset by the danger of using these procedures when the residuals are not normal, with the loss of power accompanying such a misapplication. The methods of this section are often used without any consideration given as to the validity of the normality assumption. This is a widespread but unfortunate misuse of statistical methods. A check of the normality of the residuals is necessary to assure the validity of the methods of this section.

INFERENCES ABOUT SLOPE

A method is presented in this section for testing hypotheses about the slope β of the population regression line. It may be used only if the residuals are normally

---

A TEST FOR SLOPE (NORMALLY DISTRIBUTED RESIDUALS)

Data. The data consist of n pairs of observations (X_i, Y_i), $i = 1, \ldots, n$.

Assumptions

1. (linear regression) Each Y_i is a linear function of the associated value of X_i through the equation

$$Y_i = \alpha + \beta X_i + \epsilon_i,$$

where α and β are the same for all pairs (X_i, Y_i), and where ϵ_i is a random variable.
2. The ϵ_i's are a random sample from a normal population with mean 0 and variance σ^2.

Estimation. The sample slope b provides an unbiased **point estimate** of the true slope β. A $100(1 - \alpha)$ percent **confidence interval** for β is given by β_L and β_U, where

$$\beta_L = b - t_{\alpha/2, n-2} \frac{\hat{\sigma}}{s_x \sqrt{n-1}}, \tag{12.11}$$

$$\beta_U = b + t_{\alpha/2, n-2} \frac{\hat{\sigma}}{s_x \sqrt{n-1}}, \tag{12.12}$$

and where $t_{\alpha/2, n-2}$ is the $1 - \alpha/2$ quantile of the t-distribution with $n - 2$ degrees of freedom, given in Table A3.

Null Hypothesis. $H_0: \beta = \beta_0$ (β_0 is some specified number).

Test Statistic

$$T = \frac{(b - \beta_0) s_x \sqrt{n-1}}{\hat{\sigma}} \tag{12.13}$$

Decision Rule. The decision rule depends on the alternative hypothesis of interest.

1. $H_1: \beta > \beta_0$. Reject H_0 if $T > t_{\alpha, n-2}$; otherwise accept H_0.
2. $H_1: \beta < \beta_0$. Reject H_0 if $T < -t_{\alpha, n-2}$; otherwise accept H_0.
3. $H_1: \beta \neq \beta_0$. Reject H_0 if $|T| > t_{\alpha/2, n-2}$; otherwise accept H_0.

Critical values $t_{\alpha, n-2}$ for a level of significance α are given in Table A3, for $n - 2$ degrees of freedom.

distributed. The hypothesis test outlined in this section uses a Student's
t-distribution.

A confidence interval for the unknown β is easily obtainable under the
assumption of normality of residuals. In general, a confidence interval for β
provides an interval estimate of how much the mean of Y will increase for a
given unit increase in X.

EXAMPLE

Grade point averages and ACT scores were reported for 12 college freshmen in
Problem Setting 12.2. If the University Admissions Officer in that problem
setting determines that a linear relationship exists between G.P.A. and ACT
score then this information can be used to predict changes in G.P.A. relative to
the ACT score. This relationship can be measured by the straight line fit to the
data. Suppose that a recent publication states that $\beta = .1$. This means that an
increase of 1 in the ACT score will tend to be associated with an increase of .1
in G.P.A. Clearly some type of hypothesis test is appropriate here to test the
hypothesis $\beta = \beta_0$. Equally appropriate might be a confidence interval for the
unknown value of β. The data and the worksheet of Problem Setting 12.2 are
repeated in Figure 12.18 for the reader's convenience, with an additional
column added so s_y^2 may be obtained. The calculations proceed as follows.

$$s_x^2 = [7857 - (303)^2/12]/11 = 18.750$$

$$s_y^2 = [115.8856 - (36.54)^2/12]/11 = .420$$

$$b = \frac{950.77 - 12(303/12)(36.54/12)}{7857 - 12(303/12)^2}$$

$$= .1364$$

$$a = 36.54/12 - (.1364)(303/12) = -.399$$

$$\hat{\sigma}^2 = 11[.420 - (.1364)^2(18.750)]/10 = .0783$$

Student	ACT X	G.P.A. Y	X^2	XY	Y^2
1	24	2.66	576	63.84	7.0756
2	20	2.25	400	45.00	5.0625
3	23	2.82	529	64.86	7.9524
4	27	2.91	729	78.57	8.4681
5	31	3.45	961	106.95	11.9025
6	29	3.47	841	100.63	12.0409
7	28	3.75	784	105.00	14.0625
8	26	3.33	676	86.58	11.0889
9	31	4.00	961	124.00	16.0000
10	26	3.49	676	90.74	12.1801
11	20	2.61	400	52.20	6.8121
12	18	1.80	324	32.40	3.2400
Totals	303	36.54	7857	950.77	115.8856

FIGURE 12.18 Grade
Point Average and ACT
Scores for 12 College
Freshmen

A scatterplot of the data points is given in Figure 12.19, along with the least squares regression line

$$\hat{\mu}_{y|x} = -.399 + .1364x.$$

To test the null hypothesis $\beta = .1$ against the two-sided alternative $\beta = .1$, the test statistic

$$T = \frac{(.1364 - .1)\sqrt{18.750}\sqrt{11}}{\sqrt{.0783}} = 1.8682$$

is compared with $t_{\alpha/2, n-2}$. At $\alpha = .05$ and for $n = 12$ the critical value is $t_{.025, 10} = 2.2281$, which is the .975 quantile obtained from Table A3 for 10 degrees of freedom. Since $|T| = 1.8682$, which is less than 2.2281, H_0 is accepted. The p-value associated with $T = 1.8682$ is approximately .09.

A 95 percent confidence interval is easily found for β. Since $t_{.025, 10} = 2.2281$ from Table A3, the 95 percent confidence interval is from

$$\beta_L = .1364 - 2.2281 \sqrt{.0783}/\sqrt{18.750(11)} = .0930$$

to

$$\beta_U = .1364 + 2.2281 \sqrt{.0783}/\sqrt{18.750(11)} = .1798,$$

which includes, notably, the value .1 within the interval.

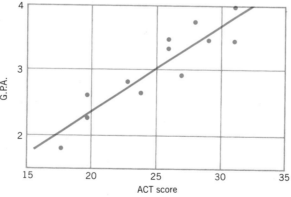

FIGURE 12.19 Scatterplot and Least Squares Regression Line for G.P.A. (Y) and ACT Score (X)

THE SPECIAL INTERPRETATION OF $\beta = 0$

If there is no linear relationship between two variables X and Y, then the slope of the sample regression line should be essentially equal to zero, except for chance variation due to sampling. That is, the true regression line should be horizontal if there is no linear relationship between X and Y. A horizontal regression line has the slope β equal to zero. Therefore, a test of the hypothesis "no linear relationship exists between X and Y" is accomplished by testing

$H_0: \beta = 0$, which is equivalent to a test of $H_0: \rho = 0$, since r is a measure of the strength of the linear relationship.

A CONFIDENCE INTERVAL FOR $\mu_{y|x}$

If the true regression equation were known, then the mean of Y given $X = x$ would be determined exactly from the relationship

$$\mu_{y|x} = \alpha + \beta x.$$

However, α and β are estimated using the observed values (X_i, Y_i), $i = 1, \ldots, n$, and hence $\mu_{y|x}$ is also estimated. A **confidence interval for $\mu_{y|x}$** is based on the point estimate $\hat{\mu}_{y|x}$ given by (12.14), and has endpoints μ_L and μ_U given by (12.15) and (12.16).

A CONFIDENCE INTERVAL FOR $\mu_{y|x}$, THE MEAN OF Y AS PREDICTED BY A GIVEN VALUE OF X

The **Data** and **Assumptions** are the same as previously given in this section.

Estimation. A **point estimate** of $\mu_{y|x}$ is provided by $\hat{\mu}_{y|x}$, where

$$\hat{\mu}_{y|x} = a + bx \qquad (12.14)$$

and where x is the given value of X. A $100(1 - \alpha)$ percent **confidence interval** for $\mu_{y|x}$ is given by μ_L and μ_U, where

$$\mu_L = \hat{\mu}_{y|x} - t_{\alpha/2, n-2}\,\hat{\sigma}\,\sqrt{\frac{1}{n} + \frac{(x - \bar{X})^2}{(n-1)\,s_x^2}} \qquad (12.15)$$

and

$$\mu_U = \hat{\mu}_{y|x} + t_{\alpha/2, n-2}\,\hat{\sigma}\,\sqrt{\frac{1}{n} + \frac{(x - \bar{X})^2}{(n-1)\,s_x^2}}. \qquad (12.16)$$

The number $t_{\alpha/2, n-2}$ is obtained from Table A3, using $n - 2$ degrees of freedom.

A PREDICTION INTERVAL FOR Y

Even if the true regression line were known, there would still be variation in the Y's because the Y's are normally distributed about the regression line. So a **prediction interval for** Y incorporates the variation of Y about the regression line and the variation in $\hat{\mu}_{y|x}$ because the true regression line is not known but is established from a sample. For this reason it is true that prediction intervals for Y are always much wider than confidence intervals for $\mu_{y|x}$.

The difference between the interpretation of a prediction interval for Y and a confidence interval for $\mu_{y|x}$ may be illustrated using the following examples. See Figure 12.20.

1. In Problem Setting 12.2 the confidence interval for $\mu_{y|x}$ is a confidence interval for the mean G.P.A. for a given ACT score, while a prediction interval for Y gives bounds for the G.P.A. that might occur for any given ACT score.

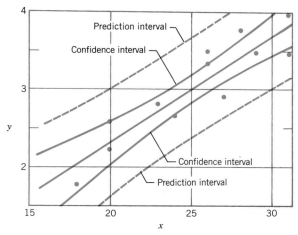

FIGURE 12.20 Bands Formed by Construction Confidence Intervals and Prediction Intervals for All Values of x

2. In Problem Setting 12.3 the confidence interval for $\mu_{y|x}$ estimates the mean production for the population of all workers who have been on the assembly line X hours. The prediction interval gives an estimate of an individual worker's production for the same length of time.

3. If Y represents the number of months of storage life of a battery and X is the number of days of conditioning, a confidence interval for $\mu_{y|x}$ gives an interval estimate of the mean storage life of a battery with a specific number of days of conditioning. Predictions for an individual battery's storage life are given by the prediction interval on Y for any given X, the number of days of conditioning.

A PREDICTION INTERVAL FOR Y GIVEN X

The **Data** and **Assumptions** are the same as previously given in this section.

Prediction. The **predicted value** $\hat{Y}$ for Y given X is

$$\hat{Y} = a + bx,$$

where x is the given value of X. A $100(1 - \alpha)$ percent **prediction interval** for Y is given by Y_L and Y_U, where

$$Y_L = \hat{Y} - t_{\alpha/2,n-2}\,\hat{\sigma}\,\sqrt{1 + \frac{1}{n} + \frac{(x - \bar{X})^2}{(n-1)\,s_x^2}} \qquad (12.17)$$

and

$$Y_U = \hat{Y} + t_{\alpha/2,n-2}\,\hat{\sigma}\,\sqrt{1 + \frac{1}{n} + \frac{(x - \bar{X})^2}{(n-1)\,s_x^2}}. \qquad (12.18)$$

The number $t_{\alpha/2,n-2}$ is obtained from Table A3 using $n - 2$ degrees of freedom.

EXAMPLE

In continuation of the previous example, a confidence interval for $\mu_{y|x}$ and a prediction interval for Y given X will be found.

Suppose a high school senior scores a 22 on the ACT exam. What freshman G.P.A. would be predicted for such an individual? There are two distinct questions of interest.

1. What is a 95 percent *confidence interval* for the expected value ($\mu_{y|x}$) of the G.P.A. next year if the ACT score is 22?

This question concerns the mean G.P.A., but does not describe the variation that accompanies any single observation, as does the next question.

2. What is a 95 percent *prediction interval* for the individual's G.P.A. (Y) next year if the individual's ACT score is 22?

The first question is answered using a confidence interval for $\mu_{y|x}$, with $X = 22$. Equations (12.14), (12.15), and (12.16) show that the point estimate for $\mu_{y|x}$ is

$$\hat{\mu}_{y|x} = -0.399 + .1364\,(22) = 2.60$$

and the 95 percent confidence interval goes from

$$\mu_L = 2.60 - 2.2281\,\sqrt{.0783}\sqrt{\frac{1}{12} + \frac{(22 - 25.25)^2}{11(18.750)}} = 2.37$$

to

$$\mu_U = 2.60 + 2.2281\,\sqrt{.0783}\sqrt{\frac{1}{12} + \frac{(22 - 25.25)^2}{11(18.750)}} = 2.83.$$

The prediction interval uses the point estimate Y as found above for $X = 22$:

$$\hat{Y} = -0.399 + .1364\,(22) = 2.60.$$

The 95 percent prediction interval goes from

$$Y_L = 2.60 - 2.2281\,\sqrt{.0783}\sqrt{1 + \frac{1}{12} + \frac{(22 - 25.25)^2}{11(18.750)}} = 1.94$$

to

$$Y_U = 2.60 + 2.2281\,\sqrt{.0783}\sqrt{1 + \frac{1}{12} + \frac{(22 - 25.25)^2}{11(18.750)}} = 3.26$$

using Equations (12.17) and (12.18). Thus the 95 percent confidence interval for the mean $\mu_{y|x}$ of the freshman G.P.A. is from 2.37 to 2.83, but an individual G.P.A. may be anywhere from 1.94 to 3.26 with probability .95. This illustrates the principle that the mean can be estimated with more precision than an individual observation can be estimated.

Exercises

12.21 Find a 95 percent confidence interval for β using the regression data given in Exercise 12.1.

12.22 Find a 95 percent confidence interval for β using the regression data given in Exercise 12.5.

12.23 Refer to the sample data given in Exercise 12.1 and test the hypothesis $H_0: \beta = 2$ versus $H_1: \beta > 2$ using the methods of this section. Let $\alpha = .05$ and state the p-value.

12.24 Refer to the sample data given in Exercise 12.5 and test the hypothesis $H_0: \beta = 10$ versus $H_1: \beta > 10$ using the methods of this section. Let $\alpha = .05$ and state the p-value.

12.25 A car owner kept track of his gasoline bills for 4 years. Fit a least squares regression line to these data by letting the 16 quarters be represented by the integers 1 to 16—that is, $X = 1, 2, \ldots, 16$. Use a level of significance of $\alpha = .05$ to test the hypothesis $H_0: \beta = 0$ versus $H_1: \beta > 0$. State the p-value.

	1977	1978	1979	1980
1st quarter	$142	$155	$178	$206
2nd quarter	83	100	112	121
3rd quarter	122	134	148	161
4th quarter	110	131	140	165

12.26 Find a 90 percent confidence interval for β using the data in Exercise 12.25.

12.27 Find a 95 percent confidence interval for $\mu_{y|x}$ for the first quarter of 1981 (i.e., $X = 17$) using the regression results in Exercise 12.25.

12.28 Find a 95 percent prediction interval for the first quarter of 1981 (i.e., $X = 17$) using the regression results in Exercise 12.25.

12.29 If the production engineer in Problem Setting 12.3 believes that the number of units assembled should increase by 50 for each hour worked, then test the $H_0: \beta = 50$ versus $H_1: \beta \neq 50$ for the data given in Figure 12.14. Let $\alpha = .05$ and recall that the necessary calculations were done in the example of the previous section.

12.30 Find a 95 percent confidence interval for the mean increase in the number of units assembled per hour for the sample data referred to in Exercise 12.29.

12.5
A Nonparametric Test for Slope

The test for slope presented in the previous section relies on the assumption of normality for the residuals. If that assumption is not satisfied, then the method of this section can be used. In reality the method of this section can also be used with normal data, but in such a case the methods of the previous section should be preferred due to their simplicity and versatility. That is, the previous section presented a test for slope, confidence intervals for β and $\mu_{y|x}$, and a prediction interval for Y. In this section only a test for slope is given. Other alternative procedures exist for analyzing nonnormal data, but they are not included in this text because they are somewhat cumbersome to use.

A TEST FOR SLOPE

If the wrong slope is used to fit a straight line to a set of bivariate data, then the fact that the wrong slope was used should be apparent from a scatterplot of the data along with the erroneous regression curve. That is, if the wrong slope is used, the line will tend to be above the points on one side of the graph and below the points on the other side of the graph. This means that the residuals will tend to be positive on one side of the graph and negative on the other. Thus the residuals will appear to be correlated with the x-coordinates of the points. Spearman's rho may be used to see if the residuals are correlated with the x-values. If r_s is significant, then the slope is probably not correct. This is the basis for this test on slope. Note that the residuals do not need to be normally distributed for a test using r_s to be valid.

NONPARAMETRIC TEST FOR SLOPE

Data. The data consist of n pairs of observations (X_i, Y_i), $i = 1, \ldots, n$.

Assumptions

1. (Linear Regression) Each Y_i is a linear function of the associated value of $X_i = x$ through the equation

$$Y_i = \alpha + \beta X_i + \epsilon_i,$$

where α and β are the same for all pairs (X_i, Y_i).
2. The ϵ_i's are a random sample from a population (not necessarily normal) with mean 0 and variance σ^2.

Null Hypothesis. H_0: $\beta = \beta_0$ (β_0 is some specified number).

Test Statistic. The test statistic is Spearman's rho computed between the ranks of X_i and the ranks of the residuals assuming the slope is β_0. That is:

1. For each pair (X_i, Y_i) compute $U_i = Y_i - \beta_0 X_i$, $i = 1, \ldots, n$.
2. Rank the U_i's from 1 to n, using average ranks in case of ties. Call these ranks R_{u_i}. Although the U's are not the residuals because a was not used in the computation, the ranks R_{u_i} are the same as the ranks of the residuals because subtraction of the constant a does not change the relative ranking.
3. Rank the X_i's from 1 to n using average ranks in case of ties. Call these ranks R_{x_i}.
4. Compute Spearman's rho

$$r_s = \frac{\Sigma R_{u_i} R_{x_i} - C}{\left[\Sigma R_{u_i}^2 - C\right]^{1/2} \left[\Sigma R_{x_i}^2 - C\right]^{1/2}} \tag{12.16}$$

between the X_i's and the hypothesized residuals U_i, where $C = n(n + 1)^2/4$.

Decision Rule. The decision rule depends on the alternative hypothesis of interest.

1. H_1: $\beta > \beta_0$. Reject H_0 if $r_s > T_{\alpha,n}^*$; otherwise accept H_0.
2. H_1: $\beta < \beta_0$. Reject H_0 if $r_s < -T_{\alpha,n}^*$; otherwise accept H_0.
3. H_1: $\beta \neq \beta_0$. Reject H_0 if $|r_s| > T_{\alpha/2,n}^*$; otherwise accept H_0.

Critical values $T_{\alpha,n}^*$ are given in Figure 11.9 for a sample size n and a level of significance α.

EXAMPLE

In a laboratory experiment involving an experimental weight loss drug, 16 adult rats have been receiving a specified amount of the experimental drug. One week the dosages received are all varied from animal to animal. The percent change in dosage and the precent change in weight are reported in Figure 12.21 for each animal.

Rat	Change in Dosage X (percent)	Change in Weight Y (percent)	X^2	XY	Y^2	Predicted Y $\hat{Y}_i = a + bX_i$	Residual $Y - \hat{Y}_i$	Standardized Residual
1	−0.6	−3.2	0.36	1.92	10.24	−0.50	−2.70	−0.53
2	2.2	−1.9	4.84	−4.18	3.61	2.62	−4.52	−0.89
3	3.4	1.9	11.56	6.46	3.61	3.96	−2.06	−0.40
4	2.9	5.0	8.41	14.50	25.00	3.40	1.60	0.31
5	−3.6	−7.6	12.96	27.36	57.76	−3.84	−3.76	−0.74
6	−0.6	12.2	0.36	−7.32	148.84	−0.50	12.70	2.49
7	−5.9	−9.8	34.81	57.82	96.04	−6.40	−3.40	−0.67
8	−7.4	−8.8	54.76	65.12	77.44	−8.07	−0.73	−0.14
9	5.7	1.5	32.49	8.55	2.25	6.52	−5.02	−0.98
10	1.1	11.7	1.21	12.87	136.89	1.40	10.30	2.02
11	5.0	6.7	25.00	33.50	44.89	5.74	0.96	0.19
12	5.7	4.9	32.49	27.93	24.01	6.52	−1.62	−0.32
13	7.9	7.3	62.41	57.67	53.29	8.97	−1.67	−0.33
14	8.2	11.0	67.24	90.20	121.00	9.30	1.70	0.33
15	1.4	1.0	1.96	1.40	1.00	1.73	−0.73	−0.14
16	1.3	−2.3	1.69	2.99	5.29	1.20	−1.02	−0.20
Totals	24.1	29.6	352.55	396.79	811.16		0.03	0.00

FIGURE 12.21 Percent Changes in Weight (Y) and Percent Changes in Dosage (X), With Worksheet

Computations yield

$$s_x^2 = [352.55 - (24.1)^2/ 16] / 15 = 21.0833$$

$$s_y^2 = [811.16 - (29.6)^2/ 16] / 15 = 50.4267$$

$$b = \frac{396.79 - 16(24.1/16)(29.6/16)}{352.55 - 16(24.1/16)^2}$$

$$= 1.1140$$

$$a = [29.6 - 1.1140(24.1)]/16 = 0.1720$$

$$\hat{\sigma}^2 = \frac{15}{14} [50.4267 - (1.1140)^2 (21.0833)] = 25.9954$$

A scatterplot of the data points and the least squares regression line

$$\hat{\mu}_{y|x} = .172 + 1.114x$$

appear in Figure 12.22. Note the two points that lie far above the sample

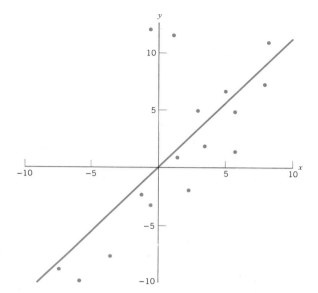

FIGURE 12.22 A Scatterplot of Percent Change in Dosage (X), Percent Changes in Weight (Y), and the Least Squares Regression Line

regression line. These two "outliers" suggest nonnormality of the residuals. To test the normality assumption of the residuals, and hence the Y's, the Lilliefors test is used on the standardized residuals, as given in Figure 12.21. The empirical distribution function of the standardized residuals falls outside the Lilliefors bounds for $n = 16$ in Figure 12.23, so normality cannot be assumed for the Y's in this case. To see if the percent weight change corresponds to percent change

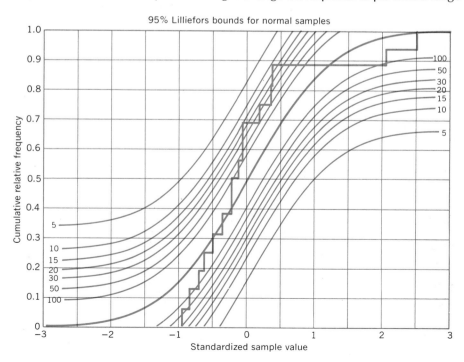

FIGURE 12.23 A Lilliefors Test of Normality of Residuals

in dosage, it might be desirable to test $H_0: \beta = 1.0$ against the two-sided alternative $H_1: \beta \neq 1.0$. The rank correlation coefficient is computed between the X's and the hypothesized residuals $Y_i - (1.0)X_i$ as shown in the worksheet in Figure 12.24. Spearman's rho

$$r_s = \frac{1231.5 - 1156}{[1496 - 1156]^{1/2}\,[1495 - 1156]^{1/2}} = .222$$

is compared with $T^*_{.025,16} = 0.500$ from Figure 11.9 for a two-sided test at $\alpha = .05$. Since $|r_s| < 0.500$, the null hypothesis $\beta = 1.0$ is accepted. The p-value is greater than .10. Therefore, it is reasonable to assume that the percent weight change noted in grams corresponds almost directly with the percent change dosage in milliliters.

Rat	X	Y	$U = Y - (1.0)X$	R_x	R_u	R_x^2	R_u^2	R_xR_u
1	−0.6	−3.2	−2.6	5.5	5	30.25	25	27.5
2	2.2	−1.9	−4.1	9	2	81	4	18
3	3.4	1.9	−1.5	11	6	121	36	66
4	2.9	5.0	2.1	10	13	100	169	130
5	−3.6	-7.6	−4.0	3	3	9	9	9
6	−0.6	12.2	12.8	5.5	16	30.25	256	88
7	−5.9	−9.8	−3.9	2	4	4	16	8
8	−7.4	−8.8	−1.4	1	7	1	49	7
9	5.7	1.5	−4.2	13.5	1	182.25	1	13.5
10	1.1	11.7	10.6	7	15	49	225	105
11	5.0	6.7	1.7	12	12	144	144	144
12	5.7	4.9	−0.8	13.5	9	182.25	81	121.5
13	7.9	7.3	−0.6	15	10	225	100	150
14	8.2	11.0	2.8	16	14	256	196	224
15	1.4	1.0	−0.4	8	11	64	121	88
16	−1.3	−2.3	−1.0	4	8	16	64	32
Totals			$C = 16(17^2)'/4 = 1156$			1495	1496	1231.5

FIGURE 12.24 Worksheet for Finding Spearman's Rho between X and U.

EXAMPLE

Laboratory tests on 10 randomly selected pieces of metal have yielded pairs of measurements on hardness and tensile strength. The data and the worksheet are given in Figure 12.25. The computations are as follows:

$$s_x^2 = [6311.08 - (243.15)^2/10]/9 = 44.3209$$

$$s_y^2 = [601.53 - (76.94)^2/10]/9 = 1.0615$$

$$b = \frac{1906.44 - 10(243.15/10)(76.94/10)}{6311.08 - 10(243.15/10)^2}$$

$$= .0894$$

$$a = [76.94 - .0894(243.15)]/10 = 5.5202$$

Tensile Strength Y	Hardness X	Y²	X²	XY
6.82	28.21	46.5124	795.8041	192.3922
8.71	21.88	75.8641	478.7344	190.5748
5.97	15.94	35.6409	254.0836	95.1618
8.23	26.16	67.7329	684.3456	215.2968
9.42	38.26	88.7364	1463.8276	360.4092
7.65	26.42	58.5225	698.0164	202.1130
6.84	18.41	46.7856	338.9281	125.9244
7.72	16.82	59.5984	282.9124	129.8504
8.41	23.14	70.7281	525.4596	194.6074
7.17	27.91	51.4089	778.9681	200.1147
Totals 76.94	243.15	601.5302	6311.0799	1906.4447

FIGURE 12.25 Results of Hardness and Tensile Strength Tests on a Metal

A scatterplot of the data points and the least squares regression line $\hat{\mu}_{y|x} = 5.5202 + 0.0894x$ appear in Figure 12.26. Since the procedure of this section is equally valid for normal as well as nonnormal residuals, the procedure is preformed on these data for purposes of illustration, without the benefit of a Lilliefors test for normality. To test H_0: $\beta = 0$ the rank correlation coefficient is computed between the X_i and the hypothesized residuals $Y_i - \beta_0 X_i$. Since $\beta_0 = 0$ in this example, the ranks of the hypothesized residuals are merely the ranks of the Y_i's themselves. Therefore, the test statistic is the rank correlation coefficient computed between the X's and the Y's. As the worksheet in Figure 12.27, shows, there are no ties, thus the simpler form for Spearman's rho, given by Equation (4.27), is used:

$$r_s = 1 - \frac{6\Sigma[R_{x_i} - R_{y_i}]^2}{n(n^2 - 1)} = 1 - \frac{6(120)}{10(99)} = .2727.$$

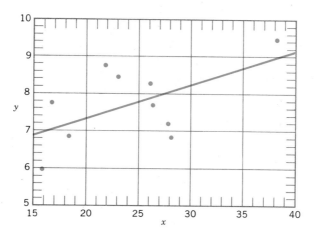

FIGURE 12.26 Scatterplot and Least Square Regression Line of Tensile Strength (Y) and Hardness (X)

A comparison of this value $r_s = .2727$ with the critical value for a two-sided test at $\alpha = .05$, $T^*_{.025,10} = 0.636$, obtained from Figure 11.9, shows that the null hypothesis $\beta = 0$ is easily accepted, hence the data do not exhibit a significant linear relationship. Note that the computations in Figure 12.25 are not needed for the hypothesis test; they are used only for finding the equation of the least squares line and for the Lilliefors test.

Tensile Strength Y	Hardness X	R_y	R_x	$(R_y - R_x)^2$
6.82	28.21	2	9	49
8.71	21.88	9	4	25
5.97	15.94	1	1	0
8.23	26.16	7	6	1
9.42	38.26	10	10	0
7.65	26.42	5	7	4
6.84	18.41	3	3	0
7.72	16.82	6	2	16
8.41	23.14	8	5	9
7.17	27.91	4	8	16
				120

FIGURE 12.27 Worksheet for Computing r_s Between Tensile Strength and Hardness

Exercises

12.31 Test the hypothesis $H_0: \beta = 2$ versus $H_1: \beta > 2$ using the methods of this section on the data referred to in Exercise 12.23. Let $\alpha = .05$ for this test. Compare your results with those obtained for Exercise 12.23.

12.32 Test the hypothesis $H_0: \beta = 10$ versus $H_1: \beta > 10$ using the methods of this section on the data referred to in Exercise 12.24. Let $\alpha = .05$ for this test. Compare your results with those obtained for Exercise 12.24.

12.33 Test the normality assumption for the residuals of the last example of this section.

12.34 Refer to Exercises 12.11 and 12.12. Test the hypothesis $H_0: \beta = 1$ versus $H_1: \beta > 1$ for the least squares regression on these data. Let $\alpha = .05$ for this test.

12.35 Refer to the sample data given in Exercise 11.3. For each budget type ($n = 8$ data pairs) test the hypothesis $H_0: \beta = 0$ versus $H_1: \beta \neq 0$. Let $\alpha = .05$ for each test. *Note:* For this special test it is only necessary to calculate Spearman's rank correlation coefficient for each budget type. (See Exercise 4.54.)

12.36 Refer to the sample data given in Exercise 11.3. Pool the data for all three budget types together ($n = 24$ data pairs) and then test $H_0: \beta = 0$ versus $H_1: \beta \neq 0$. How do you explain the difference between your answer on this exercise and the one obtained for Exercise 12.35? The same significance level should be used as in Exercise 12.35.

12.37 Use the methods of this section to work Exercise 12.29 and compare your results with those obtained on that exercise.

12.6

Serial Correlation of Residuals, and Monotone Regression

PROBLEM SETTING 12.4

Records were kept on 10 farms for 2 consecutive years regarding the amount of fertilizer applied each year to the wheat crop and the resultant yield of the crop. Data recorded are the percentage change in the amount of fertilizer applied from the first year to the second year (X) and the percentage change in the amount of the wheat yield.

Farm:	1	2	3	4	5	6	7	8	9	10
X (fertilizer):	4	62	31	−11	47	88	16	−1	74	21
Y (yield):	10	33	39	−14	37	39	18	−8	45	33

Figure 12.28 contains a scatterplot of the data and the least squares regression line, which is obtained using the techniques of the previous sections. You may want to do the calculations to check the work.

It is apparent from the scatterplot that the straight line can be used to account for some of the variation in the yield change, given the fertilizer change; but it is also apparent that some kind of a curve might be better for use as a sample regression function. In this section a method is given for determining whether a straight line provides an adequate model for fitting a set of data. An alternative method of regression is also presented here for use in many situations where straight line regression is not a reasonable assumption.

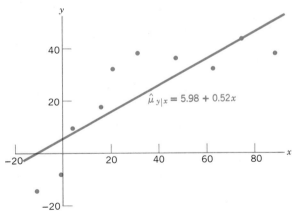

FIGURE 12.28 A Scatterplot of Percentage Change in Yield (Y) versus Percentage Change in Fertilizer (X) with the Least Squares Regression Line

SERIAL CORRELATION

Consider a sequence of observations $W_1, W_2, \ldots, W_n$. The first-order serial correlation coefficient is simply the correlation of each observation W_i with the following observation W_{i+1}, using the sample correlation coefficient, as given in Equation (12.20). Sometimes this is called the **serial correlation of *lag 1*** as opposed to **lag 2**, where observations W_i are paired with W_{i+2}, or **lag 3**, where the observations W_i are paired with W_{i+3}, and so on.

SERIAL CORRELATION COEFFICIENT (LAG 1)

The **serial correlation coefficient** r_1 *for a sequence of observations* $W_1, W_2, \ldots, W_n$ *is given by*

$$r_1 = \frac{\sum_{i=1}^{n-1} W_i W_{i+1} - \left(\sum_{i=1}^{n-1} W_i \right)\left(\sum_{i=2}^{n} W_i \right)/n}{\left[\sum_{i=1}^{n-1} W_i^2 - \left(\sum_{i=1}^{n-1} W_i \right)^2 /n \right]^{1/2} \left[\sum_{i=2}^{n} W_i^2 - \left(\sum_{i=2}^{n} W_i \right)^2 /n \right]^{1/2}}. \quad (12.20)$$

SERIAL CORRELATION OF RESIDUALS

If the points in a scatterplot appear to be scattered at random about the regression line, then the regression line probably is sufficient to use as a model. However, in Figure 12.28 the points do not appear to be scattered randomly around the regression line. Instead, negative residuals tend to follow other negative residuals such as on the left and right sides of the plot, and positive residuals tend to follow other positive residuals in the middle of the plot. One way of measuring the correlation of adjacent residuals is through a first-order serial correlation coefficient. To compute r_1 the observations need to be arranged in order of increasing values of x, as in Figure 12.29. Then the ordinary

Percent Fertilizer Change X_i	Percent Yield Change Y_i	Predicted Change In Yield $\hat{Y}_i = 5.98 + .52X$	Residuals $W_i = Y_i - \hat{Y}_i$	Residuals Lag One W_{i+1}	W_i^2	W_{i+1}^2	$W_i \times W_{i+1}$
−11	−14	0.26	−14.26	−13.46	203.35	181.17	191.94
−1	−8	5.46	−13.46	1.94	181.17	3.76	−26.11
4	10	8.06	1.94	3.69	3.76	13.62	7.16
16	18	14.31	3.69	16.09	13.62	258.89	59.37
21	33	16.91	16.09	16.89	258.89	285.27	271.76
31	39	22.11	16.89	6.57	285.27	43.16	110.97
47	37	30.43	6.57	−5.23	43.16	27.35	−34.36
62	33	38.23	−5.23	0.53	27.35	0.28	−2.77
74	45	44.47	0.53	−12.75	0.28	162.56	−6.76
88	39	51.75	−12.75[a]				
		Totals	12.76	14.27	1016.86	976.07	571.19

[a]Not used in figuring total.

FIGURE 12.29 Worksheet for Calculating Serial Correlation of Residuals

sample correlation coefficient is computed when each residual W_i is paired with its successor W_{i+1}, as indicated by Equation (12.20):

$$r_1 = \frac{571.19 - 12.76(14.29)/9}{[1016.86 - (12.76)^2/9]^{1/2} \, [976.07 - (14.27)^2/9]^{1/2}} = .565.$$

Large positive values of r_1 indicate that the model being used (linear regression in this case) could be improved. The problem is that the methods for analyzing correlation, as presented in Chapter 11, are not exact when applied to serial correlation. However, it is reasonable to use the critical values in Figure 11.6 as *approximate* standards against which r_1 may be judged. Since there are nine pairs of residuals, $n = 9$ in Figure 11.6, and the upper 5 percent critical value $T_{.05,9}$ is 0.582, which is very close to the observed $r_1 = .565$. Although the null hypothesis is accepted in this case, the p-value is very close to .05, and it may be safe to conclude that the linear model does not adequately fit the data in this case. An alternative regression model will now be introduced.

MONOTONIC REGRESSION

In Figure 12.28 the data points (X_i, Y_i) exhibit a **monotonic relationship**; that is, both X and Y tend to increase together. They do not increase together in a **linear** relationship but rather in a **monotonic** relationship. Many users of linear regression techniques realize this shortcoming but proceed with linear regression on the belief that the straight line provides a "good approximation" to the curve. Still other users of regression techniques would try to fit a curve to the data. While this latter approach may be an improvement over fitting a straight line, it is not without its drawbacks and limitations. For example, it may be possible to match the curves in Figure 12.30 by curve fitting. However, there is a very simple technique that will allow for such curves to be fit accurately and produce reliable predictions. That method consists of a linear regression computed on the rank transformed data. The only requirement is that the data exhibit a monotonic (or nearly so) relationship, hence the reason for the name **monotonic regression**.

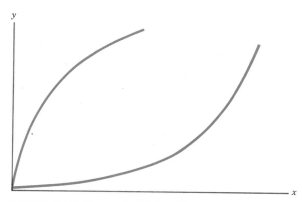

FIGURE 12.30 Monotonic Relationships

PROCEDURE FOR MONOTONIC REGRESSION

1. Replace the X_i, $i = 1, \ldots, n$, with their corresponding ranks 1 to n. Denote these ranks by R_{x_i}. Replace the Y_i, $i = 1, \ldots, n$, with their corresponding ranks 1 to n. Denote these ranks by R_{y_i}. Use the method of average ranks in case of ties.

2. Compute the usual least squares estimates a and b on the *ranks*. For ranks these calculations simplify to

$$b = \frac{\Sigma R_{x_i} R_{y_i} - n(n+1)^2/4}{\Sigma R_{x_i}^2 - n(n+1)^2/4}, \tag{12.21}$$

$$a = (n+1)/2 - b(n+1)/2. \tag{12.22}$$

3. The regression equation is expressed in terms of ranks as

$$\hat{R}_{y_i} = a + b R_{x_i}. \tag{12.23}$$

MAKING PREDICTIONS

So far the regression on ranks is at least as simple as working with the raw data (perhaps simpler). The only unusual part comes in making predictions, as the regression equation is expressed in terms of ranks, therefore the *ranks must be used* to get a predicted raw value Y_0 from a raw value X_0. This means first a rank must be associated with the value X_0 for which the prediction is to be made. This is done by comparing X_0 with the observed X_i's, and assigning a rank to X_0 that is commensurate with the ranks that belong to the other X's. The rank assigned to X_0 may be a noninteger.

ASSIGNING A RANK TO X_0

1. Order the original X's from smallest to largest

	Data	Corresponding Ranks	Ranks If there Are No Ties
(smallest)	X_1	R_{x_1}	1
	X_2	R_{x_2}	2
	. . .	. . .	. . .
(largest)	X_n	R_{x_n}	n

2. If $X_0 \le X_1$, then assign R_{x_1} to X_0.

3. If $X_0 \ge X_n$, then assign R_{x_n} to X_0.

4. If $X_0 = X_i$, then assign R_{x_i} to X_0.

5. If X_0 is between two adjacent ordered values of X, say $X_i < X_0 < X_j$, then assign R_{x_0} by linear interpolation as

$$R_{x_0} = R_{x_i} + \left(\frac{X_0 - X_i}{X_j - X_i}\right)(R_{x_j} - R_{x_i}). \tag{12.24}$$

CONVERTING THE PREDICTED RANK OF Y_0 TO $\hat{Y}_0$

Once the value R_{x_0} is determined, it is substituted into the rank regression equation to get a predicted rank for Y_0 as $\hat{R}_{y_0} = a + bR_{x_0}$. The last step involves converting the predicted $\hat{R}_{y_0}$ into a raw predicted value $\hat{Y}_0$. This is done by comparing the predicted rank $\hat{R}_{y_0}$, which may be a noninteger, with the observed ranks of the Y's, and choosing a number $\hat{Y}_0$ for $\hat{R}_{y_0}$ that is commensurate with the observed Y's and their ranks.

CONVERTING $\hat{R}_{y_0}$ INTO $\hat{Y}_0$

1. Order the original Y's from smallest to largest

	Data	Corresponding Ranks	Ranks If there Are No Ties
(smallest)	Y_1	R_{y_1}	1
	Y_2	R_{y_2}	2
	. . .	. . .	. . .
(largest)	Y_n	R_{y_n}	n

2. If $\hat{R}_{y_0} \leq R_{y_1}$, then $\hat{Y}_0 = Y_1$.
3. If $\hat{R}_{y_0} \geq R_{y_n}$, then $\hat{Y}_0 = Y_n$.
4. If $\hat{R}_{y_0} = R_{y_i}$, then $\hat{Y}_0 = Y_i$.
5. If $\hat{R}_{y_0}$ is between two adjacent ranks assigned to the Y's, say $R_{y_i} < \hat{R}_{y_0} < R_{y_j}$, then by linear interpolation,

$$\hat{Y}_0 = Y_i + \left(\frac{R_{y_0} - R_{y_i}}{R_{y_j} - R_{y_i}} \right) (Y_j - Y_i). \tag{12.25}$$

SOME APPLICATIONS OF REGRESSION ON RANKS

The technique of regression on ranks has a wide variety of applications and can be used to great advantage in many settings.

1. Although a student's G.P.A. can be expected to be higher as the students I.Q. is higher, the relationship is not likely to be linear, because a G.P.A. close to 4.0 cannot increase very much, although the corresponding I.Q. may increase considerably for very bright students. Therefore, monotonic regression methods may be more appropriate.

2. Increasing the number of checkers in a supermarket can be expected to decrease the average waiting time for customers waiting to check out. The relationship cannot be expected to be linear, however, because the larger number of tellers will increase the amount of time in which one or more checkers will be idle. Monotonic regression would be more appropriate.

3. Increasing the amount of fertilizer can be expected to increase the yield of a crop, but the gain in yield can be expected to be less for a unit of fertilizer added to a field already rich in fertilizer, than for the same unit added to a field low in fertilizer. The relationship may be monotonic but not linear.

A detailed example will now be used to demonstrate this procedure.

EXAMPLE

As a continuation of the example given in Problem Setting 12.4, wheat yield and fertilizer data are replaced by their ranks (average ranks in case of ties) in Figure 12.31. The regression on ranks involves calculating

$$b = \frac{376.5 - 10(11)^2/4}{385 - 10(11)^2/4} = .897$$

and

$$a = 11/2 - .897(11/2) = .567$$

from Equations (12.21) and (12.22) to get the regression equation on ranks

$$\hat{R}_{y_i} = .567 + .897\, R_{x_i} \tag{12.26}$$

X_i	Y_i	R_{x_i}	R_{y_i}	$R_{x_i}^2$	$R_{x_i}R_{y_i}$
−11	−14	1	1	1	1
−1	−8	2	2	4	4
4	10	3	3	9	9
16	18	4	4	16	16
21	33	5	5.5	25	27.5
31	39	6	8.5	36	51
47	37	7	7	49	49
62	33	8	5.5	64	44
74	45	9	10	81	90
88	39	10	8.5	100	85
				385	376.5

FIGURE 12.31 Ranks for Yield (Y) and Fertilizer (X)

from Equation (12.23). A scatterplot of the ranks of the X's versus the ranks of the Y's is given in Figure 12.32, along with the least squares regression equation on ranks. Note how the ranks seem to follow a linear regression even though the original points do not seem to do so. Suppose one of the farms was planning a 25 percent increase in the amount of fertilizer applied. What change in yield could it expect, as determined by regression on ranks? The steps necessary to find the answer are as follows.

1. *Convert $X_0 = 25$ to a rank.* Since $X_0 = 25$ lies between $X_i = 21$ (rank = 5) and $X_j = 31$ (rank = 6), the rank assigned to X_0 is found from Equation (12.24) to be

$$R_{x_0} = 5 + \left(\frac{25 - 21}{31 - 21}\right)(6 - 5) = 5.4.$$

2. *Predict a rank for Y_0.* Substitution of $R_{x_0} = 5.4$ into Equation (12.26) yields

$$\hat{R}_{y_0} = .567 + .897\,(5.4) = 5.41.$$

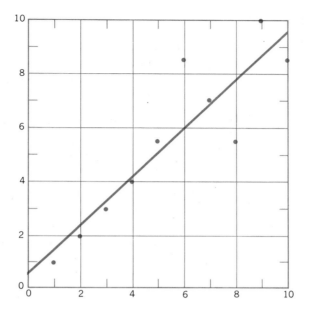

FIGURE 12.32 A Scatterplot of the Ranks of Yield (Y) Versus the Ranks of Fertilizer (X) with the Least Squares Line

3. *Convert $\hat{R}_{y_0}$ into a value for $\hat{Y}_0$.* The rank 5.41 lies between the observed values $R_{y_i} = 4$ for $Y_i = 18$ and $R_{y_j} = 5.5$ for $Y_j = 33$. Therefore, from Equation (12.25)

$$\hat{Y}_0 = 18 + \left(\frac{5.41 - 4}{5.5 - 4}\right)(33 - 18) = 32.1$$

is the predicted value for Y, given X equals 25. A 32.1 percent increase in yield is predicted from a 25 percent increase in the amount of fertilizer applied.

A list of all of the residuals is useful for this example. It enables the sizes of the residuals using regression on ranks to be compared with the sizes of the residuals obtained using ordinary linear regression on the data. It also permits the residuals to be analyzed to see if a strong serial correlation exists, as it did earlier with ordinary linear regression.

Again the points are listed in order of increasing X, in Figure 12.33, along with the predicted rank $\hat{R}_{y_i}$ of Y_i as determined by Equation (12.26) for each observed X_i. This is simple to do because the actual values $R_{x_{i'}}$ from 1 to 10, are used in the equation. For each value of $\hat{R}_{y_i}$ a predicted value $\hat{Y}_i$ is obtained for column (6), as illustrated previously. The residuals are given in column (7). The serial correlation of these residuals is found just as before,

$$r_1 = \frac{-52.74 - (-14.33)(-15.73)/9}{[252.41 - (-14.33)^2/9]^{1/2}[262.10 - (-15.73)^2/9]^{1/2}} = .335.$$

Whereas the previous analysis showed $r_1 = .565$ with a p-value close to .05, the p-value for $r_1 = .335$ is between .10 and .25, thus offering some assurance that this monotone regression model is quite adequate for this example.

(1) X_i	(2) Y_i	(3) R_{x_i}	(4) R_{y_i}	(5) $\hat{R}_{y_i}$	(6) $\hat{Y}_i$	(7) $W_i = Y_i - \hat{Y}_i$	(8) W_{i+1}	(9) W_i^2	(10) W_{i+1}^2	(11) $W_i W_{i+1}$
−11	−14	1	1	1.46	−11.24	−2.76	−6.48	7.62	41.99	17.88
−1	−8	2	2	2.36	−1.52	−6.48	−2.08	41.99	4.33	13.48
4	10	3	3	3.26	12.08	−2.08	−1.50	4.33	2.25	3.12
16	18	4	4	4.15	19.50	−1.50	4.50	2.25	20.25	−6.75
21	33	5	5.5	5.05	28.50	4.50	−12.09	20.25	146.17	−54.41
31	39	6	8.5	5.95	34.20	−12.09	0.40	146.17	0.16	−4.84
47	37	7	7	6.85	36.60	0.40	0.24	0.16	0.06	0.10
62	33	8	5.5	7.74	37.99	0.24	5.44	0.06	29.59	1.31
74	45	9	10	8.64	39.56	5.44	−4.16	29.59	17.31	−22.63
88	39	10	8.5	9.54	43.16	−4.16[a]				
					Totals	−14.33	−15.73	252.41	262.10	−52.74

[a] Not used in total.

FIGURE 12.33 Worksheet for Calculating Residuals from Rank Regression, and Serial Correlation on Residuals

CONCLUDING REMARKS

The method of monotone regression is analogous to the method of least squares presented in Section 12.2, in that no assumptions are made regarding distributions, independence, and so on. Also no inferences are made by way of hypothesis tests or confidence intervals. Both methods provide a way of estimating the unknown regression curve.

One way of comparing the two methods is by comparing the sizes of the residuals obtained from the least squares regression with the sizes of the residuals from monotone regression. The usual way of measuring the overall residual size is through the **residual sum of squares**, also called the **error sum of squares**:

$$\text{Error SS} = \Sigma(Y_i - \hat{Y}_i)^2.$$

A smaller error sum of squares implies a better fit to the data. The error sum of squares from the regression on ranks is easily found from Figure 12.33 as

$$\text{Error SS} = \sum_{i=1}^{n} W_i^2 = 269.72 \quad \text{(rank regression)},$$

which is considerably smaller than the error sum of squares for least squares regression on the data, obtained from Figure 12.29:

$$\text{Error SS} = \sum_{i=1}^{n} e_i^2 = 1179.42 \quad \text{(ordinary regression)}.$$

The smaller sum of squares for rank regression confirms what the serial correlation coefficient indicated: that the monotone regression model is more appropriate for this set of data.

For use as a predictive model, monotone regression may be better or worse than the linear regression method. A smaller error sum of squares associated with the monotone regression does not necessarily imply that the monotone regression model will be better in predicting unknown values of Y, but it serves as an indication that it *may* be better.

Exercises

12.38   A university report contains the average number of sick days (including time spent away from work on medical and dental appointments) for their employees with 1 to 20 years service with the university. These data are recorded below. Plot these data in a scatterplot.

Number of Years Service	Average Number of Sick Days
1	1.4
2	1.7
3	1.1
4	1.5
5	2.3
6	1.4
7	1.8
8	2.1
9	3.0
10	2.8
11	3.2
12	3.8
13	4.0
14	5.5
15	6.0
16	6.5
17	8.0
18	10.7
19	13.0
20	12.2

12.39   Find the least squares regression line for the data of Exercise 12.38 and add it to the scatterplot made for that exercise.

12.40   Compute the serial correlation for the least squares residuals of Exercise 12.39 and use Figure 11.6 to find an approximate standard for judging the significance of r_1. What is your conclusion regarding the adequacy of the linear fit to the data in Exercise 12.38?

12.41   Convert the regression data in Exercise 12.38 to ranks and make a scatterplot of these ranks. How does the rank scatterplot compare with the raw data scatterplot made in Exercise 12.38?

12.42   Use the ranks of Exercise 12.41 to find the rank regression equation for the data of Exercise 12.38.

12.43   Use the rank regression equation found in Exercise 12.42 to predict the average number of sick days for employees with 16.5 years of service.

12.44   Compute the serial correlation for the residuals from the monotone regression in Exercise 12.42 and use Figure 11.6 as an approximate standard for judging the significance of r_1. What is your conclusion regarding the fit on the ranks of the data compared to fit on the raw data?

12.45   Find the error sum of squares for the least squares fit to the raw data for Exercise 12.38. Also find the error sum of squares from the monotone regression fit to these same data. What conclusion is made regarding the two fits based on the error sum of squares?

12.7
Review Exercises

12.46   Find the least squares regression line for the sample data given in Exercise 2.39. (See Exercise 4.55.)

12.47   Use the sample regression equation in Exercise 12.46 to predict the average weight for individuals with a height of 72 inches. Find a 95 percent confidence interval for this prediction.

12.48   An experiment was set up to determine if the amount of wear on a new fabric is affected by the speed of the washing machine. Six pieces of fabric were tested at each of five machine speeds and a measurement of the amount of wear was recorded for each piece of fabric as follows.

Machine Speed (rpm)	Amount of Wear					
110	24.9,	24.8,	25.1,	26.4,	27.0,	26.6
130	27.4,	27.3,	26.4,	28.5,	28.7,	28.5
150	30.4,	30.3,	29.5,	31.7,	31.6,	31.4
170	37.9,	36.9,	37.2,	38.5,	38.8,	39.1
190	48.7,	42.7,	47.8,	49.6,	49.9,	49.5

Graph machine speed versus the amount of wear in a scatterplot and calculate the total sum of squares $= \Sigma(Y_i - \bar{Y})^2$ for the amount of wear.

12.49   For the sample data in Exercise 12.48 find the conditional mean of Y for each machine speed. Add these points to the scatterplot of Exercise 12.48 and connect them to form a sample regression curve.

12.50   Find the error sum of squares associated with the sample regression curve in Exercise 12.49.

12.51   Find the coefficient of determination for the sample regression curve in Exercise 12.49. What is the interpretation of this value?

12.52   Find the least squares regression equation for the sample data given in Exercise 12.48. Find the value of r^2 for these data and compare with the value found in Exercise 12.51.

12.53   Use the least squares regression equation found in Exercise 12.52 to predict the mean amount of wear on a piece of fabric washed at a machine speed of 170 rpm.

12.54   Test the normality assumption for the least squares residuals in Exercise 12.52.

12.55   Test the assumption of homogeneity of variance for the least squares residuals in Exercise 12.52.

12.56 Find a 95 percent confidence interval for the mean amount of wear predicted in Exercise 12.53.

12.57 Use a level of significance of $\alpha = .05$ to test the hypothesis $H_0: \beta = 0$ versus $H_1: \beta > 0$ for the least squares regression line found in Exercise 12.52.

12.58 Use the nonparametric test for slope to test the hypothesis in Exercise 12.57.

12.59 Convert the regression data in Exercise 12.48 to ranks and make a scatterplot of these ranks. How does the rank scatterplot compare with the raw data scatterplot made in Exercise 12.48?

12.60 Use the ranks of Exercise 12.59 to find the rank regression equation for the data in Exercise 12.48.

12.61 Use the rank regression equation found in Exercise 12.60 to predict the amount of wear on fabric that is washed at a speed of 170 rpm.

Bibliography

Additional material on the topics presented in this chapter can be found in the following publications.

Conover, W. J. (1980). *Practical Nonparametric Statistics, 2nd ed.,* John Wiley and Sons, Inc., New York.

Conover, W. J. and Iman, R. L. (1981). "Rank Transformations as a Bridge between Parametric and Nonparametric Statistics," *The American Statistician,* **35**(3), 124–129.

Iman, R. L. and Conover, W. J. (1979). "The Use of the Rank Transform in Regression," *Technometrics,* **21**(4), 499–509.

13

Analysis of Variance for One-Factor Experiments

PRELIMINARY REMARKS

The concept of an **experimental design** was introduced in Chapter 8 where matched pairs were used as a method of controlling unwanted variation in experimental results. There it was pointed out that the design, or set of plans, of an experiment is very important in order to obtain results sensitive to the purposes of the experiment. The usage of matched pairs of data in an experimental design enables one to detect differences between population means that are not possible to detect when independent samples are obtained. Since it is not always possible to collect data in matched pairs, methods of analysis are needed for independent samples also. These were given in Chapter 9. In Chapter 9, comparisons of means were made using two independent samples in another type of experimental design. These experimental designs are but two of many possible ones that exist to cover a variety of experimental requirements.

EXPERIMENTAL DESIGN	*The **experimental design** is the set of plans and instructions by which the data in an experiment are collected.*

It is very important that the experimental design be established *prior* to the collection of the data. In this way the experimenter is assured that a method of analysis is available for the data. If the data are collected prior to the consideration of an experimental design, often it is not possible to extract the desired information from the data, either because the data were collected improperly, or because no known statistical procedure exists for the analysis of the data even if the method of collection was theoretically sound.

The computations associated with the experimental design in this chapter are referred to as an **analysis of variance** for the simple reason that the analysis consists of an examination and identification of the sources of variation present in the sample data. For example, suppose the purpose of an experiment is to determine the average m.p.g. of three types of small pickup trucks. A test is set up using six Toyota, five Datsun, and four Mazda pickups. Each vehicle is driven over the same 300 mile course at 55 m.p.h. with the following m.p.g. results recorded.

Toyota	Datsun	Mazda
27.1	25.3	23.1
27.5	26.5	24.3
27.0	26.4	23.4
26.9	26.8	24.2
27.7	26.5	
27.3		

A first step in examining these sample data is to ask what is causing the variation in the results; that is since all of the 15 numbers are not identical there must be some explanation for the observed variability. One obvious response to this query is that some of the variation is due to the fact that different varieties of pickups produce different average m.p.g. readings due to different engineering. For example, in these sample data the four Mazda readings are less than all the results recorded for either Toyota or Datsun. Even among pickups of the same brand mileage will vary. That is to say, the six Toyota observations were obtained from six different Toyota pickups that may have been driven by different drivers, or perhaps the tire inflations differed from vehicle to vehicle, or maybe the carburetors were adjusted differently, or the weather conditions (windy, rainy) may have differed when the six tests were made, and so on.

The above illustration points out that experimental results contain several sources of variability. A good experimental design enables the primary sources of variability to be identified and the amount of variability due to each source to be separated out of the total variability in the data. The remaining variability in the data is attributed to randomness, and this source of variation is called **error**. A good experimental design reduces the error variation as much as possible so that the factors of interest, such as the different brands of pickup trucks, can be examined more accurately. The analysis associated with an experimental design attempts to identify all sources contributing to the variability in the results and to measure the amount of variability due to each source. One of the simplest and most frequently used experimental designs is the **completely randomized design**, which is considered in this chapter.

13.1
An Overview of Completely Randomized Designs

PROBLEM SETTING 13.1

An air quality engineer has the job of monitoring the amount of contaminant in the output of smokestacks from factories in a large city. The monitoring is done by randomly selecting times at which smokestacks are observed for factories associated with particular industries and measuring the amount of contaminants. One week, three steelmills are selected for monitoring and the amount of contaminant is measured at five randomly selected times at each factory with the results given in Figure 13.1. The air quality engineer needs to make comparisons among the factories to see if they differ with respect to the amount of contaminants produced by each.

Factory A	Factory B	Factory C
46.3	48.6	45.1
43.7	52.3	46.7
51.2	50.9	41.8
49.6	53.6	40.4
48.8	55.7	42.6

FIGURE 13.1 Measurements on the Amount of Contaminant at Each of Three Factories

PROBLEM SETTING 13.2

A consumer product evaluation group is interested in comparing the mean life in minutes of four types of batteries commonly used with children's toys. A random sample of each of the four battery types is selected and put on a continuous test where the time required for the energy output of the battery to fall below a predetermined acceptable level is measured. The test results are given in Figure 13.2. The evaluation group would like to make statements about the relative merits of the battery types as measured by average lifetime.

Battery 1	Battery 2	Battery 3	Battery 4
43	45	45	45
47	48	43	48
48	49	41	55
45	46	41	47
46	52	38	58
42	45	46	50
46	44	45	46
45	47	41	53
49		43	56
		41	

FIGURE 13.2 Results of Life Tests on Four Battery Types

PROBLEM SETTING 13.3

As input to a consumer price index, the price of a representative market basket of food is obtained at each of 10 randomly selected grocery stores in each of five cities. The prices observed in each of these stores are recorded in Figure 13.3. The prices in Figure 13.3 will be used to determine an index for food costs across the country as well as for each city. Additionally, the prices can be used to make comparisons about the relative cost of living in each of the five cities.

St. Louis	San Francisco	Omaha	Washington, D.C.	Dallas
$73.34	$76.29	$70.28	$75.81	$74.61
71.57	74.78	74.25	78.19	73.55
77.46	77.41	70.51	77.94	76.51
73.86	75.83	74.59	76.42	75.11
74.10	79.25	72.58	77.46	73.41
71.66	75.65	71.91	75.29	74.24
75.75	74.12	71.58	78.82	76.87
76.82	78.56	73.13	77.71	75.34
73.88	75.90	72.18	76.12	74.66
72.49	77.03	74.43	78.88	75.46

FIGURE 13.3 Prices of a Representative Market Basket of Food In Each of Five Cities

THE COMPLETELY RANDOMIZED DESIGN

Each of the Problem Settings 13.1, 13.2, and 13.3 involves sample data obtained as independent random samples. This is the same situation encountered in Chapter 9 where *two* independent random samples were used to make inference about the means of the corresponding populations. In this chapter the case of *two or more* independent random samples is considered to make inferences about the respective population means. Hence, the methods of this chapter represent a generalization of the methods in Chapter 9.

A design where the data consist of independent random samples from each of k populations is an **independent sampling design**, but is usually referred to as a **completely randomized design**. Strictly speaking, a completely randomized design is one in which a group of experimental units, such as people, are randomly assigned to treatments so that each experimental unit is assigned to one of the k treatments. Because the matching of experimental units with treatments is done in a completely random manner, this is called a completely randomized design. The statistical analysis is the same whether experimental units are assigned at random to k treatments, or whether random samples are taken from k populations, so both designs are treated as completely randomized designs in this book.

AN INDEPENDENT SAMPLING DESIGN (THE COMPLETELY RANDOMIZED DESIGN)	An **independent sampling design** *is one in which independent random samples are obtained from each of k populations. This design is often referred to as a* **completely randomized (CR) design**.

The main purpose of a completely randomized design is a comparison of the population means from which the sample data are obtained. Hence, for k populations the null hypothesis of interest is of the form, H_0: $\mu_1 = \mu_2 = \cdots = \mu_k$ (i.e., all k populations have the same mean). Procedures for testing this hypothesis are given in Sections 13.2 and 13.4. If H_0 is rejected, then more specific information is required with respect to the individual population means. For example, in Problem Setting 13.1, the null hypothesis takes the form H_0: $\mu_A = \mu_B = \mu_C$ (i.e., the mean amount of contaminant is the same for all factories) and if H_0 is rejected, then it is desired to know how the means μ_A, μ_B, and μ_C differ from one another. This is the subject of Section 13.3.

The remainder of this section is devoted to a graphical presentation of the sample data from a completely randomized design. This graphical presentation is meant to aid the reader in further understanding the problem of interest as well as to provide a check on the conclusions of Sections 13.2 and 13.3.

A GRAPHICAL COMPARISON OF SAMPLE DATA FROM A CR DESIGN

The empirical distribution functions associated with the two-sample data of Chapter 9 were used in Figure 9.1 to demonstrate the fact that the mean number of sickness absences tends to be higher on Mon.–Fri. than on Tue.–Wed.–Thurs. Of course, the graph did not prove in and of itself that such a difference existed in the two groups, but rather it provided a pictorial represen- tation that supported the conclusions of the analysis in Chapter 9. The e.d.f. also provides a convenient method for use with data from a completely ran- domized design. For example, consider the smokestack data of Problem Setting 13.1, which is displayed graphically in Figure 13.4.

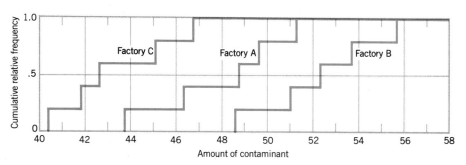

FIGURE 13.4 Empirical Distribution Functions for Smokestack Data From Figure 13.1

Figure 13.4 indicates that it would be reasonable to think in terms of reject- ing the null hypothesis H_0: $\mu_A = \mu_B = \mu_C$ since there does appear to be a clear separation of the amount of contaminants released by each factory. Addi- tionally, the comparison of the means would seem to indicate that $\mu_C < \mu_A < \mu_B$. That is, Factory C is lower than the other two factories with Factory B higher than the other two factories while Factory A is below Factory B and is above Factory C. These tentative conclusions will be examined in detail in Sections 13.2 and 13.3.

Next consider the e.d.f.'s for the battery life test data given in Figure 13.3. These e.d.f.'s appear in Figure 13.5. Figure 13.5 does not make the decision regarding the null hypothesis H_0: $\mu_1 = \mu_2 = \mu_3 = \mu_4$ (i.e., equal mean battery lives) as clear cut as was the case with Figure 13.4 for the smokestack data. It does seem safe to say from observing Figure 13.5 that the data do not seem to support H_0. However, statements about the relative ordering of the population means are more difficult to make since μ_1 and μ_2 appear to be very nearly the same and both are in between the other two. Probably the best guess at this point would be $\mu_3 < \mu_1 = \mu_2 < \mu_4$, which merely puts in symbols what the previous statement said in words. Again these tentative conclusions will be examined in detail in the next two sections.

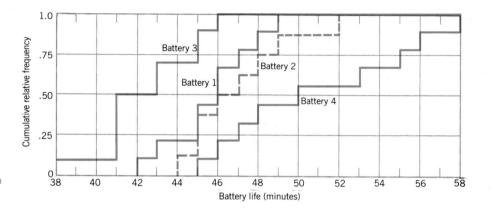

FIGURE 13.5 Empirical Distribution Functions for Battery Life Test Data from Figure 13.2

As a last example of e.d.f.'s consider the market basket costs given in Figure 13.3. These e.d.f.'s appear in Figure 13.6. The e.d.f.'s in Figure 13.6 indicate that the null hypothesis H_0: $\mu_{SL} = \mu_{SF} = \mu_O = \mu_W = \mu_D$ (i.e., equal mean market basket costs) could probably be rejected due to separation of the e.d.f.'s. However, statements regarding the means become more difficult. It would appear that the costs are higher in San Francisco and Washington, D.C.

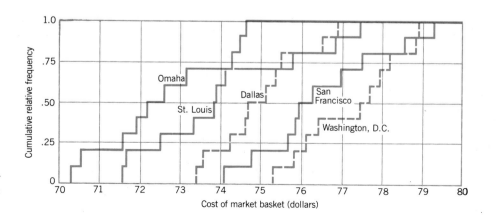

FIGURE 13.6 Empirical Distribution Functions for Market Basket Costs from Figure 13.3

than in the other three cities. At the same time the cost in Omaha appears to be less than the other cities while the e.d.f.'s for St. Louis and Dallas overlap somewhat. Hence, it may be reasonable to think in terms of the relative ordering of the population means as $\mu_O < \mu_{SL} = \mu_D < \mu_{SF} = \mu_W$. As with the previous examples these tentative conclusions will be investigated in the next two sections.

Exercises

13.1 Refer to the sample m.p.g. data given in the preliminary remarks of this chapter and graph the e.d.f.'s for each model of pickup truck.

13.2 Make a tentative guess about the acceptability of the null hypothesis H_0: $\mu_T = \mu_D = \mu_M$ from the e.d.f.'s in Exercise 13.1. If your decision is to reject H_0, then make another tentative guess about the relative ordering of the population means.

13.3 In order to compare four different methods of growing corn, a large number of plots were used where one of the methods of growing corn was randomly assigned to each plot of land. The number of bushels per acre was recorded for each plot of land with the following results. Graph the e.d.f.'s for these sample data.

Method 1	Method 2	Method 3	Method 4
83	91	101	78
91	90	100	82
94	81	91	81
89	83	93	77
89	84	96	79
96	83	95	81
91	88	94	80
92	91		81
90	89		
	84		

13.4 Make a tentative guess about the acceptability of the null hypothesis H_0: $\mu_1 = \mu_2 = \mu_3 = \mu_4$ from the e.d.f.'s in Exercise 13.3. If your decision is to reject H_0, then make another tentative guess about the relative ordering of the population means.

13.5 A random sample of the number of units of production each hour was obtained for each of five workers with the results as follows. Graph the e.d.f.'s for these sample data.

Worker 1	Worker 2	Worker 3	Worker 4	Worker 5
23	28	19	31	29
27	23	21	25	23
34	31	29	37	32
24	18	17	28	21

13.6 Make a tentative guess about the acceptability of the null hypothesis H_0: $\mu_1 = \mu_2 = \mu_3 = \mu_4 = \mu_5$ from the e.d.f.'s in Exercise 13.5. If your decision is to reject H_0, then make another tentative guess about the relative ordering of the population means.

13.7 The tar content (in milligrams) is measured on randomly selected samples of each of four brands of cigarettes. The test measurements are as follows. Graph the e.d.f.'s for these sample data.

Brand 1	Brand 2	Brand 3	Brand 4
16.4	17.7	16.1	16.6
15.5	17.1	17.9	19.2
15.4	17.2	18.1	17.3
15.8	17.3	16.5	17.9
15.1	17.1	16.8	16.6
16.8	17.3	16.1	16.9
16.8	16.5	17.2	16.2
	17.5	16.7	17.6
		17.8	17.4
			15.5

13.8 Make a tentative guess about the acceptability of the null hypothesis $H_0: \mu_1 = \mu_2 = \mu_3 = \mu_4$ from the e.d.f.'s in Exercise 13.7. If your decision is to reject H_0, then make another tentative guess about the relative ordering of the population means.

13.2

The Analysis of Variance for the Completely Randomized Design

While the graphs of the e.d.f's of the previous section aid intuition with respect to testing the null hypothesis of equality of means of k populations—$H_0: \mu_1 = \mu_2 = \cdots = \mu_k$—the graphical procedure should not be relied on entirely to make a decision regarding the acceptance or rejection of H_0 due to the subjectivity involved. Rather, what is needed is a testing procedure that removes the subjectivity such as the two-sample t-test of Chapter 9 for testing the equality of two population means ($H_0: \mu_1 = \mu_2$). Such a testing procedure is given in this section and is referred to as a one-way analysis of variance. The use of the expression "one-way" means there is only one factor used to classify the sample data as opposed to two or more factors.

IDENTIFYING THE SOURCES OF VARIATION IN THE SAMPLE DATA

As alluded to in the preliminary remarks of this chapter, an analysis of variance is quite simply an analysis of the sources of variation present in the sample data. As a demonstration the m.p.g. data presented in the preliminary remarks are considered. The sample mean for all 15 data points is $\overline{X} = 26.0$ and the total sum of squares

$$\sum_{i=1}^{15} (X_i - \overline{X})^2 = 32.940$$

represents **the total variation** present in the sample data. That is, if all 15 data points were exactly the same number, there would be no variation in the data and the total sum of squares would be equal to zero; otherwise the total sum of squares will always be greater than zero.

If the null hypothesis H_0: $\mu_T = \mu_D = \mu_M$ is true, then all three model types tend to yield the same average m.p.g. and the only source of variation in the data is the variation naturally observed from one vehicle to another for a given model type. For the m.p.g. data, the average m.p.g. observed for Toyota is $\bar{X}_T = 27.25$ and the variation for only the Toyota vehicles is given as

$$\sum_{i=1}^{6} (X_{Ti} - 27.25)^2 = 0.475.$$

Likewise the sample means for Datsun and Mazda are 26.30 and 23.75, respectively, and the corresponding measures of variation within each of these model types are

$$\sum_{i=1}^{5} (X_{Di} - 26.30)^2 = 1.340$$

and

$$\sum_{i=1}^{4} (X_{Mi} - 23.75)^2 = 1.050.$$

The sum of these three sources of variation is $0.475 + 1.340 + 1.050 = 2.865$. Since this sum is considerably less than the total variation 32.940, the data may be indicating that the null hypothesis is not true, and the population means are really unequal.

If the null hypothesis H_0: $\mu_T = \mu_D = \mu_M$ is not true, then some of the variation present in the sample data is due to differences in model types. This variation is measured by determining how much each of the three sample means (27.25, 26.30, and 23.75) differs from the overall mean (26.0) when each difference is weighted by the respective sample size. The calculation is

$$6(27.25 - 26.0)^2 + 5(26.30 - 26.0)^2 + 4(23.75 - 26.0)^2 = 30.075.$$

Since this variation between model types is almost equal to the total variation, this again may be a strong indication that the null hypothesis is not true. The variation **between** model types is added to the variation **within** each model type to produce the total variation, that is,

$$30.075 + 2.865 = 32.940.$$

A more general way of stating this result is that the total sum of squares has been partitioned (divided) into two parts that are generally referred to as the treatment (between) sum of squares and the error (within) sum of squares, or

Total sum of squares = Treatment sum of squares + Error sum of squares.

The total sum of squares may be computed using Equation (13.1) and the treatment sum of squares may be computed using Equation (13.2). The error sum of squares is easily found as the difference of the total sum of squares and the treatment sum of squares.

TOTAL SUM OF SQUARES

*The total of the squared deviations between each observation X_{ij} and the mean $\bar{X}$ of all of the observations is the **total sum of squares**. In symbols, if*

X_{ij} = *the ith observation in the jth population or treatment,*

N = *the total number of observations,*

$$\bar{X} = \frac{1}{N} \sum_i \sum_j X_{ij} = \textit{overall sample mean,}$$

then the total sum of squares is given as

$$\text{Total SS} = \sum_i \sum_j (X_{ij} - \bar{X})^2 = \sum_i \sum_j X_{ij}^2 - N\bar{X}^2 \qquad (13.1)$$

THE ONE-WAY ANALYSIS OF VARIANCE

The importance of partitioning the total sum of squares is that it leads to the development of a test statistic. For example, in the m.p.g. data almost all of the total variation (32.940) is due to differences between model types (30.075), which is perhaps strong evidence that H_0 is not true. On the other hand, if each of the model types had yielded a sample mean (26.0) exactly equal to the overall mean (26.0), the sum of the squares between model types (SST) would have been zero:

$$\text{SST} = 6(26.0 - 26.0)^2 + 5(26.0 - 26.0)^2 + 4(26.0 - 26.0)^2 = 0.$$

Hence, the contribution to the total variation (**Total SS**) from model-type differences would have been zero and H_0 would likely be true. The larger SST is relative to the Total SS or to the Error SS; the stronger is the statistical evidence that the population means are not equal. This corresponds to large values of the F statistic given by Equation (13.4). The formal testing procedure for H_0 is called a **one-way analysis of variance**.

TREATMENT SUM OF SQUARES

*The sum of squared deviations between each sample mean $\bar{X}_j$ and the overall mean $\bar{X}$, weighted by the number of observations in each sample, is called the **treatment sum of squares (SST)**. In symbols, if*

n_j = *the sample size from the jth population,*

$$\bar{X}_j = \frac{1}{n_j} \sum_{i=1}^{n_j} X_{ij} = \begin{array}{l}\textit{sample mean of the observations from the}\\ \textit{jth population,}\end{array}$$

then the treatment sum of squares is given as

$$\text{SST} = \sum_{j=1}^{k} n_j(\bar{X}_j - \bar{X})^2 = \sum_{j=1}^{k} n_j\bar{X}_j^2 - N\bar{X}^2 \qquad (13.2)$$

ERROR SUM OF SQUARES

The sum of squared deviations between each observation X_{ij} and the sample mean $\overline{X}_j$ for the population from which X_{ij} was obtained is called the **error sum of squares.**

$$\text{Error SS} = \sum_{j=1}^{k} \sum_{i=1}^{n_j} (X_{ij} - \overline{X}_j)^2 = \text{Total SS} - \text{SST} \qquad (13.3)$$

PROCEDURE FOR INFERENCE ABOUT THE DIFFERENCE IN MEANS OF k NORMAL POPULATIONS WITH EQUAL VARIANCES (THE ONE-WAY ANALYSIS OF VARIANCE)

Data. The data consist of k independent samples of sizes $n_1, n_2, \ldots, n_k$ from each of k populations. The respective population means are denoted by $\mu_1, \mu_2, \ldots, \mu_k$.

Assumptions
1. All samples are random samples from their respective populations.
2. All samples are all independent of one another.
3. All populations are normal with equal variances ($\sigma_1^2 = \sigma_2^2 = \cdots = \sigma_k^2$).

Null Hypothesis. H_0: $\mu_1 = \mu_2 = \cdots = \mu_k$

Alternative Hypothesis. H_1: At least two population means are unequal ($\mu_i \neq \mu_j$).

Test Statistic. $F = \text{MST/MSE}$ (13.4)
where $\text{MST} = \text{SST}/(k - 1)$ and $\text{MSE} = \text{Error SS}/(N - k)$ with SST and Error SS defined by Equations (13.2) and (13.3), respectively.

Decision Rule. H_0 is rejected and it is concluded that all the means are not equal to one another if $F > F_{\alpha,k-1,N-k}$ where $F_{\alpha,k-1,N-k}$ is the $1 - \alpha$ quantile from Table A5 with $k - 1$ and $N - k$ degrees of freedom.

Note: If the desired degrees of freedom for the F distribution cannot be found in Table A5, simply use the next smaller degrees of freedom given in the table.

ANALYSIS OF VARIANCE TABLE (CR DESIGN)

The sums of squares, degrees of freedom (DF), and mean squares may be summarized in an **analysis of variance table,** *along with the F statistic and its p-value, as follows.*

Source	DF	Sum of Squares	Mean Square	F Value	PR > F
TREATMENT	$k - 1$	SST	$\text{MST} = \text{SST}/(k - 1)$	MST/MSE	
ERROR	$N - k$	Error SS	$\text{MSE} = \text{Error SS}/(N - k)$		
TOTAL	$N - 1$	Total SS			

A WORD ABOUT THE TEST ASSUMPTIONS

The first two assumptions for the one-way analysis of variance are usually within the control of the experimenter and may be verified by examining the protocol by which the data were obtained. However, the assumption of normality should be tested by using the Lilliefors test on each set of sample data. If the sample sizes are small, nonnormality is difficult to detect. So the authors recommend that the Kruskal-Wallis test in Section 13.4 be used with small sample sizes whenever the experimenter doubts that the populations are normal, even if a goodness-of-fit test fails to detect the nonnormality.

The assumption of equal variances *should be checked only after the normality assumption has been checked and verified*. Good tests for equality of variances involve considerable computation and in general have low power. See the references at the end of this chapter for a recent survey article on tests for equality of variances. If the equality of variance assumption is suspect, then the procedure in Section 13.4 is recommended because it is less sensitive to unequal variances. The one-way analysis of variance for testing equality of population means is now demonstrated with three examples.

EXAMPLE

The smokestack data given in Problem Setting 13.1 are now used to test H_0: $\mu_A = \mu_B = \mu_C$ (i.e., equal mean levels of contaminants for all factories) at $\alpha = .05$. Calculations on the sample data show the test assumptions to be satisfied and reveal the following.

$$\Sigma X_i = 717.3, \qquad \overline{X} = 717.3/15 = 47.82$$

$$\Sigma X_i^2 = 34588.99, \qquad \overline{X}_A = 52.22, \quad \overline{X}_B = 47.92, \qquad \overline{X}_C = 43.32.$$

From Equation (13.1) the total sum of squares is found as

Total SS $= 34588.99 - 15(47.82)^2 = 287.704.$

Equation (13.2) yields the treatment sum of squares as

$$SST = 5(52.22)^2 + 5(47.92)^2 + 5(43.32)^2 - 15(47.82)^2$$

$$= 34499.386 - 34301.286$$

$$= 198.100.$$

And finally from Equation (13.3) the error sum of squares is

Error SS $=$ Total SS $-$ SST

$$= 287.704 - 198.100$$

$$= 89.604.$$

From Equation (13.4) MST $= 198.100/(3 - 1) = 99.05$ and MSE $= 89.604/(15 - 3) = 7.467$ from which the F ratio is given as

$F = 99.05/7.467 = 13.27.$

This value of F is much larger than the critical value for a level of significance of $\alpha = .05$, $F_{.05,2,12} = 3.885$. The results of these calculations are conveniently summarized in an analysis of variance table.

Source	DF	Sum of Squares	Mean Square	F Value	PR > F
TREATMENT	2	198.100	99.050	13.27	.0009
ERROR	12	89.604	7.467		
TOTAL	14	287.704			

The p-value is less than .05 so H_0 is rejected. It is concluded that the mean amount of contaminants is not the same for all factories. This conclusion is in agreement with tentative conclusions of Section 13.1 based on the e.d.f.'s graphed in Figure 13.4.

THE MODEL

The assumptions in the one-way analysis of variance imply a particular mathematical model for the variables involved. In this case the model is usually written in the form

$$X_{ij} = \mu_j + \epsilon_{ij}, \quad j = 1, \ldots, k, \quad i = 1, \ldots, n_j,$$

where the ϵ_{ij} are independent, identically distributed normal random variables with mean zero. Then the mean of X_{ij} is simply μ_j, the mean for the population from which X_{ij} is obtained.

USE OF THE COMPUTER WITH THE ONE-WAY ANALYSIS OF VARIANCE

Although the calculations in the previous example were relatively easy, the analysis of other CR designs involving more treatments and observations could be more time consuming. For this reason computer programs such as SAS are used to do the analysis on a CR design. Additionally, the computer program provides other output that will be useful in subsequent sections.

EXAMPLE

The battery life test data given in Problem Setting 13.2 will now be analyzed in order to test $H_0: \mu_1 = \mu_2 = \mu_3 = \mu_4$. The calculations will first be done by hand and then the computer printout will be presented as a check on the calculations. The sample data calculations provide the following:

$$\Sigma X_i = 1669, \quad \bar{X} = 1669/36 = 46.361$$

$$\Sigma X_i^2 = 78049, \quad \bar{X}_1 = 45.667, \quad \bar{X}_2 = 47.000$$

$$\bar{X}_3 = 42.400, \quad \bar{X}_4 = 50.889$$

From Equation (13.1),

Total SS $= 78049 - 36(46.361)^2 = 672.306.$

Equation (13.2) yields

SST $= 9(45.667)^2 + 8(47.000)^2 + 9(42.400)^2 + 10(50.889)^2$

$$- 36(46.361)^2$$

$$= 349.017.$$

(This value results from carrying more decimal places in the calculations than shown.) Equation (13.3) gives

Error SS $= 672.306 - 349.017$

$$= 323.289.$$

From Equation (13.4) MST $= 349.017/(4 - 1) = 116.339$ and MSE $= 323.289/(36 - 4) = 10.103$ from which the F ratio is given as

$F = 116.339/10.103 = 11.52.$

This value of F exceeds the critical value for a level of significance of $\alpha = .05$, $F_{.05,3,32} \cong 2.922$, obtained from Table A5 with 3 and 30 degrees of freedom. Therefore, H_0 is rejected and it is concluded that the different battery types do indeed differ with respect to their mean lives. The p-value is less than .001, according to Table A5. This conclusion is consistent with the tentative guess made in Section 13.1 based on the e.d.f.'s displayed in Figure 13.5. The computer printout for the analysis of variance table is as follows.

Source	DF	Sum of Squares	Mean Square	F Value	PR > F
TREATMENT	3	349.017	116.339	11.52	.0001
ERROR	32	323.289	10.103		
TOTAL	35	672.306			

EXAMPLE

The sample data on the cost of a market basket of food in five cities will now be analyzed to test $H_0: \mu_{SL} = \mu_{SF} = \mu_O = \mu_W = \mu_D$ (i.e., the mean market basket price is the same for all cities). This analysis will be done using only the computer printout. Since the p-value is .0001, the null hypothesis is soundly rejected at $\alpha = .05$, and it is concluded that the mean market basket prices are different in the five cities although it should be pointed out that the analysis thus far does not reveal which of the five cities is different. Again this conclusion agrees with the tentative conclusion based on the e.d.f.'s appearing in Figure 13.6.

Source	DF	Sum of Squares	Mean Square	F Value	PR > F
TREATMENT	4	141.514	35.379	14.73	.0001
ERROR	45	108.078	2.402		
TOTAL	49	249.592			

This section has considered testing the equality of the means from k normal populations. Remaining unanswered is the question regarding the relative ordering of the population means *when the null hypothesis is rejected*. This question will be answered in the next section.

Exercises

13.9 Test the null hypothesis H_0: $\mu_T = \mu_D = \mu_M$ for the m.p.g. sample data of the pickup trucks referred to in Exercise 13.1. Let $\alpha = .05$ and state the p-value. (If a computer is used to analyze these data, then check the calculations by hand using Equations (13.1) to (13.4).) Assume that the assumptions of the test are satisfied.

13.10 Compare your conclusion in Exercise 13.9 with your first tentative guess made in Exercise 13.2.

13.11 Test the null hypothesis H_0: $\mu_1 = \mu_2 = \mu_3 = \mu_4$ for the four methods of growing corn using the sample data given in Exercise 13.3. Let $\alpha = .05$ and state the p-value. What assumptions are you making regarding the samples?

13.12 Compare your conclusion in Exercise 13.11 with your first tentative guess made in Exercise 13.4.

13.13 Test the null hypothesis H_0: $\mu_1 = \mu_2 = \mu_3 = \mu_4 = \mu_5$ for the production of five workers using the sample data given in Exercise 13.5. Let $\alpha = .05$ and state the p-value. Assume that the assumptions of the test are satisfied. The computer output of the analysis of variance table is as follows.

Source	DF	Sum of Squares	Mean Square	F Value	PR > F
TREATMENT	4	161.5	40.375	1.47	.2610
ERROR	15	412.5	27.500		
TOTAL	19	574.0			

13.14 Compare your conclusion in Exercise 13.13 with your first tentative guess made in Exercise 13.6.

13.15 Test the null hypothesis H_0: $\mu_1 = \mu_2 = \mu_3 = \mu_4$ for mean tar content of four brands of cigarettes using the sample data given in Exercise 13.7. What assumptions are you making regarding the samples? The computer output of the analysis of variance table is as follows.

Source	DF	Sum of Squares	Mean Square	F Value	PR > F
TREATMENT	3	7.625	2.542	3.89	.0184
ERROR	30	19.603	0.653		
TOTAL	33	27.227			

13.16 Compare your conclusion in Exercise 13.15 with your first tentative guess made in Exercise 13.8.

13.17 Check the computer computations given in the third example of this section by hand using Equations (13.1) to (13.4).

13.18 Use the model $X_{ij} = \mu_j + \epsilon_{ij}$ to describe the data from the m.p.g. experiment referred to in the Preliminary Remarks of this chapter. Write a separate model for each brand of pickup truck. Interpret the parameter in each model.

13.19 Use the model $X_{ij} = \mu_j + \epsilon_{ij}$ to describe the data from Problem Setting 13.2. Write a separate model for each battery type. Interpret the parameter in each model.

13.3

Comparing Population Means in a Completely Randomized Design

The one-way analysis of variance was presented in the previous section as a method of testing for the equality of k population means—that is, $H_0: \mu_1 = \mu_2 = \cdots = \mu_k$. *If the null hypothesis is rejected*, the question still remains as to the correct relative ordering of the population means. That question is addressed in this section. For example, suppose a null hypothesis concerning only three populations ($H_0: \mu_1 = \mu_2 = \mu_3$) is rejected. A decision must now be made about the relative ordering of the population means. In this case these orderings could include the following.

$$\mu_1 < \mu_2 < \mu_3 \qquad \mu_1 < \mu_2 = \mu_3$$
$$\mu_1 < \mu_3 < \mu_2 \qquad \mu_2 < \mu_1 = \mu_3$$
$$\mu_2 < \mu_1 < \mu_3 \qquad \mu_3 < \mu_1 = \mu_2$$
$$\mu_2 < \mu_3 < \mu_1 \qquad \mu_1 = \mu_2 < \mu_3$$
$$\mu_3 < \mu_1 < \mu_2 \qquad \mu_1 = \mu_3 < \mu_2$$
$$\mu_3 < \mu_2 < \mu_1 \qquad \mu_2 = \mu_3 < \mu_1$$

With four or more populations, even more comparisons are possible. Fortunately an orderly procedure exists for making the comparisons of the population means. This is an area in statistics known as **multiple comparisons**.

FISHER'S LEAST SIGNIFICANT DIFFERENCE

There are many multiple comparisons procedures in statistics and much research effort has been devoted to the examination of multiple comparisons techniques over the years. One procedure that continually ranks among the best in studies by various researchers, and is preferred by the authors, is also one of the oldest procedures. The procedure was developed by the famed British statistician R. A. Fisher (1890–1962) and is referred to as **Fisher's Least Significant Difference (LSD)**. The LSD procedure, used only after the null hypothesis has been rejected, provides a "measuring stick" for comparing the amount of separation necessary between any two sample means (i.e., $|\bar{X}_i - \bar{X}_j|$)

"ALL RIGHT , WHO TOLD THE CAMPUS POLICE THAT
THERE WOULD BE LSD IN THE CLASSROOM TODAY?"

before a significant difference can be declared to exist between the corresponding population means. The LSD multiple comparisons procedure is now stated formally.

FISHER'S LEAST SIGNIFICANT DIFFERENCE PROCEDURE

If the null hypothesis $H_0: \mu_1 = \mu_2 = \cdots = \mu_k$ has been rejected, the population means μ_i and μ_j are declared to be significantly different at a level of significance α if

$$|\bar{X}_i - \bar{X}_j| > \text{LSD}_\alpha$$

where

$$\text{LSD}_\alpha = t_{\alpha/2, N-k} \sqrt{\text{MSE}} \sqrt{\frac{1}{n_i} + \frac{1}{n_j}}$$

and where

$\bar{X}_i$ and $\bar{X}_j$ are the sample means being compared,
n_i and n_j are the respective sample sizes,
MSE is the mean square error defined in Section 13.2,
$t_{\alpha/2, N-k}$ is the $1 - \alpha/2$ quantile from the Student's t-distribution with $N - k$ degrees of freedom obtained from Table A3.

Note that with three populations it is mathematically possible for $|\bar{X}_1 - \bar{X}_3|$ to exceed LSD_α, while neither $|\bar{X}_1 - \bar{X}_2|$ nor $|\bar{X}_2 - \bar{X}_3|$ exceeds LSD_α, and this situation often occurs in practice. In this case the conclusion is simply that μ_1 is less than μ_3, but there is not enough statistical evidence to show that either $\mu_1 < \mu_2$ or $\mu_2 < \mu_3$. The LSD multiple comparisons procedure will now be demonstrated with three examples.

EXAMPLE

The smokestack data of Problem Setting 13.1 was used to test $H_0: \mu_A = \mu_B = \mu_C$ in an example in Section 13.2. The null hypothesis was rejected in that example. The sample calculations showed the following results.

$$\bar{X}_A = 47.92, \quad \bar{X}_B = 52.22, \quad \bar{X}_C = 43.32,$$
$$n_A = 5, \quad n_B = 5, \quad n_C = 5,$$
$$\text{MSE} = 7.467.$$

For a level of significance of $\alpha = .05$, the Student's t-value from Table A3 is $t_{.025,12} = 2.1788$ and the $\text{LSD}_{.05}$ value for comparing all population means (since all sample sizes are equal) is given as

$$\text{LSD}_{.05} = 2.1788 \sqrt{7.467} \sqrt{\frac{1}{5} + \frac{1}{5}} = 3.77.$$

It is easier to make comparisons if the sample means are ordered from smallest to largest as follows.

Population:	C	A	B
Sample Mean:	43.32	47.92	52.22

The value of $|\bar{X}_C - \bar{X}_A| = 4.60$ exceeds the $LSD_{.05}$ value of 3.77, so μ_C and μ_A are declared to be significantly different with $\mu_C < \mu_A$. Additionally the value of $|\bar{X}_A - \bar{X}_B| = 4.30$ exceeds the $LDS_{.05}$ value, so μ_A and μ_B are declared to be significantly different with $\mu_A < \mu_B$. It obviously follows that $\mu_C < \mu_B$, so the correct relative ordering of sample means is

$$\mu_C < \mu_A < \mu_B.$$

The conclusion is that the mean amount of contaminant produced by the three factories is different for all three, and that their means are in the respective order shown. This result is in agreement with the tentative guess made in Section 13.1 regarding the ordering of the population means based on the e.d.f.'s appearing in Figure 13.4.

EXAMPLE

The battery life test data given in Problem Setting 13.2 was used in an example in Section 13.2 to test H_0: $\mu_1 = \mu_2 = \mu_3 = \mu_4$. The null hypothesis was rejected in that hypothesis. Calculations based on the sample data gave the following results.

$$\bar{X}_1 = 45.667, \quad \bar{X}_2 = 47.000, \quad \bar{X}_3 = 42.400, \quad \bar{X}_4 = 50.889,$$
$$n_1 = 9, \quad\quad n_2 = 8, \quad\quad n_3 = 10 \quad\quad n_4 = 9,$$
$$MSE = 10.103.$$

For a level of significance of $\alpha = .05$ the Student's t-value from Table A3 is $t_{.025,32} = 2.0369$. Since the sample sizes are unequal a separate $LSD_{.05}$ value must be computed for each possible pair of sample sizes. That is, $LSD_{.05}$ values are needed for the following pairs of sample sizes (8,9), (8,10), (9,9), and (9,10):

$$LSD_{.05,(8,9)} = 2.0369\sqrt{10.103}\,\sqrt{\frac{1}{8} + \frac{1}{9}} = 3.146,$$

$$LSD_{.05,(8,10)} = 2.0369\sqrt{10.103}\,\sqrt{\frac{1}{8} + \frac{1}{10}} = 3.071,$$

$$LSD_{.05,(9,9)} = 2.0369\sqrt{10.103}\,\sqrt{\frac{1}{9} + \frac{1}{9}} = 3.052,$$

$$LSD_{.05,(9,10)} = 2.0369\sqrt{10.103}\,\sqrt{\frac{1}{9} + \frac{1}{10}} = 2.975.$$

Again it is easier to make comparisons if the sample means are ordered from smallest to largest.

Population:	3	1	2	4
Sample Mean:	42.400	45.667	47.000	50.889

The value of $|\bar{X}_3 - \bar{X}_1| = 3.267$ is greater than $LSD_{.05,(9,10)} = 2.975$, so μ_1 and μ_3 are declared to be significantly different with $\mu_3 < \mu_1$. Comparisons of $|\bar{X}_3 -$

$\bar{X}_2| = 4.600$ and $|\bar{X}_3 - \bar{X}_4| = 8.489$ with $LSD_{.05,(8,10)} = 3.071$ and $LSD_{.05,(9,10)} = 2.975$, respectively, also show that μ_3 is less than both μ_2 and μ_1. Next, proceeding left to right with the ordered sample means the value $|\bar{X}_1 - \bar{X}_2| = 1.333$ is less than $LSD_{.05,(8,9)} = 3.146$, so no significant difference is detected between μ_1 and μ_2. The value $|\bar{X}_1 - \bar{X}_4| = 5.222$ is greater than $LSD_{.05,(9,9)} = 3.052$, so μ_1 and μ_4 are declared to be significantly different with $\mu_1 < \mu_4$. The last comparison has $|\bar{X}_2 - \bar{X}_4| = 3.889$ larger than $LSD_{.05,(8,9)} = 3.146$, so μ_2 and μ_8 are declared to be significantly different with $\mu_2 < \mu_4$. Putting all of the parts together concludes the relative ordering of population means to be

$$\mu_3 < \mu_1 = \mu_2 < \mu_4.$$

This represents the conclusion concerning which battery types have mean lives different or the same as other types. This result is in agreement with the tentative guess made in Section 13.1 about the relative ordering on the e.d.f.'s in Figure 13.5.

EXAMPLE

The cost of a market basket of food was sampled for five cities in Problem Setting 13.3. The null hypothesis $H_0: \mu_{SL} = \mu_{SF} = \mu_O = \mu_W = \mu_D$ was tested in an example in Section 13.2 and was rejected. Sample data calculations give the following results.

$$\bar{X}_{SL} = 74.093, \quad n_{SL} = 10, \quad MSE = 2.402,$$
$$\bar{X}_{SF} = 76.482, \quad n_{SF} = 10,$$
$$\bar{X}_O = 72.544, \quad n_O = 10,$$
$$\bar{X}_W = 77.264, \quad n_W = 10,$$
$$\bar{X}_D = 74.976, \quad n_D = 10,$$

For a level of significance of $\alpha = .05$ the Student's t-value from Table A3 is $t_{.025,45} = 2.0141$. The $LSD_{.05}$ value for comparing all population means (since all sample sizes are equal) is given as

$$LSD_{.05} = 2.0141\sqrt{2.402}\sqrt{\frac{1}{10} + \frac{1}{10}} = 1.396.$$

The sample means are ordered from smallest to largest to make comparisons easier.

Population:	Omaha	St. Louis	Dallas	San Francisco	Washington
Sample Mean:	72.544	74.093	74.976	76.482	77.264

The value $|\bar{X}_O - \bar{X}_{SL}| = 1.549$ exceeds the $LSD_{.05}$ value of 1.396, so μ_O and μ_{SL} are declared to be significantly different with $\mu_O < \mu_{SL}$. Obviously μ_O will also be less than all other population means. Moving from left to right across the sample means the value $|\bar{X}_{SL} - \bar{X}_D| = 0.883$ is less than the $LSD_{.05}$ value, so μ_{SL} and μ_D cannot be declared significantly different. Next, $|\bar{X}_{SL} - \bar{X}_{SF}| = 2.389$ is compared against the $LSD_{.05}$ value and it is concluded that $\mu_{SL} < \mu_{SF}$. Also, it is clear that $\mu_{SL} < \mu_W$. The value $|\bar{X}_D - \bar{X}_{SF}| = 1.506$ is found to be greater than the $LSD_{.05}$ value, so it is concluded that μ_D and μ_{SF} are significantly different with

$\mu_D < \mu_{SF}$. Again, it follows immediately that $\mu_D < \mu_W$. The last comparison shows the value $|\bar{X}_{SF} - \bar{X}_W| = 0.782$ to be less than the LSD$_{.05}$ value, so μ_{SF} and μ_W cannot be declared significantly different. In summary, this analysis concludes that the relative ordering of the population mean market basket prices is as follows:

$$\mu_O < \mu_{SL} = \mu_D < \mu_{SF} = \mu_W.$$

This result is in agreement with the tentative conclusion given in Section 13.1 based on the e.d.f.'s in Figure 13.6.

OTHER MULTIPLE COMPARISON PROCEDURES

Fisher's LSD multiple comparisons procedure is only one of many procedures available for making several comparisons among population means, based on the observed differences between sample means. Some of these procedures are intended for use only after an overall test, like the F test in this case, declares that a difference exists. Other procedures may be used without any prior overall test; they detect overall differences as well as differences between pairs of means, and some even allow combinations of means as groups to be tested against combinations of other means. Much has been written about the advantages and disadvantages of the various multiple comparisons procedures. However, the LSD procedures given here has relatively good power to detect differences in means when the differences are present, and still retains protection against a Type I error by virtue of the F test, which must be significant before this procedure is to be used. Therefore, it is recommended here as a good procedure to use for making multiple comparisons.

Exercises

13.20 Apply the LSD multiple comparisons procedure with $\alpha = .05$ to the analysis of variance problem in Exercise 13.9 to see which makes of pickup trucks differ with regard to mean gas mileage.

13.21 Compare your conclusion in Exercise 13.20 with your second tentative guess made in Exercise 13.2.

13.22 Apply the LSD multiple comparisons procedure with $\alpha = .05$ to the analysis of variance problem in Exercise 13.11 to see which methods of growing corn tend to produce higher yields.

13.23 Compare your conclusion in Exercise 13.22 with your second tentative guess made in Exercise 13.4.

13.24 Apply the LSD multiple comparisons procedure with $\alpha = .05$ to the analysis of variance problem in Exercise 13.13 to see which workers produce more than other workers.

13.25 Compare your conclusion in Exercise 13.24 with your second tentative guess made in Exercise 13.6.

13.26 See which cigarettes have lower tar content than other brands by applying the LSD multiple comparisons procedure with $\alpha = .05$ to the analysis of variance problem in Exercise 13.15.

13.27 Compare your conclusion in Exercise 13.26 with your second tentative guess made in Exercise 13.8.

13.4
A Comparison of Means for General Populations

THE KRUSKAL-WALLIS TEST

The one-way analysis of variance procedure presented in Section 13.2 has the underlying assumption of normality for testing the equality of means of k populations. If a Lilliefors test shows this assumption not to be reasonable, then the

---

PROCEDURE FOR TESTING HYPOTHESES ABOUT THE DIFFERENCE IN THE MEANS OF k GENERAL POPULATIONS (THE KRUSKAL-WALLIS TEST)

Data. The data consist of k samples of sizes $n_1, n_2, \ldots, n_k$ from each of k populations. The population means are denoted by $\mu_1, \mu_2, \ldots, \mu_k$. Assign ranks 1 to N, where $N = n_1 + n_2 + \cdots + n_k$, to the k samples jointly, assigning average ranks in case of ties. All of the computations are made on the ranks, as numbers, rather than on the data, as before. That is, all of the ranks in each sample are added and divided by the sample size to get the sample means of the ranks, $\bar{R}_1, \bar{R}_2, \ldots, \bar{R}_k$. The sums of squares are computed using Equations (13.1), (13.2), and (13.3) but the computations are made on the ranks, not the original data.

Assumptions
1. All samples are random samples from their respective populations.
2. All k samples are independent of one another.
3. Either the k population distributions are identical, or else some of the populations have larger means than other populations.

Null Hypothesis. $H_0: \mu_1 = \mu_2 = \cdots = \mu_k$

Alternative Hypothesis

H_1: At least two of the populations have unequal means ($\mu_i \neq \mu_j$).

Test Statistic. $F_R = \text{MST}/\text{MSE}$, where MST and MSE are as defined following Equation (13.4), only they are computed on the ranks of the data rather than on the raw data.

Decision Rule. H_0 is rejected if $F_R > F_{\alpha, k-1, N-k}$, where $F_{\alpha, k-1, N-k}$ is the $1 - \alpha$ quantile from Table A5 with $k - 1$ and $N - k$ degrees of freedom.

Note. If the desired degrees of freedom for the F distribution cannot be found in Table A5, simply use the next smaller degrees of freedom given in the table.

Multiple Comparisons. The multiple comparisons LSD procedure proceeds as in Section 13.3 only with all computations done on the ranks of the data including using $\bar{R}_1, \bar{R}_2, \ldots, \bar{R}_k$ rather than the sample means $\bar{X}_1, \bar{X}_2, \ldots, \bar{X}_k$. As with the analysis on raw data, the multiple comparisons procedure should only be used when H_0 is rejected.

---

procedure in this section should be preferred. Additionally, the technique presented in this section is not nearly as sensitive to the assumption of equal variances as is the F test in the one-way analysis of variance. This procedure is called the **Kruskal-Wallis test** and is calculated by performing the one-way analysis on rank transformed data. Therefore, this test is another example of a rank transformation test as were the Wilcoxon signed ranks test and Wilcoxon-Mann-Whitney test previously presented. An overall ranking is used with this test; that is, the sample data from k independent samples are replaced by their ranks from the rank 1 for the smallest observation to the rank N for the largest observation, where N is the total number of observations as before. The notation is the same as in earlier sections of this chapter, with $R_{X_{ij}}$ used to denote the rank of the observation X_{ij}.

Obviously this procedure is the same as the one-way analysis of variance except that it is applied to the ranks of the data rather than to the data. This means that existing computer programs for the one-way analysis of variance can be used to produce the Kruskal-Wallis test after the data are replaced by their ranks. This makes the Kruskal-Wallis test very easy to use since computer packages such as SAS will also rank transform the data. The Kruskal-Wallis test will now be demonstrated with an example.

EXAMPLE

The smokestack data given in Problem Setting 13.1 are used to demonstrate the Kruskal-Wallis test. The data are repeated here with the ranks appearing in parentheses.

Factory A	Factory B	Factory C
46.3(6)	48.6(8)	45.1(5)
43.7(4)	52.3(13)	46.7(7)
51.2(12)	50.9(11)	41.8(2)
49.6(10)	53.6(14)	40.4(1)
48.8(9)	55.7(15)	42.6(3)

Calculations on the ranks yield the following:

$$\sum_i \sum_j R_{X_{ij}} = 120, \qquad \bar{R} = 120/15 = 8,$$

$$\sum_i \sum_j R^2_{X_{ij}} = 1240,$$

$$\bar{R}_A = 8.2, \qquad \bar{R}_B = 12.2, \qquad \bar{R}_C = 3.6.$$

Equation (13.1) gives the total sum of squares as

$$\text{Total SS} = 1240 - 15(8)^2 = 280.$$

Equation (13.2) yields the treatment sum of squares as

$$\text{SST} = 5(8.2)^2 + 5(12.2)^2 + 5(3.6)^2 - 15(8)^2$$

$$= 185.20.$$

Equation (13.3) uses subtraction to produce the error sum of squares as

Error SS = 280 − 185.20 = 94.80.

From Equation (13.4) MST = 185.20/(3 − 1) = 92.6, MSE = 94.80/(15 − 3) = 7.90, and the F ratio is computed as

$F_R = 92.6/7.90 = 11.72.$

The value of F_R is slightly less than the value of $F = 13.27$ found from the analysis on raw data in Section 13.2, but still leads to a sound rejection of the null hypothesis of equal means at $\alpha = .05$ when compared with 3.885, the .95 quantile of the F distribution from Table A5 with 2 and 12 degrees of freedom. The computer printout summary of the analysis of variance table based on ranks appears as follows.

Source	DF	Sum of Squares	Mean Square	F Value	PR > F
TREATMENT	2	185.2	92.6	11.72	.0015
ERROR	12	94.8	7.9		
TOTAL	14	280.0			

For multiple comparisons at a level of significance of $\alpha = .05$ the value $t_{.025,12} = 2.1788$ from Table A3 is used and the $LSD_{.05}$ value for all comparisons (since all sample sizes are equal) is given as

$$LSD_{.05} = 2.1788 \sqrt{7.90} \sqrt{\frac{1}{5} + \frac{1}{5}} = 3.87.$$

The comparisons are made on rank means, which ordered from smallest to largest appear as

Population: C   A   B
Rank Mean: 3.6   8.2   12.2

Since the difference between all pairs of means exceeds the $LSD_{.05}$ value of 3.87, all mean levels of contaminant are declared to be significantly different from one another for the three factories which is in agreement with the conclusion in Section 13.3.

THE EFFECT OF OUTLIERS ON THE F TEST

In the above example the results of the analysis using the Kruskal-Wallis test agree well with the results of the analysis using the one-way analysis of variance on raw data. In the above example, both the normality assumption and the equal variance assumption are satisfied, as the reader can verify. When the assumptions of the analysis of variance procedure are satisfied, the results of the two tests tend to be in agreement with each other; but when the analysis of variance assumptions are not met, the Kruskal-Wallis test tends to have more power, especially when outliers are present in the data. This is illustrated by changing the last example slightly. The largest sample observation is the value

$X^{(15)} = 55.7$, which occurs with Factory B. What happens to the Kruskal-Wallis test as this observation gets larger? Clearly the Kruskal-Wallis test is unaffected by such a change since no matter how large $X^{(15)}$ becomes it will still be assigned a rank of 15 and the results of the Kruskal-Wallis test remain the same.

On the other hand, the F test on raw data does not remain the same as $X^{(15)}$ increases. Consider the e.d.f's in Figure 13.4 for these data. What effect does increasing $X^{(15)}$ have on this graph? Quite simply the last step for Factory B at 55.7 is moved to the right apparently increasing the distance between Factories A and B (at least as measured by their sample means this would appear to be the case). However, increasing $X^{(15)}$ causes adverse problems for the F test on raw data. To illustrate this point consider Figure 13.7, which shows what happens to the F test on raw data as $X^{(15)}$ is increased in steps of 10 from 55.7 to 95.7.

$X^{(15)}$	Total SS	SST	Error SS	F
55.7	287.70	198.10	89.60	13.27
65.7	538.64	299.43	239.21	7.51
75.7	976.24	427.43	548.81	4.67
85.7	1600.50	538.83	1061.67	3.05
95.7	2411.44	720.16	1691.28	2.55

FIGURE 13.7 A Modification of Problem Setting 13.1 To Show the Adverse Effect on the *F* Test When Assumptions are Violated

Figure 13.7 shows SST increases as $X^{(15)}$ increases. However, SST does not increase as fast as either the Total SS or the Error SS. Since the degrees of freedom are fixed the F ratio is forced to become smaller. In fact, the last two values of F in Figure 13.7 are no longer significant at the .05 level of significance. Increasing the value of $X^{(15)}$ causes problems with both the normality assumption and the equal variance assumption, and this example should serve to point out that these assumptions cannot be casually ignored. That is, when outliers are present in the data, nonnormality of the distributions is indicated, so the nonparametric Kruskal-Wallis test should be preferred because the assumptions of the F test are not satisfied, and because the power of the F test is likely to be diminished by the presence of outliers.

Exercises

13.28 Work Exercise 13.9 using the Kruskal-Wallis test and compare your answer with the one obtained in that exercise.

13.29 Work Exercise 13.20 using the LSD multiple comparisons procedure on rank transformed data. Compare your answer with the one obtained in that exercise.

13.30 Work Exercise 13.11 using the Kruskal-Wallis test and compare your answer with the one obtained in that exercise.

13.31 Work Exercise 13.22 using the LSD multiple comparisons procedure on rank trans-formed data. Compare your answer with the one obtained in that exercise.

13.32 Work Exercise 13.13 using the Kruskal-Wallis test and compare your answer with the one obtained in that exercise. The computer output of the analysis of variance table is as follows.

Source	DF	Sum of Squares	Mean Square	F Value	PR > F
TREATMENT	4	171.5	42.875	1.31	.3096
ERROR	15	489.5	32.63		
TOTAL	19	661.0			

13.33 Work Exercise 13.24 using the LSD multiple comparisons procedure on rank trans-formed data. Compare your answer with the one obtained in that exercise.

13.34 Work Exercise 13.15 using the Kruskal-Wallis test and compare your answer with the one obtained in that exercise. The computer output of the analysis of variance table is as follows.

Source	DF	Sum of Squares	Mean Square	F Value	PR > F
TREATMENT	3	922.025	307.342	3.94	.0176
ERROR	30	2442.975	78.099		
TOTAL	33	3265.000			

13.35 Work Exercise 13.26 using the LSD multiple comparisons procedure on rank trans-formed data. Compare your answer with the one obtained in that exercise.

13.5
Review Exercises

13.36 Use the Kruskal-Wallis test with the data given in Problem Setting 13.2. Compare your results with those obtained in the example of Section 13.2 that used the one-way analysis of variance on the raw data.

13.37 Use the LSD multiple comparisons procedure on the rank transformed data given in Problem Setting 13.2. Compare your results with those obtained in the example of Section 13.3 that applied the LSD procedure to the raw data.

13.38 Use the Kruskal-Wallis test with the data given in Problem Setting 13.3. Compare your results with those obtained in the example of Section 13.2 that used the one-way analysis of variance on the raw data.

13.39 Use the LSD multiple comparisons procedure on the rank transformed data given in Problem Setting 13.3. Compare your results with those obtained in the example of Section 13.3 that applied the LSD procedure to the raw data.

13.40 An experiment has been designed to assess the effect of alcoholic consumption on keypunch operators. Twenty keypunch operators were randomly selected and in turn randomly assigned to receive either 0, 1, 2, or 3 ounces of liquor one hour prior to keypunching the same computer program. The number of keypunch errors was then noted for each operator and recorded below. Graph the e.d.f.'s for these data.

0 Ounces	1 Ounce	2 Ounces	3 Ounces
0	1	5	7
1	2	8	11
1	3	8	12
2	3	9	13
3	4	10	15

13.41 Make a tentative guess about the acceptability of the null hypothesis H_0: $\mu_0 = \mu_1 = \mu_2 = \mu_3$ from the e.d.f.'s in Exercise 13.40. If your decision is to reject H_0, then make another tentative guess about the relative ordering of the population means.

13.42 Test the null hypothesis H_0: $\mu_0 = \mu_1 = \mu_2 = \mu_3$ for the sample data given in Exercise 13.40. Let $\alpha = .05$ and state the p-value. Assume that the analysis of variance assumptions are met.

13.43 Compare your answer in Exercise 13.42 with your first tentative guess in Exercise 13.41.

13.44 Apply the LSD multiple comparisons procedure with $\alpha = .05$ to the analysis of variance problem in Exercise 13.42.

13.45 Compare your conclusion in Exercise 13.44 with your second tentative guess made in Exercise 13.41.

13.46 A department examines their last 17 credit card sales and records the amount charged (nearest dollar) for each of three types of credit cards. Graph the e.d.f.'s for these charges.

Store	VISA	Master Charge
$56	$ 80	$ 73
20	51	56
37	40	123
28	72	56
	132	37
	60	44
		40

13.47 Make a tentative guess about the acceptability of the null hypothesis H_0: $\mu_S = \mu_V = \mu_{MC}$ from the e.d.f's in Exercise 13.46. If your decision is to reject H_0, then make another tentative guess about the relative ordering of the population means.

13.48 Test the null hypothesis H_0: $\mu_S = \mu_V = \mu_{MC}$ for the sample data given in Exercise 13.44. Let $\alpha = .05$ and state the p-value. What assumptions are you making about the samples and populations? The computer output of the analysis of variance table is as follows.

Source	DF	Sum of Squares	Mean Square	F Value	PR > F
TREATMENT	2	3386.09	1693.04	2.09	.1604
ERROR	14	11333.68	809.55		
TOTAL	16	14719.76			

13.49   Compare your answer in Exercise 13.48 with your first tentative guess in Exercise 13.47.

13.50   Apply the LSD multiple comparisons procedure with $\alpha = .05$ to the analysis of variance problem in Exercise 13.48.

13.51   Compare your conclusion in Exercise 13.50 with your second tentative guess made in Exercise 13.47.

Bibliography

Additional material on the topics presented in this chapter can be found in the following publications.

Carmer, S. G. and Swanson, M. R. (1973). "An Evaluation of Ten Pairwise Multiple Comparisons Procedures by Monte Carlo Methods." *Journal of the American Statistical Association*, **68**(341), 66–74.

Conover, W. J. (1980). *Practical Nonparametric Statistics, 2nd ed.*, John Wiley and Sons, Inc., New York.

Conover, W. J., and Iman, R. L. (1981). "Rank Transformations as a Bridge between Parametric and Nonparametric Statistics," *The American Statistician*, **35**(3), 124–133.

Conover, W. J., Johnson, M. E., and Johnson, M. M. (1981). "A Comparative Study of Tests for Homogeneity of Variances with Applications to Outer Continental Shelf Bidding Data," *Technometrics*, **23**(4), 351–361.

Iman, R. L., and Conover, W. J. (1980). "Multiple Comparisons Procedures Based on the Rank Transformation." Paper presented at the 140th Annual Meeting of the American Statistical Association, Houston, Texas (August).

Iman, R. L., and Davenport, J. M. (1976). "New Approximations to the Exact Distribution of the Kruskal-Wallis Test Statistic," *Communications In Statistics*, **A5**(14), 1335–1348.

A

Tables

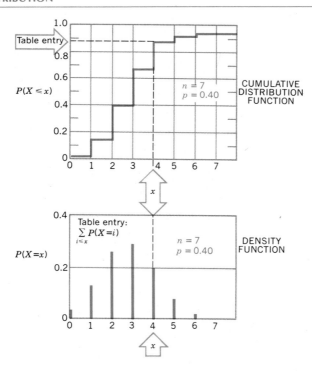

TABLE A1 Binomial Distribution

n	x	p = .05	.10	.15	.20	.25	.30	.35	.40	.45	.50	.55	.60	.65	.70	.75	.80	.85	.90	.95
1	0	.9500	.9000	.8500	.8000	.7500	.7000	.6500	.6000	.5500	.5000	.4500	.4000	.3500	.3000	.2500	.2000	.1500	.1000	.0500
	1	1.0000	1.0000	1.0000	1.0000	1.0000	1.0000	1.0000	1.0000	1.0000	1.0000	1.0000	1.0000	1.0000	1.0000	1.0000	1.0000	1.0000	1.0000	1.0000
2	0	.9025	.8100	.7225	.6400	.5625	.4900	.4225	.3600	.3025	.2500	.2025	.1600	.1225	.0900	.0625	.0400	.0225	.0100	.0025
	1	.9975	.9900	.9775	.9600	.9375	.9100	.8775	.8400	.6975	.7500	.6975	.6400	.5775	.5100	.4375	.3600	.2775	.1900	.0975
	2	1.0000	1.0000	1.0000	1.0000	1.0000	1.0000	1.0000	1.0000	1.0000	1.0000	1.0000	1.0000	1.0000	1.0000	1.0000	1.0000	1.0000	1.0000	1.0000
3	0	.8574	.7290	.6141	.5120	.4219	.3430	.2746	.2160	.1664	.1250	.0911	.0640	.0429	.0270	.0156	.0080	.0034	.0010	.0001
	1	.9928	.9720	.9392	.8960	.8438	.7840	.7182	.6480	.5748	.5000	.4252	.3520	.2818	.2160	.1562	.1040	.0608	.0280	.0072
	2	.9999	.9990	.9966	.9920	.9844	.9730	.9571	.9360	.9089	.8750	.8336	.7840	.7254	.6570	.5781	.4880	.3859	.2710	.1426
	3	1.0000	1.0000	1.0000	1.0000	1.0000	1.0000	1.0000	1.0000	1.0000	1.0000	1.0000	1.0000	1.0000	1.0000	1.0000	1.0000	1.0000	1.0000	1.0000
4	0	.8145	.6561	.5220	.4096	.3164	.2401	.1785	.1296	.0915	.0625	.0410	.0256	.0150	.0081	.0039	.0016	.0005	.0001	.0000
	1	.9860	.9477	.8905	.8192	.7383	.6517	.5630	.4752	.3910	.3125	.2415	.1792	.1265	.0837	.0508	.0272	.0120	.0037	.0005
	2	.9995	.9963	.9880	.9728	.9492	.9163	.8735	.8208	.7585	.6875	.6090	.5248	.4370	.3483	.2617	.1808	.1095	.0523	.0140
	3	1.0000	.9999	.9995	.9984	.9961	.9919	.9850	.9744	.9590	.9375	.9085	.8704	.8215	.7599	.6836	.5904	.4780	.3439	.1855
	4	1.0000	1.0000	1.0000	1.0000	1.0000	1.0000	1.0000	1.0000	1.0000	1.0000	1.0000	1.0000	1.0000	1.0000	1.0000	1.0000	1.0000	1.0000	1.0000
5	0	.7738	.5905	.4437	.3277	.2373	.1681	.1160	.0778	.0503	.0312	.0185	.0102	.0053	.0024	.0010	.0003	.0001	.0000	.0000
	1	.9774	.9185	.8352	.7373	.6328	.5282	.4284	.3370	.2562	.1875	.1312	.0870	.0540	.0308	.0156	.0067	.0022	.0005	.0000
	2	.9988	.9914	.9734	.9421	.8965	.8369	.7648	.6826	.5931	.5000	.4069	.3174	.2352	.1631	.1035	.0579	.0266	.0086	.0012
	3	1.0000	.9995	.9978	.9933	.9844	.9692	.9460	.9130	.8688	.8125	.7438	.6630	.5716	.4718	.3672	.2627	.1648	.0815	.0226
	4	1.0000	1.0000	.9999	.9997	.9990	.9976	.9947	.9898	.9815	.9688	.9497	.9222	.8840	.8319	.7627	.6723	.5563	.4095	.2262
	5	1.0000	1.0000	1.0000	1.0000	1.0000	1.0000	1.0000	1.0000	1.0000	1.0000	1.0000	1.0000	1.0000	1.0000	1.0000	1.0000	1.0000	1.0000	1.0000
6	0	.7351	.5314	.3771	.2621	.1780	.1176	.0754	.0467	.0277	.0156	.0083	.0041	.0018	.0007	.0002	.0001	.0000	.0000	.0000
	1	.9672	.8857	.7765	.6554	.5339	.4202	.3191	.2333	.1636	.1094	.0692	.0410	.0223	.0109	.0046	.0016	.0004	.0001	.0000
	2	.9978	.9842	.9527	.9011	.8306	.7443	.6471	.5443	.4415	.3438	.2553	.1792	.1174	.0705	.0376	.0170	.0059	.0013	.0001
	3	.9999	.9987	.9941	.9830	.9624	.9295	.8826	.8208	.7447	.6562	.5585	.4557	.3529	.2557	.1694	.0989	.0473	.0158	.0022
	4	1.0000	.9999	.9996	.9984	.9954	.9891	.9777	.9590	.9308	.8906	.8364	.7667	.6809	.5798	.4661	.3446	.2235	.1143	.0328
	5	1.0000	1.0000	1.0000	.9999	.9998	.9993	.9982	.9959	.9917	.9844	.9723	.9533	.9246	.8824	.8220	.7379	.6229	.4686	.2649
	6	1.0000	1.0000	1.0000	1.0000	1.0000	1.0000	1.0000	1.0000	1.0000	1.0000	1.0000	1.0000	1.0000	1.0000	1.0000	1.0000	1.0000	1.0000	1.0000
7	0	.6983	.4783	.3206	.2097	.1335	.0824	.0490	.0280	.0152	.0078	.0037	.0016	.0006	.0002	.0001	.0000	.0000	.0000	.0000
	1	.9556	.8503	.7166	.5767	.4449	.3294	.2338	.1586	.1024	.0625	.0357	.0188	.0090	.0038	.0013	.0004	.0001	.0000	.0000
	2	.9962	.9743	.9262	.8520	.7564	.6471	.5323	.4199	.3164	.2266	.1529	.0963	.0556	.0288	.0129	.0047	.0012	.0002	.0000
	3	.9998	.9973	.9879	.9667	.9294	.8740	.8002	.7102	.6083	.5000	.3917	.2898	.1998	.1260	.0706	.0333	.0121	.0027	.0002
	4	1.0000	.9998	.9988	.9953	.9871	.9712	.9444	.9037	.8471	.7734	.6836	.5801	.4677	.3529	.2436	.1480	.0738	.0257	.0038
	5	1.0000	1.0000	.9999	.9996	.9987	.9962	.9910	.9812	.9643	.9375	.8976	.8414	.7662	.6706	.5551	.4233	.2834	.1497	.0444
	6	1.0000	1.0000	1.0000	1.0000	.9999	.9998	.9994	.9984	.9963	.9922	.9848	.9720	.9510	.9176	.8665	.7903	.6794	.5217	.3017
	7	1.0000	1.0000	1.0000	1.0000	1.0000	1.0000	1.0000	1.0000	1.0000	1.0000	1.0000	1.0000	1.0000	1.0000	1.0000	1.0000	1.0000	1.0000	1.0000

n	x	p = .05	.10	.15	.20	.25	.30	.35	.40	.45	.50	.55	.60	.65	.70	.75	.80	.85	.90	.95
8	0	.6634	.4305	.2725	.1678	.1001	.0576	.0319	.0168	.0084	.0039	.0017	.0007	.0002	.0001	.0000	.0000	.0000	.0000	.0000
	1	.9428	.8131	.6572	.5033	.3671	.2553	.1691	.1064	.0632	.0352	.0181	.0085	.0036	.0013	.0004	.0001	.0000	.0000	.0000
	2	.9942	.9619	.8948	.7969	.6785	.5518	.4278	.3154	.2201	.1445	.0885	.0498	.0253	.0113	.0042	.0012	.0002	.0000	.0000
	3	.9996	.9950	.9786	.9437	.8862	.8059	.7064	.5941	.4770	.3633	.2604	.1737	.1061	.0580	.0273	.0104	.0029	.0004	.0000
	4	1.0000	.9996	.9971	.9896	.9727	.9420	.8939	.8263	.7396	.6367	.5230	.4059	.2936	.1941	.1138	.0563	.0214	.0050	.0004
	5	1.0000	1.0000	.9998	.9988	.9958	.9887	.9747	.9502	.9115	.8555	.7799	.6846	.5722	.4482	.3215	.2031	.1052	.0381	.0058
	6	1.0000	1.0000	1.0000	.9999	.9996	.9987	.9964	.9915	.9819	.9648	.9368	.8936	.8309	.7447	.6329	.4967	.3428	.1869	.0572
	7	1.0000	1.0000	1.0000	1.0000	1.0000	.9999	.9998	.9993	.9983	.9961	.9916	.9832	.9681	.9424	.8999	.8322	.7275	.5695	.3366
	8	1.0000	1.0000	1.0000	1.0000	1.0000	1.0000	1.0000	1.0000	1.0000	1.0000	1.0000	1.0000	1.0000	1.0000	1.0000	1.0000	1.0000	1.0000	1.0000
9	0	.6302	.3874	.2316	.1342	.0751	.0404	.0207	.0101	.0046	.0020	.0008	.0003	.0001	.0000	.0000	.0000	.0000	.0000	.0000
	1	.9288	.7748	.5995	.4362	.3003	.1960	.1211	.0705	.0385	.0195	.0091	.0038	.0014	.0004	.0001	.0000	.0000	.0000	.0000
	2	.9916	.9470	.8591	.7382	.6007	.4628	.3373	.2318	.1495	.0898	.0498	.0250	.0112	.0043	.0013	.0003	.0000	.0000	.0000
	3	.9994	.9917	.9661	.9144	.8343	.7297	.6089	.4826	.3614	.2539	.1658	.0994	.0536	.0253	.0100	.0031	.0006	.0001	.0000
	4	1.0000	.9991	.9944	.9804	.9511	.9012	.8283	.7334	.6214	.5000	.3786	.2666	.1717	.0988	.0489	.0196	.0056	.0009	.0000
	5	1.0000	.9999	.9994	.9969	.9900	.9747	.9464	.9006	.8342	.7461	.6386	.5174	.3911	.2703	.1657	.0856	.0339	.0083	.0006
	6	1.0000	1.0000	1.0000	.9997	.9987	.9957	.9888	.9750	.9502	.9102	.8505	.7682	.6627	.5372	.3993	.2618	.1409	.0530	.0084
	7	1.0000	1.0000	1.0000	1.0000	.9999	.9996	.9986	.9962	.9909	.9805	.9615	.9295	.8789	.8040	.6997	.5638	.4005	.2252	.0712
	8	1.0000	1.0000	1.0000	1.0000	1.0000	1.0000	.9999	.9997	.9992	.9980	.9954	.9899	.9793	.9596	.9249	.8658	.7684	.6126	.3698
	9	1.0000	1.0000	1.0000	1.0000	1.0000	1.0000	1.0000	1.0000	1.0000	1.0000	1.0000	1.0000	1.0000	1.0000	1.0000	1.0000	1.0000	1.0000	1.0000
10	0	.5987	.3487	.1969	.1074	.0563	.0282	.0135	.0060	.0025	.0010	.0003	.0001	.0000	.0000	.0000	.0000	.0000	.0000	.0000
	1	.9139	.7361	.5443	.3758	.2440	.1493	.0860	.0464	.0233	.0107	.0045	.0017	.0005	.0001	.0000	.0000	.0000	.0000	.0000
	2	.9885	.9298	.8202	.6778	.5256	.3828	.2616	.1673	.0996	.0547	.0274	.0123	.0048	.0016	.0004	.0001	.0000	.0000	.0000
	3	.9990	.9872	.9500	.8791	.7759	.6496	.5138	.3823	.2660	.1719	.1020	.0548	.0260	.0106	.0035	.0009	.0001	.0000	.0000
	4	.9999	.9984	.9901	.9672	.9219	.8497	.7515	.6331	.5044	.3770	.2616	.1662	.0949	.0473	.0197	.0064	.0014	.0001	.0000
	5	1.0000	.9999	.9986	.9936	.9803	.9527	.9051	.8338	.7384	.6230	.4956	.3669	.2485	.1503	.0781	.0328	.0099	.0016	.0001
	6	1.0000	1.0000	.9999	.9991	.9965	.9894	.9740	.9452	.8980	.8281	.7340	.6177	.4862	.3504	.2241	.1209	.0500	.0128	.0010
	7	1.0000	1.0000	1.0000	.9999	.9996	.9984	.9952	.9877	.9726	.9453	.9004	.8327	.7384	.6172	.4744	.3222	.1798	.0702	.0115
	8	1.0000	1.0000	1.0000	1.0000	1.0000	.9999	.9995	.9983	.9955	.9893	.9767	.9536	.9140	.8507	.7560	.6242	.4557	.2639	.0861
	9	1.0000	1.0000	1.0000	1.0000	1.0000	1.0000	1.0000	.9999	.9997	.9990	.9975	.9940	.9865	.9718	.9437	.8926	.8031	.6513	.4013
	10	1.0000	1.0000	1.0000	1.0000	1.0000	1.0000	1.0000	1.0000	1.0000	1.0000	1.0000	1.0000	1.0000	1.0000	1.0000	1.0000	1.0000	1.0000	1.0000
11	0	.5688	.3138	.1673	.0859	.0422	.0198	.0088	.0036	.0014	.0005	.0002	.0000	.0000	.0000	.0000	.0000	.0000	.0000	.0000
	1	.8981	.6974	.4922	.3221	.1971	.1130	.0606	.0302	.0139	.0059	.0022	.0007	.0002	.0000	.0000	.0000	.0000	.0000	.0000
	2	.9848	.9104	.7788	.6174	.4552	.3127	.2001	.1189	.0652	.0327	.0148	.0059	.0020	.0006	.0001	.0000	.0000	.0000	.0000
	3	.9984	.9815	.9306	.8389	.7133	.5696	.4256	.2963	.1911	.1133	.0610	.0293	.0122	.0043	.0012	.0002	.0000	.0000	.0000
	4	.9999	.9972	.9841	.9496	.8854	.7897	.6683	.5328	.3971	.2744	.1738	.0994	.0501	.0216	.0076	.0020	.0003	.0000	.0000
	5	1.0000	.9997	.9973	.9883	.9657	.9218	.8513	.7535	.6331	.5000	.3669	.2465	.1487	.0782	.0343	.0117	.0027	.0003	.0000
	6	1.0000	1.0000	.9997	.9980	.9924	.9784	.9499	.9006	.8262	.7256	.6029	.4672	.3317	.2103	.1146	.0504	.0159	.0028	.0001
	7	1.0000	1.0000	1.0000	.9998	.9988	.9957	.9878	.9707	.9390	.8867	.8089	.7037	.5744	.4304	.2867	.1611	.0694	.0185	.0016
	8	1.0000	1.0000	1.0000	1.0000	.9999	.9994	.9980	.9941	.9852	.9673	.9348	.8811	.7999	.6873	.5448	.3826	.2212	.0896	.0152
	9	1.0000	1.0000	1.0000	1.0000	1.0000	1.0000	.9998	.9993	.9978	.9941	.9861	.9698	.9394	.8870	.8029	.6779	.5078	.3026	.1019
	10	1.0000	1.0000	1.0000	1.0000	1.0000	1.0000	1.0000	1.0000	.9998	.9995	.9986	.9964	.9912	.9802	.9578	.9141	.8327	.6862	.4312
	11	1.0000	1.0000	1.0000	1.0000	1.0000	1.0000	1.0000	1.0000	1.0000	1.0000	1.0000	1.0000	1.0000	1.0000	1.0000	1.0000	1.0000	1.0000	1.0000

TABLE A1 Binomial Distribution (continued)

n	x	p = .05	.10	.15	.20	.25	.30	.35	.40	.45	.50	.55	.60	.65	.70	.75	.80	.85	.90	.95
12	0	.5404	.2824	.1422	.0687	.0317	.0138	.0057	.0022	.0008	.0002	.0001	.0000	.0000	.0000	.0000	.0000	.0000	.0000	.0000
	1	.8816	.6590	.4435	.2749	.1584	.0850	.0424	.0196	.0083	.0032	.0011	.0003	.0001	.0000	.0000	.0000	.0000	.0000	.0000
	2	.9804	.8891	.7358	.5583	.3907	.2528	.1513	.0834	.0421	.0193	.0079	.0028	.0008	.0002	.0000	.0000	.0000	.0000	.0000
	3	.9978	.9744	.9078	.7946	.6488	.4925	.3467	.2253	.1345	.0730	.0356	.0153	.0056	.0017	.0004	.0001	.0000	.0000	.0000
	4	.9998	.9957	.9761	.9274	.8424	.7237	.5833	.4382	.3044	.1938	.1117	.0573	.0255	.0095	.0028	.0006	.0001	.0000	.0000
	5	1.0000	.9995	.9954	.9806	.9456	.8822	.7873	.6652	.5269	.3872	.2607	.1582	.0846	.0386	.0143	.0039	.0007	.0001	.0000
	6	1.0000	.9999	.9993	.9961	.9857	.9614	.9154	.8418	.7393	.6128	.4731	.3348	.2127	.1178	.0544	.0194	.0046	.0005	.0000
	7	1.0000	.9999	.9999	.9994	.9972	.9905	.9745	.9427	.8883	.8062	.6956	.5618	.4167	.2763	.1576	.0726	.0239	.0043	.0002
	8	1.0000	1.0000	.9999	.9999	.9996	.9983	.9944	.9847	.9644	.9270	.8655	.7747	.6533	.5075	.3512	.2054	.0922	.0256	.0022
	9	1.0000	1.0000	1.0000	1.0000	1.0000	.9998	.9992	.9972	.9921	.9807	.9579	.9166	.8487	.7472	.6093	.4417	.2642	.1109	.0196
	10	1.0000	1.0000	1.0000	1.0000	1.0000	1.0000	.9999	.9997	.9989	.9968	.9917	.9804	.9576	.9150	.8416	.7251	.5565	.3410	.1184
	11	1.0000	1.0000	1.0000	1.0000	1.0000	1.0000	1.0000	1.0000	.9999	.9998	.9992	.9978	.9943	.9862	.9683	.9313	.8578	.7176	.4596
	12	1.0000	1.0000	1.0000	1.0000	1.0000	1.0000	1.0000	1.0000	1.0000	1.0000	1.0000	1.0000	1.0000	1.0000	1.0000	1.0000	1.0000	1.0000	1.0000
13	0	.5133	.2542	.1209	.0550	.0238	.0097	.0037	.0013	.0004	.0001	.0000	.0000	.0000	.0000	.0000	.0000	.0000	.0000	.0000
	1	.8646	.6213	.3983	.2336	.1267	.0637	.0296	.0126	.0049	.0017	.0005	.0001	.0000	.0000	.0000	.0000	.0000	.0000	.0000
	2	.9755	.8661	.6920	.5017	.3326	.2025	.1132	.0579	.0269	.0112	.0041	.0013	.0003	.0001	.0000	.0000	.0000	.0000	.0000
	3	.9969	.9658	.8820	.7473	.5843	.4206	.2783	.1686	.0929	.0461	.0203	.0078	.0025	.0007	.0001	.0000	.0000	.0000	.0000
	4	.9997	.9935	.9658	.9009	.7940	.6543	.5005	.3530	.2279	.1334	.0698	.0321	.0126	.0040	.0010	.0002	.0000	.0000	.0000
	5	1.0000	.9991	.9925	.9700	.9198	.8346	.7159	.5744	.4268	.2905	.1788	.0977	.0462	.0182	.0056	.0012	.0002	.0000	.0000
	6	1.0000	.9999	.9987	.9930	.9757	.9376	.8705	.7712	.6437	.5000	.3563	.2288	.1295	.0624	.0243	.0070	.0013	.0001	.0000
	7	1.0000	1.0000	.9998	.9988	.9944	.9818	.9538	.9023	.8212	.7095	.5732	.4256	.2841	.1654	.0802	.0300	.0075	.0009	.0000
	8	1.0000	1.0000	1.0000	.9998	.9990	.9960	.9874	.9679	.9302	.8666	.7721	.6470	.4995	.3457	.2060	.0991	.0342	.0065	.0003
	9	1.0000	1.0000	1.0000	1.0000	.9999	.9993	.9975	.9922	.9797	.9539	.9071	.8314	.7217	.5794	.4157	.2527	.1180	.0342	.0031
	10	1.0000	1.0000	1.0000	1.0000	1.0000	.9999	.9997	.9987	.9959	.9888	.9731	.9421	.8868	.7975	.6674	.4983	.3080	.1339	.0245
	11	1.0000	1.0000	1.0000	1.0000	1.0000	1.0000	1.0000	.9999	.9995	.9983	.9951	.9874	.9704	.9363	.8733	.7664	.6017	.3787	.1354
	12	1.0000	1.0000	1.0000	1.0000	1.0000	1.0000	1.0000	1.0000	1.0000	.9999	.9996	.9987	.9963	.9903	.9762	.9450	.8791	.7458	.4867
	13	1.0000	1.0000	1.0000	1.0000	1.0000	1.0000	1.0000	1.0000	1.0000	1.0000	1.0000	1.0000	1.0000	1.0000	1.0000	1.0000	1.0000	1.0000	1.0000
14	0	.4877	.2288	.1028	.0440	.0178	.0068	.0024	.0008	.0002	.0001	.0000	.0000	.0000	.0000	.0000	.0000	.0000	.0000	.0000
	1	.8470	.5846	.3567	.1979	.1010	.0475	.0205	.0081	.0029	.0009	.0003	.0001	.0000	.0000	.0000	.0000	.0000	.0000	.0000
	2	.9699	.8416	.6479	.4481	.2811	.1608	.0839	.0398	.0170	.0065	.0022	.0006	.0001	.0000	.0000	.0000	.0000	.0000	.0000
	3	.9958	.9559	.8535	.6982	.5213	.3552	.2205	.1243	.0632	.0287	.0114	.0039	.0011	.0002	.0000	.0000	.0000	.0000	.0000
	4	.9996	.9908	.9533	.8702	.7415	.5842	.4227	.2793	.1672	.0898	.0426	.0175	.0060	.0017	.0003	.0000	.0000	.0000	.0000
	5	1.0000	.9985	.9885	.9561	.8883	.7805	.6405	.4859	.3373	.2120	.1189	.0583	.0243	.0083	.0022	.0004	.0000	.0000	.0000
	6	1.0000	.9998	.9978	.9884	.9617	.9067	.8164	.6925	.5461	.3953	.2586	.1501	.0753	.0315	.0103	.0024	.0003	.0000	.0000
	7	1.0000	1.0000	.9997	.9976	.9897	.9685	.9247	.8499	.7414	.6047	.4539	.3075	.1836	.0933	.0383	.0116	.0022	.0002	.0000
	8	1.0000	1.0000	1.0000	.9996	.9978	.9917	.9757	.9417	.8811	.7880	.6627	.5141	.3595	.2195	.1117	.0439	.0115	.0015	.0000
	9	1.0000	1.0000	1.0000	1.0000	.9997	.9983	.9940	.9825	.9574	.9102	.8328	.7207	.5773	.4158	.2585	.1298	.0467	.0092	.0004
	10	1.0000	1.0000	1.0000	1.0000	1.0000	.9998	.9989	.9961	.9886	.9713	.9368	.8757	.7795	.6448	.4787	.3018	.1465	.0441	.0042
	11	1.0000	1.0000	1.0000	1.0000	1.0000	1.0000	.9999	.9994	.9978	.9935	.9830	.9602	.9161	.8392	.7189	.5519	.3521	.1584	.0301
	12	1.0000	1.0000	1.0000	1.0000	1.0000	1.0000	1.0000	1.0000	.9999	.9997	.9971	.9919	.9795	.9525	.8990	.8021	.6433	.4154	.1530
	13	1.0000	1.0000	1.0000	1.0000	1.0000	1.0000	1.0000	1.0000	1.0000	.9999	.9998	.9992	.9976	.9932	.9822	.9560	.8972	.7712	.5123
	14	1.0000	1.0000	1.0000	1.0000	1.0000	1.0000	1.0000	1.0000	1.0000	1.0000	1.0000	1.0000	1.0000	1.0000	1.0000	1.0000	1.0000	1.0000	1.0000

n	x	p = .05	.10	.15	.20	.25	.30	.35	.40	.45	.50	.55	.60	.65	.70	.75	.80	.85	.90	.95
15	0	.4633	.2059	.0874	.0352	.0134	.0047	.0016	.0005	.0001	.0000	.0000	.0000	.0000	.0000	.0000	.0000	.0000	.0000	.0000
	1	.8290	.5490	.3186	.1671	.0802	.0353	.0142	.0052	.0017	.0005	.0001	.0000	.0000	.0000	.0000	.0000	.0000	.0000	.0000
	2	.9638	.8159	.6042	.3980	.2361	.1268	.0617	.0271	.0107	.0037	.0011	.0003	.0001	.0000	.0000	.0000	.0000	.0000	.0000
	3	.9945	.9444	.8227	.6482	.4613	.2969	.1727	.0905	.0424	.0176	.0063	.0019	.0005	.0001	.0000	.0000	.0000	.0000	.0000
	4	.9994	.9873	.9383	.8358	.6865	.5155	.3519	.2173	.1204	.0592	.0255	.0093	.0028	.0007	.0001	.0000	.0000	.0000	.0000
	5	.9999	.9978	.9832	.9389	.8516	.7216	.5643	.4032	.2608	.1509	.0769	.0338	.0124	.0037	.0008	.0001	.0000	.0000	.0000
	6	1.0000	.9997	.9964	.9819	.9434	.8689	.7548	.6098	.4522	.3036	.1818	.0950	.0422	.0152	.0042	.0008	.0001	*.0000	.0000
	7	1.0000	1.0000	.9994	.9958	.9827	.9500	.8868	.7869	.6535	.5000	.3465	.2131	.1132	.0500	.0173	.0042	.0006	.0000	.0000
	8	1.0000	1.0000	.9999	.9992	.9958	.9848	.9578	.9050	.8182	.6964	.5478	.3902	.2452	.1311	.0566	.0181	.0036	.0003	.0000
	9	1.0000	1.0000	1.0000	.9999	.9992	.9963	.9876	.9662	.9231	.8491	.7392	.5968	.4357	.2784	.1484	.0611	.0168	.0022	.0001
	10	1.0000	1.0000	1.0000	1.0000	.9999	.9993	.9972	.9907	.9745	.9408	.8796	.7827	.6481	.4845	.3135	.1642	.0617	.0127	.0006
	11	1.0000	1.0000	1.0000	1.0000	1.0000	.9999	.9995	.9981	.9937	.9824	.9576	.9095	.8273	.7031	.5387	.3518	.1773	.0556	.0055
	12	1.0000	1.0000	1.0000	1.0000	1.0000	1.0000	.9999	.9997	.9989	.9963	.9893	.9729	.9383	.8732	.7639	.6020	.3958	.1841	.0362
	13	1.0000	1.0000	1.0000	1.0000	1.0000	1.0000	1.0000	1.0000	.9999	.9995	.9983	.9948	.9858	.9647	.9198	.8329	.6814	.4510	.1710
	14	1.0000	1.0000	1.0000	1.0000	1.0000	1.0000	1.0000	1.0000	1.0000	1.0000	.9999	.9995	.9984	.9953	.9866	.9648	.9126	.7941	.5367
	15	1.0000	1.0000	1.0000	1.0000	1.0000	1.0000	1.0000	1.0000	1.0000	1.0000	1.0000	1.0000	1.0000	1.0000	1.0000	1.0000	1.0000	1.0000	1.0000
16	0	.4401	.1853	.0743	.0281	.0100	.0033	.0010	.0003	.0001	.0000	.0000	.0000	.0000	.0000	.0000	.0000	.0000	.0000	.0000
	1	.8108	.5147	.2839	.1407	.0635	.0261	.0098	.0033	.0010	.0003	.0001	.0000	.0000	.0000	.0000	.0000	.0000	.0000	.0000
	2	.9571	.7892	.5614	.3518	.1971	.0994	.0451	.0183	.0066	.0021	.0006	.0001	.0000	.0000	.0000	.0000	.0000	.0000	.0000
	3	.9930	.9316	.7899	.5981	.4050	.2459	.1339	.0651	.0281	.0106	.0035	.0009	.0002	.0000	.0000	.0000	.0000	.0000	.0000
	4	.9991	.9830	.9209	.7982	.6302	.4499	.2892	.1666	.0853	.0384	.0149	.0049	.0013	.0003	.0000	.0000	.0000	.0000	.0000
	5	.9999	.9967	.9765	.9183	.8103	.6598	.4900	.3288	.1976	.1051	.0486	.0191	.0062	.0016	.0003	.0000	.0000	.0000	.0000
	6	1.0000	.9995	.9944	.9733	.9204	.8247	.6881	.5272	.3660	.2272	.1241	.0583	.0229	.0071	.0016	.0002	.0000	.0000	.0000
	7	1.0000	.9999	.9989	.9930	.9729	.9256	.8406	.7161	.5629	.4018	.2559	.1423	.0671	.0257	.0075	.0015	.0002	.0000	.0000
	8	1.0000	1.0000	.9998	.9985	.9925	.9743	.9329	.8577	.7441	.5982	.4371	.2839	.1594	.0744	.0271	.0070	.0011	.0001	.0000
	9	1.0000	1.0000	1.0000	.9998	.9984	.9929	.9771	.9417	.8759	.7728	.6340	.4728	.3119	.1753	.0796	.0267	.0056	.0005	.0000
	10	1.0000	1.0000	1.0000	1.0000	.9997	.9984	.9938	.9809	.9514	.8949	.8024	.6712	.5100	.3402	.1897	.0817	.0235	.0033	.0001
	11	1.0000	1.0000	1.0000	1.0000	1.0000	.9997	.9987	.9951	.9851	.9616	.9147	.8334	.7108	.5501	.3698	.2018	.0791	.0170	.0009
	12	1.0000	1.0000	1.0000	1.0000	1.0000	1.0000	.9998	.9991	.9965	.9894	.9719	.9349	.8661	.7541	.5950	.4019	.2101	.0684	.0070
	13	1.0000	1.0000	1.0000	1.0000	1.0000	1.0000	1.0000	.9999	.9994	.9979	.9934	.9817	.9549	.9006	.8029	.6482	.4386	.2108	.0429
	14	1.0000	1.0000	1.0000	1.0000	1.0000	1.0000	1.0000	1.0000	.9999	.9997	.9990	.9967	.9902	.9739	.9365	.8593	.7161	.4853	.1892
	15	1.0000	1.0000	1.0000	1.0000	1.0000	1.0000	1.0000	1.0000	1.0000	1.0000	.9999	.9997	.9990	.9967	.9900	.9719	.9257	.8147	.5599
	16	1.0000	1.0000	1.0000	1.0000	1.0000	1.0000	1.0000	1.0000	1.0000	1.0000	1.0000	1.0000	1.0000	1.0000	1.0000	1.0000	1.0000	1.0000	1.0000

TABLE A1 Binomial Distribution (*concluded*)

n	x	p = .05	.10	.15	.20	.25	.30	.35	.40	.45	.50	.55	.60	.65	.70	.75	.80	.85	.90	.95
17	0	.4181	.1668	.0631	.0225	.0075	.0023	.0007	.0002	.0000	.0000	.0000	.0000	.0000	.0000	.0000	.0000	.0000	.0000	.0000
	1	.7922	.4818	.2525	.1182	.0501	.0193	.0067	.0021	.0006	.0001	.0000	.0000	.0000	.0000	.0000	.0000	.0000	.0000	.0000
	2	.9497	.7618	.5198	.3096	.1637	.0774	.0327	.0123	.0041	.0012	.0003	.0001	.0000	.0000	.0000	.0000	.0000	.0000	.0000
	3	.9912	.9174	.7556	.5489	.3530	.2019	.1028	.0464	.0184	.0064	.0019	.0005	.0001	.0000	.0000	.0000	.0000	.0000	.0000
	4	.9988	.9779	.9013	.7582	.5739	.3887	.2348	.1260	.0596	.0245	.0086	.0025	.0006	.0001	.0000	.0000	.0000	.0000	.0000
	5	.9999	.9953	.9681	.8943	.7653	.5968	.4197	.2639	.1471	.0717	.0301	.0106	.0030	.0007	.0001	.0000	.0000	.0000	.0000
	6	1.0000	.9992	.9917	.9623	.8929	.7752	.6188	.4478	.2902	.1662	.0826	.0348	.0120	.0032	.0006	.0001	.0000	.0000	.0000
	7	1.0000	.9999	.9983	.9891	.9598	.8954	.7872	.6405	.4743	.3145	.1834	.0919	.0383	.0127	.0031	.0005	.0000	.0000	.0000
	8	1.0000	1.0000	.9997	.9974	.9876	.9597	.9006	.8011	.6626	.5000	.3374	.1989	.0994	.0403	.0124	.0026	.0003	.0000	.0000
	9	1.0000	1.0000	1.0000	.9995	.9969	.9873	.9617	.9081	.8166	.6855	.5257	.3595	.2128	.1046	.0402	.0109	.0017	.0001	.0000
	10	1.0000	1.0000	1.0000	.9999	.9994	.9968	.9880	.9652	.9174	.8338	.7098	.5522	.3812	.2248	.1071	.0377	.0083	.0008	.0000
	11	1.0000	1.0000	1.0000	1.0000	.9999	.9993	.9970	.9894	.9699	.9283	.8529	.7361	.5803	.4032	.2347	.1057	.0319	.0047	.0001
	12	1.0000	1.0000	1.0000	1.0000	1.0000	.9999	.9994	.9975	.9914	.9755	.9404	.8740	.7652	.6113	.4261	.2418	.0987	.0221	.0012
	13	1.0000	1.0000	1.0000	1.0000	1.0000	1.0000	.9999	.9995	.9981	.9936	.9816	.9536	.8972	.7981	.6470	.4511	.2444	.0826	.0088
	14	1.0000	1.0000	1.0000	1.0000	1.0000	1.0000	1.0000	.9999	.9997	.9988	.9959	.9877	.9673	.9226	.8363	.6904	.4802	.2382	.0503
	15	1.0000	1.0000	1.0000	1.0000	1.0000	1.0000	1.0000	1.0000	1.0000	.9999	.9994	.9979	.9933	.9807	.9499	.8818	.7475	.5182	.2078
	16	1.0000	1.0000	1.0000	1.0000	1.0000	1.0000	1.0000	1.0000	1.0000	1.0000	1.0000	.9998	.9993	.9977	.9925	.9775	.9369	.8332	.5819
	17	1.0000	1.0000	1.0000	1.0000	1.0000	1.0000	1.0000	1.0000	1.0000	1.0000	1.0000	1.0000	1.0000	1.0000	1.0000	1.0000	1.0000	1.0000	1.0000
18	0	.3972	.1501	.0536	.0180	.0056	.0016	.0004	.0001	.0000	.0000	.0000	.0000	.0000	.0000	.0000	.0000	.0000	.0000	.0000
	1	.7735	.4503	.2241	.0991	.0395	.0142	.0046	.0013	.0003	.0001	.0000	.0000	.0000	.0000	.0000	.0000	.0000	.0000	.0000
	2	.9419	.7338	.4797	.2713	.1353	.0600	.0236	.0082	.0025	.0007	.0001	.0000	.0000	.0000	.0000	.0000	.0000	.0000	.0000
	3	.9891	.9018	.7202	.5010	.3057	.1646	.0783	.0328	.0120	.0038	.0010	.0002	.0000	.0000	.0000	.0000	.0000	.0000	.0000
	4	.9985	.9718	.8794	.7164	.5187	.3327	.1886	.0942	.0411	.0154	.0049	.0013	.0003	.0000	.0000	.0000	.0000	.0000	.0000
	5	.9998	.9936	.9581	.8671	.7175	.5344	.3550	.2088	.1077	.0481	.0183	.0058	.0014	.0003	.0000	.0000	.0000	.0000	.0000
	6	1.0000	.9988	.9882	.9487	.8610	.7217	.5491	.3743	.2258	.1189	.0537	.0203	.0062	.0014	.0002	.0000	.0000	.0000	.0000
	7	1.0000	.9998	.9973	.9837	.9431	.8593	.7283	.5634	.3915	.2403	.1280	.0576	.0212	.0061	.0012	.0002	.0000	.0000	.0000
	8	1.0000	1.0000	.9995	.9957	.9807	.9404	.8609	.7368	.5778	.4073	.2527	.1347	.0597	.0210	.0054	.0009	.0001	.0000	.0000
	9	1.0000	1.0000	.9999	.9991	.9946	.9790	.9403	.8653	.7473	.5927	.4222	.2632	.1391	.0596	.0193	.0043	.0005	.0000	.0000
	10	1.0000	1.0000	1.0000	.9998	.9988	.9939	.9788	.9424	.8720	.7597	.6085	.4366	.2717	.1407	.0569	.0163	.0027	.0002	.0000
	11	1.0000	1.0000	1.0000	1.0000	.9998	.9986	.9938	.9797	.9463	.8811	.7742	.6257	.4509	.2783	.1390	.0513	.0118	.0012	.0000
	12	1.0000	1.0000	1.0000	1.0000	1.0000	.9997	.9986	.9942	.9817	.9519	.8923	.7912	.6450	.4656	.2825	.1329	.0419	.0064	.0002
	13	1.0000	1.0000	1.0000	1.0000	1.0000	1.0000	.9997	.9987	.9951	.9846	.9589	.9058	.8114	.6673	.4813	.2836	.1206	.0282	.0015
	14	1.0000	1.0000	1.0000	1.0000	1.0000	1.0000	1.0000	.9998	.9990	.9962	.9880	.9672	.9217	.8354	.6943	.4990	.2798	.0982	.0109
	15	1.0000	1.0000	1.0000	1.0000	1.0000	1.0000	1.0000	1.0000	.9999	.9993	.9975	.9918	.9764	.9400	.8647	.7287	.5203	.2662	.0581
	16	1.0000	1.0000	1.0000	1.0000	1.0000	1.0000	1.0000	1.0000	1.0000	.9999	.9997	.9987	.9954	.9858	.9605	.9009	.7759	.5497	.2265
	17	1.0000	1.0000	1.0000	1.0000	1.0000	1.0000	1.0000	1.0000	1.0000	1.0000	1.0000	.9999	.9996	.9984	.9944	.9820	.9464	.8499	.6028
	18	1.0000	1.0000	1.0000	1.0000	1.0000	1.0000	1.0000	1.0000	1.0000	1.0000	1.0000	1.0000	1.0000	1.0000	1.0000	1.0000	1.0000	1.0000	1.0000

n	x	p = .05	.10	.15	.20	.25	.30	.35	.40	.45	.50	.55	.60	.65	.70	.75	.80	.85	.90	.95
19	0	.3774	.1351	.0456	.0144	.0042	.0011	.0003	.0001	.0000	.0000	.0000	.0000	.0000	.0000	.0000	.0000	.0000	.0000	.0000
	1	.7547	.4203	.1985	.0829	.0310	.0104	.0031	.0008	.0002	.0000	.0000	.0000	.0000	.0000	.0000	.0000	.0000	.0000	.0000
	2	.9335	.7054	.4413	.2369	.1113	.0462	.0170	.0055	.0015	.0004	.0001	.0000	.0000	.0000	.0000	.0000	.0000	.0000	.0000
	3	.9868	.8850	.6841	.4551	.2631	.1332	.0591	.0230	.0077	.0022	.0005	.0001	.0000	.0000	.0000	.0000	.0000	.0000	.0000
	4	.9980	.9648	.8556	.6733	.4654	.2822	.1500	.0696	.0280	.0096	.0028	.0006	.0001	.0000	.0000	.0000	.0000	.0000	.0000
	5	.9998	.9914	.9463	.8369	.6678	.4739	.2968	.1629	.0777	.0318	.0109	.0031	.0007	.0001	.0000	.0000	.0000	.0000	.0000
	6	1.0000	.9983	.9837	.9324	.8251	.6655	.4812	.3081	.1727	.0835	.0342	.0116	.0031	.0006	.0001	.0000	.0000	.0000	.0000
	7	1.0000	.9997	.9959	.9767	.9225	.8180	.6656	.4878	.3169	.1796	.0871	.0352	.0114	.0028	.0005	.0000	.0000	.0000	.0000
	8	1.0000	1.0000	.9992	.9933	.9713	.9161	.8145	.6675	.4940	.3238	.1841	.0885	.0347	.0105	.0023	.0003	.0000	.0000	.0000
	9	1.0000	1.0000	.9999	.9984	.9911	.9674	.9125	.8139	.6710	.5000	.3290	.1861	.0875	.0326	.0089	.0016	.0001	.0000	.0000
	10	1.0000	1.0000	1.0000	.9997	.9977	.9895	.9653	.9115	.8159	.6762	.5060	.3325	.1855	.0839	.0287	.0067	.0008	.0000	.0000
	11	1.0000	1.0000	1.0000	1.0000	.9995	.9972	.9886	.9648	.9129	.8204	.6831	.5122	.3344	.1820	.0775	.0233	.0041	.0003	.0000
	12	1.0000	1.0000	1.0000	1.0000	.9999	.9994	.9969	.9884	.9658	.9165	.8273	.6919	.5188	.3345	.1749	.0676	.0163	.0017	.0000
	13	1.0000	1.0000	1.0000	1.0000	1.0000	.9999	.9993	.9969	.9891	.9682	.9223	.8371	.7032	.5261	.3322	.1631	.0537	.0086	.0002
	14	1.0000	1.0000	1.0000	1.0000	1.0000	1.0000	.9999	.9994	.9972	.9904	.9720	.9304	.8500	.7178	.5346	.3267	.1444	.0352	.0020
	15	1.0000	1.0000	1.0000	1.0000	1.0000	1.0000	1.0000	.9999	.9995	.9978	.9923	.9770	.9409	.8668	.7369	.5449	.3159	.1150	.0132
	16	1.0000	1.0000	1.0000	1.0000	1.0000	1.0000	1.0000	1.0000	.9999	.9996	.9985	.9945	.9830	.9538	.8887	.7631	.5587	.2946	.0665
	17	1.0000	1.0000	1.0000	1.0000	1.0000	1.0000	1.0000	1.0000	1.0000	1.0000	.9998	.9992	.9969	.9896	.9690	.9171	.8015	.5797	.2453
	18	1.0000	1.0000	1.0000	1.0000	1.0000	1.0000	1.0000	1.0000	1.0000	1.0000	1.0000	.9999	.9997	.9989	.9958	.9856	.9544	.8649	.6226
	19	1.0000	1.0000	1.0000	1.0000	1.0000	1.0000	1.0000	1.0000	1.0000	1.0000	1.0000	1.0000	1.0000	1.0000	1.0000	1.0000	1.0000	1.0000	1.0000
20	0	.3585	.1216	.0388	.0115	.0032	.0008	.0002	.0000	.0000	.0000	.0000	.0000	.0000	.0000	.0000	.0000	.0000	.0000	.0000
	1	.7358	.3917	.1756	.0692	.0243	.0076	.0021	.0005	.0001	.0000	.0000	.0000	.0000	.0000	.0000	.0000	.0000	.0000	.0000
	2	.9245	.6769	.4049	.2061	.0913	.0355	.0121	.0036	.0009	.0002	.0000	.0000	.0000	.0000	.0000	.0000	.0000	.0000	.0000
	3	.9841	.8670	.6477	.4114	.2252	.1071	.0444	.0160	.0049	.0013	.0003	.0000	.0000	.0000	.0000	.0000	.0000	.0000	.0000
	4	.9974	.9568	.8298	.6296	.4148	.2375	.1182	.0510	.0189	.0059	.0015	.0003	.0000	.0000	.0000	.0000	.0000	.0000	.0000
	5	.9997	.9887	.9327	.8042	.6172	.4164	.2454	.1256	.0553	.0207	.0064	.0016	.0003	.0000	.0000	.0000	.0000	.0000	.0000
	6	1.0000	.9976	.9781	.9133	.7858	.6080	.4166	.2500	.1299	.0577	.0214	.0065	.0015	.0003	.0000	.0000	.0000	.0000	.0000
	7	1.0000	.9996	.9941	.9679	.8982	.7723	.6010	.4159	.2520	.1316	.0580	.0210	.0060	.0013	.0002	.0000	.0000	.0000	.0000
	8	1.0000	.9999	.9987	.9900	.9591	.8867	.7624	.5956	.4143	.2517	.1308	.0565	.0196	.0051	.0009	.0001	.0000	.0000	.0000
	9	1.0000	1.0000	.9998	.9974	.9861	.9520	.8782	.7553	.5914	.4119	.2493	.1275	.0532	.0171	.0039	.0006	.0000	.0000	.0000
	10	1.0000	1.0000	1.0000	.9994	.9961	.9829	.9468	.8725	.7507	.5881	.4086	.2447	.1218	.0480	.0139	.0026	.0002	.0000	.0000
	11	1.0000	1.0000	1.0000	.9999	.9991	.9949	.9804	.9435	.8692	.7483	.5857	.4044	.2376	.1133	.0409	.0100	.0013	.0001	.0000
	12	1.0000	1.0000	1.0000	1.0000	.9998	.9987	.9940	.9790	.9420	.8684	.7480	.5841	.3990	.2277	.1018	.0321	.0059	.0004	.0000
	13	1.0000	1.0000	1.0000	1.0000	1.0000	.9997	.9985	.9935	.9786	.9423	.8701	.7500	.5834	.3920	.2142	.0867	.0219	.0024	.0000
	14	1.0000	1.0000	1.0000	1.0000	1.0000	1.0000	.9997	.9984	.9936	.9793	.9447	.8744	.7546	.5836	.3828	.1958	.0673	.0113	.0003
	15	1.0000	1.0000	1.0000	1.0000	1.0000	1.0000	1.0000	.9997	.9985	.9941	.9811	.9490	.8818	.7625	.5852	.3704	.1702	.0432	.0026
	16	1.0000	1.0000	1.0000	1.0000	1.0000	1.0000	1.0000	1.0000	.9997	.9987	.9951	.9840	.9556	.8929	.7748	.5886	.3523	.1330	.0159
	17	1.0000	1.0000	1.0000	1.0000	1.0000	1.0000	1.0000	1.0000	1.0000	.9998	.9991	.9964	.9879	.9645	.9087	.7939	.5951	.3231	.0755
	18	1.0000	1.0000	1.0000	1.0000	1.0000	1.0000	1.0000	1.0000	1.0000	1.0000	.9999	.9995	.9979	.9924	.9757	.9308	.8244	.6083	.2642
	19	1.0000	1.0000	1.0000	1.0000	1.0000	1.0000	1.0000	1.0000	1.0000	1.0000	1.0000	1.0000	.9998	.9992	.9968	.9885	.9612	.8784	.6415
	20	1.0000	1.0000	1.0000	1.0000	1.0000	1.0000	1.0000	1.0000	1.0000	1.0000	1.0000	1.0000	1.0000	1.0000	1.0000	1.0000	1.0000	1.0000	1.0000

For n larger than 20, the rth quantile x_r of a binomial random variable may be approximated using $x_r = np + z_r \sqrt{np(1-p)}$, where z_r is the rth quantile of a standard normal random variable, obtained from Table A2.

Source: Table generated by W. J. Conover.

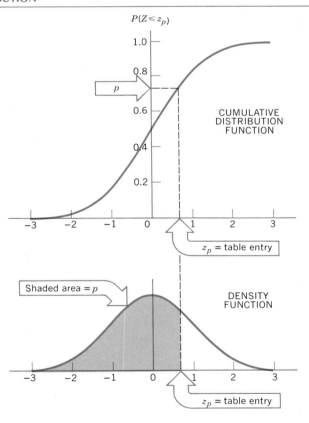

$P(Z \leqslant z_p)$

CUMULATIVE DISTRIBUTION FUNCTION

z_p = table entry

Shaded area = p

DENSITY FUNCTION

z_p = table entry

Selected quantiles: $z_{.0001} = -3.7190$ $z_{.0005} = -3.2905$ $z_{.01} = -2.3263$ $z_{.025} = -1.9600$
$z_{.05} = -1.6449$ $z_{.10} = -1.2816$ $z_{.90} = 1.2816$ $z_{.95} = 1.6449$
$z_{.975} = 1.9600$ $z_{.99} = 2.3263$ $z_{.9995} = 3.2905$ $z_{.9999} = 3.7190$

Note that the value of p to two decimal places determines which row to use; the third decimal place of p determines which column to use to find z_p.

TABLE A2 Normal Distribution

p	.000	.001	.002	.003	.004	.005	.006	.007	.008	.009
.00		−3.0902	−2.8782	−2.7478	−2.6521	−2.5758	−2.5121	−2.4573	−2.4089	−2.3656
.01	−2.3263	−2.2904	−2.2571	−2.2262	−2.1973	−2.1701	−2.1444	−2.1201	−2.0969	−2.0749
.02	−2.0537	−2.0335	−2.0141	−1.9954	−1.9774	−1.9600	−1.9431	−1.9268	−1.9110	−1.8957
.03	−1.8808	−1.8663	−1.8522	−1.8384	−1.8250	−1.8119	−1.7991	−1.7866	−1.7744	−1.7624
.04	−1.7507	−1.7392	−1.7279	−1.7169	−1.7060	−1.6954	−1.6849	−1.6747	−1.6646	−1.6546
.05	−1.6449	−1.6352	−1.6258	−1.6164	−1.6072	−1.5982	−1.5893	−1.5805	−1.5718	−1.5632
.06	−1.5548	−1.5464	−1.5382	−1.5301	−1.5220	−1.5141	−1.5063	−1.4985	−1.4909	−1.4833
.07	−1.4758	−1.4684	−1.4611	−1.4538	−1.4466	−1.4395	−1.4325	−1.4255	−1.4187	−1.4118
.08	−1.4051	−1.3984	−1.3917	−1.3852	−1.3787	−1.3722	−1.3658	−1.3595	−1.3532	−1.3469
.09	−1.3408	−1.3346	−1.3285	−1.3225	−1.3165	−1.3106	−1.3047	−1.2988	−1.2930	−1.2873
.10	−1.2816	−1.2759	−1.2702	−1.2646	−1.2591	−1.2536	−1.2481	−1.2426	−1.2372	−1.2319
.11	−1.2265	−1.2212	−1.2160	−1.2107	−1.2055	−1.2004	−1.1952	−1.1901	−1.1850	−1.1800
.12	−1.1750	−1.1700	−1.1650	−1.1601	−1.1552	−1.1503	−1.1455	−1.1407	−1.1359	−1.1311
.13	−1.1264	−1.1217	−1.1170	−1.1123	−1.1077	−1.1031	−1.0985	−1.0939	−1.0893	−1.0848
.14	−1.0803	−1.0758	−1.0714	−1.0669	−1.0625	−1.0581	−1.0537	−1.0494	−1.0450	−1.0407
.15	−1.0364	−1.0322	−1.0279	−1.0237	−1.0194	−1.0152	−1.0110	−1.0069	−1.0027	−.9986
.16	−.9945	−.9904	−.9863	−.9822	−.9782	−.9741	−.9701	−.9661	−.9621	−.9581
.17	−.9542	−.9502	−.9463	−.9424	−.9385	−.9346	−.9307	−.9269	−.9230	−.9192
.18	−.9154	−.9116	−.9078	−.9040	−.9002	−.8965	−.8927	−.8890	−.8853	−.8816
.19	−.8779	−.8742	−.8705	−.8669	−.8633	−.8596	−.8560	−.8524	−.8488	−.8452
.20	−.8416	−.8381	−.8345	−.8310	−.8274	−.8239	−.8204	−.8169	−.8134	−.8099
.21	−.8064	−.8030	−.7995	−.7961	−.7926	−.7892	−.7858	−.7824	−.7790	−.7756
.22	−.7722	−.7688	−.7655	−.7621	−.7588	−.7554	−.7521	−.7488	−.7454	−.7421
.23	−.7388	−.7356	−.7323	−.7290	−.7257	−.7225	−.7192	−.7160	−.7128	−.7095
.24	−.7063	−.7031	−.6999	−.6967	−.6935	−.6903	−.6871	−.6840	−.6808	−.6776
.25	−.6745	−.6713	−.6682	−.6651	−.6620	−.6588	−.6557	−.6526	−.6495	−.6464
.26	−.6433	−.6403	−.6372	−.6341	−.6311	−.6280	−.6250	−.6219	−.6189	−.6158
.27	−.6128	−.6098	−.6068	−.6038	−.6008	−.5978	−.5948	−.5918	−.5888	−.5858
.28	−.5828	−.5799	−.5769	−.5740	−.5710	−.5681	−.5651	−.5622	−.5592	−.5563
.29	−.5534	−.5505	−.5476	−.5446	−.5417	−.5388	−.5359	−.5330	−.5302	−.5273
.30	−.5244	−.5215	−.5187	−.5158	−.5129	−.5101	−.5072	−.5044	−.5015	−.4987
.31	−.4959	−.4930	−.4902	−.4874	−.4845	−.4817	−.4789	−.4761	−.4733	−.4705
.32	−.4677	−.4649	−.4621	−.4593	−.4565	−.4538	−.4510	−.4482	−.4454	−.4427
.33	−.4399	−.4372	−.4344	−.4316	−.4289	−.4261	−.4234	−.4207	−.4179	−.4152
.34	−.4125	−.4097	−.4070	−.4043	−.4016	−.3989	−.3961	−.3934	−.3907	−.3880
.35	−.3853	−.3826	−.3799	−.3772	−.3745	−.3719	−.3692	−.3665	−.3638	−.3611
.36	−.3585	−.3558	−.3531	−.3505	−.3478	−.3451	−.3425	−.3398	−.3372	−.3345
.37	−.3319	−.3292	−.3266	−.3239	−.3213	−.3186	−.3160	−.3134	−.3107	−.3081
.38	−.3055	−.3029	−.3002	−.2976	−.2950	−.2924	−.2898	−.2871	−.2845	−.2819
.39	−.2793	−.2767	−.2741	−.2715	−.2689	−.2663	−.2637	−.2611	−.2585	−.2559
.40	−.2533	−.2508	−.2482	−.2456	−.2430	−.2404	−.2378	−.2353	−.2327	−.2301
.41	−.2275	−.2250	−.2224	−.2198	−.2173	−.2147	−.2121	−.2096	−.2070	−.2045
.42	−.2019	−.1993	−.1968	−.1942	−.1917	−.1891	−.1866	−.1840	−.1815	−.1789
.43	−.1764	−.1738	−.1713	−.1687	−.1662	−.1637	−.1611	−.1586	−.1560	−.1535
.44	−.1510	−.1484	−.1459	−.1434	−.1408	−.1383	−.1358	−.1332	−.1307	−.1282
.45	−.1257	−.1231	−.1206	−.1181	−.1156	−.1130	−.1105	−.1080	−.1055	−.1030
.46	−.1004	−.0979	−.0954	−.0929	−.0904	−.0878	−.0853	−.0828	−.0803	−.0778
.47	−.0753	−.0728	−.0702	−.0677	−.0652	−.0627	−.0602	−.0577	−.0552	−.0527
.48	−.0502	−.0476	−.0451	−.0426	−.0401	−.0376	−.0351	−.0326	−.0301	−.0276
.49	−.0251	−.0226	−.0201	−.0175	−.0150	−.0125	−.0100	−.0075	−.0050	−.0025

TABLE A2 Normal Distribution (*continued*)

p	.000	.001	.002	.003	.004	.005	.006	.007	.008	.009
.50	.0000	.0025	.0050	.0075	.0100	.0125	.0150	.0175	.0201	.0226
.51	.0251	.0276	.0301	.0326	.0351	.0376	.0401	.0426	.0451	.0476
.52	.0502	.0527	.0552	.0577	.0602	.0627	.0652	.0677	.0702	.0728
.53	.0753	.0778	.0803	.0828	.0853	.0878	.0904	.0929	.0954	.0979
.54	.1004	.1030	.1055	.1080	.1105	.1130	.1156	.1181	.1206	.1231
.55	.1257	.1282	.1307	.1332	.1358	.1383	.1408	.1434	.1459	.1484
.56	.1510	.1535	.1560	.1586	.1611	.1637	.1662	.1687	.1713	.1738
.57	.1764	.1789	.1815	.1840	.1866	.1891	.1917	.1942	.1968	.1993
.58	.2019	.2045	.2070	.2096	.2121	.2147	.2173	.2198	.2224	.2250
.59	.2275	.2301	.2327	.2353	.2378	.2404	.2430	.2456	.2482	.2508
.60	.2533	.2559	.2585	.2611	.2637	.2663	.2689	.2715	.2741	.2767
.61	.2793	.2819	.2845	.2871	.2898	.2924	.2950	.2976	.3002	.3029
.62	.3055	.3081	.3107	.3134	.3160	.3186	.3213	.3239	.3266	.3292
.63	.3319	.3345	.3372	.3398	.3425	.3451	.3478	.3505	.3531	.3558
.64	.3585	.3611	.3638	.3665	.3692	.3719	.3745	.3772	.3799	.3826
.65	.3853	.3880	.3907	.3934	.3961	.3989	.4016	.4043	.4070	.4097
.66	.4125	.4152	.4179	.4207	.4234	.4261	.4289	.4316	.4344	.4372
.67	.4399	.4427	.4454	.4482	.4510	.4538	.4565	.4593	.4621	.4649
.68	.4677	.4705	.4733	.4761	.4789	.4817	.4845	.4874	.4902	.4930
.69	.4959	.4987	.5015	.5044	.5072	.5101	.5129	.5158	.5187	.5215
.70	.5244	.5273	.5302	.5330	.5359	.5388	.5417	.5446	.5476	.5505
.71	.5534	.5563	.5592	.5622	.5651	.5681	.5710	.5740	.5769	.5799
.72	.5828	.5858	.5888	.5918	.5948	.5978	.6008	.6038	.6068	.6098
.73	.6128	.6158	.6189	.6219	.6250	.6280	.6311	.6341	.6372	.6403
.74	.6433	.6464	.6495	.6526	.6557	.6588	.6620	.6651	.6682	.6713
.75	.6745	.6776	.6808	.6840	.6871	.6903	.6935	.6967	.6999	.7031
.76	.7063	.7095	.7128	.7160	.7192	.7225	.7257	.7290	.7323	.7356
.77	.7388	.7421	.7454	.7488	.7521	.7554	.7588	.7621	.7655	.7688
.78	.7722	.7756	.7790	.7824	.7858	.7892	.7926	.7961	.7995	.8030
.79	.8064	.8099	.8134	.8169	.8204	.8239	.8274	.8310	.8345	.8381
.80	.8416	.8452	.8488	.8524	.8560	.8596	.8633	.8669	.8705	.8742
.81	.8779	.8816	.8853	.8890	.8927	.8965	.9002	.9040	.9078	.9116
.82	.9154	.9192	.9230	.9269	.9307	.9346	.9385	.9424	.9463	.9502
.83	.9542	.9581	.9621	.9661	.9701	.9741	.9782	.9822	.9863	.9904
.84	.9945	.9986	1.0027	1.0069	1.0110	1.0152	1.0194	1.0237	1.0279	1.0322
.85	1.0364	1.0407	1.0450	1.0494	1.0537	1.0581	1.0625	1.0669	1.0714	1.0758
.86	1.0803	1.0848	1.0893	1.0939	1.0985	1.1031	1.1077	1.1123	1.1170	1.1217
.87	1.1264	1.1311	1.1359	1.1407	1.1455	1.1503	1.1552	1.1601	1.1650	1.1700
.88	1.1750	1.1800	1.1850	1.1901	1.1952	1.2004	1.2055	1.2107	1.2160	1.2212
.89	1.2265	1.2319	1.2372	1.2426	1.2481	1.2536	1.2591	1.2646	1.2702	1.2759
.90	1.2816	1.2873	1.2930	1.2988	1.3047	1.3106	1.3165	1.3225	1.3285	1.3346
.91	1.3408	1.3469	1.3532	1.3595	1.3658	1.3722	1.3787	1.3852	1.3917	1.3984
.92	1.4051	1.4118	1.4187	1.4255	1.4325	1.4395	1.4466	1.4538	1.4611	1.4684
.93	1.4758	1.4833	1.4909	1.4985	1.5063	1.5141	1.5220	1.5301	1.5382	1.5464
.94	1.5548	1.5632	1.5718	1.5805	1.5893	1.5982	1.6072	1.6164	1.6258	1.6352
.95	1.6449	1.6546	1.6646	1.6747	1.6849	1.6954	1.7060	1.7169	1.7279	1.7392
.96	1.7507	1.7624	1.7744	1.7866	1.7991	1.8119	1.8250	1.8384	1.8522	1.8663
.97	1.8808	1.8957	1.9110	1.9268	1.9431	1.9600	1.9774	1.9954	2.0141	2.0335
.98	2.0537	2.0749	2.0969	2.1201	2.1444	2.1701	2.1973	2.2262	2.2571	2.2904
.99	2.3263	2.3656	2.4089	2.4573	2.5121	2.5758	2.6521	2.7478	2.8782	3.0902

Source. Table Generated by R. L. Iman.

$P(T \leqslant x)$

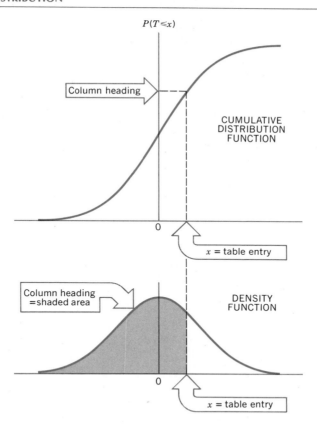

Column heading

CUMULATIVE
DISTRIBUTION
FUNCTION

0

x = table entry

Column heading
=shaded area

DENSITY
FUNCTION

0

x = table entry

TABLE A3 Student's *t*-Distribution

DF					Quantiles					
	.60	.75	.90	.95	.975	.99	.995	.999	.9995	.9999
1	.3249	1.0000	3.0777	6.3138	12.706	31.821	63.657	318.31	636.62	3183.1
2	.2887	.8165	1.8856	2.9200	4.3027	6.9646	9.9248	22.327	31.599	70.700
3	.2767	.7649	1.6377	2.3534	3.1824	4.5407	5.8409	10.215	12.924	22.204
4	.2707	.7407	1.5332	2.1318	2.7764	3.7469	4.6041	7.1732	8.6103	13.034
5	.2672	.7267	1.4759	2.0150	2.5706	3.3649	4.0321	5.8934	6.8688	9.6776
6	.2648	.7176	1.4398	1.9432	2.4469	3.1427	3.7074	5.2076	5.9588	8.0248
7	.2632	.7111	1.4149	1.8946	2.3646	2.9980	3.4995	4.7853	5.4079	7.0634
8	.2619	.7064	1.3968	1.8595	2.3060	2.8965	3.3554	4.5008	5.0413	6.4420
9	.2610	.7027	1.3830	1.8331	2.2622	2.8214	3.2498	4.2968	4.7809	6.0101
10	.2602	.6998	1.3722	1.8125	2.2281	2.7638	3.1693	4.1437	4.5869	5.6938
11	.2596	.6974	1.3634	1.7959	2.2010	2.7181	3.1058	4.0247	4.4370	5.4528
12	.2590	.6955	1.3562	1.7823	2.1788	2.6810	3.0545	3.9296	4.3178	5.2633
13	.2586	.6938	1.3502	1.7709	2.1604	2.6503	3.0123	3.8520	4.2208	5.1106
14	.2582	.6924	1.3450	1.7613	2.1448	2.6245	2.9768	3.7874	4.1405	4.9850
15	.2579	.6912	1.3406	1.7531	2.1314	2.6025	2.9467	3.7328	4.0728	4.8800
16	.2576	.6901	1.3368	1.7459	2.1199	2.5835	2.9208	3.6862	4.0150	4.7909
17	.2573	.6892	1.3334	1.7396	2.1098	2.5669	2.8982	3.6458	3.9651	4.7144
18	.2571	.6884	1.3304	1.7341	2.1009	2.5524	2.8784	3.6105	3.9216	4.6480
19	.2569	.6876	1.3277	1.7291	2.0930	2.5395	2.8609	3.5794	3.8834	4.5899
20	.2567	.6870	1.3253	1.7247	2.0860	2.5280	2.8453	3.5518	3.8495	4.5385
21	.2566	.6864	1.3232	1.7207	2.0796	2.5176	2.8314	3.5272	3.8193	4.4929
22	.2564	.6858	1.3212	1.7171	2.0739	2.5083	2.8188	3.5050	3.7921	4.4520
23	.2563	.6853	1.3195	1.7139	2.0687	2.4999	2.8073	3.4850	3.7676	4.4152
24	.2562	.6848	1.3178	1.7109	2.0639	2.4922	2.7969	3.4668	3.7454	4.3819
25	.2561	.6844	1.3163	1.7081	2.0595	2.4851	2.7874	3.4502	3.7251	4.3517
26	.2560	.6840	1.3150	1.7056	2.0555	2.4786	2.7787	3.4350	3.7066	4.3240
27	.2559	.6837	1.3137	1.7033	2.0518	2.4727	2.7707	3.4210	3.6896	4.2987
28	.2558	.6834	1.3125	1.7011	2.0484	2.4671	2.7633	3.4082	3.6739	4.2754
29	.2557	.6830	1.3114	1.6991	2.0452	2.4620	2.7564	3.3962	3.6594	4.2539
30	.2556	.6828	1.3104	1.6973	2.0423	2.4573	2.7500	3.3852	3.6460	4.2340
31	.2555	.6825	1.3095	1.6955	2.0395	2.4528	2.7440	3.3749	3.6335	4.2155
32	.2555	.6822	1.3086	1.6939	2.0369	2.4487	2.7385	3.3653	3.6218	4.1983
33	.2554	.6820	1.3077	1.6924	2.0345	2.4448	2.7333	3.3563	3.6109	4.1822
34	.2553	.6818	1.3070	1.6909	2.0322	2.4411	2.7284	3.3479	3.6007	4.1672
35	.2553	.6816	1.3062	1.6896	2.0301	2.4377	2.7238	3.3400	3.5911	4.1531
36	.2552	.6814	1.3055	1.6883	2.0281	2.4345	2.7195	3.3326	3.5821	4.1399
37	.2552	.6812	1.3049	1.6871	2.0262	2.4314	2.7154	3.3256	3.5737	4.1275
38	.2551	.6810	1.3042	1.6860	2.0244	2.4286	2.7116	3.3190	3.5657	4.1158
39	.2551	.6808	1.3036	1.6849	2.0227	2.4258	2.7079	3.3128	3.5581	4.1047
40	.2550	.6807	1.3031	1.6839	2.0211	2.4233	2.7045	3.3069	3.5510	4.0942
41	.2550	.6805	1.3025	1.6829	2.0195	2.4208	2.7012	3.3013	3.5442	4.0843
42	.2550	.6804	1.3020	1.6820	2.0181	2.4185	2.6981	3.2960	3.5377	4.0749
43	.2549	.6802	1.3016	1.6811	2.0167	2.4163	2.6951	3.2909	3.5316	4.0659
44	.2549	.6801	1.3011	1.6802	2.0154	2.4141	2.6923	3.2861	3.5258	4.0574
45	.2549	.6800	1.3006	1.6794	2.0141	2.4121	2.6896	3.2815	3.5203	4.0493
46	.2548	.6799	1.3002	1.6787	2.0129	2.4102	2.6870	3.2771	3.5150	4.0416
47	.2548	.6797	1.2998	1.6779	2.0117	2.4083	2.6846	3.2729	3.5099	4.0343
48	.2548	.6796	1.2994	1.6772	2.0106	2.4066	2.6822	3.2689	3.5051	4.0272
49	.2547	.6795	1.2991	1.6766	2.0096	2.4049	2.6800	3.2651	3.5004	4.0205
50	.2547	.6794	1.2987	1.6759	2.0086	2.4033	2.6778	3.2614	3.4960	4.0140

TABLE A3 Student's *t*-Distribution (*continued*)

DF	.60	.75	.90	.95	.975	.99	.995	.999	.9995	.9999
					Quantiles					
51	.2547	.6793	1.2984	1.6753	2.0076	2.4017	2.6757	3.2579	3.4918	4.0079
52	.2546	.6792	1.2980	1.6747	2.0066	2.4002	2.6737	3.2545	3.4877	4.0020
53	.2546	.6791	1.2977	1.6741	2.0057	2.3988	2.6718	3.2513	3.4838	3.9963
54	.2546	.6791	1.2974	1.6736	2.0049	2.3974	2.6700	3.2481	3.4800	3.9908
55	.2546	.6790	1.2971	1.6730	2.0040	2.3961	2.6682	3.2451	3.4764	3.9856
56	.2546	.6789	1.2969	1.6725	2.0032	2.3948	2.6665	3.2423	3.4729	3.9805
57	.2545	.6788	1.2966	1.6720	2.0025	2.3936	2.6649	3.2395	3.4696	3.9757
58	.2545	.6787	1.2963	1.6716	2.0017	2.3924	2.6633	3.2368	3.4663	3.9710
59	.2545	.6787	1.2961	1.6711	2.0010	2.3912	2.6618	3.2342	3.4632	3.9664
60	.2545	.6786	1.2958	1.6706	2.0003	2.3901	2.6603	3.2317	3.4602	3.9621
61	.2545	.6785	1.2956	1.6702	1.9996	2.3890	2.6589	3.2293	3.4573	3.9579
62	.2544	.6785	1.2954	1.6698	1.9990	2.3880	2.6575	3.2270	3.4545	3.9538
63	.2544	.6784	1.2951	1.6694	1.9983	2.3870	2.6561	3.2247	3.4518	3.9499
64	.2544	.6783	1.2949	1.6690	1.9977	2.3860	2.6549	3.2225	3.4491	3.9461
65	.2544	.6783	1.2947	1.6686	1.9971	2.3851	2.6536	3.2204	3.4466	3.9424
66	.2544	.6782	1.2945	1.6683	1.9966	2.3842	2.6524	3.2184	3.4441	3.9389
67	.2544	.6782	1.2943	1.6679	1.9960	2.3833	2.6512	3.2164	3.4417	3.9354
68	.2543	.6781	1.2941	1.6676	1.9955	2.3824	2.6501	3.2145	3.4394	3.9321
69	.2543	.6781	1.2939	1.6672	1.9949	2.3816	2.6490	3.2126	3.4372	3.9288
70	.2543	.6780	1.2938	1.6669	1.9944	2.3808	2.6479	3.2108	3.4350	3.9257
71	.2543	.6780	1.2936	1.6666	1.9939	2.3800	2.6469	3.2090	3.4329	3.9226
72	.2543	.6779	1.2934	1.6663	1.9935	2.3793	2.6459	3.2073	3.4308	3.9197
73	.2543	.6779	1.2933	1.6660	1.9930	2.3785	2.6449	3.2057	3.4289	3.9168
74	.2543	.6778	1.2931	1.6657	1.9925	2.3778	2.6439	3.2041	3.4269	3.9140
75	.2542	.6778	1.2929	1.6654	1.9921	2.3771	2.6430	3.2025	3.4250	3.9113
76	.2542	.6777	1.2928	1.6652	1.9917	2.3764	2.6421	3.2010	3.4232	3.9086
77	.2542	.6777	1.2926	1.6649	1.9913	2.3758	2.6412	3.1995	3.4214	3.9061
78	.2542	.6776	1.2925	1.6646	1.9908	2.3751	2.6403	3.1980	3.4197	3.9036
79	.2542	.6776	1.2924	1.6644	1.9905	2.3745	2.6395	3.1966	3.4180	3.9011
80	.2542	.6776	1.2922	1.6641	1.9901	2.3739	2.6387	3.1953	3.4163	3.8988
81	.2542	.6775	1.2921	1.6639	1.9897	2.3733	2.6379	3.1939	3.4147	3.8964
82	.2542	.6775	1.2920	1.6636	1.9893	2.3727	2.6371	3.1926	3.4132	3.8942
83	.2542	.6775	1.2918	1.6634	1.9890	2.3721	2.6364	3.1913	3.4116	3.8920
84	.2542	.6774	1.2917	1.6632	1.9886	2.3716	2.6356	3.1901	3.4102	3.8899
85	.2541	.6774	1.2916	1.6630	1.9883	2.3710	2.6349	3.1889	3.4087	3.8878
86	.2541	.6774	1.2915	1.6628	1.9879	2.3705	2.6342	3.1877	3.4073	3.8857
87	.2541	.6773	1.2914	1.6626	1.9876	2.3700	2.6335	3.1866	3.4059	3.8837
88	.2541	.6773	1.2912	1.6624	1.9873	2.3695	2.6329	3.1854	3.4045	3.8818
89	.2541	.6773	1.2911	1.6622	1.9870	2.3690	2.6322	3.1843	3.4032	3.8799
90	.2541	.6772	1.2910	1.6620	1.9867	2.3685	2.6316	3.1833	3.4019	3.8780
91	.2541	.6772	1.2909	1.6618	1.9864	2.3680	2.6309	3.1822	3.4007	3.8762
92	.2541	.6772	1.2908	1.6616	1.9861	2.3676	2.6303	3.1812	3.3994	3.8745
93	.2541	.6771	1.2907	1.6614	1.9858	2.3671	2.6297	3.1802	3.3982	3.8727
94	.2541	.6771	1.2906	1.6612	1.9855	2.3667	2.6291	3.1792	3.3971	3.8710
95	.2541	.6771	1.2905	1.6611	1.9853	2.3662	2.6286	3.1782	3.3959	3.8694
96	.2541	.6771	1.2904	1.6609	1.9850	2.3658	2.6280	3.1773	3.3948	3.8678
97	.2540	.6770	1.2903	1.6607	1.9847	2.3654	2.6275	3.1764	3.3937	3.8662
98	.2540	.6770	1.2902	1.6606	1.9845	2.3650	2.6269	3.1755	3.3926	3.8646
99	.2540	.6770	1.2902	1.6604	1.9842	2.3646	2.6264	3.1746	3.3915	3.8631
100	.2540	.6770	1.2901	1.6602	1.9840	2.3642	2.6259	3.1737	3.3905	3.8616
∞	.2533	.6745	1.2816	1.6449	1.9600	2.3263	2.5758	3.0902	3.2905	3.7190

Source: Table generated by R. L. Iman.

$P(T \leq x)$ for 10 d.f.

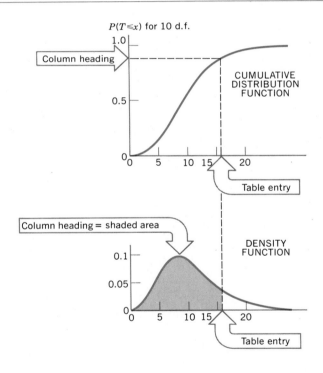

TABLE A4 Chi-Square Distribution

DF	.60	.75	.90	.95	.975	.99	.995	.999	.9995	.9999
					Quantiles					
1	.708	1.323	2.706	3.841	5.024	6.635	7.879	10.83	12.12	15.14
2	1.833	2.773	4.605	5.991	7.378	9.210	10.60	13.82	15.20	18.42
3	2.946	4.108	6.251	7.815	9.348	11.34	12.84	16.27	17.73	21.11
4	4.045	5.385	7.779	9.488	11.14	13.28	14.86	18.47	20.00	23.51
5	5.132	6.626	9.236	11.07	12.83	15.09	16.75	20.52	22.11	25.74
6	6.211	7.841	10.64	12.59	14.45	16.81	18.55	22.46	24.10	27.86
7	7.283	9.037	12.02	14.07	16.01	18.48	20.28	24.32	26.02	29.88
8	8.351	10.22	13.36	15.51	17.53	20.09	21.95	26.12	27.87	31.83
9	9.414	11.39	14.68	16.92	19.02	21.67	23.59	27.88	29.67	33.72
10	10.47	12.55	15.99	18.31	20.48	23.21	25.19	29.59	31.42	35.56
11	11.53	13.70	17.28	19.68	21.92	24.72	26.76	31.26	33.14	37.37
12	12.58	14.85	18.55	21.03	23.34	26.22	28.30	32.91	34.82	39.13
13	13.64	15.98	19.81	22.36	24.74	27.69	29.82	34.53	36.48	40.87
14	14.69	17.12	21.06	23.68	26.12	29.14	31.32	36.12	38.11	42.58
15	15.73	18.25	22.31	25.00	27.49	30.58	32.80	37.70	39.72	44.26
16	16.78	19.37	23.54	26.30	28.85	32.00	34.27	39.25	41.31	45.92
17	17.82	20.49	24.77	27.59	30.19	33.41	35.72	40.79	42.88	47.57
18	18.87	21.60	25.99	28.87	31.53	34.81	37.16	42.31	44.43	49.19
19	19.91	22.72	27.20	30.14	32.85	36.19	38.58	43.82	45.97	50.80
20	20.95	23.83	28.41	31.41	34.17	37.57	40.00	45.31	47.50	52.39
21	21.99	24.93	29.62	32.67	35.48	38.93	41.40	46.80	49.01	53.96
22	23.03	26.04	30.81	33.92	36.78	40.29	42.80	48.27	50.51	55.52
23	24.07	27.14	32.01	35.17	38.08	41.64	44.18	49.73	52.00	57.07
24	25.11	28.24	33.20	36.42	39.36	42.98	45.56	51.18	53.48	58.61
25	26.14	29.34	34.38	37.65	40.65	44.31	46.93	52.62	54.95	60.14
26	27.18	30.43	35.56	38.89	41.92	45.64	48.29	54.05	56.41	61.66
27	28.21	31.53	36.74	40.11	43.19	46.96	49.64	55.48	57.86	63.16
28	29.25	32.62	37.92	41.34	44.46	48.28	50.99	56.89	59.30	64.66
29	30.28	33.71	39.09	42.56	45.72	49.59	52.34	58.30	60.73	66.15
30	31.32	34.80	40.26	43.77	46.98	50.89	53.67	59.70	62.16	67.63
31	32.35	35.89	41.42	44.99	48.23	52.19	55.00	61.10	63.58	69.11
32	33.38	36.97	42.58	46.19	49.48	53.49	56.33	62.49	65.00	70.57
33	34.41	38.06	43.75	47.40	50.73	54.78	57.65	63.87	66.40	72.03
34	35.44	39.14	44.90	48.60	51.97	56.06	58.96	65.25	67.80	73.48
35	36.47	40.22	46.06	49.80	53.20	57.34	60.27	66.62	69.20	74.93
36	37.50	41.30	47.21	51.00	54.44	58.62	61.58	67.99	70.59	76.36
37	38.53	42.38	48.36	52.19	55.67	59.89	62.88	69.35	71.97	77.80
38	39.56	43.46	49.51	53.38	56.90	61.16	64.18	70.70	73.35	79.22
39	40.59	44.54	50.66	54.57	58.12	62.43	65.48	72.05	74.73	80.65
40	41.62	45.62	51.81	55.76	59.34	63.69	66.77	73.40	76.09	82.06
41	42.65	46.69	52.95	56.94	60.56	64.95	68.05	74.74	77.46	83.47
42	43.68	47.77	54.09	58.12	61.78	66.21	69.34	76.08	78.82	84.88
43	44.71	48.84	55.23	59.30	62.99	67.46	70.62	77.42	80.18	86.28
44	45.73	49.91	56.37	60.48	64.20	68.71	71.89	78.75	81.53	87.68
45	46.76	50.98	57.51	61.66	65.41	69.96	73.17	80.08	82.88	89.07
46	47.79	52.06	58.64	62.83	66.62	71.20	74.44	81.40	84.22	90.46
47	48.81	53.13	59.77	64.00	67.82	72.44	75.70	82.72	85.56	91.84
48	49.84	54.20	60.91	65.17	69.02	73.68	76.97	84.04	86.90	93.22
49	50.87	55.27	62.04	66.34	70.22	74.92	78.23	85.35	88.23	94.60
50	51.89	56.33	63.17	67.50	71.42	76.15	79.49	86.66	89.56	95.97

TABLE A4 Chi-Square Distribution *(continued)*

DF					Quantiles					
	.60	.75	.90	.95	.975	.99	.995	.999	.9995	.9999
51	52.92	57.40	64.30	68.67	72.62	77.39	80.75	87.97	90.89	97.34
52	53.94	58.47	65.42	69.83	73.81	78.62	82.00	89.27	92.21	98.70
53	54.97	59.53	66.55	70.99	75.00	79.84	83.25	90.57	93.53	100.1
54	55.99	60.60	67.67	72.15	76.19	81.07	84.50	91.87	94.85	101.4
55	57.02	61.66	68.80	73.31	77.38	82.29	85.75	93.17	96.16	102.8
56	58.04	62.73	69.92	74.47	78.57	83.51	86.99	94.46	97.47	104.1
57	59.06	63.79	71.04	75.62	79.75	84.73	88.24	95.75	98.78	105.5
58	60.09	64.86	72.16	76.78	80.94	85.95	89.48	97.04	100.1	106.8
59	61.11	65.92	73.28	77.93	82.12	87.17	90.72	98.32	101.4	108.2
60	62.13	66.98	74.40	79.08	83.30	88.38	91.95	99.61	102.7	109.5
61	63.16	68.04	75.51	80.23	84.48	89.59	93.19	100.9	104.0	110.8
62	64.18	69.10	76.63	81.38	85.65	90.80	94.42	102.2	105.3	112.2
63	65.20	70.16	77.75	82.53	86.83	92.01	95.65	103.4	106.6	113.5
64	66.23	71.23	78.86	83.68	88.00	93.22	96.88	104.7	107.9	114.8
65	67.25	72.28	79.97	84.82	89.18	94.42	98.11	106.0	109.2	116.2
66	68.27	73.34	81.09	85.96	90.35	95.63	99.33	107.3	110.5	117.5
67	69.29	74.40	82.20	87.11	91.52	96.83	100.6	108.5	111.7	118.8
68	70.32	75.46	83.31	88.25	92.69	98.03	101.8	109.8	113.0	120.1
69	71.34	76.52	84.42	89.39	93.86	99.23	103.0	111.1	114.3	121.4
70	72.36	77.58	85.53	90.53	95.02	100.4	104.2	112.3	115.6	122.8
71	73.38	78.63	86.64	91.67	96.19	101.6	105.4	113.6	116.9	124.1
72	74.40	79.69	87.74	92.81	97.35	102.8	106.6	114.8	118.1	125.4
73	75.42	80.75	88.85	93.95	98.52	104.0	107.9	116.1	119.4	126.7
74	76.44	81.80	89.96	95.08	99.68	105.2	109.1	117.3	120.7	128.0
75	77.46	82.86	91.06	96.22	100.8	106.4	110.3	118.6	121.9	129.3
76	78.48	83.91	92.17	97.35	102.0	107.6	111.5	119.9	123.2	130.6
77	79.51	84.97	93.27	98.48	103.2	108.8	112.7	121.1	124.5	131.9
78	80.53	86.02	94.37	99.62	104.3	110.0	113.9	122.3	125.7	133.2
79	81.55	87.08	95.48	100.7	105.5	111.1	115.1	123.6	127.0	134.5
80	82.57	88.13	96.58	101.9	106.6	112.3	116.3	124.8	128.3	135.8
81	83.59	89.18	97.68	103.0	107.8	113.5	117.5	126.1	129.5	137.1
82	84.61	90.24	98.78	104.1	108.9	114.7	118.7	127.3	130.8	138.4
83	85.63	91.29	99.88	105.3	110.1	115.9	119.9	128.6	132.0	139.7
84	86.65	92.34	101.0	106.4	111.2	117.1	121.1	129.8	133.3	140.9
85	87.67	93.39	102.1	107.5	112.4	118.2	122.3	131.0	134.5	142.2
86	88.68	94.45	103.2	108.6	113.5	119.4	123.5	132.3	135.8	143.5
87	89.70	95.50	104.3	109.8	114.7	120.6	124.7	133.5	137.0	144.8
88	90.72	96.55	105.4	110.9	115.8	121.8	125.9	134.7	138.3	146.1
89	91.74	97.60	106.5	112.0	117.0	122.9	127.1	136.0	139.5	147.4
90	92.76	98.65	107.6	113.1	118.1	124.1	128.3	137.2	140.8	148.6
91	93.78	99.70	108.7	114.3	119.3	125.3	129.5	138.4	142.0	149.9
92	94.80	100.8	109.8	115.4	120.4	126.5	130.7	139.7	143.3	151.2
93	95.82	101.8	110.9	116.5	121.6	127.6	131.9	140.9	144.5	152.4
94	96.84	102.8	111.9	117.6	122.7	128.8	133.1	142.1	145.8	153.7
95	97.85	103.9	113.0	118.8	123.9	130.0	134.2	143.3	147.0	155.0
96	98.87	104.9	114.1	119.9	125.0	131.1	135.4	144.6	148.2	156.3
97	99.89	106.0	115.2	121.0	126.1	132.3	136.6	145.8	149.5	157.5
98	100.9	107.0	116.3	122.1	127.3	133.5	137.8	147.0	150.7	158.8
99	101.9	108.1	117.4	123.2	128.4	134.6	139.0	148.2	151.9	160.1
100	102.9	109.1	118.5	124.3	129.6	135.8	140.2	149.4	153.2	161.3
z_p	.2533	.6745	1.2816	1.6449	1.9600	2.3263	2.5758	3.0902	3.2905	3.7190

For $DF > 100$ use the approximation $x_p = (1/2)(z_p + \sqrt{2k-1})^2$, or, the more accurate, $x_p = k(1 - 2/(9k) + z_p\sqrt{2/(9k)})^3$, where $k = DF$ and z_p is the value from the standardized normal distribution shown in the bottom of the table.
Source: Table generated by R. L. Iman.

$P(F \leq x)$ for (10,10) d.f.

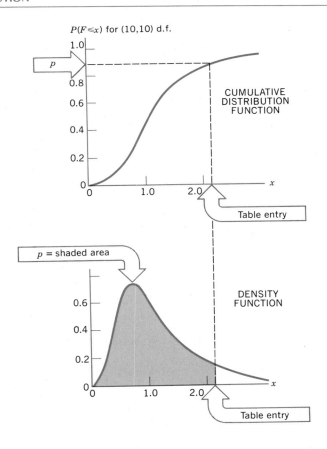

CUMULATIVE
DISTRIBUTION
FUNCTION

Table entry

p = shaded area

DENSITY
FUNCTION

Table entry

TABLE A5 The F Distribution

k_2	p	$k_1:$ 1	2	3	4	5	6	7	8	9	10	12	15	20	24	30	40	60	120
1	.75	5.828	7.500	8.200	8.581	8.820	8.983	9.102	9.192	9.263	9.320	9.406	9.493	9.581	9.625	9.670	9.714	9.759	9.804
	.90	39.86	49.50	53.59	55.83	57.24	58.20	58.91	59.44	59.86	60.19	60.71	61.22	61.74	62.00	62.26	62.53	62.79	63.06
	.95	161.4	199.5	215.7	224.6	230.2	234.0	236.8	238.9	240.5	241.9	243.9	245.9	248.0	249.1	250.1	251.1	252.2	253.3
	.975	647.8	799.5	864.2	899.6	921.8	937.1	948.2	956.7	963.3	968.6	976.7	984.9	993.1	997.2	1001.	1006.	1010.	1014.
	.99	4052.	5000.	5403.	5625.	5764.	5859.	5928.	5981.	6022.	6056.	6106.	6157.	6209.	6235.	6261.	6287.	6313.	6339.
2	.75	2.571	3.000	3.153	3.232	3.280	3.312	3.335	3.353	3.366	3.377	3.393	3.410	3.426	3.435	3.443	3.451	3.459	3.468
	.90	8.526	9.000	9.162	9.243	9.293	9.326	9.349	9.367	9.381	9.392	9.408	9.425	9.441	9.450	9.458	9.466	9.475	9.483
	.95	18.51	19.00	19.16	19.25	19.30	19.33	19.35	19.37	19.38	19.40	19.41	19.43	19.45	19.45	19.46	19.47	19.48	19.49
	.975	38.51	39.00	39.17	39.25	39.30	39.33	39.36	39.37	39.39	39.40	39.41	39.43	39.45	39.46	39.46	39.47	39.48	39.49
	.99	98.50	99.00	99.17	99.25	99.30	99.33	99.36	99.37	99.39	99.40	99.42	99.43	99.45	99.46	99.47	99.47	99.48	99.49
	.999	998.5	999.0	999.2	999.2	999.3	999.3	999.4	999.4	999.4	999.4	999.4	999.4	999.4	999.5	999.5	999.5	999.5	999.5
3	.75	2.024	2.280	2.356	2.390	2.409	2.422	2.430	2.436	2.441	2.445	2.450	2.455	2.460	2.463	2.465	2.467	2.470	2.472
	.90	5.538	5.462	5.391	5.343	5.309	5.285	5.266	5.252	5.240	5.230	5.216	5.200	5.184	5.176	5.168	5.160	5.151	5.143
	.95	10.13	9.552	9.277	9.117	9.013	8.941	8.887	8.845	8.812	8.786	8.745	8.703	8.660	8.639	8.617	8.594	8.572	8.549
	.975	17.44	16.04	15.44	15.10	14.88	14.73	14.62	14.54	14.47	14.42	14.34	14.25	14.17	14.12	14.08	14.04	13.99	13.95
	.99	34.12	30.82	29.46	28.71	28.24	27.91	27.67	27.49	27.35	27.23	27.05	26.87	26.69	26.60	26.50	26.41	26.32	26.22
	.999	167.0	148.5	141.1	137.1	134.6	132.8	131.6	130.6	129.9	129.2	128.3	127.4	126.4	125.9	125.4	125.0	124.5	124.0
4	.75	1.807	2.000	2.047	2.064	2.072	2.077	2.079	2.080	2.081	2.082	2.083	2.083	2.083	2.083	2.082	2.082	2.082	2.081
	.90	4.545	4.325	4.191	4.107	4.051	4.010	3.979	3.955	3.936	3.920	3.896	3.870	3.844	3.831	3.817	3.804	3.790	3.775
	.95	7.709	6.944	6.591	6.388	6.256	6.163	6.094	6.041	5.999	5.964	5.912	5.858	5.803	5.774	5.746	5.717	5.688	5.658
	.975	12.22	10.65	9.979	9.605	9.364	9.197	9.074	8.980	8.905	8.844	8.751	8.657	8.560	8.511	8.461	8.411	8.360	8.309
	.99	21.20	18.00	16.69	15.98	15.52	15.21	14.98	14.80	14.66	14.55	14.37	14.20	14.02	13.93	13.84	13.75	13.65	13.56
	.999	74.14	61.25	56.18	53.44	51.71	50.53	49.66	49.00	48.47	48.05	47.41	46.76	46.10	45.77	45.43	45.09	44.75	44.40
5	.75	1.692	1.853	1.884	1.893	1.895	1.894	1.894	1.892	1.891	1.890	1.888	1.885	1.882	1.880	1.878	1.876	1.874	1.872
	.90	4.060	3.780	3.619	3.520	3.453	3.405	3.368	3.339	3.316	3.297	3.268	3.238	3.207	3.191	3.174	3.157	3.140	3.123
	.95	6.608	5.786	5.409	5.192	5.050	4.950	4.876	4.818	4.772	4.735	4.678	4.619	4.558	4.527	4.496	4.464	4.431	4.398
	.975	10.01	8.434	7.764	7.388	7.146	6.978	6.853	6.757	6.681	6.619	6.525	6.428	6.329	6.278	6.227	6.175	6.123	6.069
	.99	16.26	13.27	12.06	11.39	10.97	10.67	10.46	10.29	10.16	10.05	9.888	9.722	9.553	9.466	9.379	9.291	9.202	9.112
	.999	47.18	37.12	33.20	31.09	29.75	28.83	28.16	27.65	27.24	26.92	26.42	25.91	25.39	25.13	24.87	24.60	24.33	24.06
6	.75	1.621	1.762	1.784	1.787	1.785	1.782	1.779	1.776	1.773	1.771	1.767	1.762	1.757	1.754	1.751	1.748	1.744	1.741
	.90	3.776	3.463	3.289	3.181	3.108	3.055	3.014	2.983	2.958	2.937	2.905	2.871	2.836	2.818	2.800	2.781	2.762	2.742
	.95	5.987	5.143	4.757	4.534	4.387	4.284	4.207	4.147	4.099	4.060	4.000	3.938	3.874	3.841	3.808	3.774	3.740	3.705
	.975	8.813	7.260	6.599	6.227	5.988	5.820	5.695	5.600	5.523	5.461	5.366	5.269	5.168	5.117	5.065	5.012	4.959	4.904
	.99	13.75	10.92	9.780	9.148	8.746	8.466	8.260	8.102	7.976	7.874	7.718	7.559	7.396	7.313	7.229	7.143	7.057	6.969
	.999	35.51	27.00	23.70	21.92	20.80	20.03	19.46	19.03	18.69	18.41	17.99	17.56	17.12	16.90	16.67	16.44	16.21	15.98
7	.75	1.573	1.701	1.717	1.716	1.711	1.706	1.701	1.697	1.693	1.690	1.684	1.678	1.671	1.667	1.663	1.659	1.655	1.650
	.90	3.589	3.257	3.074	2.961	2.883	2.827	2.785	2.752	2.725	2.703	2.668	2.632	2.595	2.575	2.555	2.535	2.514	2.493
	.95	5.591	4.737	4.347	4.120	3.972	3.866	3.787	3.726	3.677	3.637	3.575	3.511	3.445	3.410	3.376	3.340	3.304	3.267
	.975	8.073	6.542	5.890	5.523	5.285	5.119	4.995	4.899	4.823	4.761	4.666	4.568	4.467	4.415	4.362	4.309	4.254	4.199
	.99	12.25	9.547	8.451	7.847	7.460	7.191	6.993	6.840	6.719	6.620	6.469	6.314	6.155	6.074	5.992	5.908	5.824	5.737
	.999	29.25	21.69	18.77	17.20	16.21	15.52	15.02	14.63	14.33	14.08	13.71	13.32	12.93	12.73	12.53	12.33	12.12	11.91

k_2	p	k_1: 1	2	3	4	5	6	7	8	9	10	12	15	20	24	30	40	60	120
8	.75	1.538	1.657	1.668	1.664	1.658	1.651	1.645	1.640	1.635	.631	1.624	1.617	1.609	1.604	1.600	1.595	1.589	1.584
	.90	3.458	3.113	2.924	2.806	2.726	2.668	2.624	2.589	2.561	2.538	2.502	2.464	2.425	2.404	2.383	2.361	2.339	2.316
	.95	5.318	4.459	4.066	3.838	3.687	3.581	3.500	3.438	3.388	3.347	3.284	3.218	3.150	3.115	3.079	3.043	3.005	2.967
	.975	7.571	6.059	5.416	5.053	4.817	4.652	4.529	4.433	4.357	4.295	4.200	4.101	3.999	3.947	3.894	3.840	3.784	3.728
	.99	11.26	8.649	7.591	7.006	6.632	6.371	6.178	6.029	5.911	5.814	5.667	5.515	5.359	5.279	5.198	5.116	5.032	4.946
	.999	25.41	18.49	15.83	14.39	13.48	12.86	12.40	12.05	11.77	11.54	11.19	10.84	10.48	10.30	10.11	9.919	9.727	9.532
9	.75	1.512	1.624	1.632	1.625	1.617	1.609	1.602	1.596	1.591	1.586	1.579	1.570	1.561	1.556	1.551	1.545	1.539	1.533
	.90	3.360	3.006	2.813	2.693	2.611	2.551	2.505	2.469	2.440	2.416	2.379	2.340	2.298	2.277	2.255	2.232	2.208	2.184
	.95	5.117	4.256	3.863	3.633	3.482	3.374	3.293	3.230	3.179	3.137	3.073	3.006	2.936	2.900	2.864	2.826	2.787	2.748
	.975	7.209	5.715	5.078	4.718	4.484	4.320	4.197	4.102	4.026	3.964	3.868	3.769	3.667	3.614	3.560	3.505	3.449	3.392
	.99	10.56	8.022	6.992	6.422	6.057	5.802	5.613	5.467	5.351	5.257	5.111	4.962	4.808	4.729	4.649	4.567	4.483	4.398
	.999	22.86	16.39	13.90	12.56	11.71	11.13	10.70	10.37	10.11	9.894	9.570	9.238	8.898	8.724	8.548	8.369	8.187	8.001
10	.75	1.491	1.598	1.603	1.595	1.585	1.576	1.569	1.562	1.556	1.551	1.543	1.534	1.523	1.518	1.512	1.506	1.499	1.492
	.90	3.285	2.924	2.728	2.605	2.522	2.461	2.414	2.377	2.347	2.323	2.284	2.244	2.201	2.178	2.155	2.132	2.107	2.082
	.95	4.965	4.103	3.708	3.478	3.326	3.217	3.135	3.072	3.020	2.978	2.913	2.845	2.774	2.737	2.700	2.661	2.621	2.580
	.975	6.937	5.456	4.826	4.468	4.236	4.072	3.950	3.855	3.779	3.717	3.621	3.522	3.419	3.365	3.311	3.255	3.198	3.140
	.99	10.04	7.559	6.552	5.994	5.636	5.386	5.200	5.057	4.942	4.849	4.706	4.558	4.405	4.327	4.247	4.165	4.082	3.996
	.999	21.04	14.91	12.55	11.28	10.48	9.926	9.517	9.204	8.956	8.754	8.445	8.129	7.804	7.638	7.469	7.297	7.122	6.944
11	.75	1.475	1.577	1.580	1.570	1.560	1.550	1.542	1.535	1.528	1.523	1.514	1.504	1.493	1.487	1.481	1.474	1.466	1.459
	.90	3.225	2.860	2.660	2.536	2.451	2.389	2.342	2.304	2.274	2.248	2.209	2.167	2.123	2.100	2.076	2.052	2.026	2.000
	.95	4.844	3.982	3.587	3.357	3.204	3.095	3.012	2.948	2.896	2.854	2.788	2.719	2.646	2.609	2.570	2.531	2.490	2.448
	.975	6.724	5.256	4.630	4.275	4.044	3.881	3.759	3.664	3.588	3.526	3.430	3.330	3.226	3.173	3.118	3.061	3.004	2.944
	.99	9.646	7.206	6.217	5.668	5.316	5.069	4.886	4.744	4.632	4.539	4.397	4.251	4.099	4.021	3.941	3.860	3.776	3.690
	.999	19.69	13.81	11.56	10.35	9.578	9.047	8.655	8.355	8.116	7.922	7.626	7.321	7.008	6.847	6.684	6.518	6.348	6.175
12	.75	1.461	1.560	1.561	1.550	1.539	1.529	1.520	1.512	1.505	1.500	1.490	1.480	1.468	1.461	1.454	1.447	1.439	1.431
	.90	3.177	2.807	2.606	2.480	2.394	2.331	2.283	2.245	2.214	2.188	2.147	2.105	2.060	2.036	2.011	1.986	1.960	1.932
	.95	4.747	3.885	3.490	3.259	3.106	2.996	2.913	2.849	2.796	2.753	2.687	2.617	2.544	2.505	2.466	2.426	2.384	2.341
	.975	6.554	5.096	4.474	4.121	3.891	3.728	3.607	3.512	3.436	3.374	3.277	3.177	3.073	3.019	2.963	2.906	2.848	2.787
	.99	9.330	6.927	5.953	5.412	5.064	4.821	4.640	4.499	4.388	4.296	4.155	4.010	3.858	3.780	3.701	3.619	3.535	3.449
	.999	18.64	12.97	10.80	9.633	8.892	8.379	8.001	7.710	7.480	7.292	7.005	6.709	6.405	6.249	6.090	5.928	5.762	5.593
13	.75	1.450	1.545	1.545	1.534	1.521	1.511	1.501	1.493	1.486	1.480	1.470	1.459	1.447	1.440	1.432	1.425	1.416	1.408
	.90	3.136	2.763	2.560	2.434	2.347	2.283	2.234	2.195	2.164	2.138	2.097	2.053	2.007	1.983	1.958	1.931	1.904	1.876
	.95	4.667	3.806	3.411	3.179	3.025	2.915	2.832	2.767	2.714	2.671	2.604	2.533	2.459	2.420	2.380	2.339	2.297	2.252
	.975	6.414	4.965	4.347	3.996	3.767	3.604	3.483	3.388	3.312	3.250	3.153	3.053	2.948	2.893	2.837	2.780	2.720	2.659
	.99	9.074	6.701	5.739	5.205	4.862	4.620	4.441	4.302	4.191	4.100	3.960	3.815	3.665	3.587	3.507	3.425	3.341	3.255
	.999	17.82	12.31	10.21	9.073	8.354	7.856	7.489	7.206	6.982	6.799	6.519	6.231	5.934	5.781	5.626	5.467	5.305	5.138
14	.75	1.440	1.533	1.532	1.519	1.507	1.495	1.485	1.477	1.470	1.463	1.453	1.441	1.428	1.421	1.414	1.405	1.397	1.387
	.90	3.102	2.726	2.522	2.395	2.307	2.243	2.193	2.154	2.122	2.095	2.054	2.010	1.962	1.938	1.912	1.885	1.857	1.828
	.95	4.600	3.739	3.344	3.112	2.958	2.848	2.764	2.699	2.646	2.602	2.534	2.463	2.388	2.349	2.308	2.266	2.223	2.178
	.975	6.298	4.857	4.242	3.892	3.663	3.501	3.380	3.285	3.209	3.147	3.050	2.949	2.844	2.789	2.732	2.674	2.614	2.552
	.99	8.862	6.515	5.564	5.035	4.695	4.456	4.278	4.140	4.030	3.939	3.800	3.656	3.505	3.427	3.348	3.266	3.181	3.094
	.999	17.14	11.78	9.729	8.622	7.922	7.436	7.077	6.802	6.583	6.404	6.130	5.848	5.557	5.407	5.254	5.098	4.938	4.773

TABLE A5 The F Distribution (continued)

k_2	p	k_1: 120	60	40	30	24	20	15	12	10	9	8	7	6	5	4	3	2	1
15	.75	1.370	1.380	1.389	1.397	1.405	1.413	1.426	1.438	1.449	1.456	1.463	1.472	1.482	1.494	1.507	1.520	1.523	1.432
	.90	1.787	1.817	1.845	1.873	1.899	1.924	1.972	2.017	2.059	2.086	2.119	2.158	2.208	2.273	2.361	2.490	2.695	3.073
	.95	2.114	2.160	2.204	2.247	2.288	2.328	2.403	2.475	2.544	2.588	2.641	2.707	2.790	2.901	3.056	3.287	3.682	4.543
	.975	2.461	2.524	2.585	2.644	2.701	2.756	2.862	2.963	3.060	3.123	3.199	3.293	3.415	3.576	3.804	4.153	4.765	6.200
	.99	2.959	3.047	3.132	3.214	3.294	3.372	3.522	3.666	3.805	3.895	4.004	4.142	4.318	4.556	4.893	5.417	6.359	8.683
	.999	4.475	4.638	4.796	4.950	5.101	5.248	5.535	5.812	6.081	6.256	6.471	6.741	7.092	7.567	8.253	9.335	11.34	16.59
16	.75	1.354	1.365	1.374	1.383	1.391	1.399	1.413	1.426	1.437	1.443	1.451	1.460	1.471	1.483	1.497	1.510	1.514	1.425
	.90	1.751	1.782	1.811	1.839	1.866	1.891	1.940	1.985	2.028	2.055	2.088	2.128	2.178	2.244	2.333	2.462	2.668	3.048
	.95	2.059	2.106	2.151	2.194	2.235	2.276	2.352	2.425	2.494	2.538	2.591	2.657	2.741	2.852	3.007	3.239	3.634	4.494
	.975	2.383	2.447	2.509	2.568	2.625	2.681	2.788	2.889	2.986	3.049	3.125	3.219	3.341	3.502	3.729	4.077	4.687	6.115
	.99	2.845	2.933	3.018	3.101	3.181	3.259	3.409	3.553	3.691	3.780	3.890	4.026	4.202	4.437	4.773	5.292	6.226	8.531
	.999	4.226	4.388	4.545	4.697	4.846	4.992	5.274	5.547	5.812	5.984	6.195	6.460	6.805	7.272	7.944	9.006	10.66	16.12
17	.75	1.341	1.351	1.361	1.370	1.379	1.387	1.401	1.414	1.426	1.433	1.441	1.450	1.460	1.473	1.487	1.502	1.506	1.419
	.90	1.719	1.751	1.781	1.809	1.836	1.862	1.912	1.958	2.001	2.028	2.061	2.102	2.152	2.218	2.308	2.437	2.645	3.026
	.95	2.011	2.058	2.104	2.148	2.190	2.230	2.308	2.381	2.450	2.494	2.548	2.614	2.699	2.810	2.965	3.197	3.592	4.451
	.975	2.315	2.380	2.442	2.502	2.560	2.616	2.723	2.825	2.922	2.985	3.061	3.156	3.277	3.438	3.665	4.011	4.619	6.042
	.99	2.746	2.835	2.920	3.003	3.084	3.162	3.312	3.455	3.593	3.682	3.791	3.927	4.102	4.336	4.669	5.185	6.112	8.400
	.999	4.016	4.177	4.332	4.484	4.631	4.775	5.054	5.324	5.584	5.754	5.962	6.223	6.562	7.022	7.683	8.727	10.39	15.72
18	.75	1.328	1.340	1.350	1.359	1.368	1.376	1.391	1.404	1.416	1.423	1.431	1.441	1.452	1.464	1.479	1.494	1.499	1.413
	.90	1.691	1.723	1.754	1.783	1.810	1.837	1.887	1.933	1.977	2.005	2.038	2.079	2.130	2.196	2.286	2.416	2.624	3.007
	.95	1.968	2.017	2.063	2.107	2.150	2.191	2.269	2.342	2.412	2.456	2.510	2.577	2.661	2.773	2.928	3.160	3.555	4.414
	.975	2.256	2.321	2.384	2.445	2.503	2.559	2.667	2.769	2.866	2.929	3.005	3.100	3.221	3.382	3.608	3.954	4.560	5.978
	.99	2.660	2.749	2.835	2.919	2.999	3.077	3.227	3.371	3.508	3.597	3.705	3.841	4.015	4.248	4.579	5.092	6.013	8.285
	.999	3.836	3.996	4.151	4.301	4.447	4.590	4.866	5.132	5.390	5.558	5.763	6.021	6.355	6.808	7.459	8.487	10.16	15.38
19	.75	1.317	1.329	1.339	1.349	1.358	1.367	1.382	1.395	1.407	1.414	1.423	1.432	1.444	1.457	1.472	1.487	1.493	1.408
	.90	1.666	1.699	1.730	1.759	1.787	1.814	1.865	1.912	1.956	1.984	2.017	2.058	2.109	2.176	2.266	2.397	2.606	2.990
	.95	1.930	1.980	2.026	2.071	2.114	2.155	2.234	2.308	2.378	2.423	2.477	2.544	2.628	2.740	2.895	3.127	3.522	4.381
	.975	2.203	2.270	2.333	2.394	2.452	2.509	2.617	2.720	2.817	2.880	2.956	3.051	3.172	3.333	3.559	3.903	4.508	5.922
	.99	2.584	2.674	2.761	2.844	2.925	3.003	3.153	3.297	3.434	3.523	3.631	3.765	3.939	4.171	4.500	5.010	5.926	8.185
	.999	3.680	3.840	3.994	4.143	4.288	4.430	4.704	4.967	5.222	5.388	5.590	5.845	6.175	6.622	7.265	8.280	9.953	15.08
20	.75	1.307	1.319	1.330	1.340	1.349	1.358	1.374	1.387	1.399	1.407	1.415	1.425	1.437	1.450	1.465	1.481	1.487	1.404
	.90	1.643	1.677	1.708	1.738	1.767	1.794	1.845	1.892	1.937	1.965	1.999	2.040	2.091	2.158	2.249	2.380	2.589	2.975
	.95	1.896	1.946	1.994	2.039	2.082	2.124	2.203	2.278	2.348	2.393	2.447	2.514	2.599	2.711	2.866	3.098	3.493	4.351
	.975	2.156	2.223	2.287	2.349	2.408	2.464	2.573	2.676	2.774	2.837	2.913	3.007	3.128	3.289	3.515	3.859	4.461	5.871
	.99	2.517	2.608	2.695	2.778	2.859	2.938	3.088	3.231	3.368	3.457	3.564	3.699	3.871	4.103	4.431	4.938	5.849	8.096
	.999	3.544	3.703	3.856	4.005	4.149	4.290	4.562	4.823	5.075	5.239	5.440	5.692	6.019	6.461	7.096	8.098	9.953	14.82
21	.75	1.298	1.311	1.322	1.332	1.341	1.350	1.366	1.380	1.392	1.400	1.409	1.419	1.430	1.444	1.459	1.475	1.482	1.400
	.90	1.623	1.657	1.689	1.719	1.748	1.776	1.827	1.875	1.920	1.948	1.982	2.023	2.075	2.142	2.233	2.365	2.575	2.961
	.95	1.866	1.916	1.965	2.010	2.054	2.096	2.176	2.250	2.321	2.366	2.420	2.488	2.573	2.685	2.840	3.072	3.467	4.325
	.975	2.114	2.182	2.246	2.308	2.368	2.425	2.534	2.637	2.735	2.798	2.874	2.969	3.090	3.250	3.475	3.819	4.420	5.827
	.99	2.457	2.548	2.636	2.720	2.801	2.880	3.030	3.173	3.310	3.398	3.506	3.640	3.812	4.042	4.369	4.874	5.780	8.017
	.999	3.424	3.583	3.736	3.884	4.027	4.167	4.437	4.696	4.946	5.109	5.308	5.557	5.881	6.318	6.947	7.938	9.772	14.59

TABLE A5 The F Distribution (continued)

k_2	p	k_1: 1	2	3	4	5	6	7	8	9	10	12	15	20	24	30	40	60	120
22	.75	1.396	1.477	1.470	1.454	1.438	1.424	1.413	1.402	1.394	1.386	1.374	1.359	1.343	1.334	1.324	1.314	1.303	1.290
	.90	2.949	2.561	2.351	2.219	2.128	2.060	2.008	1.967	1.933	1.904	1.859	1.811	1.759	1.731	1.702	1.671	1.639	1.604
	.95	4.301	3.443	3.049	2.817	2.661	2.549	2.464	2.397	2.342	2.297	2.226	2.151	2.071	2.028	1.984	1.938	1.889	1.838
	.975	5.786	4.383	3.783	3.440	3.215	3.055	2.934	2.839	2.763	2.700	2.602	2.498	2.389	2.331	2.272	2.210	2.145	2.076
	.99	7.945	5.719	4.817	4.313	3.988	3.758	3.587	3.453	3.346	3.258	3.121	2.978	2.827	2.749	2.667	2.583	2.495	2.403
	.999	14.38	9.612	7.796	6.814	6.191	5.758	5.438	5.190	4.993	4.832	4.583	4.326	4.058	3.919	3.776	3.629	3.476	3.317
23	.75	1.393	1.473	1.466	1.449	1.433	1.419	1.407	1.397	1.388	1.380	1.368	1.353	1.337	1.327	1.318	1.307	1.295	1.282
	.90	2.937	2.549	2.339	2.207	2.115	2.047	1.995	1.953	1.919	1.890	1.845	1.796	1.744	1.716	1.686	1.655	1.622	1.587
	.95	4.279	3.422	3.028	2.796	2.640	2.528	2.442	2.375	2.320	2.275	2.204	2.128	2.048	2.005	1.961	1.914	1.865	1.813
	.975	5.750	4.349	3.750	3.408	3.183	3.023	2.902	2.808	2.731	2.668	2.570	2.466	2.357	2.299	2.239	2.176	2.111	2.041
	.99	7.881	5.664	4.765	4.264	3.939	3.710	3.539	3.406	3.299	3.211	3.074	2.931	2.781	2.702	2.620	2.535	2.447	2.354
	.999	14.20	9.469	7.669	6.696	6.078	5.649	5.331	5.085	4.890	4.730	4.483	4.227	3.961	3.822	3.680	3.533	3.380	3.222
24	.75	1.390	1.470	1.462	1.445	1.428	1.414	1.402	1.392	1.383	1.375	1.362	1.347	1.331	1.321	1.311	1.300	1.289	1.275
	.90	2.927	2.538	2.327	2.195	2.103	2.035	1.983	1.941	1.906	1.877	1.832	1.783	1.730	1.702	1.672	1.641	1.607	1.571
	.95	4.260	3.403	3.009	2.776	2.621	2.508	2.423	2.355	2.300	2.255	2.183	2.108	2.027	1.984	1.939	1.892	1.842	1.790
	.975	5.717	4.319	3.721	3.379	3.155	2.995	2.874	2.779	2.703	2.640	2.541	2.437	2.327	2.269	2.209	2.146	2.080	2.010
	.99	7.823	5.614	4.718	4.218	3.895	3.667	3.496	3.363	3.256	3.168	3.032	2.889	2.738	2.659	2.577	2.492	2.403	2.310
	.999	14.03	9.339	7.554	6.589	5.977	5.550	5.235	4.991	4.797	4.638	4.393	4.139	3.873	3.735	3.593	3.447	3.295	3.136
25	.75	1.387	1.466	1.458	1.441	1.424	1.410	1.398	1.387	1.378	1.370	1.357	1.342	1.325	1.316	1.306	1.294	1.282	1.269
	.90	2.918	2.528	2.317	2.184	2.092	2.024	1.971	1.929	1.895	1.866	1.820	1.771	1.718	1.689	1.659	1.627	1.593	1.557
	.95	4.242	3.385	2.991	2.759	2.603	2.490	2.405	2.337	2.282	2.236	2.165	2.089	2.007	1.964	1.919	1.872	1.822	1.768
	.975	5.686	4.291	3.694	3.353	3.129	2.969	2.848	2.753	2.677	2.613	2.515	2.411	2.300	2.242	2.182	2.118	2.052	1.981
	.99	7.770	5.568	4.675	4.177	3.855	3.627	3.457	3.324	3.217	3.129	2.993	2.850	2.699	2.620	2.538	2.453	2.364	2.270
	.999	13.88	9.223	7.451	6.493	5.885	5.462	5.148	4.906	4.713	4.555	4.312	4.059	3.794	3.657	3.515	3.369	3.217	3.058
30	.75	1.376	1.452	1.443	1.424	1.407	1.392	1.380	1.369	1.359	1.351	1.337	1.321	1.303	1.293	1.282	1.270	1.257	1.242
	.90	2.881	2.489	2.276	2.142	2.049	1.980	1.927	1.884	1.849	1.819	1.773	1.722	1.667	1.638	1.606	1.573	1.538	1.499
	.95	4.171	3.316	2.922	2.690	2.534	2.421	2.334	2.266	2.211	2.165	2.092	2.015	1.932	1.887	1.841	1.792	1.740	1.683
	.975	5.568	4.182	3.589	3.250	3.026	2.867	2.746	2.651	2.575	2.511	2.412	2.307	2.195	2.136	2.074	2.009	1.940	1.866
	.99	7.562	5.390	4.510	4.018	3.699	3.473	3.304	3.173	3.067	2.979	2.843	2.700	2.549	2.469	2.386	2.299	2.208	2.111
	.999	13.29	8.773	7.054	6.125	5.534	5.122	4.817	4.581	4.393	4.239	4.001	3.753	3.493	3.357	3.217	3.072	2.920	2.760
40	.75	1.363	1.435	1.424	1.404	1.386	1.371	1.357	1.345	1.335	1.327	1.312	1.295	1.276	1.265	1.253	1.240	1.225	1.208
	.90	2.835	2.440	2.226	2.091	1.997	1.927	1.873	1.829	1.793	1.763	1.715	1.662	1.605	1.574	1.541	1.506	1.467	1.425
	.95	4.085	3.232	2.839	2.606	2.449	2.336	2.249	2.180	2.124	2.077	2.003	1.924	1.839	1.793	1.744	1.693	1.637	1.577
	.975	5.424	4.051	3.463	3.126	2.904	2.744	2.624	2.529	2.452	2.388	2.288	2.182	2.068	2.007	1.943	1.875	1.803	1.724
	.99	7.314	5.179	4.313	3.828	3.514	3.291	3.124	2.993	2.888	2.801	2.665	2.522	2.369	2.288	2.203	2.114	2.019	1.917
	.999	12.61	8.251	6.595	5.698	5.128	4.731	4.436	4.207	4.024	3.874	3.642	3.400	3.145	3.011	2.872	2.727	2.574	2.410
50	.75	1.355	1.425	1.413	1.393	1.374	1.358	1.344	1.332	1.321	1.312	1.297	1.280	1.259	1.248	1.235	1.221	1.205	1.186
	.90	2.809	2.412	2.197	2.061	1.966	1.895	1.840	1.796	1.760	1.729	1.680	1.627	1.568	1.536	1.502	1.465	1.424	1.379
	.95	4.034	3.183	2.790	2.557	2.400	2.286	2.199	2.130	2.073	2.026	1.952	1.871	1.784	1.737	1.687	1.634	1.576	1.511
	.975	5.340	3.975	3.390	3.054	2.833	2.674	2.553	2.458	2.381	2.317	2.216	2.109	1.993	1.931	1.866	1.796	1.721	1.639
	.99	7.171	5.057	4.199	3.720	3.408	3.186	3.020	2.890	2.785	2.698	2.562	2.419	2.265	2.183	2.098	2.007	1.909	1.803
	.999	12.22	7.956	6.336	5.459	4.901	4.512	4.222	3.998	3.818	3.671	3.443	3.204	2.951	2.817	2.679	2.533	2.378	2.211

TABLE A5 The F Distribution (continued)

k_2	p	k_1: 1	2	3	4	5	6	7	8	9	10	12	15	20	24	30	40	60	120
60	.75	1.349	1.419	1.405	1.385	1.366	1.349	1.335	1.323	1.312	1.303	1.287	1.269	1.248	1.236	1.223	1.208	1.191	1.172
	.90	2.791	2.393	2.177	2.041	1.946	1.875	1.819	1.775	1.738	1.707	1.657	1.603	1.543	1.511	1.476	1.437	1.395	1.348
	.95	4.001	3.150	2.758	2.525	2.368	2.254	2.167	2.097	2.040	1.993	1.917	1.836	1.748	1.700	1.649	1.594	1.534	1.467
	.975	5.286	3.925	3.343	3.008	2.786	2.627	2.507	2.412	2.334	2.270	2.169	2.061	1.944	1.882	1.815	1.744	1.667	1.581
	.99	7.077	4.977	4.126	3.649	3.339	3.119	2.953	2.823	2.718	2.632	2.496	2.352	2.198	2.115	2.028	1.936	1.836	1.726
	.999	11.97	7.768	6.171	5.307	4.757	4.372	4.086	3.865	3.687	3.541	3.315	3.078	2.827	2.694	2.555	2.409	2.252	2.082
70	.75	1.346	1.414	1.400	1.379	1.360	1.343	1.329	1.316	1.305	1.296	1.280	1.262	1.240	1.228	1.214	1.199	1.181	1.161
	.90	2.779	2.380	2.164	2.027	1.931	1.860	1.804	1.760	1.723	1.691	1.641	1.587	1.526	1.493	1.457	1.418	1.374	1.325
	.95	3.978	3.128	2.736	2.503	2.346	2.231	2.143	2.074	2.017	1.969	1.893	1.812	1.722	1.674	1.622	1.566	1.505	1.435
	.975	5.247	3.890	3.309	2.975	2.754	2.595	2.474	2.379	2.302	2.237	2.136	2.028	1.910	1.847	1.779	1.707	1.628	1.539
	.99	7.011	4.922	4.074	3.600	3.291	3.071	2.906	2.777	2.672	2.585	2.450	2.306	2.150	2.067	1.980	1.886	1.785	1.672
	.999	11.80	7.637	6.057	5.201	4.656	4.275	3.992	3.773	3.596	3.452	3.227	2.991	2.741	2.608	2.469	2.322	2.164	1.991
80	.75	1.343	1.411	1.396	1.375	1.355	1.338	1.324	1.311	1.300	1.291	1.275	1.256	1.234	1.222	1.208	1.192	1.174	1.152
	.90	2.769	2.370	2.154	2.016	1.921	1.849	1.793	1.748	1.711	1.680	1.629	1.574	1.513	1.479	1.443	1.403	1.358	1.307
	.95	3.960	3.111	2.719	2.486	2.329	2.214	2.126	2.056	1.999	1.951	1.875	1.793	1.703	1.654	1.602	1.545	1.482	1.411
	.975	5.218	3.864	3.284	2.950	2.730	2.571	2.450	2.355	2.277	2.213	2.111	2.003	1.884	1.820	1.752	1.679	1.599	1.508
	.99	6.963	4.881	4.036	3.563	3.255	3.036	2.871	2.742	2.637	2.551	2.415	2.271	2.115	2.032	1.944	1.849	1.746	1.630
	.999	11.67	7.540	5.972	5.123	4.582	4.204	3.923	3.705	3.530	3.386	3.162	2.927	2.677	2.545	2.406	2.258	2.099	1.924
90	.75	1.341	1.408	1.393	1.372	1.352	1.335	1.320	1.307	1.296	1.287	1.270	1.252	1.229	1.217	1.202	1.186	1.168	1.145
	.90	2.762	2.363	2.146	2.008	1.912	1.841	1.785	1.739	1.702	1.670	1.620	1.564	1.503	1.468	1.432	1.391	1.346	1.293
	.95	3.947	3.098	2.706	2.473	2.316	2.201	2.113	2.043	1.986	1.938	1.861	1.779	1.688	1.639	1.586	1.528	1.465	1.391
	.975	5.196	3.844	3.265	2.931	2.711	2.552	2.432	2.336	2.259	2.194	2.092	1.983	1.864	1.800	1.731	1.657	1.576	1.483
	.99	6.925	4.849	4.007	3.535	3.228	3.009	2.845	2.715	2.611	2.524	2.389	2.244	2.088	2.004	1.916	1.820	1.716	1.598
	.999	11.57	7.466	5.908	5.064	4.526	4.150	3.870	3.653	3.479	3.336	3.113	2.879	2.629	2.497	2.357	2.209	2.049	1.871
100	.75	1.339	1.406	1.391	1.369	1.349	1.332	1.317	1.304	1.293	1.283	1.267	1.248	1.226	1.213	1.198	1.182	1.163	1.140
	.90	2.756	2.356	2.139	2.002	1.906	1.834	1.778	1.732	1.695	1.663	1.612	1.557	1.494	1.460	1.423	1.382	1.336	1.282
	.95	3.936	3.087	2.696	2.463	2.305	2.191	2.103	2.032	1.975	1.927	1.850	1.768	1.676	1.627	1.573	1.515	1.450	1.376
	.975	5.179	3.828	3.250	2.917	2.696	2.537	2.417	2.321	2.244	2.179	2.077	1.968	1.849	1.784	1.715	1.640	1.558	1.463
	.99	6.895	4.824	3.984	3.513	3.206	2.988	2.823	2.694	2.590	2.503	2.368	2.223	2.067	1.983	1.893	1.797	1.692	1.572
	.999	11.50	7.408	5.857	5.017	4.482	4.107	3.829	3.612	3.439	3.296	3.074	2.840	2.591	2.458	2.319	2.170	2.009	1.829
120	.75	1.336	1.402	1.387	1.365	1.345	1.328	1.313	1.300	1.289	1.279	1.262	1.243	1.220	1.207	1.192	1.175	1.156	1.131
	.90	2.748	2.347	2.130	1.992	1.896	1.824	1.767	1.722	1.684	1.652	1.601	1.545	1.482	1.447	1.409	1.368	1.320	1.265
	.95	3.920	3.072	2.680	2.447	2.290	2.175	2.087	2.016	1.959	1.910	1.834	1.750	1.659	1.608	1.554	1.495	1.429	1.352
	.975	5.152	3.805	3.227	2.894	2.674	2.515	2.395	2.299	2.222	2.157	2.055	1.945	1.825	1.760	1.690	1.614	1.530	1.433
	.99	6.851	4.787	3.949	3.480	3.174	2.956	2.792	2.663	2.559	2.472	2.336	2.192	2.035	1.950	1.860	1.763	1.656	1.533
	.999	11.38	7.321	5.781	4.947	4.416	4.044	3.767	3.552	3.379	3.237	3.016	2.783	2.534	2.402	2.262	2.113	1.950	1.767
∞																			

Solutions to Selected Exercises

1.1 A statistical population is the entire body of information, or entire set of facts, that is the object of study. A subset of the population is a sample that should be obtained so that useful inferences can be made about the characteristics of the population from which the sample came.

1.3 *Planning.* Decide how many of the rooms in the dormitory will be contacted. If there is more than one occupant in the room, which one will you interview? Or, will you interview all occupants? Decide what to do if no one answers or if an individual refuses to cooperate.

Data Collection. Word your question carefully and give the respondent time to reply since the answer will take some thought. Perhaps you should consider asking the respondent to keep track of his or her viewing hours for a week and then make arrangements to get the results to you.

Data Analysis. Methods will have to be used to tabulate the results so that the "average" number of hours of TV viewing can be determined. (The method of determining the average will be explained later in the text along with methods for explaining how close this estimate may be to the actual average for the dormitory.)

Conclusions. The conclusions will state the estimated average number of hours spent viewing TV.

1.5 A few carefully chosen sales outlets thought to be representative of all sales outlets could provide the needed information.

1.7 Pick one of the digits 0, 1, 2, . . . , 9 at random and use it as a starting point to select every 10th account number. For example, if 7 is picked at random, the sample would be 07, 17, 27, . . . , 97.

1.8 The stratification allows all segments of the voting population to be represented rather than concentrating on only some segment of the voters.

1.9   Digit: 0   1   2   3   4   5   6   7   8   9

Frequency: 13 8 14 15 14 19 17 19 20 11

1.10 The target population is the nation's high schools. The sampled population is the high schools in Washington, D.C.

1.11 (a) Systematic;   (b) convenience;   (c) random;   (d) stratified;   (e) cluster.

1.12 Neither option (a) nor (b) is entirely satisfactory since the door may be answered by a small child who isn't capable of answering the questions or supplying a list of residents; however, it is generally accepted that a randomly chosen respondent produces a more valid cross section of opinion.

1.13 Return to the home and thus avoid sloppy execution of the survey in the field.

1.15 Record the response in the respondent's exact words as nearly as possible.

1.16 (a) is sampling error;   (b) and (c) are nonsampling error.

1.17 (a) Processing error;   (b) adequacy of respondent;   (c) concealment of the truth.

1.19 Quantitative information on (c), (d), (e), (i), and (j).

1.21 No. 1 provides information as to whether or not this is a first-time purchase. Nos. 2 and 4 provide guidance on targeting advertising to the most likely buyers. No. 3 tells if demonstrations are effective sales methods. No. 5 identifies seasons of likely sales increases. No. 6 provides informaton on how to promote the product.

1.25 (a) Assign each employee a number from 001 to 147. Next select three-digit numbers from a table of random numbers such as given in Figure 1.2 discarding 000 if it should be selected as well as any three-digit number over 147. Continue drawing numbers until ten unique numbers are obtained. These ten numbers correspond to the ten employees selected for the sample.

(b) The solution in part (a) may be difficult to apply here since there are $83 \times 144 = 11952$ parts and it may be difficult to associate numbers with the parts. However, an approximate random sample may be obtained by using the method in part (a) to select 40 of the 83 boxes and then repeating the procedure on (a) or a blind draw to select one part from each of the selected boxes.

(c) This problem differs from part (b) in that the class sizes are unequal. The best solution would be to assign all N students a number from 1 to N and repeat the procedure outlined in (a).

(d) Repeat the procedure outlined in part (a).

(e) Repeat the procedure outlined in part (a).

1.26 This method has the advantage of being easy to use, but has the drawback of being likely to select only those customers who are extremely pleased or displeased.

1.28 Information would have to be available on production for previous years that would provide information on the relationship between July figures and the total fall harvest. Then sample information obtained in July would allow a forecast to be made for fall based on this relationship.

1.29 This has the advantage of not allowing rigs to gear up for the safety inspection and thus provide more reliable information about safety conditions. The disadvantage is that some rigs will not be inspected.

2.1

12	5	8	1	3	6		
13	3	4					
14	8	3					
15	3	1	9	8	3	0	
16	2	5	6	2	6	8	0
17	8	4	8	5	0		
18	6	7	6	8	4	0	5
19							
20	1	0	2	8			
21	8	5					

2.2

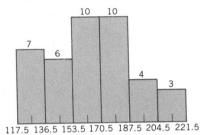

2.4 (a) 7; (b) .5 to 5.5, 5.5 to 10.5, 10.5 to 15.5, 15.5 to 20.5, 20.5 to 25.5, 25.5 to 30.5, 30.5 to 35.5; (c) 5; (d) 3, 8, 13, 18, 23, 28, 33; (e) 5, 30, 64, 67, 49, 12, 3; (f) .022, .130, .278, .291, .213, .052, .013; (g) 15.2 percent; (h) 27.8 percent.

2.6

0	95	20	68	41	45	
1	60	60	43	80	34	31
2	25	11	34	30	57	07
3	01	55				
4	70	76				
5	44					
6	85	79				
7	30					

2.7 (a)

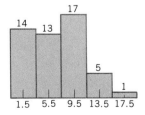

(b) 7.5 to 11.5
(c) 9.5

2.9 The class width varies from class to class, but the graph gives the impression that all class widths are equal.

2.12 The 2000-year projection has a larger percentage in all classes representing 35 and older.

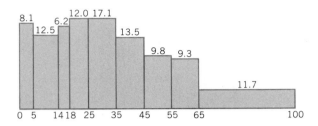

2.13

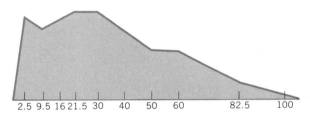

2.15 The population was 50 million in 1880 and 130 million in 1940. The population increased above 100 million between the 1910 census and the 1920 census.

2.17 Company D is the fastest growing manufacturer of integrated circuits over the 10-year period starting with 1970.

2.19

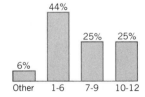

2.21

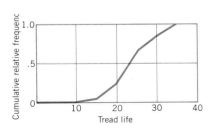

2.23

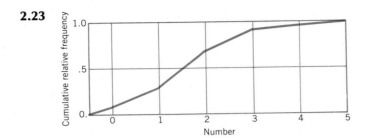

2.25 Two items is the maximum number a person can have and use the fast lane.

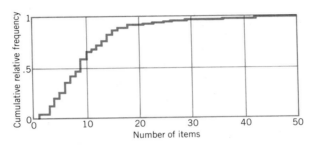

2.28 The scatterplot appears as three distinct and similarly shaped clusters of points. All points within each cluster are close to lying on a straight line except for those points corresponding to Cincinnati and Milwaukee.

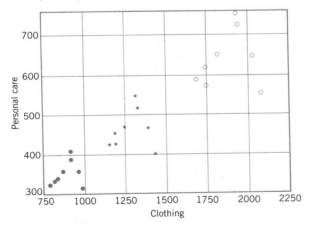

2.29 Tuition and fees get smaller as the percent of revenue from government appropriations increases.

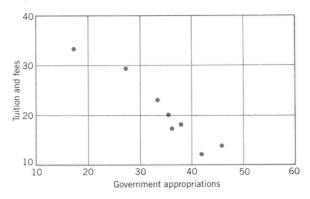

2.31 In the graph the largest ranks are associated with the highest percentages and the smallest ranks with the lowest percentages. Bonded whiskey shows the greatest decrease in percent of the market while vodka shows the greatest increase. The 45 degree line is added to make changes readily identifiable as if there were no change from 1959 to 1974 all points would fall on this line.

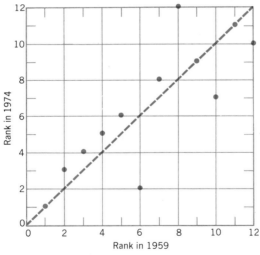

2.33

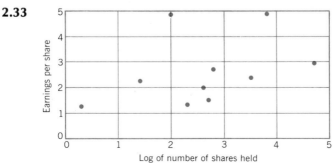

2.35

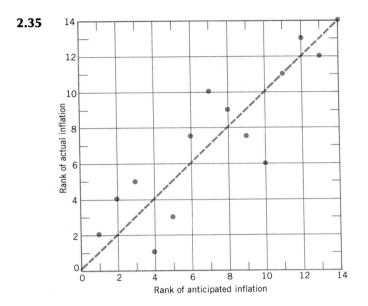

2.37

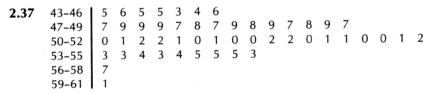

43–46	5 6 5 5 3 4 6
47–49	7 9 9 9 7 8 7 9 8 9 7 8 9 7
50–52	0 1 2 2 1 0 1 0 0 2 2 0 1 1 0 0 1 2
53–55	3 3 4 3 4 5 5 5 3
56–58	7
59–61	1

Note that the stems appear as equal width intervals.

2.38

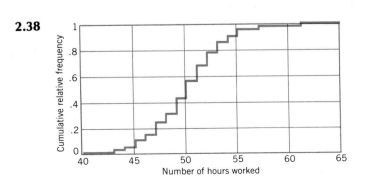

2.40

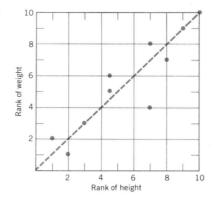

2.42

3.6–3.8	6 8 8
3.9–4.1	9 1 1 1 0 1 0 9 0 1 9 0 9 1 0
4.2–4.4	2 4 2 3 4 3 3 2 3 4 2 4 4 2 2 4 4
4.5–4.7	7 7 6 6 5 5 5 7 6 6
4.8–5.0	8 9 8 9 8

Note the stems appear as equally spaced intervals.

2.43

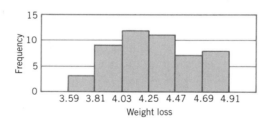

3.1 (a) The experiment consists of a student taking the exam and receiving a grade based on the number of correct responses.

(b) Answers will vary.

(c) {0, 5, 10, 15, . . . , 100}

(d) {75, 80, 85, 90, 95, 100}

3.2 (a) 1.0   (b) .83   (c) .08   (d) .48

3.3 (a)

	Married	Single	Total
Plan 1	.071	.179	.250
Plan 2	.500	.036	.536
Plan 3	.161	.054	.214
Total	.732	.268	1.000

(b) .250   (c) .732

(d) $P(\text{Married}) = .732$, $P(\text{Plan 1}) = .250$
$P(\text{Married and Plan 1}) = .071$
Since $.732 \times .250 = .183 \neq .071$, these events are not independent.

(e) $P(\text{Plan 1}|\text{Married}) = \dfrac{P(\text{Plan 1 and Married})}{P(\text{Married})} = \dfrac{.071}{.732} = .097$

(f) $P(\text{Plan 1}|\text{Single}) = \dfrac{P(\text{Plan 1 and Single})}{P(\text{Single})} = \dfrac{.179}{.268} = .668$

(g) $P(\text{Plan 2}|\text{Married}) = \dfrac{P(\text{Plan 2 and Married})}{P(\text{Married})} = \dfrac{.500}{.732} = .683$

(h) $P(\text{Plan 2}|\text{Single}) = \dfrac{P(\text{Plan 2 and Single})}{P(\text{Single})} = \dfrac{.036}{.268} = .134$

(i) No; plan selection depends to a great deal on marital status as the probabilities on (e)–(h) demonstrate.

3.5 (a)

	Married	Single	Total
Delinquent	.04	.08	.12
Non-Delinquent	.62	.26	.88
Total	.66	.34	1.00

(b) .26

(c) $P(\text{Delinquent}|\text{Single}) = \dfrac{P(\text{Delinquent and Single})}{P(\text{Single})} = \dfrac{.08}{.34} = .24$

(d) No, since $P(\text{Married}) \times P(\text{Delinquent}) = .66 \times .12 = .07$ while $P(\text{Married and Delinquent}) = .04$.

3.7 (a)

	30	30–40	41–50	>50	Total
News	.12	.10	.11	.14	.47
Sport	.10	.07	.08	.06	.31
Hobby	.01	.03	.05	.13	.22
Total	.23	.20	.24	.33	1.00

(b) $P(\text{Over 40}) = .24 + .33 = .57$

(c) $P(\text{Over 50}|\text{Hobby}) = \dfrac{P(\text{Over 50 and Hobby})}{P(\text{Hobby})} = \dfrac{.13}{.22} = .59$

(d) $P(\text{Over 40}|\text{Hobby}) = \dfrac{P(\text{Over 40 and Hobby})}{P(\text{Hobby})} = \dfrac{.05 + .13}{.22} = .82$

(e) $P(<40 \text{ and Sport}) = .10 + .07 = .17$

3.9

x	1	2	3	4	5	>5
$P(X = x)$	.40	.23	.11	.09	.08	.09

(a) .40   (b) .74   (c) .23   (d) .74   (e) .26   (f) 0

3.11

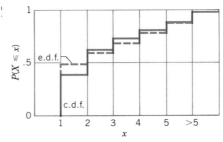

3.13 (a) $P(X \leq 10) = \sqrt{10/10} = 1$

(b) $P(X \leq 5) = \sqrt{5/10} = .71$

(c) $P(X < 5) = \sqrt{5/10} = .71$

(d) $P(X > 5) = 1 - P(X \leq 5) = 1 - .71 = .29$

(e) $P(1 < X < 5) = P(X < 5) - P(X < 1) = .71 - \sqrt{1/10} = .39$

(f) $P(X = 4) = 0$

3.15

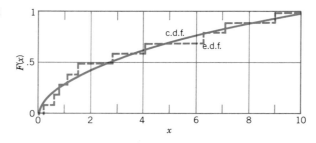

3.17 (a)

	Opposed	Undecided	In Favor	Total
Staff	18	19	38	75
Faculty	63	12	50	125
Administrator	19	4	27	50
Total	100	35	115	250

	Opposed	Undecided	In Favor	Total
Staff	.072	.076	.152	.300
Faculty	.252	.048	.200	.500
Administrator	.076	.016	.108	.200
Total	.400	.140	.460	1.000

(b) .460   (c) .200

(d) $P(\text{Favorable}|\text{Faculty}) = \dfrac{P(\text{Favorable and Faculty})}{P(\text{Faculty})} = \dfrac{.200}{.500} = .4$

3.19 (a) .41 (b) .54 (c) .18 (d) .18 (e) .05 (f) .14

3.21

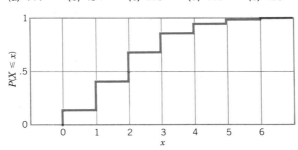

3.23 $P(\text{Male}|\text{Colorblind}) = \dfrac{P(\text{Male and Colorblind})}{P(\text{Colorblind})} = \dfrac{.025}{.030} = .83$

$P(\text{Female}|\text{Colorblind}) = \dfrac{P(\text{Female and Colorblind})}{P(\text{Colorblind})} = \dfrac{.005}{.030} = .17$

or $1 - .83 = .17$

3.25 (a) $P(A|\text{No Defects}) = \dfrac{P(A \text{ and No Defects})}{P(\text{No Defects})} = \dfrac{.1250}{.1875} = .6667$

(b) $P(\text{No Defects}|A) = \dfrac{P(\text{No Defects and } A)}{P(A)} = \dfrac{.1250}{.5000} = .2500$

(c) $(A|2 \text{ or more Defects}) = \dfrac{P(A \text{ and 2 or more Defects})}{P(2 \text{ or more Defects})} = \dfrac{.3125}{.6875} = .4545$

4.1 $\bar{X} = 75.5/5 = 15.1$

4.3 Data expressed in years: $4, \frac{8}{12}, 2, 27, 1\frac{2}{12}$. $\bar{X} = 34.83/5 = 6.97$

4.5 $\bar{X} = 334/43 = 7.77$ for ungrouped data.
For grouped data $\bar{X} = [10(2) + 9(8) + 8(16) + 7(12) + 6(5)]/43 = 7.77$.

4.8 $\bar{X} = \$215.31/6 = \$35.89$. This value is not near the center of the sample as it has been influenced greatly by the one large purchase of \$183.79.

4.10 $\mu = 50(.138) + 150(.158) + 250(.109) + 350(.079) + 450(.050)$
$+ 750(.168) + 1250(.069) + 1750(.040) + 2250(.030) + 3750(.059)$
$+ 7500(.040) + 30{,}000(.050) + 75{,}000(.010) = 3229$

4.13 $\mu = 1(.10) + 2(.30) + 3(.10) + 4(.20) + 5(.08) + 6(.11) + 7(.03) + 8(.08) = 3.71$

4.15

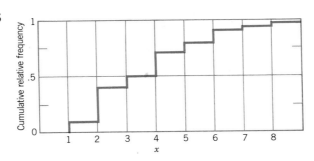

4.17 From Equation (4.9): $s = \sqrt{\frac{1}{4}(3.42)} = .92$.

From Equation (4.10): $s = \sqrt{\frac{1}{4}[1143.47 - (15.1)^2/5]} = .92$.

4.19 $\bar{X} = 6669/40 = 166.73$

$$s = \sqrt{\frac{1}{39}[1{,}138{,}039 - (166.73)^2/40]} = 25.9$$

$\bar{X} \pm s$ is from 140.8 to 192.6 and contains 27 observations, which is 68 percent of the total compared to the rule of thumb 67 percent, while $\bar{X} \pm 2s$ is from 114.9 to 218.5 and contains all 40 observations compared with the rule of thumb 95 percent.

4.21 z-scores for production data.

	Quarter			
	1	2	3	4
1978	.15	−.31	−1.01	1.31
1979	.52	−.92	−1.35	1.53
1980	.33	−.28	−1.12	1.18

The z-scores make it clear that production is below the average level in the second and third quarters each year.

4.23 $\bar{X} = 1011/102 = 9.91$ from Exercise 4.7.

$\Sigma f_i m_i^2 = 14{,}989$

$$s = \sqrt{\frac{1}{101}[14{,}989 - (9.91)^2\, 102]} = 7.01$$

4.24 Ungrouped data: $\bar{X} = 334/43 = 7.77$ from Exercise 4.5.

$\Sigma X_i^2 = 2640$

$$s = \sqrt{\frac{1}{42}\left[2640 - \left(\frac{334}{43}\right)^2 43\right]} = 1.04$$

Grouped data: $\Sigma f_i m_i^2 = 2(10)^2 + 8(9)^2 + 16(8)^2 + 12(7)^2 + 5(6)^2$

$$= 2640$$

$$s = \sqrt{\frac{1}{42}\left[2640 - \left(\frac{334}{43}\right)^2 43\right]} = 1.04$$

4.27 From Exercise 4.11:

$\mu = 5.84$

$\sigma = [4^2(11/56) + 5^2(10/56) + 6^2(12/56) + 7^2(23/56) - (5.84)^2]^{1/2} = 1.16$

4.29 From Exercise 4.14:

$\mu = 2.73$

$\sigma = [0^2(.10) + 1^2(.20) + \cdots + 7^2(.06) - (2.73)^2]^{1/2} = 1.95$

4.30 81, 83, and 85 are all modes.

4.32 (a) The median is $X_{.50} = (80 + 81)/2 = 80.5$.

(b) $X_{.25} = 71$ (c) $X_{.75} = 85$ (d) $X_{.75} - X_{.25} = 85 - 71 = 14$

(e) $X_{.90} = 95$ (f) $X_{.80} = 86$

4.34 Data Set 1: Median = 7.2, Mean = 7.2.
Data set 2: Median = $(7.2 + 8.4)/2 = 7.8$, Mean = 17.8.
The median is not influenced by the outlier 70.8, while the mean is very sensitive to extreme observations.

4.35 Median = $(377 + 544)/2 = 460.5$.
Interquartile range = $X_{.75} - X_{.25} = 3452 - 99 = 3353$.
Mean = 6508.4.
Standard deviation = 16532.

Both the mean and standard deviation are influenced by the outliers in the data.

4.37 From the ogive graphed below the median is found as approximately $430. Note that all classes of $1000 or more do not affect the calculation of the median from the graph, hence all of these classes have been combined to make the graphing easier.

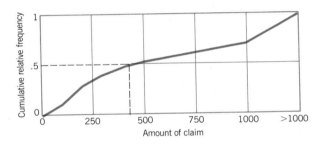

4.39 From Exercise 4.15 the population median is found as $(3 + 4)/2 = 3.5$.

4.41 From Exercise 4.16 the population interquartile range is $4 - 1 = 3$, which is slightly larger than the population standard deviation of 1.95.

4.43 $\Sigma X = 100$, $\Sigma Y = 100$, $\Sigma XY = 1285.9$, $\Sigma X^2 = 1965.2$, $\Sigma Y^2 = 1235.9$, $r = .670$.

4.45 $\Sigma X = 1200$, $\Sigma Y = 153$, $\Sigma XY = 9236$, $\Sigma X^2 = 77,020$, $\Sigma Y^2 = 1711$, $r = .034$. Typing speed and the number of errors are unrelated.

4.48 $\Sigma X_L = 7133$, $\Sigma Y_L = 2818$, $\Sigma X_L Y_L = 2,515,644$
$\Sigma X_L^2 = 6,397,387$, $\Sigma Y_L^2 = 1,000,838$, $r_L = .174$
$\Sigma X_I = 10,255$, $\Sigma Y_I = 3700$, $\Sigma X_I Y_I = 4,748,928$
$\Sigma X_I^2 = 13,217,861$, $\Sigma Y_I^2 = 1,727,374$, $r_I = .176$
$\Sigma X_H = 14,988$, $\Sigma Y_H = 5104$, $\Sigma X_H Y_H = 9,577,145$
$\Sigma X_H^2 = 28,227,310$, $\Sigma Y_H^2 = 3,292,924$, $r_H = .202$
$\Sigma X = 32376$, $\Sigma Y = 11622$, $\Sigma XY = 16,841,717$
$\Sigma X^2 = 47,842,558$, $\Sigma Y^2 = 6,021,136$, $r = .909$

Refer to the scatterplot made for Exercise 2.28. The outlying observations from Cincinnati and Milwaukee within each budget type cause the correlations for each budget type to be small; however, when all 24 pairs of data points are taken together, a strong relationship exists.

4.49 $\Sigma R_x = 36$, $\Sigma R_y = 36$, $\Sigma R_x R_y = 122$, $\Sigma R_x^2 = 204$, $\Sigma R_y^2 = 204$, $r_s = -.952$. The rank correlation coefficient is almost identical to the correlation coefficient calculated on raw data.

4.51 $\Sigma R_x = 210$, $\Sigma R_y = 210$, $\Sigma R_x R_y = 2142$, $\Sigma R_x^2 = 2869.5$, $\Sigma R_y^2 = 2857.5$. From Equation (4.25), $r_s = -.096$. From Equation (4.27), $r_s = -.085$. Equation (4.27) is not correct to use when ties are present.

4.53 $\Sigma R_x = 55$, $\Sigma R_y = 55$, $\Sigma R_x R_y = 351$, $\Sigma R_x^2 = 385$, $\Sigma R_y^2 = 385$, $r_s = .588$. The sample correlation coefficient $r = .174$ was strongly influenced by outliers.

4.54 $\Sigma R_{x_L} = 36$, $\Sigma R_{y_L} = 36$, $\Sigma R_{x_L} R_{y_L} = 170.5$
$\Sigma R_{x_L}^2 = 204$, $\Sigma R_{y_L}^2 = 203.5$, $r_s = .204$
$\Sigma R_{x_I} = 36$, $\Sigma R_{y_I} = 36$, $\Sigma R_{x_I} R_{y_I} = 168.5$
$\Sigma R_{x_I}^2 = 204$, $\Sigma R_{y_I}^2 = 203.5$, $r_s = .156$
$\Sigma R_{x_H} = 36$, $\Sigma R_{y_H} = 36$, $\Sigma R_{x_H} R_{y_H} = 168$
$\Sigma R_{x_H}^2 = 204$, $\Sigma R_{y_H}^2 = 204$, $r_s = .143$
$\Sigma R_x = 300$, $\Sigma R_y = 300$, $\Sigma R_x R_y = 4795$
$\Sigma R_x^2 = 4900$, $\Sigma R_y^2 = 4899$, $r_s = .909$

The rank correlations are similar to the raw correlations and the interpretation on the problem is the same as in Exercise 4.40.

4.55 $\Sigma X = 693$, $\Sigma Y = 1588$, $\Sigma XY = 110,896$, $\Sigma X^2 = 48163$, $\Sigma Y^2 = 257,974$, $r = .9471$

4.56 $\Sigma R_x = 55$, $\Sigma R_y = 55$, $\Sigma R_x R_y = 377.5$, $\Sigma R_x^2 = 384$, $\Sigma R_y^2 = 385$, $r_s = .9147$

4.59 $X_{.50} = 50$, $X_{.75} - X_{.25} = 52 - 48 = 4$

4.61 $\Sigma X = 214.8$, $\Sigma X^2 = 927.66$, $\overline{X} = 4.2960$, $s = .3156$

4.63 $X_{.50} = 4.3$, $X_{.75} - X_{.25} = 4.5 - 4.1 = 4.0$

4.66 $\mu = 0(.408) + (1/3)(.017) + (2/3)(.025) + 1(.550) = .572$

4.67 $\sigma = [0^2(.408) + (1/3)^2(.017) + (2/3)^2(.025) + 1^2(.550) - (.572)^2]^{1/2} = .486$

4.68 Population median $= 1$

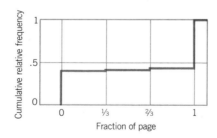

5.1 $n = 20$, $p = .4$, from Table A1, $P(X \le 9) = .7553$.

5.3 $1/5$; expected score $= 20(1/5) = 4$. $n = 20$, $p = .2$; from Table A1,

$$P(X \ge 11) = 1 - P(X \le 10) = 1 - .9994 = .0006$$

From Table A1,

$$P(X > 5) = 1 - P(X \le 5) = 1 - .8042 = .1958$$
$$P(2 \le X \le 6) = P(X \le 6) - P(X \le 1) = .9133 - .0692 = .8441$$

5.4 $n = 20, p = .1$; expected number of defectives $= 20(.1) = 2$.

$P(X \geq 4) = 1 - P(X \leq 3) = 1 - .8670 = .1330$

$$f(0) = \binom{5}{0}\left(\frac{1}{3}\right)^0\left(\frac{2}{3}\right)^5 = \frac{32}{243} = P(X = 0)$$

$$f(1) = \binom{5}{1}\left(\frac{1}{3}\right)^1\left(\frac{2}{3}\right)^4 = \frac{80}{243} = P(X = 1)$$

$$f(2) = \binom{5}{2}\left(\frac{1}{3}\right)^2\left(\frac{2}{3}\right)^3 = \frac{80}{243} = P(X = 2)$$

$$f(3) = \binom{5}{3}\left(\frac{1}{3}\right)^3\left(\frac{2}{3}\right)^2 = \frac{40}{243} = P(X = 3)$$

$$f(4) = \binom{5}{4}\left(\frac{1}{3}\right)^4\left(\frac{2}{3}\right)^1 = \frac{10}{243} = P(X = 4)$$

$$f(5) = \binom{5}{5}\left(\frac{1}{3}\right)^5\left(\frac{2}{3}\right)^0 = \frac{1}{243} = P(X = 5)$$

(a) $P(X = 0) = 32/243$

(b) $P(X = 5) = 1/243$

(c) $P(X \geq 3) = 40/243 + 10/243 + 1/243 = 51/243$

(d) Either 1 or 2 as $P(X = 1) = P(X = 2) = 80/243$.

5.7 $n = 8, p = .05, P(X = 0) = .3585$ from Table A1.

5.8 $n = 15, p = .7, P(X \leq 8) = .1311$ from Table A1.

5.10 $z_{.01} = -2.3263, z_{.05} = -1.6449, z_{.95} = 1.6449, z_{.99} = 2.3263$

5.11 $X_{.025} = 10 + 2z_{.025} = 10 + 2(-1.9600) = 6.0800$
$X_{.975} = 10 + 2z_{.975} = 10 + 2(1.9600) = 13.9200$

5.12 (a) .01,   (b) .05,   (c) .95,   (d) .99

5.14 (a) $P(X \leq 160) = .196$ as $z = (160 - 163)/3.5 = -.8571$

(b) $P(X > 170) = 1 - P(X \leq 170) = 1 - .977 = .023$ as $z = (170 - 163)/3.5 = 2.0000$

(c) $P(160 \leq X \leq 165) = P(X \leq 165) - P(X \leq 160)$ and $z_1 = (160 - 163)/3.5 = -.8571$
and $z_2 = (165 - 163)/3.5 = .5714$. From Table A2, $.716 - .196 = .520$.

(d) A number x is needed such that $P(X \geq x) = .80$ or, equivalently, $P(X \leq x) = .20$, (i.e., $x_{.20}$) and $x_{.20} = 163 + 3.5z_{.20} = 163 + 3.5(-.8416) = 159.05$.

5.15 (a) $z = (.5 - .8)/.2 = -1.5000$. From Table A2, $P(X \leq .5) = .067$.

(b) $P(X \geq 1.2) = 1 - P(X \leq 1.2) = 1 - .977 = .023$ as $z = (1.2 - .8)/.2 = 2.0000$

(c) $P(.75 \leq X \leq 1.25) = P(X \leq 1.25) - P(X \leq .75) = .988 - .401 = .587$ as $z_1 = (.75 - .8)/.2 = -.2500$ and $z_2 = (1.25 - .8)/.2 = 2.2500$

5.17 $X_{.95} = 125 + 10z_{.95} = 125 + 10(1.6449) = 141.45$; therefore, his score of 140 is not quite high enough to get in the top 5 percent.

5.18 $X_{.80} = 63 + 5z_{.80} = 63 + 5(.8416) = 67.21$

5.19 $P(X \leq 9) \approx P(Z \leq (9 - 20(.4) + .5)/\sqrt{20(.4)(.6)}) = P(Z \leq .6847) = .753$, which compares with the exact answer of .7553.

5.21 From Table A1 $P(1 \leq X \leq 3) = P(X \leq 3) - P(X \leq 0) = .9961 - .3164 = .6797$ as the exact answer.

$$P(1 \leq X \leq 3) \approx P\left(\frac{1 - 4(.25) - .5}{\sqrt{4(.25)(.75)}} \leq Z \leq \frac{3 - 4(.25) + .5}{\sqrt{4(.25)(.75)}}\right)$$

$$= P(-.5774 \leq Z \leq 2.8868) = P(Z \leq 2.8868) - P(Z \leq -.5774)$$

$$= .998 - .282 = .716$$

Error is somewhat large since np and nq are not both greater than 5.

5.23 (a) $50(.6) = 30$

(b) $P(26 \leq X \leq 35) \approx P\left(\frac{26 - 30 - .5}{\sqrt{50(.6)(.4)}} \leq Z \leq \frac{35 - 30 + .5}{\sqrt{50(.6)(.4)}}\right)$

$$= P(-1.2990 \leq Z \leq 1.5877) = P(Z \leq 1.5877) - P(Z \leq -1.2990)$$

$$= .944 - .097 = .847$$

(c) $P(X \leq 25) \approx P\left(Z \leq \frac{25 - 30 + .5}{\sqrt{50(.6)(.4)}}\right) = P(Z \leq -1.2990) = .097$

5.25 $\mu = 1(.2) + 2(.2) + 3(.2) + 4(.2) + 5(.2) = 3$
$\sigma = [1^2(.2) + 2^2(.2) + 3^2(.2) + 4^2(.2) + 5^2(.2) - 3^2]^{1/2}$
$= \sqrt{2}$

5.27 $P(\bar{X} > 18m) = 1 - P(\bar{X} \leq 18m) = 1 - P\left(Z \leq \frac{18m - 16m}{5m/\sqrt{5}}\right)$

$$= 1 - P(Z \leq .8944) = 1 - .814 = .186$$

5.30

$\bar{x}$		$P(\bar{X} = \bar{x})$
1.0	.3(.3)	= .0900
1.5	2(.3)(.15)	= .0900
2.0	2(.3)(.1) + .15(.15)	= .0825
2.5	2(.3)(.15) + 2(.15)(.1)	= .1200
3.0	2(.3)(.3) + 2(.15)(.15) + .1 (.1)	= .2350
3.5	2(.15)(.3) + 2(.1)(.15)	= .1200
4.0	2(.1)(.3) + .15(.15)	= .0825
4.5	2(.15)(.3)	= .0900
5.0	.3(.3)	= .0900

5.31 The graph is heavier in the tails than Figure 5.18 and lacks the triangular appearance of Figure 5.18.

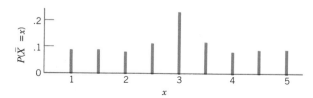

5.32 $\mu_{\bar{x}} = 1(.09) + 1.5(.09) + 2(.0825) + 2.5(.12) + 3(.2350) + 3.5(.12) + 4(.0825)$
$+ 4.5(.09) + 5(.09) = 3$ which is in agreement with Equation (5.14).
$\sigma_{\bar{x}} = [1^2(.09) + (1.5)^2(.09) + 2^2(.0825) + (2.5)^2(.12) + 3^2(.2350)$
$+ (3.5)^2(.12) + 4^2(.0825) + (4.5)^2(.09) + 5^2(.09) - 3^2]^{1/2}$
$= 1.1619$, which is in agreement with Equation (5.16) since $\sigma_{\bar{x}} = 1.6432/\sqrt{2} =$
1.1619 (see Exercise 5.29).

5.34 $\bar{X} = 9.4$, standardized sample values: .426, .532, .851, .957, .957, 1.064, 1.064, 1.277, 1.383, 1.489. The graph of the e.d.f. falls outside of the bounds for $n = 10$, hence the assumption of an exponential distribution does not seem reasonable for these data.

5.35 (a) $\bar{X} = 5$

(b) Standardized sample values: .06, .20, .36, .54, .64, .90, .96, 1.24, 1.96, 3.14. The graph of the e.d.f. stays within the bounds for $n = 10$, so it seems reasonable to regard these data as coming from an exponential distribution.

(c) $\hat{\lambda} = 1/5 = .2$

(d) $P(X \leq 1) = 1 - e^{-.2(1)} = .18$; $P(X \geq 10) = e^{-.2(10)} = .14$

5.37 $\hat{\lambda} = 1/1 = 1$, $P(X \geq 5) = e^{-1(5)} = .01$

5.39 $\hat{\lambda} = 1/300 = .0033$, $P(X \leq 90) = 1 - e^{-.0033(90)} = .26$. Therefore, 26 percent can be expected back within 90 days if an exponential distribution is used to describe the time to repair.

5.41 $P(Y > 3) = 1 - P(Y = 0) - P(Y = 1) - P(Y = 2) - P(Y = 3)$
$P(Y = 0) = e^{-2.16} (2.16)^0/0! = .115$
$P(Y = 1) = e^{-2.16} (2.16)^1/1! = .249$
$P(Y = 2) = e^{-2.16} (2.16)^2/2! = .269$
$P(Y = 3) = e^{-2.16} (2.16)^3/3! = .194$
Therefore, $P(Y > 3) = 1 - .115 - .249 - .269 - .194 = .173$.

5.43 $P(X < 23,500) = P\left(Z \leq \dfrac{23500 - 27500}{3000}\right) = P(Z \leq -1.33) = .092$

$P(X > 30,000) = 1 - P\left(Z \leq \dfrac{30000 - 27500}{3000}\right) = 1 - P(Z \leq .83)$

$= 1 - .797 = .203$

5.45 (a) $P(\bar{X} \leq 1230) = P\left(Z \leq \dfrac{1230 - 1500}{900/\sqrt{25}}\right) = P(Z \leq -1.50) = .067$

(b) $P(\bar{X} \leq 1600) = P\left(Z \leq \dfrac{1600 - 1500}{900/\sqrt{25}}\right) = P(Z \leq -.56) = .712$

(c) $P(1230 \leq \bar{X} \leq 1600) = P(\bar{X} \leq 1600) - P(\bar{X} \leq 1230) = .712 - .067 = .645$

5.47 $P(X < 16) = P\left(Z \leq \dfrac{16 - 16.3}{.15}\right) = P(Z \leq -2) = .023$

5.49 $\lambda = 1$, $P(X \leq 14/12) = 1 - e^{-1(14/12)} = .69$

5.51 $\bar{X} = 62.99$, standardized sample values: .59, .68, .71, .75, .77, .82, .89, .92, .98, 1.00, 1.05, 1.20, 1.29, 1.46, 1.91. Based on the e.d.f. of these standardized values plotted on the Lilliefors graph, it does not seem reasonable to assume an exponential distribution for these data.

5.52   This graph is skewed to the left.

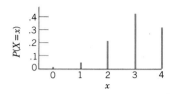

6.1   Medians: 61.55, 59.25, 58.10, 61.90, 58.65. The standard deviation of these medians is 1.73, which would indicate that the sample mean has smaller variance associated with it for these data.

6.2   (a) and (b)

Sample No.	Sample Mean	Sample Median
1	4.6	5
2	5.2	6
3	2.8	3
4	4.0	4
5	5.6	5
6	2.8	3
7	4.8	3
8	4.6	5
9	6.6	8
10	2.8	2
11	2.2	3
12	5.4	7
13	1.8	1
14	4.2	4
15	2.2	2

(c)

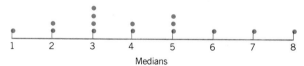

(d) 1.46,   (e) 1.97,   (f) the sample mean

6.3 $\$2100 \pm 1.9600(400)/\sqrt{50}$ or $1989 to $2211. There is 95 percent confidence that the interval from $1989 to $2211 contains the true average daily balance.

6.5   The required sample size is given as $n = [2(2.5758)(500)/400]^2$ or $n = 41.5$, which would need to be rounded up to 42 to be within $200.

6.7 The required sample size is given as $n = [2(1.9600)(25)/3]^2$ or $n = 1067.1$, which should be rounded up to 1068.

6.9 $\bar{X} = 808.2$ and $s = 113.4$; 90 percent confidence interval: $808.2 \pm 2.1318(113.4)/\sqrt{5}$ or 700 to 916. There is 90 percent confidence that the mean of the population is contained in this interval.

6.11 Since the e.d.f. of the standardized values does not fall outside of the Lilliefors bounds for $n = 5$, it is reasonable to assume these data come from a normal distribution. $\bar{X} = 71$ and $s = 12.65$; 95 percent confidence interval: $71 \pm 2.7764(12.65)/\sqrt{5}$ or 55.3 to 86.7. There is 95 percent confidence that the population mean score of all exams falls in this interval.

6.13 $L = .10 + (.15 - .10)\left(\dfrac{.9750 - .9936}{.9581 - .9936}\right) = .126$

$U = .55 + (.60 - .55)\left(\dfrac{.0250 - .0537}{.0203 - .0537}\right) = .593$

6.15 $(6/18) \pm 1.9600 \sqrt{\dfrac{(6/18)(12/18)}{18}}$ or .116 to .551

This interval is slightly smaller than the one found using the exact method.

6.17 $(6/50) \pm 1.9600 \sqrt{\dfrac{(6/50)(44/50)}{50}}$ or .030 to .210

6.19 $.57 \pm 2.5758 \sqrt{\dfrac{(.57)(.43)}{1200}}$ or .533 to .607

6.21

Median	90 Percent Confidence Interval
61.55	56.8 to 63.8
59.25	57.0 to 60.4
58.10	54.7 to 63.5
60.45	58.6 to 64.4
58.65	54.8 to 61.6

6.23 Median = 24.43; the 95 percent confidence is constructed from Figure 6.5, where $S_1 = 6$ and $S_2 = 15$. Therefore, the confidence interval is from $X^{(6)}$ to $X^{(15)}$, or 24.14 to 24.62.

6.25 From Figure 6.5, $S_1 = 5$ and $S_2 = 16$. Therefore, the 99 percent confidence interval is from $X^{(5)}$ to $X^{(16)}$ or 3.11 to 6.35.

6.27 $S_1^* = (40 - 1.9600\sqrt{40})/2 = 13.80$, so $S_1 = 14$ and $S_2 = 40 - 14 + 1 = 27$. Therefore, the 95 percent confidence interval is from $X^{(14)}$ to $X^{(27)}$ or 75 to 84.

6.29 Median = 20. $S_1^* = (30 - 1.9600\sqrt{30})/2 = 9.63$, or $S_1 = 10$ and $S_2 = 30 - 10 + 1 = 21$. Therefore, the 95 percent confidence interval is from $X^{(10)}$ to $X^{(21)}$ or 17 to 22.

6.31 $84.2 \pm t_{.05,9}(12.2)/\sqrt{10}$ or $84.2 \pm (1.8331)(12.2)/\sqrt{10}$ or 77.1 to 91.3

6.33 $L = .15 + (.20 - .15)\left(\dfrac{.9750 - .9765}{.9183 - .9765}\right) = .151$

$U = .60 + (.65 - .60)\left(\dfrac{.0250 - .0583}{.0229 - .0583}\right) = .687$

6.35 Median = 125. The 95 percent confidence interval is constructed from Figure 6.5, where $S_1 = 6$ and $S_2 = 15$. Therefore, the confidence interval is from $X^{(6)}$ to $X^{(15)}$ or 103 to 142.

6.36 $n = [2(1.9600)(.02)/.01]^2 = 61.5$ or 62

7.1 (a) Employee typing speeds are increased as a result of taking a specialized training course.

(b) Daily sales are increased after the start of TV commercials.

(c) The average number of sickness absentees for Monday and Friday is higher than the average number for Tuesday, Wednesday, and Thursday.

(d) The proportion of viewers watching the local news on Channel 4 is more than 40 percent.

(e) The median grade point average at this university is not 2.613.

7.3 (a) It is concluded that smoking is harmful to your health when it really is not.

(b) It is concluded that smoking is not harmful to your health when it really is.

7.5 (a) H_1: The proportion of accounts qualifying for the "500 Club" is greater than 40 percent.

(b) Let T = the number of accounts qualifying for the "500 Club." Accept H_0 if $T \leq 11$ and reject H_0 if $T > 11$. The exact level of significance for this test is $\alpha = .0565$.

(c) H_0 is accepted.

(d) The p-value is $P(T \leq 11) = .0565$ from Table A1 with $n = 20$ and $p = .4$. The probability of falsely rejecting H_0 when $T = 11$ is .0565.

7.6 (a) Let $T_1 = \dfrac{\bar{X} - 16}{.5/\sqrt{100}}$. Reject H_0 if $T_1 < -z_{.05} = -1.6449$.

(b) For $\bar{X} = 15.5$, $T_1 = -10.00$; therefore, reject H_0.
For $\bar{X} = 15.95$, $T_1 = -1.00$; therefore, accept H_0.
For $\bar{X} = 16.1$, $T_1 = 2.00$; therefore, accept H_0.
For $\bar{X} = 15.90$, $T_1 = -2.00$; therefore, reject H_0.

7.7 (a) $\bar{X} = 23.90$ and $s = .89$. Standardized sample values: -1.24, -1.02, $-.11$, $.34$, $.68$, 1.36. The Lilliefors graph shows the assumption of normality to be reasonable.

(b) $T_2 = \dfrac{23.9 - 25}{.89/\sqrt{6}} = -3.04$. Reject H_0 if $T_2 < -t_{.05,5} = -2.0150$,

or if $T_2 > t_{.05,5} = 2.0150$. H_0 is rejected.

(c) The p-value is between .02 and .05; therefore, the sample indicates the mean weight very likely has changed from 25.0 grams.

7.9 $\bar{X} = 26.9$ and $T_1 = \dfrac{26.9 - 25}{4/\sqrt{10}} = 1.50$. Reject H_0 if $T_1 > z_{.05} = 1.6449$.

H_0 is accepted. The p-value is approximately .067, so this small sample size has not indicated an increase in ACT scores.

7.11 $T_1 = \dfrac{42.8 - 40}{6.89/\sqrt{200}} = 5.747$. Reject H_0 if $T_1 > z_{.01} = 2.3263$. H_0 is rejected.

The p-value is less than .001, so the evidence is quite strong that the average age is greater than 40.

7.13 $T_2 = \dfrac{6 - 6.3}{1.5/\sqrt{25}} = -1$. Reject H_0 if $T_2 < -t_{.05,24} = -1.7109$.

H_0 is accepted. The p-value is between .10 and .25, so the sample evidence does not indicate that the mean number of cubic yards of dirt is less than 6.3.

7.15 Reject H_0 if $X < 3$. Since $X = 2$, H_0 is rejected. the p-value is .0498. The evidence is sufficient to show the percentage of drivers wearing seat belts is less than 60 percent.

7.17 (a) $\alpha = P(X < 5) + P(X > 9)$ when $n = 10$ and $p = .7$. This is found in Table A1 as .0473 + .0282 = .0755.

(b) Power $= P(X < 5) + P(X > 9)$ when $n = 10$ and $p = .5$, which is found in Table A1 as .3770 + .0010 = .3780.

7.19 $H_0: p = .78$ versus $H_1: p \neq .78$. Reject H_0 if $X > 92(.78) + z_{.025}\sqrt{92(.78)(.22)} = 79.55$ or if $X < 92(.78) - z_{.025}\sqrt{92(.78)(.22)} = 63.97$. Since $X = 64$, H_0 is accepted. The p-value is $2P(X \leq 64) = 2P(Z < (64 - 92(.78) + .5)/\sqrt{92(.78)(.22)}) = 2P(Z < 1.8278) = 2(.034) = .068$, so the sample evidence supports the claim.

7.21

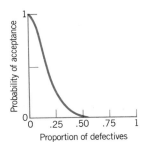

7.23 (a) $\alpha = P(X > 23) = 1 - P(X \leq 23) = 1 - P\left(Z < \dfrac{23 - 30(.7) + .5}{\sqrt{30(.7)(.3)}}\right)$

$= 1 - P(Z < .9960) = .160$

(b) $P(X > 23) = 1 - P(X \leq 23) = 1 - P\left(Z < \dfrac{23 - 30(.8) + .5}{\sqrt{30(.8)(.2)}}\right)$

$= 1 - P(Z < -.2282) = .590$

7.25 Reject H_0 if $T_1 \leq (40 - 2.3263\sqrt{40})/2 = 12.64$. H_0 is rejected since $T_1 = 9$. The p-value is

$$P(T_1 \leq 9) \approx P\left(Z < \dfrac{9 - 40(.5) + .5}{\sqrt{40(.5)(.5)}}\right)$$

$$= P(Z < -3.3204)$$

which is less than .001.

7.27 Reject H_0 if $T_1 \leq 1$ as $P(T_1 \leq 1) = .0107$ from Table A1 with $n = 20$ and $p = .5$. H_0 is rejected since $T_1 = 1$. The p-value is also .0107, which strongly indicates that median earnings are greater than $2.00.

7.29 Reject H_0 if $T_1 \leq 4$ as $P(T_1 \leq 4) = .1334$ from Table 1 with $n = 13$ and $p = .5$. H_0 is accepted since $T_1 = 6$. The p-value is $P(T_1 \leq 6) = .5000$, so H_0 cannot reasonably be rejected.

7.31 Reject H_0 if $T_2 \geq T_U$, where $T_U = (30 + 1.6449/\sqrt{30})/2 = 19.5$. H_0 is rejected since $T_2 = 20$. The p-value is .05 and this is the smallest value of α that will lead to rejection of H_0.

7.33 $T = \dfrac{1498.9 - 1492}{11.4/\sqrt{16}} = 2.44$. Reject H_0 if $T > t_{.05,15} = 1.7531$.

H_0 is rejected and the p-value is between .025 and .01, so it is strongly concluded that the melting point is greater than 1492.

7.35 Power $= P(\bar{X} > 30 + 2.3263(15.7)/\sqrt{30}) = P(\bar{X} > 36.668)$

$$= P\left(Z > \frac{36.668 - 35}{15.7/\sqrt{30}}\right) = P(Z > .582) = 1 - P(Z < .582)$$

$$= 1 - .720 = .280$$

7.38

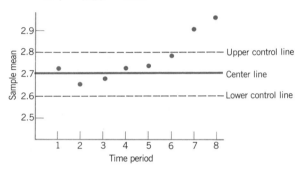

7.39 Process is judged to be out of control as indicated in the plot in Exercise 7.38.

8.1 The data (i.e., the D_i) pass the Lilliefors test for normality. $H_0: \mu_X = \mu_Y$ versus $H_1: \mu_X > \mu_Y$. Reject H_0 if $T > t_{.025,9} = 2.2622$. $T = \dfrac{4.5}{5.48/\sqrt{10}} = 2.5957$, so H_0 is rejected. The p-value is between .01 and .025. 95 percent confidence interval is $4.5 \pm 2.2622 \, (5.48)/\sqrt{10}$, or .58 to 8.42.

8.3 The data (i.e., the D_i) pass the Lilliefors test for normality. $H_0: \mu_X = \mu_Y$ versus $H_1: \mu_X > \mu_Y$. Reject H_0 if $T > t_{.05,19} = -1.7291$. $T = \dfrac{2.4}{5.88/\sqrt{20}} = 1.8255$, so H_0 is rejected. The p-value is between .025 and .05.

8.5 The data (i.e., the D_i) pass the Lilliefors test for normality. $H_0: \mu_X = \mu_Y$ versus $H_1: \mu_X \neq \mu_Y$. Reject H_0 if $T < -t_{.025,14} = -2.1448$ or if $T > t_{.025,14} = 2.1448$. $T = \dfrac{-.04}{16/\sqrt{5}} = -0.9045$, so H_0 is accepted. The p-value is between .20 and .50.

8.7 The data (i.e., the D_i) pass the Lilliefors test for normality. $H_0: \mu_X = \mu_Y$ versus $H_1: \mu_X \neq \mu_Y$. Reject H_0 if $T < -t_{.025,19} = -2.0930$ or if $T > t_{.025,19} = 2.0930$. $T = \dfrac{-4.15}{19.25/\sqrt{20}} = .9643$, so H_0 is accepted. The p-value is between .20 and .50.

8.9 The technique described is a randomization technique that is used to avoid any bias on the part of the drivers. $H_0: \mu_X = \mu_Y$ versus $H_1: \mu_X \neq \mu_Y$. Reject H_0 if $T < -t_{.005,99} = -2.6264$ or if $T > t_{.005,99} = 2.6264$. $T = \dfrac{43}{1.8/\sqrt{100}} = 2.3889$, so H_0 is accepted. The p-value is between .01 and .02.

8.10 $H_0: \mu_X = \mu_Y$ versus $H_1: \mu_X > \mu_Y$. Reject H_0 if $T_R > t_{.025,9} = 2.2622$. $T_R = \dfrac{3.90}{5.08/\sqrt{10}} = 2.4270$, so H_0 is rejected in agreement with the paired t-test. The p-value is between .01 and .025.

8.12 $H_0: \mu_X = \mu_Y$ versus $H_1: \mu_X > \mu_Y$. Reject H_0 if $T_R > t_{.05,19} = 1.7291$. $T_R = \dfrac{4.8}{11.23/\sqrt{20}} = 1.9123$, so H_0 is rejected in agreement with the paired t-test. The p-value is between .025 and .05.

8.14 $H_0: \mu_X = \mu_Y$ versus $H_1: \mu_X \neq \mu_Y$. Reject H_0 if $T_R < -t_{.025,14} = -2.1448$ or if $T_R > t_{.025,14} = 2.1448$. $T_R = -3.33/(8.74/\sqrt{15}) = -1.4768$, so H_0 is accepted in agreement with the paired t-test. The p-value is between .10 and .20.

8.16 $H_0: \mu_X = \mu_Y$ versus $H_1: \mu_X \neq \mu_Y$. Reject H_0 if $T_R < -t_{.025,19} = -2.0930$ or if $T_R > t_{.025,19} = 2.0930$. $T_R = -2.60/(11.99/\sqrt{20}) = -.9695$, so H_0 is accepted in agreement with the paired t-test. The p-value is between .20 and .50.

8.17 (a) $H_0: \mu_X = \mu_Y$ versus $H_1: \mu_X \neq \mu_Y$. Reject H_0 if $T_R < -t_{.025,12} = -2.1788$ or if $T_R > t_{.025,12} = 2.1788$. $T_R = 4.69/(6.63/\sqrt{13}) = 2.5499$, so H_0 is rejected. The p-value is between .02 and .05.

  (b) $\bar{D} = 7$ and $s_D = 11.66$. Standardized sample values: $-.77, -.77, -.69, -.60, -.51, -.51, -.43, -.34, -.26, -.26, 1.46, 1.63, 2.06$. The normality assumption is rejected for the differences.

  (c) $T = 7/(11.66/\sqrt{13}) = 2.1642$, so H_0 is accepted with the p-value close to .05. This is in disagreement with the WSR test and this can be attributed to the nonnormality of the D_i; that is, the paired t-test has no meaning when the normality assumption is not satisfied.

8.18 Reject H_0 if $T > t_{.05,11} = 1.7959$. $T = 3.67/(6.65/\sqrt{12}) = 1.9096$, so H_0 is rejected. The p-value is between .025 and .05 and the Lilliefors test indicates the assumption of normality to be reasonable for these data.

8.19 Reject H_0 if $T_R > t_{.05,11} = 1.7959$. $T_R = 3.50/(6.76/\sqrt{12}) = 1.7947$, so H_0 is accepted with a p-value of .05.

8.23 $\bar{D} = .108$, $s = .387$. Standardized sample values: $-1.05, -.54, -.54, -.54, -.54, -.28, -.28, -.02, -.02, .24, .75, 2.82$. The Lilliefors test shows the assumption of normality to not be reasonable for these data.

8.24 The appropriate test is the Wilcoxon signed ranks test. $H_0: \mu_N = \mu_J$ versus $H_1: \mu_N \neq \mu_J$. Reject H_0 if $T_R > t_{.025,11} = 2.2010$ or if $T_R < -t_{.025,11} = -2.2010$. $T_R = 1.17/(7.48/\sqrt{12}) = .5402$, so H_0 is accepted with a p-value between .50 and .80.

8.25

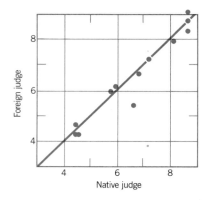

8.26 The techniques of this chapter cannot be used to find a meaningful 95 percent confidence interval since normality was a requirement for small sample sizes and Exercise 8.23 showed normality not to be a reasonable assumption.

9.1 $(18000 - 12000) \pm 1.9600 \sqrt{\frac{(2100)^2}{30} + \frac{(820)^2}{30}}$ or 5193 to 6807. The confidence is 95

percent that the true difference is contained in this interval.

9.3 $H_0: \mu_X = \mu_Y$ versus $H_1: \mu_X \neq \mu_Y$. Reject H_0 if $Z < -z_{.025} = -1.9600$ or if $Z > z_{.025} = 1.9600$.

$$Z = (.90 - .70) \bigg/ \sqrt{\frac{.0081}{30} + \frac{.0256}{30}} = 5.97$$

H_0 is soundly rejected with a p-value less than .001. 95 percent confidence interval:

$$(.90 - .70) \pm 1.9600 \sqrt{\frac{.0081}{30} + \frac{.0256}{30}} = .13 \text{ to } .27 \text{ hours.}$$

9.5 $H_0: \mu_X = \mu_Y$ versus $H_1: \mu_X \neq \mu_Y$. Reject H_0 if $Z < -z_{.05} = -1.6449$ or if $Z > z_{.05} = 1.6449$.

$$Z = (15.88 - 15.58) \bigg/ \sqrt{\frac{1.5625}{45} + \frac{1.4641}{45}} = 1.1568$$

H_0 is accepted and the p-value is about .248.

9.6 $H_0: \mu_X = \mu_Y$ versus $H_1: \mu_X \neq \mu_Y$. Reject H_0 if $Z < -z_{.005} = -2.5758$ or if $Z > z_{.005} = 2.5758$.

$$Z = (23.8 - 23.4) \bigg/ \sqrt{\frac{2.89}{50} + \frac{3.61}{50}} = 3.0769$$

H_0 is rejectd with a p-value of about .002. The paired comparisons would be the better way to do the comparison as both gasoline types are tried on the same drivers, otherwise there is no control over the variation that might occur from two different groups of drivers.

9.7 $H_0:\mu_X = \mu_Y$ versus $H_1:\mu_X \neq \mu_Y$. Reject H_0 if $Z < -z_{.025} = -1.9600$ or if $Z > z_{.025} = 1.9600$.

$$Z = (28.3 - 26.7) \Big/ \sqrt{\frac{10.89}{30} + \frac{24.01}{40}} = 1.6302$$

H_0 is accepted and the p-value is about .104.

9.9 The Lilliefors test shows the assumption of normality to be reasonable for both sets of data. Reject H_0 if $T < -t_{.05,51} = -1.6753$.

$$\bar{D} = -8.42, \quad s_1^2 = 310.63, \quad s_2^2 = 217.34, \quad s_p^2 = 257.58, \quad s_{\bar{D}} = 4.45$$
$$T = -8.42/4.45 = -1.8924$$

H_0 is rejected and the p-value is between .025 and .05.

9.10
$$-8.42 \pm 2.0076\,(16.0493)\,\sqrt{\frac{1}{23} + \frac{1}{30}} \quad \text{or} \quad -17.35 \text{ to } .51.$$

9.11 The test for equal variances shows $F = 310.63/217.34 = 1.43$, which is compared with $F_{.05,22,29} = 1.984$, so the variances are not declared to be significantly different. Note the normality was checked in Exercise 9.9. The only difference between this exercise and Exercise 9.9 is in the degrees of freedom used with the Student's t random variable. The previous degrees of freedom 51 is replaced by

$$f = \frac{(310.63/23 + 217.34/30)^2}{\left(\dfrac{(310.63/23)^2}{22} + \dfrac{(217.34/30)^2}{29}\right)} = 42.6$$

Interpolation in Table A3 yields $-t_{.05,42.6} = -1.6815$, which is only slightly different from the value of -1.6753 used in Exercise 9.9. The decision is the same as in Exercise 9.9.

9.12 $H_0: \mu_X = \mu_Y$ versus $H_1: \mu_X \neq \mu_Y$. Reject H_0 if $T < -t_{.05,13} = -1.7709$ or if $T > t_{.05,13} = 1.7709$.

$$s_p = 11.95, \quad T = (52.33 - 45.17)/11.95 \, \sqrt{\frac{1}{9} + \frac{1}{6}} = 1.1368$$

H_0 is accepted and the p-value is about .149, which is about half the p-value used with the large sample test in Exercise 9.2.

9.15 $H_0: \mu_X = \mu_Y$ versus $H_1: \mu_X \neq \mu_Y$. Reject H_0 if $T < -t_{.025,38} = -2.0244$ or if $T > t_{.025,38} = 2.0244$.

$$T = (335 - 375)/21.8 \, \sqrt{\frac{1}{20} + \frac{1}{20}} = -2.9034$$

H_0 is rejected and the p-value is between .002 and .01.

9.17 (a) $H_0: \mu_X = \mu_Y$ versus $H_1: \mu_X \neq \mu_Y$. Reject H_0 if $T_R < -t_{.025,18} = -2.1009$ or if $T_R > t_{.025,18} = 2.1009$.

$$T_R = (7.89 - 12.64)/3.71 \, \sqrt{\frac{1}{9} + \frac{1}{11}} = -2.8469$$

H_0 is rejected and the p-value is between .005 and .01.

(b) Sample 1: $\bar{X} = 649.7$, $s_X = 96.53$. Standardized sample values: -1.09, $-.65$, $-.60$, $-.38$, $-.26$, $-.21$, $-.13$, 1.32, 2.00. The assumption of normality is rejected based on the Lilliefors graph.

Sample 2: $\bar{Y} = 691.8$, $s_Y = 92.17$. Standardized sample values: $-.91$, $-.76$, $-.62$, $-.52$, $-.46$, $-.30$, $-.26$, $-.26$, $.24$, 1.75, 2.10. The assumption of normality is rejected based on the Lilliefors graph.

$$T = (649.7 - 691.8)/94.13 \sqrt{\frac{1}{9} + \frac{1}{11}} = -.9950$$

H_0 would not be rejected with the two-sample t-test; however, this test has no meaning since the assumption of normality did not seem reasonable for these data.

(c)

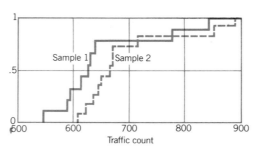

The graph of the e.d.f.'s supports the WMW decision on part (a) as Sample 2 is shifted to right of Sample 1.

(d) This would have removed time of day as a source of variation and probably would have made a more accurate comparison, particularly since the sample sizes are so small.

9.19 H_0: $\mu_X = \mu_Y$ versus H_1: $\mu_X \neq \mu_Y$. Reject H_0 if $T_R < -t_{.05,491} = -1.6449$ or if $T_R > t_{.05,491} = 1.6449$. The test parallels the example in the text due to the large number of ties.

$$n_X = 233, \quad \bar{R}_X = 241.6, \quad s_{R_X}^2 = 19891.3$$

$$n_Y = 260, \quad \bar{R}_Y = 251.8, \quad s_{R_Y}^2 = 20116.7, \quad s_p = 141.5$$

$$T_R = (241.6 - 251.8)/141.5 \sqrt{\frac{1}{233} + \frac{1}{260}} = -.8022$$

H_0 is accepted and the p-value is about .422.

9.21 H_0: $\mu_X = \mu_Y$ versus H_1: $\mu_X \neq \mu_Y$. Reject H_0 if $T_R < -t_{.05,13} = -1.7709$ or if $T_R > t_{.05,13} = 1.7709$.

$$n_X = 9, \quad \bar{R}_X = 8.89, \quad s_{R_X}^2 = 17.61$$

$$n_Y = 6, \quad \bar{R}_Y = 6.67, \quad s_{R_Y}^2 = 23.77, \quad s_p = 4.47$$

$$T_R = (8.89 - 6.67)/4.47 \sqrt{\frac{1}{9} + \frac{1}{6}} = .9424$$

H_0 is accepted with a p-value between .20 and .50, which is in agreement with the previous test.

9.23 $H_0: \mu_X = \mu_Y$ versus $H_1: \mu_X \neq \mu_Y$. Reject H_0 if $T_R < -t_{.025,18} = -2.1009$ or if $T_R > t_{.025,18} = 2.1009$.

$$\bar{R}_X = 7.90, \quad s^2_{R_X} = 30.99$$

$$\bar{R}_Y = 13.10, \quad s^2_{R_Y} = 27.66, \quad s_p = 5.42$$

$$T_R = (7.90 - 13.10)/5.42 \sqrt{\frac{1}{10} + \frac{1}{10}} = -2.1473$$

H_0 is rejected and the p-value is between .02 and .05, which is in agreement with the results from the two-sample t-test.

9.25 Reject H_0 if $T > t_{.05,18} = 2.5524$.

$$T = \frac{1.24 - 1.00}{.0474 \sqrt{\frac{1}{10} + \frac{1}{10}}} = 11.3137$$

so H_0 is soundly rejected. 95 percent confidence interval:

$$(1.24 - 1.00) \pm 2.1009(.0474)\sqrt{\frac{1}{10} + \frac{1}{10}} \text{ or } .20 \text{ to } .28.$$

9.27 $F = 9/5 = 1.8$; reject the null hypothesis of equal variances if $F > F_{.05,39,59} = 1.637$ from Table A5. So the null hypothesis is rejected with a p-value close to .025.

9.29 $F = 1.19/.22 < 5.33$; reject the null hypothesis of equal variances if $F > F_{.05,15,12} = 2.475$ from Table A5. So the null hypothesis is rejected with a p-value between .001 and .01.

9.31 The following graph, which shows the distribution of octanes at 1200 feet that shift to the right, would seem to justify the decision of the approximate test in Exercise 9.30.

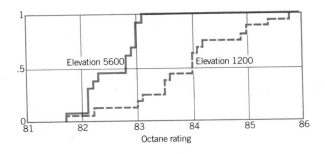

9.33 The separation of the graphs of the e.d.f.'s would seem to justify the decision in Exercise 9.32.

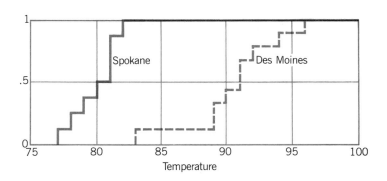

10.1

	Men	Women	Total
Union	48	22	70
Non-union	15	15	30
Total	63	37	100

Reject H_0 if $\phi > z_{.05}/\sqrt{100} = .1645$.

$$\phi = \frac{48 \cdot 15 - 15 \cdot 22}{\sqrt{63 \cdot 37 \cdot 30 \cdot 70}} = .1763$$

H_0 is rejected and the p-value is about .039.

10.3 Reject H_0 if $T > \chi^2_{.05,1} = 3.841$.

$$T = \frac{200 (5 \cdot 85 - 15 \cdot 95)^2}{20 \cdot 180 \cdot 100 \cdot 100} = 5.556$$

H_0 is rejected and the p-value is between .01 and .025.

10.5 Reject H_0 if $T > \chi^2_{.10,1} = 2.706$.

$$T = \frac{40 (8 \cdot 9 - 9 \cdot 14)^2}{17 \cdot 23 \cdot 22 \cdot 18} = .753$$

H_0 is accepted and the p-value is between .25 and .40.

10.7 $\phi = \dfrac{8 \cdot 9 - 14 \cdot 9}{\sqrt{22 \cdot 18 \cdot 17 \cdot 23}} = -.137$

10.9 $\displaystyle\sum_{\substack{\text{all} \\ \text{cells}}} \frac{(O_{ij} - E_{ij})^2}{E_{ij}} = 52.28$ from Equation (10.2), which is in agreement with Equation

(10.3); however, Equation (10.3) is easier to compute.

10.11 Reject H_0 if $T > \chi^2_{.05,4} = 9.488$.

$$T = \sum_{\substack{all \\ cells}} \frac{O_{ij}^2}{E_{ij}} - N = 242.05 - 200 = 42.05$$

H_0 is soundly rejected and the p-value is less than .0001.

10.13 Reject H_0 if $T > \chi^2_{.01,2} = 9.210$.

$$T = \sum_{\substack{all \\ cells}} \frac{O_{ij}^2}{E_{ij}} - N = 308.28 - 306 = 2.28$$

H_0 is accepted and the p-value is between .25 and .40.

10.15 Reject H_0 if $T > \chi^2_{.10,2} = 4.605$.

$$T = \sum_{\substack{all \\ cells}} \frac{O_{ij}^2}{E_{ij}} - N = 261.87 - 260 = 1.87$$

H_0 is accepted and the p-value is between .25 and .40.

10.17 Reject H_0 if $T > \chi^2_{.05,9} = 16.92$.

$$T = \sum_{i=1}^{k} \frac{O_i^2}{15} - 150 = 158.80 - 150 = 8.80$$

H_o is accepted and the p-value is more than .40.

10.19 Reject H_0 if $T > \chi^2_{.01,7} = 18.48$.

$$T = \sum_{i=1}^{k} \frac{O_i^2}{149} - 1192 = 1216.15 - 1192 = 24.15$$

H_0 is rejected and the p-value is about .001.

10.21 Reject H_0 if $T > \chi^2_{.05,6} = 12.59$.

$$T = \sum_{i=1}^{k} \frac{O_i^2}{Np_i} - N = 108.95 - 100 = 8.95$$

H_0 is accepted and the p-value is between .10 and .25.

10.23 Reject H_0 if $T > \chi^2_{.01,11} = 24.72$.

$$T = \sum_{i=1}^{k} \frac{(O_i - E_i)^2}{E_i} = 15.08$$

H_0 is accepted and the p-value is between .10 and .25.

10.25 Reject H_0 if $T > \chi^2_{.05,2} = 5.991$.

$$T = \sum_{\substack{all \\ cells}} \frac{O_{ij}^2}{E_{ij}} - N = 303.78 - 300 = 3.78$$

H_0 is accepted and the p-value is between .10 and .25.

10.27   Reject H_0 if $T > \chi^2_{.05,3} = 7.815$.

$$T = \sum_{\substack{\text{all} \\ \text{cells}}} \frac{O^2_{ij}}{E_{ij}} - N = 1004.56 - 1000 = 4.56$$

H_0 is accepted with a p-value between .10 and .25.

10.29   Reject H_0 if $T > \chi^2_{.01,6} = 16.81$.

$$T = \sum_{i=1}^{k} \frac{O^2_i}{E_i} - N = 154.00 - 148 = 6.00$$

H_0 is accepted with a p-value of approximately .40.

10.31

	> 110	< 110	Total
Completed Course	38	6	44
Didn't Complete Course	8	13	21
Total	46	19	65

Reject the hypothesis of independence of aptitude test score and course if $\phi > z_{.05}/\sqrt{n} = 1.6449/\sqrt{65} = .2040$.

$$\phi = \frac{38 \cdot 13 - 8 \cdot 6}{\sqrt{46 \cdot 19 \cdot 21 \cdot 44}} = .4963$$

H_0 is soundly rejected with a p-value less than .001.

10.33

	A	B	C	D	E	F	Total
Defective	6	7	17	8	10	4	52
Not Defective	94	93	83	92	90	96	548
Total	100	100	100	100	100	100	600

Reject H_0 if $T > \chi^2_{.05,5} = 11.07$.

$$T = \sum_{\substack{\text{all} \\ \text{cells}}} \frac{O^2_{ij}}{E_{ij}} - N = 613.054 - 600 = 13.054$$

H_0 is rejected with a p-value between .01 and .025.

11.1   Reject H_0 if $r < -T_{.05,8} = -.621$ from Figure 11.6. From Exercise 4.34 $r = -.963$ and H_0 is rejected with a p-value of less than .005. The decision means that tuition and fees decrease as government appropriations increase.

11.3   For each budget type reject H_0 if $r > T_{.025,8} = .707$ or if $r < -T_{.025,8} = -.707$ from Figure 11.6. From Exercise 4.48 $r_L = .174$, $r_I = .176$, and $r_H = .202$, so H_0 is accepted for each of the three budget types and the p-values in each case are much greater than .10.

For all data pooled together H_0 is rejected if $r > T_{.025,24} = .404$ or if $r < -T_{.025,24} = -.404$ from Figure 11.6. From Exercise 4.48 $r = .909$ and H_0 is soundly rejected with a p-value much less than .01.

If the scatterplot of Exercise 2.28 is consulted, a strong linear trend is apparent for all points taken together; however, the two outliers (Cincinnati and Milwaukee) hold down the correlation within each budget type.

11.5   Reject H_0 if $r > T_{.05,10} = .549$ or if $r < -T_{.05,10} = -.549$ from Figure 11.6. From Exercise 4.47 $r = .174$ and H_0 is accepted with a p-value much greater than .10. However, the assumption of normality is easily rejected for the X values; hence, the bivariate normality of these data is not justified and the results of this test are without meaning.

11.7   From Exercise 4.45 $r = .034$ and $w \approx .034$ from Figure 11.5. $w_L = .034 - 1.9600/\sqrt{20} = -.404$ and $w_U = .034 + 1.9600/\sqrt{20} = .472$ and $r_L = -.38$ and $r_U = .44$ from Figure 11.5 so the confidence interval is from $-.38$ to $.44$. There is 95 percent confidence that this interval contains the true value of the population correlation coefficient.

11.9   From Exercise 4.53 $r_s = .588$ and $r = .174$ from Exercise 4.47. The correlation coefficient on raw data was influenced by the outliers in the data. Reject H_0 if $|r_s| > T^*_{.05,10} = .552$ from Figure 11.9, so H_0 is rejected with a p-value between .05 and .10.

11.11   Reject H_0 if $r_s > T^*_{.005,9} = .817$ from Figure 11.9.

$$\Sigma R_X = 45, \quad \Sigma R_Y = 45, \quad \Sigma R_X R_Y = 253$$
$$\Sigma R_X^2 = 285, \quad \Sigma R_Y^2 = 285, \quad r_s = .467$$

H_0 is accepted and the p-value is more than .05.

11.13 H_0:The order of finish in the general election is independent of the finish in the primary election.

H_1:The order of finish in the general election agrees with the order of finish in the primary election.

Reject H_0 if $r_s > T^*_{.05,6} = .800$ from Figure 11.9.

$$\Sigma R_X = 21, \quad \Sigma R_Y = 21, \quad \Sigma R_X R_Y = 87$$
$$\Sigma R_X^2 = 91, \quad \Sigma R_Y^2 = 91, \quad r_s = .771$$

H_0 is accepted but the sample size is small and the p-value is slightly larger than .05.

11.15   Reject H_0 if $r_s > T^*_{.05,5} = .800$ from Figure 11.9.

$$\Sigma R_X = 15, \quad \Sigma R_Y = 15, \quad \Sigma R_X R_Y = 51$$
$$\Sigma R_X^2 = 55, \quad \Sigma R_Y^2 = 55, \quad r_s = .600$$

H_0 is accepted and the p-value is greater than .05.

11.17   Reject H_0 if $r > T_{.05,10} = .549$ from Figure 11.6. From Exercise 4.55 $r = .9471$ and H_0 is rejected with a p-value less than .005.

11.19   Reject H_0 if $r > T_{.05,10} = .549$ from Figure 11.6.

$$\Sigma X = 672.2, \quad \Sigma Y = 558.7, \quad \Sigma XY = 40601.49$$
$$\Sigma X^2 = 48227.06, \quad \Sigma Y^2 = 34733.27, \quad r = .9310$$

H_0 is soundly rejected with a p-value less than .005.

11.20   Reject H_0 if $r_s > T^*_{.05,10} = .552$ from Figure 11.9.

$\Sigma R_X = 55,$ $\Sigma R_Y = 55,$ $\Sigma R_X R_Y = 376$

$\Sigma R_X^2 = 385,$ $\Sigma R_Y^2 = 385,$ $r_s = .8906$

H_0 is soundly rejected with a p-value less than .005.

11.21   Reject H_0 if $r_s > T^*_{.05,10} = .552$ from Figure 11.9.

$\Sigma R_X = 55,$ $\Sigma R_Y = 55,$ $\Sigma R_X R_Y = 376$

$\Sigma R_X^2 = 385,$ $\Sigma R_Y^2 = 385,$ $r_s = .8424$

H_0 is soundly rejected with a p-value less than .005.

11.23   Age at inauguration. $\overline{X} = 54.29$, $s = 5.97$; standardized sample values: -2.06, -1.89, -1.39, -1.22, -1.05, $-.89$, $-.89$, $-.72$, $-.72$, $-.55$, $-.55$, $-.55$, $-.55$, $-.38$, $-.05$, $-.05$, $-.05$, $-.05$, .12, .12, .12, .29, .29, .45, .45, .45, .45, .62, .96, 1.12, 1.12, 1.29, 1.63, 1.80, 2.30. Based on the Lilliefors graph, the assumption of normality seems reasonable for these data.

Age at death. $\overline{Y} = 68.80$, $s = 11.26$; standardized sample values: -2.03, -1.76, -1.40, -1.14, -1.14, -1.05, $-.96$, $-.78$, $-.78$, $-.52$. $-.52$, $-.43$, $-.43$, $-.34$, $-.25$, $-.16$ $-.16$, $-.16$, $-.07$, .11, .20, .20, .28, .37, .46, .73, .82, .82, .91, .99, 1.26, 1.44., 1.71, 1.88, 1.88. Based on the Lilliefors graph, the assumption of normality seems reasonable for these data.

$$r = \Sigma \frac{Z_I Z_D}{n-1} = .5379$$

12.1   Total sum of squares $= 84.83$

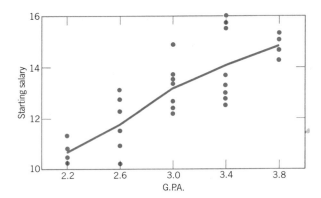

12.2

G.P.A.:	2.2	2.6	3.0	3.4	3.8
$\overline{X}$:	10.63	11.75	13.14	14.05	14.84

12.3

G.P.A.:	2.2	2.6	3.0	3.4	3.8
s^2:	.33	1.17	.84	2.09	.15

Error SS: $.99 + 5.86 + 5.06 + 14.60 + .59 = 27.09$

12.4   Total SS $= 84.83$, $r^2 = 1 - 27.09/84.83 = .68$, which implies that 68 percent of the variation in starting salaries is explained by the regression curve using G.P.A.

12.9 $r^2 = .826$, which means that 82.6 percent of the variation in G.P.A. is explained by the ACT score.

12.11 $\Sigma X = 293.5, \Sigma Y = 434.3, \Sigma XY = 10938.52, \Sigma X^2 = 7307.31, \Sigma Y^2 = 16575.11, r^2 = .906$

12.13 $\Sigma X = 91.6, \Sigma Y = 391.6, \Sigma XY = 1216.92, \Sigma X^2 = 287.60, \Sigma Y^2 = 5196.52, r^2 = .672,$ $\hat{\mu}_{Y|X} = 4.86 + 2.683X$

12.14 $\hat{\mu}_{Y|X} = 4.86 + 2.683(3.45) = 14.1$ thousand

12.17 $\hat{\sigma} = [2.925 - (2.683)^2(.273)]^{1/2}[29/28]^{1/2} = .998$. Standardized residuals $(Y_i - \hat{Y}_i)/\hat{\sigma}$: $-1.64, -1.49, -1.19, -1.09, -.94, -.81, -.76, -.68, -.66, -.61, -.46, -.36, -.34,$ $-.31, -.28, -.26, -.06, .04, .24, .36, .39, .49, .59, .64, .86, 1.17, 1.52, 1.72, 1.89,$ 2.02.

12.19 The variance associated with the residuals for the smallest 17 values of X (G.P.A. of 3.0 or less) is $s_1^2 = .7740$ and the variance associated with the residuals for the largest 13 values of X (G.P.A. of 3.4 or greater) is $s_2^2 = 1.2866$. The F ratio is $F = 1.2866/.7740 = 1.6622$. This value is compared with the .95 quantile of an F distribution with 12 and 16 degrees of freedom, which is $F_{.05,12,16} = 2.425$, so the assumption of homogeneity of variance seems reasonable for these regression data.

12.21 From the calculations in Exercise 12.13 $s_X^2 = .273$ and $s_Y^2 = 2.925$. From Equation (12.9) $\hat{\sigma}^2 = \frac{29}{28}[2.925 - (2.683)^2(.273)] = .995$ and $\hat{\sigma} = .998$. From Equations (12.11) and (12.12),

$$\beta_L = 2.683 - t_{.025,28} .998/\sqrt{.273(29)} = 1.956$$
$$\beta_U = 2.683 + t_{.025,28} .998/\sqrt{.273(29)} = 3.410$$

The 95 percent confidence interval is from 1.956 to 3.410.

12.23 Reject H_0 if $T > t_{.05,28} = 1.7011$.

$$T = \frac{(2.683 - 2)\sqrt{.273(29)}}{.998} = 1.926$$

H_0 is rejected and the p-value is between .025 and .05.

12.25 Reject H_0 if $T > t_{.05,14} = 1.7613$. $\Sigma X = 136, \Sigma Y = 2208, \Sigma XY = 19932, s_X = 4.761,$ $s_Y = 31.016, \Sigma X^2 = 1496, \Sigma Y^2 = 319,134, \hat{\sigma} = 27.314, \hat{\mu}_{Y|X} = 108.9 + 3.424X$.

$$T = \frac{(3.424 - 0)(4.761)\sqrt{15}}{27.314} = 2.3115$$

H_0 is rejected and the p-value is between .01 and .025.

12.26 From Equation (12.11) and (12.12),

$$\beta_L = 3.424 - t_{.05,14} 27.314/(4.761)\sqrt{15} = .815$$
$$\beta_U = 3.424 + t_{.05,14} 27.314/(4.761)\sqrt{15} = 6.033$$

The 90 percent confidence interval is from .815 to 6.033.

12.27 $\hat{\mu}_{Y|X} = 108.9 + 3.424(17) = 167.1$. From Equations (12.15) and (12.16),

$$\mu_L = 167.1 - t_{.025,14} 27.314\sqrt{\frac{1}{16} + \frac{(17 - 8.5)^2}{15(4.761)^2}} = 136.4$$

$$\mu_U = 167.1 + t_{.025,14}27.314\sqrt{\frac{1}{16} + \frac{(17 - 8.5)^2}{15(4.761)^2}} = 197.8$$

The 95 percent confidence interval is from 136.4 to 197.8.

12.28 From Equations (12.17) and (12.18),

$$Y_L = 167.1 - t_{.025,14}27.314\sqrt{1 + \frac{1}{16} + \frac{(17 - 8.5)^2}{15(4.761)^2}} = 101.0$$

$$Y_U = 167.1 + t_{.025,14}27.314\sqrt{1 + \frac{1}{16} + \frac{(17 - 8.5)^2}{15(4.761)^2}} = 233.2$$

The 95 percent prediction interval is from 101.0 to 233.2.

12.31 Reject H_0 if $r_s > T^*_{05,30} = .306$ from Figure 11.9. $r_s = .140$, so H_0 is accepted with a p-value greater than .05. This is in contrast to Exercise 12.23, which rejected H_0 with a p-value between .025 and .05.

12.33 $\hat{\sigma} = [1.0615 - (.08925)^2(44.3231)]^{1/2}[9/8]^{1/2} = .893$. Standardized residuals $(Y_i - \hat{Y}_i)/\hat{\sigma}$: $-1.37, -1.08, -.95, -.37, -.26, .42, .54, .78, .92, 1.38$. The assumption of normality seems reasonable for these residuals based on the Lilliefors test.

12.35 Reject H_0 if $|r_s| > T^*_{.025,8} = .714$ from Figure 11.9 for each budget type. From Exercise 4.54 the rank correlation coefficients for each budget type are respectively .204, .156, and .143. In each case H_0 would be accepted with a p-value much greater than .10.

12.36 Reject H_0 if $|r_s| > T^*_{.025,24} = .406$ from Figure 11.9. From Exercise 4.54 $r_s = .909$, so H_0 is soundly rejected with a p-value less than .005. From the scatterplot of these data given in Exercise 2.28, a strong linear relationship is indicated for all data taken together, but within each budget type the outliers represented by Cincinnati and Milwaukee detracts from the linear relationship exhibited by the other six points.

12.39 $\Sigma X = 210, \Sigma Y = 92, \Sigma XY = 1341.8, \Sigma X^2 = 2870, \Sigma Y^2 = 685.56, \hat{\mu}_{Y|X} = -1.334 + .565X$

12.40 The serial correlation on the residuals is $r_1 = .824$. From Figure 11.6 the approximate critical value for $\alpha = .005$ is .561, so a serial correlation of .824 could safely be considered to be significant. Hence, the linear fit to the data is inadequate.

12.42 $\Sigma R_x = 210, \Sigma R_y = 210, \Sigma R_x R_y = 2848.5, \Sigma R_x^2 = 2870, \Sigma R_y^2 = 2869.5, \hat{R}_y = .339 + .968R_x$

12.43 From Equation (12.24) $R_{x_0} = 16.5$ and $\hat{R}_y = .339 + .968(16.5) = 16.31$.

From Equation (12.25) $\hat{Y}_0 = 6.5 + \left(\frac{16.31 - 16}{17 - 16}\right)(8.0 - 6.5) = 6.96$.

12.44 The serial correlation on the rank residuals is $r_1 = .249$. From Figure 11.6 the p-value is much greater than .05, so a linear fit on the ranks seem adequate.

12.45 For raw data, Error SS $= \Sigma(Y_i - \hat{Y}_i)^2 = 50.03$. For ranks, Error SS $= \Sigma(Y_i - \hat{Y}_i)^2 = 3.42$ from using Equation (12.25) to get $\hat{Y}_i$.

12.47 $\hat{\mu}_{Y|X} = -266.53 + 6.14(72) = 175.4$

$\hat{\sigma} = [644.40 - (6.14)^2(15.34)]^{1/2}[9/8]^{1/2} = 8.641$

$$\mu_L = 175.4 - t_{.025,8}8.641\sqrt{\frac{1}{10} + \frac{(72 - 69.3)^2}{9(15.34)}} = 167.6$$

$$\mu_U = 175.4 - t_{.025,8}8.641\sqrt{\frac{1}{10} + \frac{(72 - 69.3)^2}{9(15.34)}} = 183.2$$

The 95 percent confidence interval is from 167.6 to 183.2.

12.49 $\bar{Y}_1 = 25.8, \bar{Y}_2 = 27.8, \bar{Y}_3 = 30.83, \bar{Y}_4 = 38.07, \bar{Y}_5 = 48.03$

12.51 $R^2 = 1 - 53.77/2029.17 = .97$, which means that 97 percent of the variation in the amount of wear is explained by machine speed.

12.53 $\hat{\mu}_{Y|X} = -6.95 + .27(170) = 39.6$

12.55   The variance associated with the smallest 18 values of X (110, 130, 150) is $s_1^2 = .85$, and the variance associated with the residuals for the largest 12 values of X (170, 190) is $s_2^2 = 1.12$. The F ratio is $1.12/.85 = 1.31$. This value is compared with the .95 quantile of an F distribution with 11 and 17 degrees of freedom, which is $F_{.05,11,17} = 2.381$, so the assumption of homogeneity of variance seems reasonable for these regression data.

12.57   Reject H_0 if $T > t_{.05,28} = 1.7011$.

$$T = \frac{.274(28.768)\sqrt{29}}{2.877} = 14.737$$

H_0 is soundly rejected with a p-value less than .0001.

12.60 $\Sigma R_x = 465, \Sigma R_y = 465, \Sigma R_x R_y = 9352.5, \Sigma R_x^2 = 9454, \Sigma R_y^2 = 9367.5, \hat{R}_y = .70 + .955R_x$

12.61   The rank of 170 is 15.5. Therefore, $\hat{R}_y = .70 + .955(15.5) = 15.5$ and, from Equation (12.25),

$$\hat{Y} = 30.4 + \left(\frac{15.5 - 15}{16 - 15}\right)(31.4 - 30.4) = 30.9$$

13.1

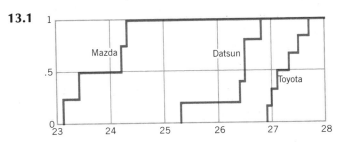

13.2   Based on the e.d.f.'s in Exercise 13.1, it would seem reasonable to reject H_0 with $\mu_M < \mu_D < \mu_T$.

13.5

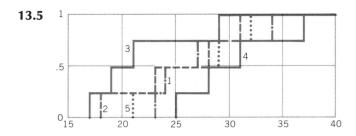

13.6   There is so much overlap present in the e.d.f.'s in Exercise 13.5 that the null hypothesis seems reasonable.

13.9   The test assumptions are reasonable for these data. The data yield the following calculations.

$$\Sigma X = 390, \quad \Sigma X^2 = 10172.94, \quad \bar{X} = 26, \quad \bar{X}_T = 27.25, \quad \bar{X}_D = 26.30, \quad \bar{X}_M = 23.75$$

Based on the following computer printout the null hypothesis is rejected with a p-value of .0001.

Source	DF	Sum of Squares	Mean Square	F Value	PR > F
TREATMENT	2	30.075	15.03750	62.98	.0001
ERROR	12	2.865	.23875		
TOTAL	14	32.940			

13.10   The conclusion in Exercise 13.9 to reject H_0 is in agreement with the first tentative guess made in Exercise 13.2.

13.13   The test assumptions are reasonable for these data. The data yield the following calculations:

$$\Sigma X = 520, \quad \Sigma X^2 = 14094, \quad \bar{X} = 26, \quad \bar{X}_1 = 27, \quad \bar{X}_2 = 25,$$
$$\bar{X}_3 = 21.5, \quad \bar{X}_4 = 30.25, \quad \bar{X}_5 = 26.25$$

Based on the following computer printout, the null hypothesis is accepted with a p-value of .2610.

Source	DF	Sum of Squares	Mean Square	F Value	PR > F
TREATMENT	4	161.5	40.375	1.47	.2610
ERROR	15	412.5	27.500		
TOTAL	19	574.0			

13.14   The conclusion in Exercise 13.13 to accept H_0 is in agreement with the first tentative guess made in Exercise 13.6.

13.17 $\Sigma X = 3753.59$, $\Sigma X^2 = 282038.3497$, $\bar{X} = 75.0718$, $\bar{X}_{SL} = 74.093$, $\bar{X}_{SF} = 76.482$, $\bar{X}_O = 72.544$, $\bar{X}_W = 77.264$, $\bar{X}_D = 74.976$

From Equation (13.1),

Total SS $= 282038.3497 - 50(75.0718)^2 = 249.5919$

From Equation (13.2),

$$SST = 10(74.093)^2 + 10(76.482)^2 + 10(72.544)^2 + 10(77.264)^2$$
$$+ 10(74.976)^2 - 50(75.0718)^2 = 141.5141$$

From Equation (13.3),

SSE $= 249.5919 - 141.5141 = 108.0778$

13.18 Toyota: $X_{i1} = \mu_1 + \epsilon_{i1}$, $i = 1, \ldots, n_1$
Datsun: $X_{i2} = \mu_2 + \epsilon_{i2}$, $i = 1, \ldots, n_2$
Mazda: $X_{i3} = \mu_3 + \epsilon_{i3}$, $i = 1, \ldots, n_3$

The values μ_1 μ_2, and μ_3 are the respective mean m.p.g. for Toyota, Datsun, and Mazda. In each case the ϵ_{ij} are independent, identically distributed normal random variables with mean zero.

13.19 Battery 1: $X_{i1} = \mu_1 + \epsilon_{i1}$, $i = 1, \ldots, n_1$
Battery 2: $X_{i2} = \mu_2 + \epsilon_{i2}$, $i = 1, \ldots, n_2$
Battery 3: $X_{i3} = \mu_3 + \epsilon_{i3}$, $i = 1, \ldots, n_3$
Battery 4: $X_{i4} = \mu_4 + \epsilon_{i4}$, $i = 1, \ldots, n_4$

The values μ_1, μ_2, μ_3, and μ_4 are the respective mean lives for Batteries 1, 2, 3, and 4. In each case the ϵ_{ij} are independent, identically distributed normal random variables with mean zero.

13.20 Refer to the computations in Exercise 13.9. For Toyota and Datsun,

$$LSD_{.05} = t_{.025,12} \sqrt{.23875} \, (\tfrac{1}{6} + \tfrac{1}{5})^{1/2} = .645 \quad \text{and} \quad \bar{X}_T - \bar{X}_D = .95$$

which exceeds .645, so these means are declared to be significantly different.
For Toyota and Mazda,

$$LSD_{.05} = t_{.025,12} \sqrt{.23875} \, (\tfrac{1}{6} + \tfrac{1}{4})^{1/2} = .687 \quad \text{and} \quad \bar{X}_T - \bar{X}_M = 3.5$$

which exceeds .687, so these means are declared to be significantly different.
For Datsun and Mazda,

$$LSD_{.05} = t_{.025,12} \sqrt{.23875} \, (\tfrac{1}{5} + \tfrac{1}{4})^{1/2} = .714 \quad \text{and} \quad \bar{X}_D - \bar{X}_M = 2.55$$

which exceeds .714, so these means are declared to be significantly different. In summary, $\mu_M < \mu_D < \mu_T$.

13.21 The conclusion in Exercise 13.20 is the same as the ordering made on the second tentative guess in Exercise 13.2.

13.24 Since the null hypothesis was not rejected on Exercise 13.13, multiple comparisons should not be made.

13.25 No multiple comparisons were made in Exercise 13.24 and this is in agreement with the conclusion in Exercise 13.6 that the null hypothesis seems reasonable.

13.28 $\Sigma R_x = 120$, $\Sigma R_x^2 = 1239.5$, $\bar{R} = 8$, $\bar{R}_T = 12.5$, $\bar{R}_D = 7$, $\bar{R}_M = 2.5$. Based on the following computer printout the null hypothesis is rejected with a p-value of .0001. This is in agreement with Exercise 13.9.

Source	DF	Sum of Squares	Mean Square	F Value	PR > F
TREATMENT	2	247.5	123.75	46.41	.0001
ERROR	12	32.0	2.67		
TOTAL	14	249.5			

13.29 Refer to the computations in Exercise 13.28. The LSD values are given by the formula $\text{LSD}_{.05} = t_{.025,12}\sqrt{2.67}\,(1/n_i + 1/n_j)^{1/2}$. These values are as follows.

	n_i	n_j	$\text{LSD}_{.05}$	$|\bar{R}_i - \bar{R}_j|$
T,D	6	5	2.154	5.5
T,M	6	4	2.297	10.5
D,M	5	4	2.387	4.5

Based on these LSD values, all pairs of means are declared to be significantly different and the summary is $\mu_M < \mu_D < \mu_T$. This is the same as the conclusion in Exercise 13.20.

13.32 $\Sigma R_x = 210$, $\Sigma R_x^2 = 2866$, $\bar{R} = 10.5$, $\bar{R}_1 = 11.5$, $\bar{R}_2 = 9.5$, $\bar{R}_3 = 5.75$, $\bar{R}_4 = 14.75$, $\bar{R}_5 = 11$. Based on the following computer printout, the null hypothesis is accepted with a p-value of .3096. This is in agreement with Exercise 13.13.

Source	DF	Sum of Squares	Mean Square	F Value	PR > F
TREATMENT	4	171.5	42.875	1.31	.3096
ERROR	15	489.5	32.633		
TOTAL	19	661.0			

13.33 Since the null hypothesis was not rejected in Exercise 13.32, multiple comparisons should not be made. This is in agreement with the decision in Exercise 13.24.

13.36 $\Sigma R_x = 666$, $\Sigma R_x^2 = 16156.5$, $\bar{R} = 18.5$, $\bar{R}_1 = 18.056$, $\bar{R}_2 = 21.313$, $\bar{R}_3 = 7.9$, $\bar{R}_4 = 28.222$. Based on the following computer printout, the null hypothesis is rejected with a p-value of .0001. This is in agreement with the analysis on the raw data.

Source	DF	Sum of Squares	Mean Square	F Value	PR > F
TREATMENT	3	2039.353	679.784	12.11	.0001
ERROR	32	1796.147	56.130		
TOTAL	35	3835.500			

13.37 Refer to the computations in Exercise 13.36. The LSD values are given by the formula $LSD_{.05} = t_{.025,32}\sqrt{56.130}\ (1/n_i + 1/n_j)^{1/2}$. These values are as follows.

i,j	n_i	n_j	$LSD_{.05}$	$|\bar{R}_i - \bar{R}_j|$
1,2	9	8	7.415	3.257
1,3	9	10	7.012	10.156
1,4	9	9	7.194	10.166
2,3	8	10	7.239	13.413
2,4	8	9	7.415	6.909
3,4	10	9	7.012	20.322

Based on these LSD values, the summary of ordered means is $\mu_3 < \mu_1$ and $\mu_3 < \mu_4$. Also, $\mu_1 = \mu_2$ and $\mu_1 < \mu_4$. So far this is in agreement with the analysis on raw data. However, the analysis on ranks shows μ_2 and μ_4 to be not quite significantly different, while the analysis on raw data just showed a significant difference for these means.

13.40

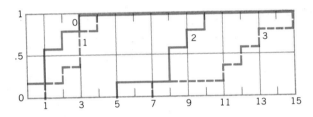

13.41 Based on the separation of the e.d.f.'s, it would seem reasonable to reject H_0 with $\mu_0 = \mu_1 < \mu_2 < \mu_3$.

13.42 The test assumptions are reasonable for these data. The data yield the following calculations.

$$\Sigma X = 118, \quad \Sigma X^2 = 1096, \quad \bar{X} = 5.9, \quad \bar{X}_0 = 1.4, \quad \bar{X}_1 = 2.6, \quad \bar{X}_2 = 8, \quad \bar{X}_3 = 11.6$$

Based on the following computer printout, the null hypothesis is rejected with a p-value of .0001.

Source	DF	Sum of Squares	Mean Square	F Value	PR > F
TREATMENT	3	340.2	113.400	30.44	.0001
ERROR	16	59.6	3.725		
TOTAL	19	399.8			

13.43 The decision in Exercise 13.42 is in agreement with the first tentative guess in Exercise 13.41.

13.44 Refer to the computations in Exercise 13.42. The LSD value for all pairs of comparisons is $LSD_{.05} = t_{.025,16} \sqrt{3.725} \, (\tfrac{1}{5} + \tfrac{1}{5})^{1/2} = 2.588$. The pairs of comparisons are as follows.

i,j	$|\bar{X}_i - \bar{X}_j|$	i,j	$|\bar{X}_i - \bar{X}_j|$
1,2	1.2	2,3	5.4
1,3	6.6	2,4	9.0
1,4	10.2	3,4	3.6

The summary is $\mu_0 = \mu_1 < \mu_2 < \mu_3$.

13.45 The decision in Exercise 13.44 is in agreement with the second tentative guess in Exercise 13.41.

INDEX

Glossary of Symbols (and page number)

$a = y -$ intercept of sample regression line (361)
$b =$ slope of sample regression line (361)
$c =$ number of columns in contingency table (302)
$CV =$ coefficient of variation (103)
$C_j =$ column total in contingency table (296)
$D_i =$ difference between matched observations (246)
$DF =$ degrees of freedom (408)
$e_i =$ sample residual in regression (368)
$E(X) =$ expected value of X (93)
$E_{ij} =$ expected cell count in contingency table (303)
Error $SS =$ error sum of square (356)
$f =$ degrees of freedom using Satterthwaite's approximation (277)
$f(x) =$ probability function (134)
$f_i =$ the frequency in the i th interval (89)
$F =$ ratio of two sample variances (275)
$F(x) =$ cumulative distribution function (75)
$H_o =$ null hypothesis (209)
$H_1 =$ alternative hypothesis (209)
$k =$ number of intervals (35, 89, 312)
$k_1, k_2 =$ parameters in the F distribution (274)
$LSD =$ least significant difference (413)
$m_i =$ midpoint of the i th interval (89)
$MSE =$ mean squared error (408)
$MST =$ mean square for treatments (408)

$n =$ number of observations in a sample (35, 88, 158)
$n =$ parameter in the binomial distribution (133)
$\binom{n}{x} =$ binomial coefficient (134, 175)
$n! = n(n-1)(n-2) \ldots (2)(1)$, "n factorial," (175)
$N =$ number of items in a population (93, 158)
$O_{ij} =$ observed cell count in contingency table (303)
$p =$ parameter in the binomial distribution (133)
$p =$ population proportion (117)
$\hat{p} =$ sample proportion (116, 194)
$p_i =$ probability associated with x_i (95)
$p_{ij} =$ cell probability in contingency table (295)
$p_{.j}, p_{i.} =$ marginal probabilities (295)
$P(A) =$ probability of an event A (65)
$P(A|B) =$ conditional probability of A given B (69)
$P(A$ and $B) = P(AB) =$ joint probability of A and B (67)
$q =$ smaller of r and c in contingency table (308)
$q = 1-p$ in binomial distribution (133)
$r =$ sample correlation coefficient (121)
$r =$ number of rows in a contingency table (302)
$r_s =$ Spearman's rho, rank correlation coefficient (126)
$r_i =$ serial correlation, lag i (388)
$\bar{R} =$ average of several ranks (281)
$R^2 =$ coefficient of determination (356)
$R_i =$ row total in contingency table (296)